A PRIMER IN FLUID MECHANICS

Dynamics of Flows in One Space Dimension

William B. Brower, Jr.

A PRIMER IN FLUID MECHANICS

Dynamics of Flows in One Space Dimension

CRC Press
Taylor & Francis Group
Boca Raton London New York

CRC Press is an imprint of the
Taylor & Francis Group, an **informa** business

CRC Press
Taylor & Francis Group
6000 Broken Sound Parkway NW, Suite 300
Boca Raton, FL 33487-2742

ISBN 13: 978-0-8493-9368-6 (hbk)

Visit the Taylor & Francis Web site at
http://www.taylorandfrancis.com

and the CRC Press Web site at
http://www.crcpress.com

Library of Congress Cataloging-in-Publication Data
Brower William B.
   A primer in fluid mechanics : flow in one space dimension /
William B. Brower, Jr.
            p.   cm.
      Includes bibliographical references and index.
      ISBN 0-8493-9368-X (alk. paper)
      1. Fluid dynamics.   I. Title.
   TA357.B75   1998   98-34265
   620.1'06--dc21   CIP

Library of Congress Card Number   98-34265

Dedication

---

to
The memory of my teacher, colleague, and friend
Joseph V. Foa

Acknowledgment

In the preparation of the seemingly endless number of drafts of this text I wish to acknowledge the meticulous work of Marjorie Baldwin, Mary Ellen Frank, Marilyn De Maria, and Victoria Lee.

The figures were computerized by the skills of Jamison Pond.

I had frequent occasion to call on the Reference Staff of the RPI Folsom Library, and they never failed to produce a requested book, technical paper, or other reference, for which I express my sincere gratitude.

A generation or more of RPI aeronautical engineering students suffered through the early drafts. I received from them a number of useful criticisms, which have resulted in improvements, and I was also the beneficiary of notification by them of a variety of existing errors, the bane of most authors.

The late Professor Joseph V. Foa read the early version of Chapters 3 through 5 and provided much informed criticism, all of which has been incorporated in the final version.

I wish also to thank my wife Yolanda who provided long-term encouragement in this endeavor and furnished much-needed support in getting out the final manuscript.

Epigraph

Although in classical and medieval times there was a certain amount of interest in hydrodynamics and more especially in hydrostatics, as is testified by the law of hydrostatics, and although this knowledge was advanced by the work of Stevin, Galileo and Newton, it is Leonhard Euler who is justly recognized as the father of hydrodynamics. To him we owe our clear ideas on fluid pressure, and it was through his grasp of this fundamental concept that he later propounded the equations of motion bearing his name. He was a great theoretical mathematician, yet he brought his genius to bear on such technical matters as the construction of turbines.

But the great growth in technical achievement which began in the nineteenth century left scientific knowledge far behind. The multitudinous problems of practice could not be answered by the hydrodynamics of Euler; they could not even be discussed. This was chiefly because, starting from Euler's equations of motion, the science had become more and more a purely academic analysis of the hypothetical frictionless "ideal fluid." This theoretical development is associated with the names of Helmholtz, Kelvin, Lamb and Rayleigh.

The analytical results obtained by means of this so-called "classical hydrodynamics" usually do not agree at all with the practical phenomena. To such extremely important questions as the magnitude of the pressure drop in pipes, or the resistance of a body moving through the field, theoretical hydrodynamics can only answer that both pressure drop and resistance are zero! Hydrodynamics thus has little significance for the engineer because of the great mathematical knowledge required for an understanding of it and the negligible possibility of applying the results. Therefore the engineers—such as Bernoulli, Hagen, Weissbach, Darcy, Bazin and Boussinesq—put their trust in a mass of empirical data collectively known as the "science of hydraulics," a branch of knowledge which grew more and more unlike hydrodynamics....

...Theoretical hydrodynamics seemed to lose all contact with reality; simplifying assumptions were made which were not permissible even as approximations. Hydraulics disintegrated into a collection of unrelated problems; each individual problem was solved by assuming a formula containing some undetermined coefficients and then determining these to fit the facts as well as possible. Hydraulics seemed to become more and more a science of coefficients.

Toward the end of the nineteenth century and the beginning of the twentieth century a new critical spirit appeared in both sciences...there grew up, among the [students of fluid mechanics] a more realistic attitude...having for its object a restoration of the unity between pure and applied science....

[Here] the relation between theory and practice will always be kept in the foreground. Theory will not be detached from the facts of experience but will be put in its proper relation to them. Experimental results, on the other hand, will be considered in their relation to the fundamental laws and underlying theory.

From the Introduction to *Fundamentals of Hydro- & Aeromechanics*, Vol. 1, by L. Prandtl and O. G. Tietjens, 1934.

Preface

When I first began taking graduate courses in fluid mechanics, in the 1950s, I observed that the books I used as an undergraduate student were never referred to by the instructors and, in fact, were almost useless since they provided no bridge to the more-advanced topics.

This lack highlighted a wider gap. The undergraduate program existed in a historical vacuum. Concepts, theories, crucial experiments were often presented as though they were stand-alone entities, created "yesterday." Mention of the creators of these scientific and engineering disciplines was at most incidental, and provided little or no insight into how they developed.

Therefore, some years ago, while commencing work on the manuscript for this text, I determined to provide a moderate historical context that would be appreciated by the more curious student. Thus, I have identified many references—some of them the great engineering or science classics—to which the individual can immediately turn for further background, if desired. A second, particular goal is to provide to the working engineer a quick refresher or a ready introduction to the subjects within, and a source for references to standard works at an advanced level.

Although this is an introductory text devoted to flows in one space dimension, I have attempted, where possible, to tie these simpler flow models to their more complex counterparts in two or three dimensions, so that the interested individual can seek them out in the advanced literature. For example, in Chapter 1, on fluid properties, the molecular nature of fluids and its relation to the macroscopic view is dealt with on the lowest and most elemental level, but the student is also referred to the classic, and eminently readable, works of Sir James Jeans, and other more modern works, for proper treatments of the subject.

Inevitably, in an introductory book, most of the major areas covered are conventional. However, I have not hesitated to include advanced applications where warranted. In Chapter 2, on fluid statics, the analysis of the stability of a floating parabolic section is new, and provides an analytical challenge even for the superior student. In the same chapter I have included a treatment of the *U.S. Standard Atmosphere* because of its interest in applications related to the space program. Chapter 2 concludes with a short discussion of surface tension.

Some introductory books on fluid mechanics ambitiously start with the derivation of the basic equations in full tensor notation. (G. I. Taylor, one of the great names in fluid mechanics of the 20th century, once referred to such authors as the "i,j,k-guys," as contrasted to the "x,y,z-guys.") However, perusal of the applications presented in such texts reveals that most could have been just as effectively treated, and much more simply, by the one-dimensional approach followed herein. This should not be taken as a criticism of the importance of tensors, however.

Chapter 3 introduces the equations of fluid dynamics in some generality, although in one space dimension. The substantial derivative is introduced early as one of the time derivatives most frequently encountered in fluid mechanics, with emphasis on its physical significance. A number of examples are provided. The one-dimensional approach lends itself to the combining of the first law of thermodynamics with the dynamic equation to produce the energy equation, in a straightforward way. The equations for the specialized cases of constant-density flow and compressible flow of a perfect gas follow readily, and without the suggestion that they are really two different subjects. Special attention is given to the question of viscous dissipation in a flow.

Chapter 4 deals with conventional applications in constant-density flow including several that employ the integrated momentum equation, one of which involves the more-complicated nonsteady form. Several deal with the concept of efficiency of fluid machinery.

In the fifth chapter, after a discussion of units, dimensions, and standards, I explore the subject of dimensional analysis starting with a real-life example from the work of Theodore von Kármán. Several examples are then used to show how some of the more commonly encountered π-ratios appear routinely in a dimensional analysis. The associated topic of inspectional analysis is then introduced to show that when a differential equation governing a physical phenomenon is available, an inspectional analysis is capable of demonstrating dependency upon the appropriate π-ratio "naturally" and without necessarily integrating the equation, an important advance. Examples are given that illustrate the virtues of correlating data in a nondimensional format and further demonstrate certain limitations on model testing.

In Chapter 6, a brief history is given of the early work on viscous flow in pipes, emphasizing the fact that over a period longer than two millennia hydraulics engineers had achieved considerable practical success in building water-distribution systems, with access to only elemental, or empirical, methods of quantitative flow prediction. Starting with the formulation of the Darcy–Weisbach law for head loss, we follow the development of the subject: the Navier–Stokes equations; the experiments of Hagen and Poiseuille on laminar flow and its theoretical confirmation by Stokes; the demonstration of transition from laminar to turbulent flow, and the criterion therefore, by Reynolds; the use of dimensional analysis to correlate pipe flow date by Blasius; Prandtl's law of turbulent flow in smooth pipes; the remarkable experiments of Nikuradse on artificial pipe roughness and velocity profiles in turbulent flow; and finally the correlation of data on commercial pipes by Moody. Two sections deal with the problem of determining pumping power in hydraulic systems. The treatment of viscous flow in pipes and fittings concludes with a prescription reducing the analysis of flow in hydraulic loops to a routine.

In 1904 Prandtl formulated the concept of the boundary layer flow that enabled researchers to deal with a flow in which the viscous effects are restricted to a small region near the body surface. The current text deals with boundary layers only in a qualitative way.

For the curious student I have included an analysis of a flow in which the duct contour interacts with the flow—the case of a pressurized flow in an elastic tube. This problem introduces the concept of an approximate solution by expanding a function in a series involving a small parameter, a technique that is further exploited in Chapter 8.

Chapter 7 starts with a review of elementary thermodynamic relations applied to gases. The first major flow discussed celebrates once again the keen insight of Reynolds who resolved the perceived paradox of a gas flow from a reservoir through an orifice which attained a maximum flow rate independent of the value of the pressure applied on the downstream side of the orifice, as long as the pressure was lower than some fraction of the reservoir pressure. His analysis shows that at a station where the flow speed equals or exceeds the local speed of sound the downstream regime can no longer "communicate" imposed pressure changes to the upstream flow regime.

This leads naturally to the study of the de Laval convergent–divergent nozzle and, eventually, to the appearance of shock waves—which are treated as finite discontinuities in the flow—in the divergent section of the nozzle. All the applicable relations are appreciably simplified by the introduction of the Mach number as a flow parameter. For the more perceptive student who might wonder about the consistency of discontinuities appearing in a supposedly continuous substance, I have included the analysis of Becker for the calculation of the (very thin!) thickness of a shock when treated by continuum theory. This is followed by discussions of Fanno flow and isothermal flow.

In the final chapter, Chapter 8, the behavior of nonsteady flows is explored. The first group of applications deals with the draining of "inviscid" liquids from ducts of fixed geometry. Certain of these problems go back to Newton himself (who got it wrong) and to other pioneers in fluid mechanics, especially Bernoulli (who got it right). Several of these solutions are presented here for the first time, and comparisons with experiment are made in several cases.

In most of these examples the analysis involves a second-order, ordinary differential equation, usually nonlinear, which must be integrated. In the case of draining a slender, conical reservoir, the equation is highly nonlinear. Its solution is approximate, and requires the use of the powerful technique involving a singular perturbation analysis.

The remainder of the applications deals with compressible flows in uniform ducts. Since the governing equations are a pair of coupled nonlinear partial differential equations, it is necessary to introduce the theory of hyperbolic functions for functions of two variables. With the advent of the computer age and the use of packaged software programs, it is vital for the analyst to be able to distinguish among hyperbolic, parabolic, and elliptical differential equations. The criterion depends on whether or not the characteristics—lines along which disturbances are transmitted—are "real."

The techniques involved are spelled out in detail, with elementary applications. These include the analysis of an isentropic expansion pulse generated by the impulsive motion of a piston and the propagation of an isentropic finite-amplitude pulse that leads to the inception of a shock wave and a calculation for the length of the duct required for the shock to form.

The chapter concludes with a short section on acoustical theory. The acoustic equations are deduced from the preceding characteristic equations of gas dynamics, by imposing approximations that require that flow disturbances satisfy certain "smallness" criteria. The resulting governing relation turns out to be the wave equation of classical physics, which is linear. One elementary application is given, and some of the technical standards used in evaluating acoustical phenomena are described.

An extensive group of problems for the first seven chapters is provided. There are few routine exercises. Most of the problems are intended to provoke thought from students and to require ulitization of a variety of analytical skills in the process of solution. Some of the problems stem from real-life situations, and, where appropriate, a modest number of them have been annotated.

To teachers who may contemplate adoption of this text, I note that for a number of years I utilized xeroxed copies of the preliminary manuscript (lacking Chapter 8) for a one-semester, 3-credit course, at Rensselaer Polytechnic Institute. The course outline included Chapters 1 through 6, omitting certain sections, e.g., Sections 2.14 and 6.14, which varied somewhat as the manuscript grew and the needs changed.

W.B. Brower, Jr.
Troy, New York
April, 1998

Table of Contents

Chapter 2

1 Fluid Properties — Kinetic Theory of Gases

At last you must admit, confess
That tiny things there are of nature indivisible.
Since this is so, this, too you must allow
That these atomic shapes are made of solid stuff,
and will for evermore endure.

Lucretius (98–55 B.C.E.)
Set in English verse by
Allen Dewes Winspear (1955)

If, in some cataclysm, all of scientific knowledge were to be destroyed, and only one
sentence passed on to the next generation of creatures, what statement would contain
the most information in the fewest words? I believe it is the atomic hypothesis...that
all things are made of atoms, little particles that move around in perpetual motion,
attracting each other when they are a little distance apart, but repelling upon being
squeezed into one another.

Richard P. Feynman et al. (1963)

1.1 INTRODUCTION

In order to discuss fluid mechanics it is desirable first to indicate in an unambiguous
way what is meant by a fluid. Intuitively, we all recognize the difference between
gases and liquids, which are both fluids and which are two of the three commonly
identified states of matter,* solids being the third.

FLUID DEFORMATION

The crucial distinction between a solid and a fluid is that a fluid will deform
continuously when subject to a shear force, however small. In Figure 1.1-1 a quantity
of matter, originally rectangular in cross section, is subject to a shear force such that
the lower boundary remains fixed. Equilibrium, of course, requires that an equal and
opposite shear force is applied on the lower boundary. The deformation is measured
by the angle ϑ between the left boundary and the vertical.

In an elastic solid (for example, a rectangular sponge to the top and bottom of
which are glued thin plates) ϑ would achieve a fixed angle and not vary thereafter
as long as the force system remains in place. However, a fluid would continue to

---

* A *plasma*, consisting of a collection of charged particles in the gaseous state, is sometimes said to be
the fourth state of matter.

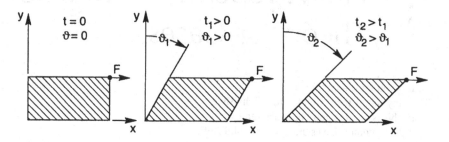

FIGURE 1.1-1 Example of continuous deformation of a fluid element under a shearing force.

deform as long as the forces are imposed. The time derivative of the angle $d\vartheta/dt \equiv \dot{\theta}$ is the *rate of shear deformation*.

NONSIMPLE FLUIDS

But there are fluids whose behavior is complex and for which the preceding elementary criterion becomes inadequate. For example. there are *thixotropic* materials such as some paints and jellies, which, having stood for some time, behave as elastic bodies. That is, as long as the forces are not too great, the bodies will return to their original shape when the shearing force is removed. However, if the material is stirred up, it behaves like a fluid.

On the other hand, there are other substances, ordinarily considered as solids, that may also exhibit fluid behavior. For example, at ordinary temperatures, pitch will fracture if struck sufficiently hard. Yet, if it is subjected to shearing action, it tends to deform, albeit very slowly. The rate of deformation is so small it may take weeks or months to detect the deformation by eye. Similarly, glass deforms under shearing action, although it may take centuries to achieve a visually observable deformation.

Modern chemistry has produced a number of remarkable substances — *polymers*, which may display simultaneously certain characteristics of solids and fluids. Nor surprisingly, the formulation of a mathematical model that can account for their behavior becomes more difficult than for more elementary substances.

SIMPLE FLUIDS

This text will concentrate on *simple fluids*, which are defined as materials that, when subjected to a shear force, however small, respond by exhibiting a finite rate of deformation. For engineering purposes this covers an extremely wide range of applications. For example, at moderate temperatures and pressures, air and water can be treated as simple fluids. Even for fluids not covered by this definition, knowledge of the behavior of a simple fluid is useful as a point of departure for studying the behavior of more complex fluids.

To understand why fluids differ from solids, and how gases differ from liquids, it is necessary to recognize that matter is molecular in nature. The rest of this chapter gives a qualitative description of how fluids differ from solids from the molecular viewpoint, followed by a cursory treatment of how *macroscopic* properties of a gas, such as pressure, temperature, and viscosity, can be explained from a molecular viewpoint.

1.2 THE MACROSCOPIC VIEW OF MATTER

Although the ancients recognized that matter consisted of extremely small particles, this was not conclusively established until the 19th century by various investigators employing a wide range of experimental techniques.

THE SOLID STATE

In the solid state, for example, the atoms of a crystal, are densely packed, often with their centers at the intersections of a regular lattice. It is known that the atoms are held in position by the combined effects of the force fields of the surrounding atoms. Yet, the individual atoms are not in fact at rest. Alder and Wainwright (1959) illustrated this fact by a calculated plot — a portion of which is shown in Figure 1.2-1 — in which the traces of the pathlines of 32 "hard" atoms (conceptually elastic spheres) in a square-grid lattice lying in a plane were plotted over a time interval involving about 3000 "collisions." The pathline for each atom repeatedly self-intersects at or near the mean position. For a sufficiently long time interval the plot of each path depicted would devolve into a smear, or blot, centered on the particle mean location in the crystal grid.

Recent developments in electron microscopy have since permitted this fact to be conclusively demonstrated. Figure 1.2-2a is a time-exposure photograph of an extremely thin slice of a zirconium oxide (ZrO_2) crystal, at a magnification of 300,000 times, taken on the Atomic Resolution Microscope by R. Gronsky at the Lawrence Berkeley Laboratory. The large, white, almost circular areas are the zirconium atoms. At the intermediate lattice points are the oxygen atoms, which are much less bright. The diffuseness of the image can be attributed partly to the resolution limit of the microscope, which is 0.18 nm, but, more importantly, to the motion of the molecules about their mean positions. Figure 1.2-2b indicates the scale of the lattice structure. For a description of the microscope see Gronsky (1984).

THE INTERMOLECULAR FORCE FIELD

The intermolecular force field between any pair of atoms depends on the magnitude of the separation distance of the pair. The force is mutually attractive when the separation d exceeds some (extremely small) distance d_0 and the attractive force magnitude is approximately proportional to d^{-7}. When the separation distance is such that $d < d_0$ the force becomes extremely repulsive with magnitude approximately proportional to d^{-13}. Deductions from calculations compared with experiment indicate that the distance d_0 is the order of 3 to 4×10^{-10} m.

As the crystal is heated, the average speed of random motion of the ensemble of molecules increases. Thus, on the molecular level, the acquisition of heat results in an increase of the kinetic energy of random motion of the molecules. If the crystal continues to be heated, the average molecular speed continues to increase. However, at any instant, the speed of a given molecule may vary from zero to some speed

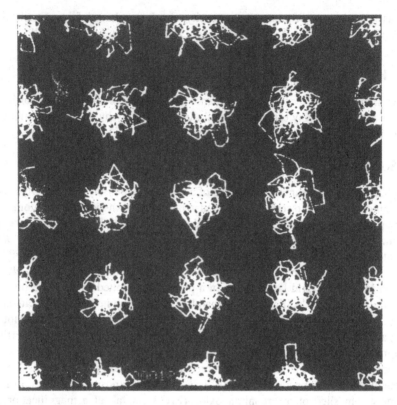

FIGURE 1.2-1 Plots of the theoretical pathlines of atoms in a crystal. (Calculated by Alder, B.J., and Wainwright, T., *J. Chem. Phys.*, 31(2), 459–466, 1959, and reprinted with permission.)

above which a molecule may have sufficient energy to move away from its position in the lattice. As the heating continues, this must occur and the crystal begins to melt.

1.3 THE LIQUID STATE

Melting

In the melting process heat is added, but the crystal temperature remains constant. For most substances the volume of the material increases, leading to a decrease in density. Water is exceptional in that the opposite occurs over a small temperature range above the melting temperature.

In the liquid state (the melt must now be confined by walls) there remain certain similarities to solids and certain crucial distinctions. In mechanical equilibrium (i.e., subject to no shearing forces) the molecules remain partially ordered. The packing spacing remains approximately the same as for the solid state. That is, a molecule may vibrate randomly about a mean position, but there is a degree of mobility that allows for possible interchange or rearrangement of individual molecules or even of large groups of molecules acting together. Thus, there is a continual reordering of

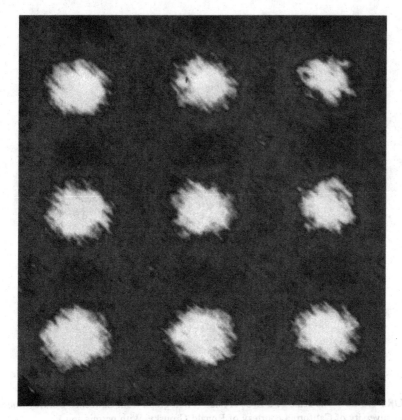

FIGURE 1.2-2a Electron microscope photograph of a slice of a crystal of zirconium oxide. (Photograph from Lawrence Berkley Laboratory, University of California courtesy of Ronald Gronsky. With permission.)

the structure. In both the liquid and solid state the average spacing of the molecules is the order of d_0.

EVAPORATION

If the heating process continues beyond the point at which the substance is fully liquefied, molecules near the free surface* may undergo a series of collisions** with their neighbors in which they acquire sufficient energy to escape from the force field exerted by the main body of a fluid. In such a case, these molecules may fly off into the space above the fluid. In the macroscopic sense this process is called *evaporation*. If the liquid is in an enclosed chamber, the liquid phase establishes, for any specified temperature, an equilibrium with its gas phase above the free surface, in which the rate at which molecules are emitted by the free surface in evaporation just equals

---

* The free surface is the boundary between the gas and the liquid.

** The reader is cautioned that no pretense is made herein to deal with the structure of a molecule on the level of quantum mechanics. For the purpose of this book it is adequate to think of a molecule as an elastic sphere of extremely small diameter.

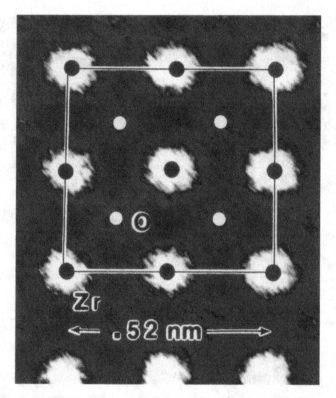

FIGURE 1.2-2b The scale of a ZrO_2 crystal. (Photograph from Lawrence Berkley Laboratory, University of California courtesy of Ronald Gronsky. With permission.)

the rate at which molecules are received from the gas phase in *condensation*. As is well known, once the equilibrium temperature is established, the equilibrium pressure is also automatically fixed at a unique value.

Viscosity

From the macroscopic view, a liquid, when subject to shearing action, tends to deform as long as the shearing force remains imposed. For a fixed shear force the rate of deformation depends on the liquid temperature; that is, as the temperature increases, the rate of deformation increases, which means that the viscosity decreases.* The state of the kinetic theory of liquids is not sufficiently developed, however, to allow complete prediction of the fluid properties from the molecular structure of the liquid. This question will be revisited in a later section.

As we continue to add heat to the liquid, its temperature and pressure rise, its density and viscosity decrease, and the fraction of the substance evaporated increases. Eventually, all the material is evaporated into the vapor phase and we are then dealing solely with a gas.

---

* A quantitative expression for viscosity is developed in Section 1.8.

1.4 ON THE KINETIC THEORY OF GASES

The kinetic theory of gases was extensively developed in the 19th century and the first half of the 20th by a host of investigators, which includes some of the greatest names in science. For a proper treatment of the subject the reader is referred to Jeans (1952), Loeb (1934), and Hirschfelder et al. (1954), in ascending order of comprehensiveness. A comprehensive, modern review of gas kinetic theory can be found in Touloukian et al. (1975). For the nonspecialist, Hildebrand (1963) provides an excellent introduction. A more recent book by Tabor (1991) covers a broader range of the properties of matter, also on a level conducive to the nonspecialist. In what follows I take the liberty to make occasional gross oversimplifications since space does not permit rigorous treatment. For the average reader this will not present a serious problem if he or she keeps in mind that the primary objective is to show that many of the fundamental concepts of fluid mechanics (of gases) can be considered to have a mechanical origin.

1.5 MASS DENSITY

DEFINITION OF DENSITY

Consider a control volume which is a cube of side ℓ with permeable walls. At a fixed instant of time inside the volume, let there be N molecules each of mass m. Then the average *number density* is defined* as

$$\nu \equiv N/\ell^3. \tag{1.5-1}$$

The average *mass density* is denoted ρ and is given by

$$\rho \equiv Nm/\ell^3 = \nu \overline{m} u, \tag{1.5-2}$$

where u is the *unified atomic mass unit*.** For future reference we note that the quantity *specific volume*, denoted v, is the inverse of the density, i.e.,

$$v \equiv 1/\rho. \tag{1.5-3}$$

CONCEPT OF A CONTINUUM

In a fluid mechanical system we shall find it desirable to introduce the concept of a *continuum*, i.e., the idea that in treating ordinary flow problems we can ignore the fact that a fluid consists of finite particles in which the masses are concentrated, separated by distances that are large relative to molecular diameter. In fact, we shall

---

* The symbol $\equiv$ denotes a definition as distinguished from an equals sign, which refers to an equation.
** In the International System of Units (SI) the unified atomic mass is fixed at $1u = 1.660\ 53 \times 10^{-27}$ kg. The mass of a molecule, then, is given by $m = (1.660\ 53 \times 10^{-27})\overline{m}$ kg, where $\overline{m}$ is the *molecular weight* (actually a pure number).

postulate, under conditions to be stated, that density is an analytic point function of space and time, i.e., $\rho = \rho(\mathbf{r}, t)$, where $\mathbf{r}$ is the position vector. But the continuum density cannot be obtained from Equation 1.5-2 by letting the control volume — centered at an arbitrary, fixed point — shrink to zero; i.e., we cannot put

$$\rho(r, t) \equiv \lim_{\ell \to 0} Nm/\ell^3.$$

To see this, imagine an enormous control volume of 10 km on the side, with centroid at 5 km. According to the *U.S. Standard Atmosphere, 1976* (Dublin, et al. 1976; see Table 2.16-2), the number densities at three different heights are

Altitude (km)	v (m^{-3})
Sea level	2.547×10^{25}
5	1.531×10^{25}
10	0.860×10^{25}

The nonlinearity of the number distribution requires that the integrated average of v over 10 km must exceed that at the median altitude of 5 km. As we shrink the cube side, keeping the centroid at 5 km, the plot of the number density (and equally true, the density) must appear something like that shown in Figure 1.5-1. As ℓ decreases to some value denoted ℓ', the volume becomes sufficiently small that the instantaneous calculation of the density starts to display a time-dependent effect due to the randomness of the number of molecules within the cube. The smaller the volume, the larger the possible deviations from the mean value become. Eventually, the macroscopic concept of density breaks down and we must resort to an alternative limiting condition.

To get a preliminary, though crude, idea of the conditions under which it is reasonable to express the density as an analytic function — which is to say that we are dealing with a continuum — we suppose that within a cube of side ℓ' there must be a minimum of N molecules, where N is to be determined by some as yet unspecified criterion. For a fluid of number density v the side of the cube would be $\ell' = (N/v)^{1/3}$.

If $N = 10^6$ molecules, the corresponding value of ℓ' in air at sea level is about 3.4×10^{-7} m. At an altitude of 10^3 km, according to the *U.S. Standard Atmosphere, 1976*, where $v = 5.422 \times 10^4$ m^{-3}, then $\ell' = 1.2 \times 10^{-2}$ m = 1.2 cm. By contrast, in liquids, the cube edge necessary to contain 10^6 molecules is only about $\ell' = 3.5 \times 10^{-8}$ m.

Suppose that we want to install a pressure tap on some aerodynamic surface. We have the instinctual feeling if there is a sufficient number of molecules present, the tap should yield valid readings, from the continuum viewpoint. Perhaps we should compare the tap diameter to ℓ'. The diameter of a really fine pressure tap or sensing element could be the order of 0.012 5 cm = 1.25×10^{-4} m. Continuing our example, we see that the pressure tap diameter is much larger than ℓ' for the sea-level measurement and much smaller at 10^3 km.

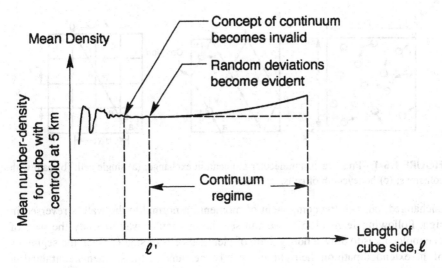

FIGURE 1.5-1 Qualitative explanation of breakdown of density function as analytic function of space and time.

However, the procedure involving 10^6 molecules is completely arbitrary. There is no way to formulate a valid criterion depending on a specific number of molecules. In Section 1.7, therefore, a different approach is used, which has a sound scientific foundation and which allows us to quantify the continuum criterion. The principal virtue, of course, is that it allows us to consider the density function — all state functions, in fact — to be treated as analytic. The sole exception occurs for shock waves, which are treated as discontinuities across which the density, pressure, velocity, etc. may involve finite jumps. Even so, as shown in Section 7.14, the concept of a shock as an infinitely thin discontinuity is a convenient analytical fiction. Of course, the idea of a continuum is itself a fiction, and aerodynamicists must keep in mind that continuum theories must not be applied to a physical body in violation of the criterion developed in the following pages.

1.6 PRESSURE IN A GAS

Consider a cubical container, Figure 1.6-1, with rigid walls of side ℓ in which there is a total of N molecules. Each molecule is considered to be an elastic sphere of mass m. These molecules are supposed to be moving randomly, and we denote the components of the average speeds* of the molecules as u, v, w, for the x-, y-, z-directions, respectively.

If there is true randomness, then, at any instant, $N/2$ of the molecules have an x-component of velocity to the right, the other half to the left. When a molecule (see Figure 1.6-1b) undergoes an elastic collision with a wall, we know from elementary mechanics that the tangential components of momentum remains

---

* The symbol u for the x-component of velocity should not be confused with the atomic mass unit, which is designated by the same symbol.

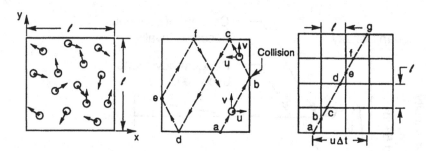

FIGURE 1.6-1 Pressure as a molecular momentum exchange: (a) single cell; (b) path with collisions; (c) honeycomb of cells.

unchanged, but that the component of momentum normal to the wall is reversed in sign. Following Jeans (1952), we can see that a simpler way to study the path of one molecule is to draw a honeycomb of identical cells of side ℓ. Then, the segments of the extended path on the right are exactly the same as the segments contained in the center sketch.

We can now make the following calculations:

x-distance traveled to right in time Δt	$= u\Delta t,$
Number of collisions with wall perpendicular to x-direction, for one molecule	$= u\Delta t/\ell,$
Total number of collisions with wall perpendicular to x-direction, for N molecules	$= Nu\Delta t\,/\,\ell,$
Total number of collisions on wall facing left	$= \frac{1}{2}\,Nu\Delta t\,/\,\ell,$
Total change of momentum of all molecules on wall facing left	$= \left(\frac{1}{2}\,Nu\Delta t\,/\,\ell\right)2mu,$
	$= Nmu^2\Delta t\,/\,\ell,$

$$\text{Force on left-facing wall} \quad = \frac{\text{Tot. chge. momentum}}{\Delta t} = \frac{Nmu^2}{\ell}.$$

Thus, the pressure on the left-facing wall is defined as

$$p_x \equiv \frac{\text{Force}}{\text{Area}} = \frac{Nmu^2}{\ell^3},$$

which is the same as for the right-facing walls. By an identical procedure

$$p_y = \frac{Nmv^2}{\ell^3}, \quad p_z = \frac{Nmw^2}{\ell^3}.$$

The average pressure within the container is defined as the average of these three quantities, i.e.,

$$p \equiv \frac{1}{3}\left(p_x + p_y + p_z\right) = \frac{1}{3}\frac{Nm}{\ell^3}\left(u^2 + v^2 + w^2\right). \qquad (1.6\text{-}1)$$

DEFINITION OF MEAN-FREE-SPEED

The *mean-free-molecular speed* C is defined as

$$C^2 \equiv u^2 + v^2 + w^2. \qquad (1.6\text{-}2)$$

In fact, in any realistic container the number of molecules is truly enormous (about 2.55×10^{25}) m^{-3} $\approx 7.21 \times 10^{23}$ ft^{-3} in air at standard atmospheric conditions). Further, there is actually a distribution of speeds for a gas in equilibrium ranging from a mean free speed of zero to infinity, according to a statistical law discovered by, and named for, the British physicist Maxwell. A rigorous analysis shows that C is the speed with the highest probability of occurrence.

Combining Equations 1.6-1 and 1.6-2, we have

$$p = \frac{2}{3}\frac{N\left(\frac{1}{2}mC^2\right)}{\ell^3}, \qquad (1.6\text{-}3)$$

which demonstrates that pressure in a (monatomic) gas can be interpreted* as

$$p = \frac{2}{3}\frac{\text{Kinetic Energy of Random Motion}}{\text{Volume}}. \qquad (1.6\text{-}4)$$

Furthermore, since kinetic energies are additive, the combined pressure of the mixture of two gases must equal the sum of the pressures exerted by the constituents of the mixture separately. This is *Dalton's law* of partial pressures. We further see that pressure varies inversely with the volume, which is *Boyle's law*. By introducing the definition of density, Equation 1.5-2,

$$p = \tfrac{1}{3}\rho C^2. \qquad (1.6\text{-}5)$$

EQUIPARTITION OF ENERGY

It can be shown rigorously, for a mixture of gases, that the average kinetic energy of a particle is the same for each constituent. Furthermore, since there is no reason to expect that there is any preferred direction in a container, the three average velocity components of the same gas must be equal, i.e.,

---

* Equation 1.6-4 is strictly valid only for monatomic gases.

$$u^2 = v^2 = w^2 = C^2/3. \tag{1.6-6}$$

One third of the kinetic energy of random motion is associated with the mean velocity component for each of the three orthogonal space coordinates. This lack of preferred direction is called the principle of *equipartition of energy.*

EQUATION OF STATE

If Equation 1.6-6 is compared with the *equation of state for a perfect gas,**

$$p = \rho RT, \tag{1.6-7}$$

which results from the thermodynamic requirement (see Chapter 7) that a gas be thermally perfect, then

$$C^2 = 3RT. \tag{1.6-8}$$

Clearly, then, we see that temperature is a measure of average kinetic energy of random motion of a particle.

1.7 THE MOLECULAR MEAN-FREE-PATH LENGTH

THE MOLECULAR DIAMETER

It was previously noted that in the process of evaporation a molecule flies off into the space above a liquid. In the gas phase the average spacing between gas molecules arranged in a cubical lattice is $v^{-1/3}$, the inverse cube root of the number density, a length which is generally very large compared with the molecular diameter. For molecules in motion the lattice spacing has little significance; a more meaningful quantity is the mean-free-path, which is described in the following paragraphs.

That a molecule has an effective diameter can easily be established for some substances. One way is to pour a known volume of a liquid onto the surface of another, immiscible liquid that is more dense. If the volume of the less-dense liquid is sufficiently small, it will spread out in a layer one molecular-diameter thick, which does not occupy the entire cross section of the container. This area is then measured. Its value, divided into the original volume, produces the desired molecular diameter.

More sophisticated techniques are required for gases. If it is assumed that the molecules are elastic spheres, then it is possible to deduce a theoretical value for the coefficient of viscosity and similarly, the coefficient of self-diffusion, both of which depend on the postulated molecular diameter. These macroscopic quantities

---

* The specific gas constant R is related to the universal gas constant $R*$ such that $R = R*/\overline{m}k$, where $\overline{m}$ is the molecular weight (actually a dimensionless number) and where k is the *molar mass,* of unit magnitude, but of units consistent with those of the universal gas constant, e.g., kg/kmol, gm/gm mol, slug/slug mol, etc. See Section 7.2 for further explanation and see Tables 2.16-2 and 7.2-1, respectively, for values of R and $\overline{m}$ for various gases.

TABLE 1.7-1
Molecular Diameters* Deduced from
Viscosity and Self-Diffusion Calculations

	Molecular diameter $\sigma \times 10^8$ (cm)	
Gas	From Viscosity	From Diffusivity
Argon	3.64	3.47
Neon	2.58	2.42
Nitrogen	3.75	3.48

* This table should be read as an equation: $\sigma \times 10^8 =$ Value from table, and similarly throughout.

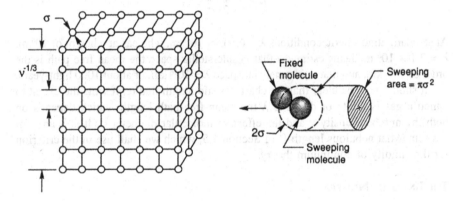

FIGURE 1.7-1 The calculation of mean-free-path length.

can be measured, allowing the effective diameter to be calculated. A few values taken from Hirschfelder et al. (1954), p. 545, are listed in Table 1.7-1.

MEAN-FREE-PATH LENGTH

In an ensemble of moving molecules, the *mean-free-path length* is defined as the average length a molecule travels between collisions with other molecules. If molecules are treated as elastic spheres the calculation is relatively simple. However, if molecules are treated as point masses, obeying the attractive/repulsive law of intermolecular forces previously mentioned, the word *encounter* better describes the interaction process, and the calculation gets quite involved.

Referring to Figure 1.7-1 we can get an approximate idea of the mean-free-path length as follows. Consider a volume of side ℓ containing N molecules of diameter σ. Suppose, for the current purpose, that the molecules are not in motion but are fixed at the lattice points of a cubical grid. If another molecule, also of diameter σ, moves through the volume to sweep it out without duplication, then we can consider a collision to occur any time the center of a fixed molecule falls within the sweeping circle of diameter 2σ. Obviously, the total number of "collisions" is the same as the number of molecules in the lattice.

The sweeping area is $\pi\sigma^2$ and the total length required to sweep the volume is $\ell^3/\pi\sigma^2$. Thus, approximately, λ, the mean-free-path length between collisions, is given by

$$\lambda \equiv \frac{\ell^3/\pi\sigma^2}{N} = \frac{1}{\pi v \sigma^2}. \qquad (1.7\text{-}1)$$

This result is surprisingly close to the result of rigorous analysis, which involves an infinity of molecules in random motion, and whose velocities vary according to Maxwell's law. The accurate calculation yields

$$\lambda = \frac{1}{\pi\sqrt{2}v\sigma^2}. \qquad (1.7\text{-}2)$$

At standard atmospheric conditions $\lambda = 6.633 \times 10^{-8}$ m, whereas at a height of 10^6 m, $\lambda = 3.1 \times 10^6$ m. Jeans estimates that in interstellar space the mean free path is the order of 10^{11} km, and that in internebular space $\lambda = 10^{16}$ km, or about 1000 light-years.

It should now be clear that the characteristic length that represents the extent to which a gas is dilute or dense is λ the mean-free-path length, which depends on both the number density v and the effective molecular diameter σ. It is λ and not the somewhat nebulous length ℓ' of Section 1.5, which we shall use in the criterion for the validity of continuum theory.

THE KNUDSEN NUMBER

The criterion is very simple. The mean-free-path length is compared with L, the characteristic dimension of a body immersed in a flow. If the body dimension is very large in comparison with λ, then the continuum concept (which implies that there is a distribution of molecular velocities in time/space governed by Maxwell's law and that collisions between molecules are important) is valid. If the mean-free-path length is equal to, or larger than the body dimension, then free-molecule theory, rather than continuum theory, must be employed. In free-molecule theory, collisions between molecules need not be considered. Density is computed (over a large volume) in the same way as before, but it is not considered to be an analytic function.

Thus, the mean-free-path and characteristic lengths are combined into the dimensionless parameter

$$\text{Kn} \equiv \frac{\text{Mean-free-path length}}{\text{Characteristic body dimension}} = \frac{\lambda}{L}, \qquad (1.7\text{-}3)$$

named for M. Knudsen, who made a number of original and fundamental investigations of free-molecule flow, early in the 20th century; see Knudsen (1933).

The characteristic body dimension is usually apparent from the nature of the flow. If the flow is through a tube or hole, then L = diameter. For a space vehicle, then L = vehicle diameter, and so forth. The criterion used by the aerodynamicist divides the flow into regimes; i.e.,

$$\text{Kn} < 0.1, \quad \text{Continuum regime;}$$

$$0.1 \leq \text{Kn} \leq 1, \quad \text{Transition regime, slip flow;} \qquad (1.7\text{-}4)$$

$$\text{Kn} > 1, \quad \text{Free - molecule regime.}$$

For example, if a piece of hypodermic tubing is used as a pressure probe, it might have an internal diameter as small as $L = 0.005$ in. $= 0.0127$ cm $\approx 1.3 \times 10^{-4}$ m. By using the data cited in Table 2.16-3, this would result in a Knudsen number $\text{Kn} = \lambda/L$ $= 6.6 \times 10^{-8}$ at sea level and $\text{Kn} = 2 \times 10^{-3}/1.3 \times 10^{-4} \approx 15.0$ at an altitude of 75 km, which is generally taken to be the *reentry altitude* (i.e., the altitude above which the aerodynamic forces of the atmosphere on a vehicle are usually sufficiently small that they can be neglected). Thus, at sea level we expect continuum theory to be applicable for flow into the tube, whereas at 75 km free-molecule theory must be employed.

The vast literature on gas flow has emphasized the continuum regime. Patterson (1956; 1971) and Bird (1976) have applied nonequilibrium (non-Maxwellian) theory to a wide range of problems in the transition regime as well as in the free-molecule regime. In the transition regime the average tangential component of the molecular velocity is not zero, hence the name *slip flow*. Neither slip flow nor free-molecule theory is further considered in this work.

1.8 RELATION BETWEEN FLUID VISCOSITY AND MEAN-FREE-PATH LENGTH

In Book II, Section 9 of the *Principia*, Newton (1686) gave the first quantitative formulation of a law describing the action of fluid viscosity on a body. Newton considered a circular cylinder enclosed by a concentric cylindrical container with a liquid lying in the annular space between. He postulated that if the cylinder were rotated about its axis, it would generate a resistance (i.e., a torque) proportional to the difference in velocity between cylinder and wall. This idea has been extended to the idea of planar motion in a fluid of infinite lateral extent, resulting in what is called *Newton's law of resistance*.

THE MACROSCOPIC LAW

Consider a layer of fluid resting on a horizontal surface; see Figure 1.8-1. On the top of the fluid is another, solid surface of which is shown only a portion, area A, which is dragged across the fluid* thereby exerting a horizontal (shear) force F on the fluid surface. The speed of the surface relative to the fixed bottom is Δu. The fluid in between has some velocity variation $u = u(y)$, which we initially treat as a linear profile.

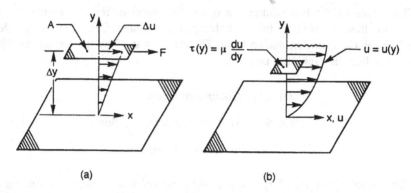

(a)   (b)

FIGURE 1.8-1   Relation of shear stress to velocity gradient in a flow: (a) linear velocity profile; (b) nonlinear velocity profile.

It is important to note that the flow velocity is a *directed velocity* on the macroscopic level, as distinguished from the microscopic, or molecular, random velocity of the molecules that is superimposed on the directed velocity. From the macroscopic view, the molecular velocity is not observable except in that the random motion produces certain macroscopic effects, such as pressure and, as it turns out, flow resistance.

Newton's law of resistance states that the resisting force is proportional to the area of the surface times the change in velocity divided by the thickness of the fluid layer, i.e.,

$$F \sim A \frac{\Delta u}{\Delta y}. \tag{1.8-1}$$

If we introduce the concept of *shear stress* (τ) and a factor of proportionality μ, called the *coefficient of dynamic viscosity*, then

$$\tau \equiv \frac{F}{A} = \mu \frac{\Delta u}{\Delta y}. \tag{1.8-2}$$

The units of μ are discussed in Section 1.10.

For the case in which the (directed) velocity profile is not linear, then the shear stress also becomes a function of the coordinate y. The local value can then be computed as a derivative,* i.e.,

$$\tau \equiv \lim_{\Delta y \to 0} \mu \frac{\Delta u}{\Delta y} = \mu \frac{du}{dy}. \tag{1.8-3}$$

---

* It has been demonstrated by countless experiments that on a material surface the relative, fluid mean velocity is zero except in the case of extremely rarified gases. This circumstance is referred to as the *no-slip condition*.

* This can be compared with the expression for shear stress for the more general case not restricted to motion parallel to the x-axis; then $\tau_{xy} = \mu(\partial v / \partial x + \partial u / \partial y)$.

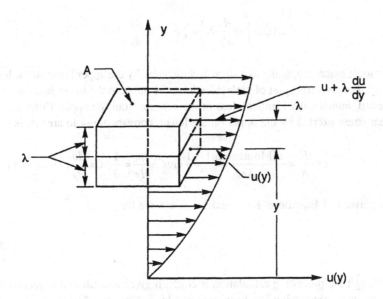

FIGURE 1.8-2 Schematic for calculation of viscosity coefficient.

A fluid that obeys Equation 1.8-3 is said to be a *newtonian* fluid. In a three-dimensional flow equation, Equation 1.8-3 must be replaced by a general relation between shear stress and the rate of strain in the medium (rate of fluid deformation).

Newton's law of friction is a *phenomenological** statement of the macroscopic behavior of the fluid. It is possible to formulate a comparable law from the molecular viewpoint. By comparing the two expressions, it is then possible to obtain an expression that relates the coefficient of dynamic viscosity to the molecular mean speed.

THE MOLECULAR LAW

In a flow over a surface, let the directed velocity profile be given as $u = u(y)$. Now consider two adjacent layers of fluid each of thickness λ, the local mean-free-path length, see Figure 1.8-2. Then, the average directed velocity in each of the two layers is u, $u + (du/dy)\lambda$, respectively. Let the local number density be v and the mean free speed be C. If we consider a pair of rectangular parallelepipeds of area A and height λ, interfacing at the common surface, then the time interval required for one molecule to traverse the height is $\Delta t = \lambda/C$. Since, on the average, one sixth of the molecules are moving toward any one face, the total number of molecules moving upward through A in interval Δt is

$$\tfrac{1}{6} vA\lambda = \tfrac{1}{6} vAC\Delta t.$$

After these molecules have interacted (collided) with the molecules in the upper layer, they must acquire the same average directed x-momentum as the upper layer. This change in momentum, which is equal to

---

* A relation depending only on quantities directly evident to the observer.

$$\tfrac{1}{6} v A C \Delta t \left(m \frac{du}{dy} \lambda \right) = \tfrac{1}{6} m v C \lambda \frac{du}{dy} A \Delta t,$$

exhibits itself macroscopically as a shear force exerted by the upper layer on the lower. Similarly, there is a transport of molecules from the upper to the lower layer resulting in a second, numerically identical, loss of momentum from the upper. Consequently, the shear stress exerted by the upper on the lower, corresponding to area A, is

$$\tau = \frac{F}{A} = \frac{\Delta(\text{Momentum})}{A\Delta t} = \frac{1}{3} m v C \lambda \frac{du}{dy} = \frac{1}{3}\rho C\lambda \frac{du}{dy}. \tag{1.8-4}$$

By comparison of Equations 1.8-3 and 1.8-4, we see that

$$\mu = \frac{1}{3}\rho C\lambda. \tag{1.8-5}$$

Although the preceding calculation is crude, it gives a result that is surprisingly good. The exact expression has been obtained by S. Chapman* as

$$\mu = 0.499\rho C\lambda. \tag{1.8-6}$$

If we eliminate λ by means of Equation 1.7-2, then

$$\mu = \frac{0.499\, mC}{\pi\sqrt{2}\sigma^2}. \tag{1.8-7}$$

Since, according to Equation 1.6-8, $C \sim T^{1/2}$, then, for a perfect gas,

$$\mu \sim T^{1/2}. \tag{1.8-8}$$

The surprising fact that viscosity depends on temperature but not on pressure or density was first demonstrated by Robert Boyle in 1660** who found that the rate at which the oscillations of a pendulum suspended in a gas chamber died away remained the same even after the chamber was partially evacuated. The increase in viscosity of a gas with temperature is also confirmed by a vast array of experiments. In Section 1.11 a brief description is given of how Equation 1.8-8 is modified to account for the behavior of real gases.

1.9 ON HEAT CONDUCTION

In the preceding section we saw, from a molecular viewpoint, that the action of viscosity can be interpreted as the transport of momentum between layers of a fluid

---

* See Jeans (1952, p. 163).
** See Jeans (1952), p. 164, for the reference.

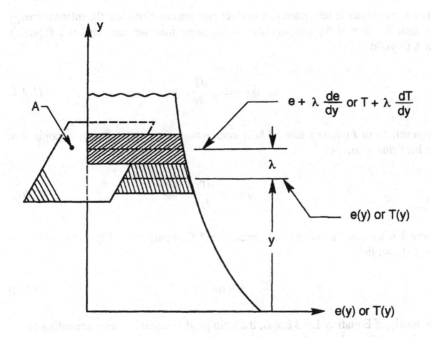

FIGURE 1.9-1   Schematic for relating heat flux term to temperature gradient.

in which gradients exist in the directed, or macroscopic, velocity. Similarly, we can show that if temperature gradients exist in a gas there is transport of thermal energy, or kinetic energy of random motion. The macroscopic term for this type of energy transport is *heat conduction*. Consider two adjacent layers of the gas, each of thickness λ, centered at y, $y + \lambda$, respectively, Figure 1.9-1.

From thermodynamic theory, the specific internal energy function for a thermally and calorically perfect gas (i.e., a gas which obeys Equation 1.6-7), and for which the coefficient of specific heat at constant volume c_v is constant) is given by $e = e(T) = c_v T$. Then, for a gas in which there is temperature gradient (or equivalently a gradient of the internal energy function) in the y-direction, the energy per molecule transported by molecular activity from a lower layer to an upper layer of mean separation λ is $-m(de/dy)\lambda$, m being the molecular mass. For all molecules transported across an area A, in both directions, over interval Δt, the energy gain of the upper layer is

$$Q = \frac{1}{3} v C(\Delta t) \left(-m \frac{de}{dy} \lambda \right) A.$$

The *energy flux* $\dot{q}$, equivalently the *heat flux*, is defined as the energy transported per unit area per unit time; i.e.,

$$\dot{q} \equiv \frac{Q}{A\Delta t} = -\frac{1}{3} m v\, C\lambda \frac{de}{dy} = -\frac{1}{3} \rho C\lambda \frac{de}{dy}.$$

Since our analysis is restricted to a perfect gas, we can eliminate the internal energy function in favor of the temperature. At the same time we can introduce Equation 1.8-5 to yield

$$\dot{q} = -\mu c_v \frac{dT}{dy}.$$

(1.9-1)

However, from *Fourier's law of heat conduction*, an empirical, macroscopic law, the heat flux is simply

$$\dot{q} = -k \frac{dT}{dy},$$

(1.9-2)

where k is the *coefficient of heat conduction*.* Comparison of Equations 1.9-1 and 1.9-2 shows that

$$k = \mu c_v.$$

(1.9-3)

Obviously, if Equation 1.9-3 holds, the ratio $\mu c_v/k$ is equal to unity, according to the simple model employed. It is precisely at this juncture that the elementary kinetic theory is least accurate. The accurate theory takes into account that a gas with a gradient in the directed velocity is not in equilibrium, and that is necessary to abandon the principle of equipartition of energy in favor of a more careful accounting of the distribution of energy between the translational and rotational degrees of freedom. This has resulted in *Eucken's formula*:**

$$k = \frac{9\gamma - 5}{4} \mu c_v,$$

(1.9-4)

where $\gamma \equiv c_p/c_v$ is the *ratio of specific heats* and is constant for a perfect gas.

It is next convenient to introduce a dimensionless parameter, the *Prandtl number* Pr, which appears in many problems in heat transfer. Its definition is

$$\text{Pr} \equiv \mu c_p/k = \gamma \mu c_v/k,$$

(1.9-5)

which can be combined with Equation 1.9-4 to yield

$$\text{Pr} = \frac{4\gamma}{9\gamma - 5}.$$

(1.9-6)

---

* Some authors use the symbol k to denote the ratio of specific heats c_p/c_v.
** See Jeans (1952), p. 190, for the details.

TABLE 1.9-1
Prandtl Number for Selected Gases at Standard Temperature and Pressure

Gas	γ (Theo.)	Prandtl Number Eq. 1.9-6	Expt.
He	5/3	0.667	0.694
Ar			0.669
H_2	7/5	0.737	0.712
N_2			0.735
O_2			0.732
CO			0.763
CO_2			0.819
H_2S	4/3	0.762	0.929
SO_2			0.833
CH_4			0.777

Source: Adapted from Loeb, 1934, p. 250.

Table 1.9-1, adapted from Loeb (1934, p. 250), compares Equation 1.9-6 using the theoretical value of γ for a perfect gas (see the table), with experiment, at standard temperature and pressure.

Apparently Eucken's formula is subject to some inaccuracies, particularly as the molecules become more complex. Its accuracy tends to increase at elevated temperature, but in the lower ranges, particularly near the critical temperature, it becomes less reliable. Nevertheless, the prediction that the Prandtl number is a constant — which is valid for every gas over some temperature range — is very useful, since it leads to important analytical simplifications in the theory of compressible flow and heat transfer.

On the Ratio of Specific Heats of a Perfect Gas

If a monatomic molecule is treated as an elastic sphere, then the capacity of a molecule to store kinetic energy is restricted to three degrees of freedom, all in translation. In this molecular model, collisions do not result in rotation of a molecule about any axis. From another point of view the mass moment of inertia of a monatomic molecule is so small that, relative to a translational degree of freedom, the energy stored in rotation is effectively negligible.

However, in a diatomic, or rigid dumbbell, molecule there are two degrees of freedom in rotation in which the mass moment of inertia of the molecule is not insignificant. These two degrees of freedom are for rotations about any pair of axes orthogonal to the axis through the atoms, as well as to each other. According to the principle of equipartition of energy for a gas in equilibrium, the average energy stored in each possible mode must be the same. Thus, for a diatomic molecule there are five degrees of freedom in all. For triatomic particles there are two possible situations. If the molecules are linear, such as CO_2, then there are only five degrees of freedom of mechanical motion. However, polar molecules, water, for example, have three nonnegligible moments of inertia, hence three degrees of freedom in rotation, for a total of six.

The translational and rotational degrees of freedom of a molecule are often referred to as the *classical* degrees of freedom, in contrast to the *vibrational modes*. For example, as the temperature of a diatomic gas is increased, a point is reached where the atoms start to vibrate along the axis of the dumbbell. This represents a "soaking up" of energy from the surrounding molecules and causes significant deviations in the thermodynamic properties predicted by the elementary theory of perfect gases.

In the absence of vibrational motion (which is typical at less-elevated temperatures), kinetic theory predicts that the ratio of specific heats is given by

$$\gamma = \frac{n+2}{n},$$
(1.9-6a)

where n is the number of classical degrees of freedom. Thus, $\gamma = \frac{5}{3}$, $\frac{7}{5}$, $\frac{4}{3}$, for monatomic, diatomic, and polar triatomic gases, respectively. For most gases there is a range of temperatures and pressures where Equation 1.9-6a gives very satisfactory predictions for γ.

1.10 ON THE UNITS OF VISCOSITY

The dimensions of μ, usually referred to as the *dynamic viscosity*, can be obtained from Newton's law of resistance, Equation 1.8-3, relating stress to velocity gradient. In SI* the units of stress, which are the same as pressure, are given in *pascal* (Pa); i.e., 1 Pa = 1 N/m^2 = 1 $kg/m \cdot s^2$, the last being in SI base units. The SI units for velocity gradient are $(m/s)/m = s^{-1}$. Consequently, to obtain the SI units of viscosity, we must have

$$\text{Units of } \mu = \frac{\text{Units of } \tau}{\text{Units of } du/dy} = \frac{\text{Units of (force/area)}}{\text{Units of (velocity/length)}} = Pa \cdot s. \quad (1.10\text{-}1)$$

Thus, the units of dynamic viscosity in SI are *pascal second*.

Many engineers, however, are still wedded to experimental data obtained in either the now-superseded cgs system, in which viscosity is measured in *poise* (1 poise = 1 g/cm sec), or to the equally superseded British system of engineering units, or to one of its variants encountered in the U.S. In the engineering system of units, one common way of expressing viscosity units is in terms of lb_f sec/ft^2 = slug/ft sec. In some important references, viscosity data are given in terms of lb_m/ft sec. To be consistent, in the latter case, stress, computed from Newton's law of resistance would turn out to be poundal/ft^2, where 1 lb_f = 32.174 poundal. In a practical sense, the poundal is as extinct as the dodo.

---

* SI is the abbreviation for *Système International*, from the French; see Chapter 5.

TABLE 1.10-1
Conversion of Viscosity-Related Units

To Convert From	To	Multiply By
poise	Pa·s	10^{-1}
$lb_f\ sec/ft^2$	ʺ	47.880
$lb_m/ft\ sec$	ʺ	1.488 16
stoke	m^2/s	10^{-4}
ft^2/sec	ʺ	9.290×10^{-2}
lb_f	N	4.448 22
slug	kg	14.594
lb_f/ft^2	Pa	47.880
bar	ʺ	10^5
atm	ʺ	101.325×10^3

A similar situation prevails for the *kinematic viscosity* $v \equiv \mu/\rho$,* where, in SI

$$\text{Units of } v = \frac{\text{Units of } \mu}{\text{Units of } \rho} = m^2/s. \qquad (1.10\text{-}2)$$

The basic cgs unit for kinematic viscosity is called the *stoke*, (1 stoke = 1 cm²/sec). The corresponding engineering unit is given in ft²/sec. Table 1.10-1 is useful for converting data between systems.

1.11 ON THE VISCOSITY OF REAL GASES

Equation 1.8-8 indicates that the dynamic viscosity of a perfect gas is proportional to the square root of the absolute temperature, according to the elementary kinetic theory that treats a molecule as an elastic sphere. The simplest frame in which to analyze a collision of two such molecules is that of an observer located at the instantaneous mass center, which lies on the line through the mass centers of the two particles at collision. During collision, the kinetic energy of the particles is absorbed by the molecules as strain energy of deformation. During rebound, the strain energy is reconverted to kinetic energy such that for each molecule the component of velocity along the line of mass centers is reversed. The component of velocity of each molecule perpendicular to the line of mass centers remains unchanged. It is further assumed that the deformation is negligible compared with the diameter of the molecule and, effectively, that the reversal takes place instantaneously. For these reasons the model is often referred to as one of rigid elastic spheres.

The elastic sphere model has inherent deficiencies. For example, in such a collision the trajectories of the colliding molecules do not depend on the molecular

---

* This use of v for kinematic viscosity should not be confused with the same symbol used for particle number density.

diameter, or on the actual velocities, as long as the two velocities are scaled up by the same constant.

This can be contrasted to the (more accurate, but more complex) model of molecules treated as point masses surrounded by force fields. In the latter the nature of the collision (i.e., the encounter) depends significantly on the magnitudes of the molecular velocities. The greater the relative speed of the mass centers, the greater the penetration of each into the other's repulsive force field. In such case the calculation of the trajectories becomes enormously more complicated.

A second deficiency of the elementary theory is that it assumes that only binary collisions (interactions) are of importance. In other words, the molecules are so far separated from each other that simultaneous collisions of three or more molecules are so rare as to be negligible. This assumption, as borne out by experiment, is valid as long as the gas density is sufficiently low. Thus, we refer to the viscosity of *dilute gases*. Strictly speaking, the theory applies in the limit as $\rho \to 0$.

SUTHERLAND'S VISCOSITY LAW FOR DILUTE GASES

To overcome the deficiency of the rigid elastic sphere model, Sutherland proposed* for dilute gases a model in which the effective molecular diameter decreases with temperature, and in which each molecule is surrounded by a force field that attracts at "large" separations and repels when two molecules are very close. This results in the law:

$$\frac{\mu}{\mu_r} = \frac{T_r + S}{T + S} \left(\frac{T}{T_r} \right)^{3/2}. \tag{1.11-1}$$

The parameter S, which varies from gas to gas, is known as *Sutherland's constant*; it has the dimension of temperature. Its value is proportional to the potential energy of mutual attraction when the pair of molecules is in contact. The quantity μ_r corresponds to a measured value of viscosity at temperature T_r, which is usually taken in the middle of the range of temperatures at which Equation 1.11-1 applies.

To illustrate the accuracy of Sutherland's law, consider a comparison with the recent tabulations of Stephan and Lucas (1979) on nitrogen, which were obtained by a comprehensive correlation of the most-advanced kinetic theory with available data, yielding "smoothed" values over an impressive range of pressure and temperature. Since Sutherland's law applies only to dilute gases, the comparison is made for the lowest pressure tabulated by Stephan and Lucas, which is for 1 bar $\equiv 10^5$ Pa ≈ 0.987 atm. The discrepancies are so small that a plot would not be adequate to show the differences, and it is necessary to tabulate the values (see Table 1.11-1).

The value of Sutherland's constant was $S = 106.7$ K, from Hilsenrath (1955). The reference values were chosen to force agreement at $T_r = 600$ K, $\mu_r = 29.08 \times$

---

* For a comprehensive treatment of the mathematical theory of viscosity, thermal conduction, and diffusion, see the treatise by Chapman and Cowling (1953), who cite the original references.

TABLE 1.11-1
Comparison of Sutherland's Relation for the
Viscosity of Nitrogen with the Calculations of
Stephan and Lucas (S&L), 1979 at 1 bar pressure

	$10^6\mu$ (Pa·s)			$10^6\mu$ (Pa·s)	
$T(K)$	Eq. 1.11-1	S&L	$T(K)$	Eq. 1.11-1	S&L
80	5.36	5.52	500	25.77	25.77
90	6.07	6.20	600	29.08	29.08
100	6.76	6.88	700	32.10	32.10
150	10.01	10.06	800	34.90	34.91
200	12.90	12.92	900	37.50	37.53
250	15.50	15.49	1000	39.95	41.50
300	17.87	17.82	1100	42.28	42.32
400	22.08	22.04	1200	44.48	44.53
450	23.98	23.96	1300	46.59	46.62

10^{-6} Pa·s. It turns out, except at the two lowest temperatures, the discrepancies between Sutherland's law and the Stephan and Lucas values are less than the deviations of their computed values from the experimental data on which they are based. The Stephan and Lucas values deviate from experiment by not more than 2.5% at a pressure of 1 bar.

On the Viscosity of Anomalous Gases

For polar molecules and for molecules formed of long chains of atoms, the rigid elastic sphere model introduces inaccuracies that may result in substantial errors for some substances. In spite of this, it may still be possible to apply Sutherland's law over restricted ranges of temperature if Sutherland's "constant" is treated as temperature dependent and if the required accuracy is not extreme. See Hirschfelder et al. (1954, p. 567), for example.

On the Viscosity of Dense Gases

There are branches of fluid mechanics — for example, subsonic aerodynamics — where the use of the dilute gas relations is a good approximation to the facts. However, at high pressures (thus, high densities) and/or low temperatures, the simple model of binary molecular interactions becomes inadequate. In particular, in the neighborhood of the critical point, where the distinction between the gaseous and liquid state vanishes, it is essential to account for the fact that the molecules are so closely packed that the diameter of a molecule is no longer negligible compared with the mean-free-path length. This complicates the theory enormously.

The whole problem is additionally complicated by the fact that the experimental difficulties in making viscosity measurements at these extreme state parameters are also greatly enhanced, particularly near the critical point. Thus, the limit of confidence in making any comparison between theory and experiment decreases accordingly.

Stephan and Lucas note that the viscosity of some pure, dense gases can be approximated by the expression

$$\mu(T,\rho) = \mu_0(T) + f(\rho). \tag{1.11-2}$$

The first term on the right of Equation 1.11-2 is the dilute gas contribution. The term $f(\rho)$ is an additive contribution, depending on density only, and is called the residual viscosity. Even in the best-behaved gases, however, the accuracy of Equation 1.11-2 is fair at best. In other cases Equation 1.11-2 is replaced by

$$\mu(T,\rho) = \mu_0(T) + \mu_1(T) \cdot \rho + \Delta\mu(\rho,T), \tag{1.11-3}$$

where $\mu_1(T)$ is the first density correction, and $\Delta\mu(\rho,T)$ a remainder, which is generally much smaller than the other two terms. Equation 1.11-3 is so complicated in practice that it requires determination of 12 empirical constants for each gas.

The forms of Equations 1.11-2 and 1.11-3 imply a format of data tabulation as a function of density and temperature. However, users generally need viscosity data as a function of pressure and temperature. Because of the complexity of the pressure–temperature–density relations for a dense gas, it becomes apparent that to produce viscosity tables for specified values of pressure and temperature requires an iterative calculation.

Figure 1.11-1 is a plot of the viscosity of nitrogen, taken from Stephan and Lucas (1979),* which delineates the limitations of the dilute gas theory (the lowest curve on the plot). At the left, the negative slope of the curves indicates that near the critical point viscosity decreases with temperature, which is characteristic of liquids. The uncertainties in some of these curves may reach 10% in the critical region, but for the most part are 6% or less, particularly at the higher temperatures.

In the high-temperature range, effects of pressure become relatively small. On the other hand, at a temperature sufficiently high, a diatomic gas begins to dissociate so that the gas becomes a mixture of diatomic molecules and atoms. This effect adds still another complication to the process of computing fluid properties — viscosity included — from the kinetic viewpoint. For nitrogen, the temperature below which dissociation effects can be neglected is about 1000 K. At the other extreme, in the neighborhood of the critical temperature (T_c = 126.4 K for N_2), pressure increases result in extremely large increases in the viscosity, especially above the critical pressure (p_c = 33.98 bar for N_2).

---

* In the original volume, which covers 50 pure fluids, an additional, expanded plot is shown for the high-temperature range for nitrogen. Stephan and Lucas also have an extensive bibliography of references for each gas.

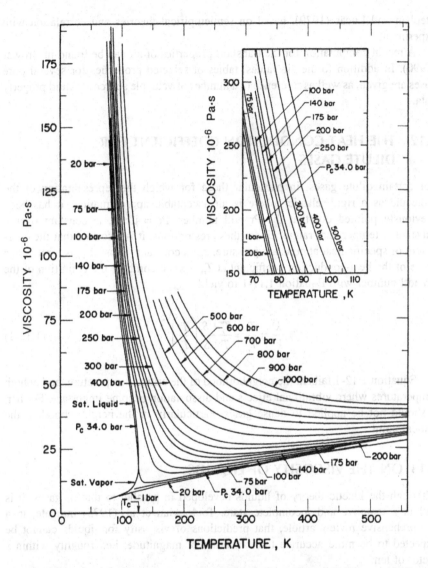

FIGURE 1.11-1   Viscosity of nitrogen. (From Stephan, K., and Lucas, K., *Viscosity of Dense Fluids*, Plenum Press, New York, 1979. With permission.)

VISCOSITY OF GAS MIXTURES

The techniques of kinetic theory can also be extended to mixtures of gases. If the viscosity dependence of any pair of pure gases is known, then the mixture viscosity can be calculated according to rules outlined in the literature, e.g., Hirschfelder et al. (1954, p. 528). For some dilute–gas mixtures Sutherland's law can be applied successfully, where a mean value of S must be applied. For dense-gas mixtures, in particular air, resort must be made to tabulations or plots of data such as given by

Stephan and Lucas (1979), based on semiempirical theories and correlation with experiment.

A recent computation for the transport properties of air can be found in Brower (1990). In addition to the air values, tables of selected properties for several pure gases are given, as well as reference to a number of valuable sources of fluid property data.

1.12 THE HEAT CONDUCTION COEFFICIENT FOR DILUTE GASES

For certain dilute gases, in particular those for which the representation of the molecule as a rigid, elastic sphere is an acceptable approximation, it has been previously pointed out that the Prandtl number, $Pr = \mu c_p/k$, is constant over a substantial temperature range. Under these restrictions, it is also true that the coefficient of specific heat at constant pressure, c_p, is constant. Thus, if k_r is a reference value of the heat-conduction coefficient at T_r, we can combine the definition of the Prandtl number with Equation 1.11-1 to yield

$$\frac{k}{k_r} = \frac{T}{T_x} = \frac{T_r+S}{T+S}\left(\frac{T}{T_r}\right)^{3/2}. \qquad (1.12-1)$$

Equation 1.12-1 fails in the neighborhood of the critical temperature, and at high temperatures where vibrational effects and dissociation become important. Further, it should not be applied to dense gases, particularly in the neighborhood of the critical point.

1.13 ON THE VISCOSITY OF LIQUIDS

Although the kinetic theory of liquids developed in parallel to that of gases, it is still in a tentative and incomplete state. Touloukian et al. (1975) indicate, in a comprehensive review article, that predictions of viscosity for liquids cannot be expected to be more accurate than an order of magnitude, i.e., roughly within a factor of ten.

Since possible errors of this magnitude are unacceptable for engineering purposes, it is necessary to rely on tabulations of experimental data, or on semiempirical analytical expressions correlated with experiment. Despite the fact that the work of Touloukian et al. is massive, comprising 1803 sets of data on 59 pure fluids and 129 fluid mixtures, including a number of dilute gases, it is not hard to find fluids or experimental states (temperatures and/or pressures) that are not included. A more recent work by Viswanath and Natarajan (1989) also has an extensive compilation of data. The lack of comprehensive data is really a symptom of the fact that reliable experimental data are expensive to compile, hence the vast gaps in the literature.

Another substantial compilation of viscosity values can be found in Vargaftik (1980), a volume that includes tables of thermophysical properties, including heat-

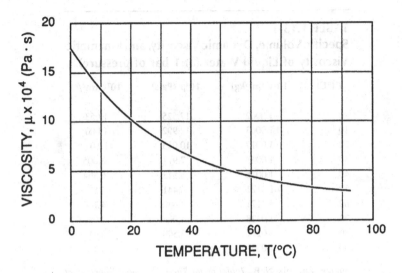

FIGURE 1.13-1 Viscosity of liquid water at 1 bar pressure. (From Vargaftik, N.B., *Tables on the Thermophysical Properties of Liquids and Gases*, Hemisphere, Washington, D.C., 1980. With permission.)

conduction coefficients and specific heats, among others. These tables are obviously based on correlations of experimental data, but the original sources are not listed so that confidence estimates of the type given by Touloukian et al. are not available.

One example suffices to illustrate the nature of viscosity dependence on temperature of a newtonian liquid. In Figure 1.13-1, a plot of the viscosity of liquid water at 1 bar of pressure is given. Near the low end of the temperature range the ice state is reached. Near the high end, the boiling temperature is reached which is slightly lower than 100°C because of the slightly lower than atmospheric pressure.

In liquid, viscous flow problems it is more typical that the investigator needs the kinematic viscosity $v \equiv \mu/\rho$ rather than the dynamic viscosity. However, the original metric system — i.e., the cgs system — was designed such that the numerical value of density in gram per cubic centimeter units, and its inverse, specific volume $v = 1/\rho$, is unity at the temperature (3.98°C) at which density reaches a maximum value, at standard atmospheric pressure. Consequently, since the density of water at this pressure varies little with temperature over the range it remains liquid the magnitudes (excepting some power of 10) of the dynamic and kinematic viscosities (but not their units, of course) are approximately equal in the same range. When converted to SI units the same behavior is observed if we compare $10^4 \mu$ and $10^7 v$, as shown in Table 1.13-1.

1.14 AVAILABILITY AND FUTURE SOURCES OF FLUID PROPERTY DATA

When individuals have conducted an extensive search for sources of fluid data, they are likely to reach certain apparently inconsistent conclusions. In the first place, the

TABLE 1.13-1
Specific Volume, Dynamic Viscosity, and Kinematic
Viscosity of Liquid Water (at 1 bar of pressure)

T (°C)	$10^3 v$ (m³/kg)	$10^4 \mu$ (Pa·s)	$10^7 v$ (m²/s)
0	1.0002	17.525	17.53
10	1.0003	12.992	13.00
20	1.0017	10.015	10.03
30	1.0043	7.971	8.005
40	1.0078	6.513	6.564
50	1.0121	5.441	5.507
60	1.0171	4.630	4.709
70	1.0228	4.004	4.095
80	1.0292	3.509	3.611
90	1.0361	3.113	3.225

Source: Vargaftik, N. B., *Tables on the Thermophysical Properties of Liquids and Gases*, Hemisphere, 1980. With permission.

library shelves have vast collections of data. The data are based on calculations and/or measurements, but often the information is piecemeal, and may be inconsistent with that of other investigators. The data may lack traceability to the original source, or, if based on calculations, the governing relations may not be cited. Often, the accuracy is not specified, and if data over large ranges of temperature and pressure are required, there may be enormous gaps.

Olien (1979) deals with these questions in a very interesting review article. As industry becomes more sophisticated, and builds ever larger and more expensive plants (for example, coal gasification plants), the need for accurate data on fluid properties becomes intense. He points out that there now exist a number of computer packages* that permit the calculation of fluid properties, such as density, transport constants, reaction rate constants, and phase-equilibrium data, over a wide range of pressure and temperature. The packages rely on intricate equations correlated against experimental data.

Olien indicates that in the two decades ending in the year 2000, the capital requirements for the U.S. supplemental gas industry alone are estimated to be on the order of $175 billion in 1977 dollars. By use of these packaged programs, which allow rapid calculation of fluid properties, Olien indicates that savings of tens of millions of dollars per year in the design and operation of these plants should be possible. At the same time, the packages may represent a serious pitfall for the

* The recent work by McCarty (1980) is a good example. For a selected group of cryogens (hydrogen, helium, neon, nitrogen, oxygen, argon, and methane) he gives a FORTRAN IV program to calculate the thermodynamic and transport properties from the triple point to some upper level of pressure and temperature. The output includes pressure, temperature, density, entropy, enthalpy for all these fluids, and the specific heats, speed of sound, viscosity, and heat-conduction coefficients for most of them.

unwary user in that the output may be accepted uncritically. In fact, it is possible to use interactive programming so that the user is not aware of what the actual predictions of fluid properties are, on a day-to-day basis. If a situation calls for data that lie outside of the range at which the reliability of the package is well established, good engineering judgment might reject the computations, or at least demand a critical assessment of them, if the data were brought to the attention of a competent engineer. Thus, Olien indicates that it is vital for every package to be "accompanied by well documented and justifed confidence limits, the existence and use of which is clearly understood by the users....In addition, the package should contain provisions for notifying the user when calculations are requested which are outside the range of the particular method used."

These computer packages, which are themselves products of years of experimental and theoretical research, are the wave of the future for use with large-scale chemical plants. Nevertheless, Olien also points out that their availability tends to stifle research. The inclination — particularly on the part of technical management — to consider their availability as a sign that the subject is complete may result in the withdrawal of funding in areas where the data are sparse or nonexistent.

Most recently, enormous computing power, which was unthinkable a decade previous, has become available for desktop computers and to the individual engineer. For example, a program named Gas Flex* uses the Benedict–Webb–Rubin equations of state to predict gas properties over a wide range of operating conditions. After a user inputs the pressure and temperature, the program computes the state properties such as enthalpy, entropy, saturation point, specific volume, sonic velocity, c_p, heat-conduction coefficient, and dynamic viscosity. The database includes more than 35 gases including the common hydrocarbons, nitrogen, and water vapor.

PROBLEMS

1.1. If containers of the gases neon and argon (see Table 7.2-1 for molecular weights) were established at standard atmospheric temperature and pressure (see Table 2.16-2), which would have the higher mean free speed, kinetic energy of random motion per unit volume, and number density? Compute the values for both gases.

1.2. For the gases of Problem 1.1 determine the mean-free-path lengths for each and the value of the dynamic viscosity based on elementary kinetic theory. Compare the latter result with any tabulated data, and give the reference.

1.3. Consider a gas in a cubical container at a known pressure and temperature (state 1). The gas is somehow compressed into one eighth the volume and

---

* Developer: Flexware, Inc., 168 Braxton Pl. S.W. Calgary, Alberta T2W 1C8, 403-238-1093, E-mail: flexware@agt.net.

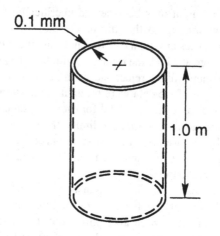

then allowed to cool to its original temperature (state 2). Let any variable f in the two states be related by $f_2 = nf_1$ where n is a pure number (e.g., $p_2 = np_1$). Compute n for each of the following variables:

a. Pressure
b. Mass density
c. Number density
d. Mean-free-path length
e. Kinetic energy per unit volume
f. Ratio of specific heats
g. Coefficient of specific heat at constant pressure
h. Dynamic viscosity
i. Knudsen number for a fixed-diameter spherical molecule
j. Speed of sound

1.4. Give a qualitative explanation of why viscosity of a liquid decreases with temperature whereas the viscosity of a gas increases with temperature.

1.5.

a. Estimate the number of air molecules within a room $10 \times 16 \times 8$ ft, under standard atmospheric conditions (see Table 2.16-2). Suppose that the room is heated to 80°F at the same pressure; compare the second number of molecules with the first.

b. Compute the total kinetic energy of random motion of the molecules in the first situation. From this value compute the specific heat at constant volume for air, convert to SI, and compare the result with the value from Table 2.16-2. If the values differ significantly, explain.

c. Compare the particle mean free speed at sea level with the corresponding speed of sound. Suggest a plausible reason why they are different and why the latter must always be higher.

1.6. A circular cylinder 0.5 m in diameter rotates inside a cylindrical container at 1000 rpm. The radial clearance between the cylinder and the wall is 0.1 mm.

a. If the fluid between the walls is air at standard atmospheric conditions, compute the wall shear stress.

b. Suppose the density were doubled. How would the shear stress change? Explain.

c. The preceding computations are based on continuum theory. If the apparatus were tested in an environment consistent with an arbitrarily high altitude, at what simulated altitude would you expect agreement between theory and experiment to break down? Justify your conclusion.

d. If the cylinder is 1 m high, determine the torque necessary to maintain the motion specified in part (a).

1.7. There are two categories of gaseous hydrogen that differ only in the direction of the electron spin. In a para-hydrogen molecule the two electrons spin in opposite directions. Consider the data from McCarty (1980) for para-hydrogen at standard atmospheric pressure (and, thus, a dilute gas):

T (K)	400	800	1200
$10^7 \mu$ (Pa·s)	92.97	136.66	176.49
$10^3 k$ (J/s·m·K)	197.45	293.77	511.15

a. Using the data for $T = 400$ K and 800 K, calculate a value for S in Sutherland's law in Equation 1.11-1; then, using the calculated S, compute μ at 1200 K and compare with the tabulated value.

b. Then, treating H_2 as a perfect gas, estimate the values of the heat-conduction coefficients k at the three temperatures and compare with the tabulated values.

1.8. Sutherland's viscosity law, $\mu = \mu(T)$ is given as

$$\frac{\mu}{\mu_r} = \frac{S+T_r}{S+T}\left(\frac{T}{T_r}\right)^{3/2},$$

where S is Sutherland's constant with dimension of absolute temperature and μ_r is a reference viscosity at reference temperature T_r. Strictly speaking, Sutherland's law should be applied only to dilute gases, i.e., in the limit as $\rho \to 0$. Following are data abstracted from Stephan and Lucas (1979) that represent their "best" smoothed values for the viscosity of air at 1 bar of pressure.

T (K)	200	300	400	500	600	700	800	900
$10^6\mu$ (Pa·s)	13.4	18.4	22.7	26.6	30.1	33.4	36.3	39.0

a. Show that, at $T_r = 500$ K, S can be determined with only one other datum. Determine S using the value of viscosity at $T_r = 900$ K.

b. Make calculations of μ at representative values of T and compare the results with the tabulated values.

c. Convert the law to the lb_f-ft-sec-°R system and compare the resulting expression to that of Equation A3 of NACA Rep. 1135, p. 19:

$$10^8\,\mu = 2.270\;\frac{T^{3/2}}{T+198.6}, \quad \text{in lb sec/ft}^2 .$$

1.9. A special electrical discharge lamp uses xenon ($\sigma = 4.06 \times 10^{-8}$ cm) as its working conductor. The lamp bulb is filled at standard temperature $T_1 = T_a$ and at $p_1 = 0.1$ atm pressure.

a. Determine the following, and give units for each response: specific gas constant of xenon, molecular number density ν_1, mass density ρ_1, mean-free-path length λ_1, initial mean free speed C_1.

b. If the operating temperature stabilizes at 800°F, determine operating pressure p_2, density ρ_2, mean free speed C_2, viscosity μ_2.

1.10. A cone viscosimeter design is proposed that consists of an inverted cone within an inverted conical reservoir (see figure) such that there is maintained a small, constant clearance d between the rotating cone surface and that of the reservoir. The cone, of dimensions shown, rotates at a fixed angular velocity $\Omega = e_z\Omega$.

a. Assuming that the fluid velocity varies linearly in the direction normal to the cone surfaces, determine an expression for the shear stress on the cone wall as a function of z, based on Newton's law of resistance.

b. Then determine expressions for the differential shearing force on a differential area dA, in cylindrical coordinates (r, ϑ, z), see diagram. From this, determine the corresponding differential moment $dM = e_z dM$.

c. Compute the integrated moment generated on the cone for the case $\Omega = 5$ rpm, $h = 6$ in., $\alpha = 30°$, $d = 0.03$ in., if the fluid viscosity is $\mu = 4 \times 10^{-5}$ lb sec/ft².

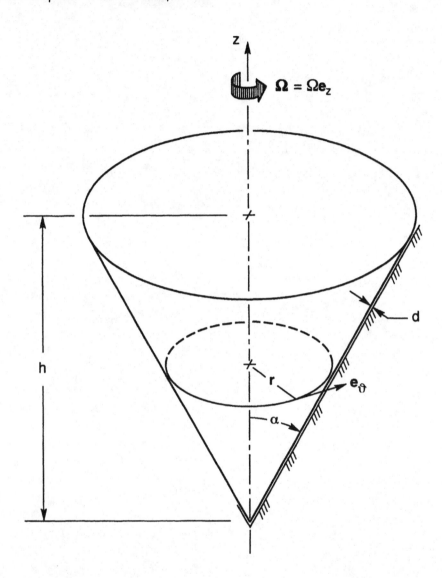

Figure Problem 1.10

2 Fluid Statics

Let it be supposed that a fluid is of such a character that, its parts lying evenly and being continuous, that part which is thrust the less is driven along by that which is thrust the more; and that each of its parts is thrust by the fluid which is above it in a perpendicular direction if the fluid be sunk in anything and compressed by anything else.

From *On Floating Bodies*
by Archimedes (287–212 B.C.E.),
translated from the Greek by
T.L. Heath (1897).

2.1 INTRODUCTION

Fluid statics, along with its parent subject statics, is undoubtedly the branch of physical science that was placed on a sound theoretical basis at an earlier date than any other. This was due almost entirely to the work of Archimedes who had the misfortune to perish under a Roman sword at the fall of Syracuse in 212 B.C.E. While ignorant of differential and integral calculus, Archimedes had knowledge of the underlying ideas on which differential calculus is founded, and he succeeded in discovering the essential principles governing buoyancy of floating or immersed bodies, including the difficult question of stability.

Although ancient in origin, fluid statics has many important applications in the modern era in such diverse fields as meteorology, astronautics, naval architecture, hydraulics, instrumentation, aeronautics, oceanography, and space technology.

2.2 EQUILIBRIUM AND PRESSURE IN A FLUID

The starting point in fluid statics, as for any statical system, is the concept of *equilibrium*, in which it is postulated that the sum of all external forces acting on an isolated system is zero; i.e., the *equation of equilibrium* is

$$\Sigma F = e_x \Sigma F_x + e_y \Sigma F_y + e_z \Sigma F_z = 0. \qquad (2.2\text{-}1)$$

Individually, therefore, each of these summations is zero. Likewise, for a system in equilibrium, the sum of the moments of these forces about any point is zero.

In a fluid mechanical system with a large extent of fluid, a small but finite element of fluid is not isolated but is in contact with surrounding fluid on all sides. However, using the same techniques of isolation employed in mechanics, it is possible to visualize a fluid element in equilibrium if, when the surrounding fluid is "removed," the forces that are exerted on the element are replaced by equivalent surface forces.

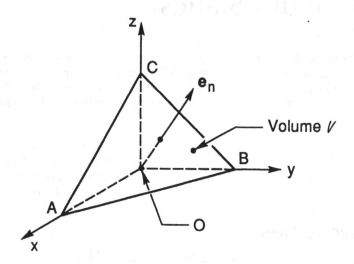

FIGURE 2.2-1   Equilibrium of a fluid tetrahedron.

The idea that pressure in a fluid acts normal to a surface, independently of the surface orientation, is due to Pascal. To demonstrate this, consider a fluid in statical equilibrium, in particular the isolated tetrahedron $OABC$, Figure 2.2-1, with the z-axis chosen vertically up. Since we have specified a statical condition, there can be no shear stresses; otherwise, by definition of a fluid, deformation would occur.

Let $A = Ae_n$ be the area of ABC and let A_x, A_y, A_z be the areas* of the three orthogonal faces. Also, let p, p_x, p_y, p_z be the average pressures acting on ABC, A_x, A_y, A_z, respectively. Postulating that the pressure be an analytic function $p = p\,(x, y, z)$, then, by the theorem of the mean, there is always at least one point on each face where the actual pressure is equal to the respective average value. Furthermore, pressure is defined to act in such a way as to produce a force in the direction opposed to the outer normal.

From the equilibrium relation, Equation 2.2-1, we have

$$p_x A_x - pA \, \cos\!\left(e_n, e_x\right) = 0,$$

$$p_y A_y - pA \, \cos\!\left(e_n, e_y\right) = 0, \qquad\qquad (2.2\text{-}2)$$

$$p_z A_z - pA \, \cos\!\left(e_n, e_z\right) - \rho g V = 0,$$

where $\cos\,(e_n, e_x)$ is the direction cosine of the unit vector with the x-axis, etc., and $-\rho g V$ is the force of gravity exerted on the fluid element, V being the volume.

But, since $A_x = A \, \cos\!\left(e_n, \, e_x\right)$, etc., we have from the first two relations $p = p_x = p_y$. Dividing the third equation by A_z and noting that $V/A_z = (\tfrac{1}{3})\overline{OC}$, we

---

* From elementary vector theory $A_x \equiv A \cdot e_x = Ae_n \cdot e_x = A \cos\,(e_n, e_x)$, where (e_n, e_x) denotes the angle between the e_n and the e_x vectors.

then let $\overline{OA}$, $\overline{OB}$, and $\overline{OC} \to 0$ simultaneously. The gravitational term vanishes, and we are left with *Pascal's law*:

$$p = p_x = p_y = p_z. \qquad (2.2\text{-}3)$$

Since the direction of e_n was chosen arbitrarily, Equation 2.2-3 applies regardless of the orientation of the surface.

2.3 CLASSIFICATION OF FORCES

There are only two essentially different kinds of forces that can be experienced by a fluid: surface forces and field forces.

SURFACE FORCES

Surface forces can be further subdivided into viscous forces and pressure forces. The former may have an arbitrary orientation on a surface and thus have components normal and tangential to the surface. As it turns out, viscous forces enter only when velocity gradients are present in a fluid. Therefore, discussion of viscous effects is reserved for the chapters on fluid dynamics. On the other hand, pressure forces exist as a consequence of molecular agitation and must be considered in fluid statics. Pressure is a scalar quantity, defined to be positive when the associated force acts on a surface in a direction opposed to the outer normal, i.e., in this case toward the inferior of the fluid element being isolated. Surface tension forces also act at interfaces between two different materials. These are treated in Section 2.21.

FIELD FORCES

Forces resulting from the presence of fields, such as electromagnetic or electrostatic fields, are termed *field forces*. In magnetohydrodynamics, a subject of research interest, the coupling of electromagnetic theory and fluid dynamics presents challenging theoretical and experimental difficulties. Electromagnetic forces depend on the charge distribution and field strength. A more prosaic example of a field force is the gravitational force, already employed in Section 2.2. Because gravitational force is proportional to the mass of a body (i.e., the fluid mass), it is generally referred to as a *body force*. Applications in the following sections are restricted to body forces.

2.4 THE FUNDAMENTAL EQUATION OF FLUID STATICS

The mode of thinking used in this section is one which we shall use frequently in the following chapters. We assume without proof that certain functions used to describe fluid behavior — pressure, for example — are analytic. Hence, there is a continuous variation of this quantity in the neighborhood of an arbitrary point, so

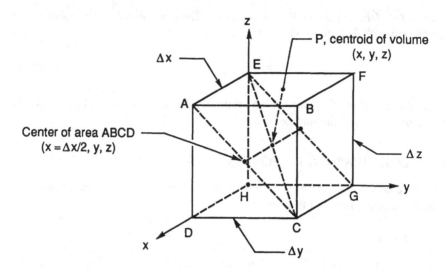

FIGURE 2.4-1   Fluid element in equilibrium.

that we can use Taylor's expansion to obtain the value for a neighboring point in terms of the original function, and its derivatives, evaluated at the *original* point. This means that the molecular nature of matter is ignored and the fluid is treated as a continuum, a concept that applies only if the characteristic body dimension is large compared with the molecular mean-free-path length.

Although the analysis can be worked out for an element of fluid of arbitrary shape, it is simpler to deal with a specific geometry, namely, the rectangular parallelpiped $AB...GH$, as the fluid element that is to be placed in equilibrium, with sides of length Δx, Δy, Δz, respectively (see Figure 2.4-1).

The centroid P of the fluid element has the coordinates (x, y, z); the center of area of the face $ABCD$ is $(x + \Delta x/2, y, z)$, and so on. Since Equation 2.2-1 can be decomposed into three independent equations, the equation for the x-direction is

$$\Sigma F_x = F_{x_{\text{surface}}} + F_{x_{\text{field}}} = 0. \tag{2.4-1}$$

PRESSURE FORCES

To compute the net x-component of the pressure force it is necessary to take into account only faces $ABCD$ and $EFGH$ since the others can contribute no force components parallel to the x-axis. The pressure, as just noted, is assumed to be a continuous well-behaved mathematical function $p(x, y, z)$, which, by definition, is independent of time in a statical system. Thus, pressure can be represented as an analytical function having an infinite number of derivatives, none of which can ever be infinite in the region under consideration.

According to Taylor's theorem, the pressure at any point in the vicinity of the centroid can be expressed as a function of the pressure and its derivatives *at the point P* itself. Therefore, expanding the pressure in a Taylor's series at $(x + \Delta x/2, y, z)$, we have

$$p(x + \Delta x/2, y, z) = p(x, y, z) + \frac{\partial p}{\partial x}\frac{(\Delta x/2)}{1!} + \frac{\partial^2 p}{\partial x^2}\frac{(\Delta x/2)^2}{2!} + \frac{\partial^3 p}{\partial x^3}\frac{(\Delta x/2)^3}{3!} + \cdots$$

$$= p + \Delta x\left[\frac{1}{2}\frac{\partial p}{\partial x} + O(\Delta x)\right],$$

where in Equation 2.4-2 we have introduced the symbol $O(\Delta c)$, which is read "of order Δx," to denote a catchall function which involves derivatives of at least the second order or higher, multiplied by Δx to at least the first power. It is important to bear in mind that these derivatives can never be infinite because of the assumed analyticity of $p(x, y, z)$; consequently, $O(\Delta x)$ is always finite.

It is now assumed that on $ABCD$ the pressure force is approximately given by the product of the area $\Delta y\Delta z$, and the pressure given by Equation 2.4-2. Since we shall let the volume shrink toward zero prior to obtaining the final result, the error involved in this approximation ultimately vanishes.

On $ABCD$, the force acts in the negative x-direction and is given by

$$-\left\{p + \Delta x\left[\frac{1}{2}\frac{\partial p}{\partial x} + O(\Delta x)\right]\right\}\Delta y\Delta z.$$

Similarly, on $EFGH$ with center of area $(x - \Delta x/2, y, z)$, the force is

$$\left\{p - \Delta x\left[\frac{1}{2}\frac{\partial p}{\partial x} + O(\Delta x)\right]\right\}\Delta y\Delta z.$$

These combine to give the algebraic sum

$$F_{x_{\text{surface}}} = -\left[\frac{\partial p}{\partial x} + O(\Delta x)\right]\Delta V \qquad (2.4\text{-}3)$$

where the elemental volume is $\Delta V \equiv \Delta x\Delta y\Delta z$.

Body Force

Before computing the force in a particular situation, it is necessary to specify the nature of the field or fields. For convenience, in the case of a body force, we define a function f_x, resulting from a field, such that

$$F_{x_{\text{field}}} = (\text{mass})f_x = \rho\,\Delta V f_x \qquad (2.4\text{-}4)$$

where f_x is the force per unit mass due to the (unspecified) field, acting in the x-direction. Combining Equations 2.4-3 and 2.4-4 in Equation 2.4-1 produces

$$\left[\rho f_x - \frac{\partial p}{\partial x} + O(\Delta x)\right]\Delta V = 0.$$

Next, we perform a ritual which will be used on other occasions. Since the elemental volume is finite, it can be divided out, leaving

$$\rho f_x - \frac{\partial p}{\partial x} + O(\Delta x) = 0. \qquad (2.4\text{-}5)$$

The last term in Equation 2.4-5 is supposed to contain all the higher-order derivatives, every one of which is multiplied at least by Δx, Δy, or Δz.

At this point we let the volume shrink to zero, i.e., put $\Delta x \rightarrow 0$, $\Delta y \rightarrow 0$, $\Delta z \rightarrow 0$; therefore,

$$\frac{\partial p}{\partial x} = \rho f_x, \qquad (2.4\text{-}6a)$$

for the x-direction. The catchall function, whose terms involve factors of powers of Δx, becomes negligible in the limit of vanishing volume. Similarly,

$$\frac{\partial p}{\partial y} = \rho f_y, \quad \frac{\partial p}{\partial z} = \rho f_z, \qquad (2.4\text{-}6b)$$

for the other directions.

Vector Form

Multiplying the x-, y-, and z-equations by the unit vectors e_x, e_y, and e_z, respectively, and adding, results in

$$e_x \frac{\partial p}{\partial x} + e_y \frac{\partial p}{\partial y} + e_z \frac{\partial p}{\partial z} = \rho\left(e_x f_x + e_y f_y + e_z f_z\right).$$

The left-hand side is recognized to be the *gradient* of the pressure, and, therefore, the fundamental equation of statics can be written

$$\text{grad } p \equiv \nabla p = \rho f, \qquad (2.4\text{-}7a)$$

where we employ the abbreviation

$$f \equiv e_x f_x + e_y f_y + e_z f_z \qquad (2.4\text{-}7b)$$

for the vector field force per unit mass.

In Equation 2.4-7a the gradient has been expressed in terms of the vector operator *del*, which in rectangular cartesian coordinates is

$$\nabla \equiv e_x \frac{\partial}{\partial x} + e_y \frac{\partial}{\partial y} + e_z \frac{\partial}{\partial z}. \qquad (2.4\text{-}8)$$

POTENTIAL BODY FORCES

In many practical cases in solid and fluid mechanics, the body force term is found to be derivable* from a potential function. In such a case we mean that

$$f_x = -\frac{\partial W}{\partial x}, \quad f_y = -\frac{\partial W}{\partial y}, \quad f_z = -\frac{\partial W}{\partial z} \qquad (2.4\text{-}9a)$$

or

$$f = -\text{grad } W, \qquad (2.4\text{-}9b)$$

where the scalar $W = W(x, y, z)$ is a point function of the space coordinates. Equation 2.4-7, therefore, becomes

$$\text{grad } p + \rho \text{grad } W = \nabla p + \rho \nabla W = 0. \qquad (2.4\text{-}10)$$

2.5 EQUATION OF FLUID STATICS FOR A UNIFORM GRAVITATIONAL FIELD

By far the most important field force in fluid statics is that caused by the gravitational field. A precise treatment of gravity would imply use of Newton's inverse-square law of gravitational attraction. For many purposes, however, it is sufficiently accurate, and much simpler, to employ the approximation of a uniform gravitational field, with $g = 9.806\ 65\ \text{m/s}^2 = 32.174\ \text{ft/sec}^2$.** In such a case, Equation 2.4-9a, with the z-axis directed vertically up, becomes

$$f_x = -\frac{\partial W}{\partial x} = 0, \quad f_y = -\frac{\partial W}{\partial y} = 0, \quad f_z = -\frac{\partial W}{\partial z} = -g. \qquad (2.5\text{-}1)$$

From the first two relations of Equation 2.5-1 it is seen that $W = W(z)$ only and, therefore,

---

* In any good text on engineering mechanics it is shown that the necessary conditions that Equation 2.4-9 be valid are $\partial f_y/\partial y = \partial f_z/\partial z$, $\partial f_z/\partial z = \partial f_x/\partial x$, $\partial f_x/\partial x = \partial f_y/\partial y$, or in vector form $\nabla \times f = 0$.

** This is the standard value of g agreed upon by the General Conference on Weights and Measures in 1913. In general, however, for the purpose of engineering calculations, any result rounded off to more than three significant figures should be viewed with suspicion.

$$\frac{\partial W}{\partial z} \to \frac{dW}{dz} = g,$$

or

$$W = gz, \tag{2.5-2}$$

where W has arbitrarily been chosen zero on the surface $z = 0$. The function W is recognized to be a potential energy function with dimensions of energy per unit mass, and its change between any two values of z represents the potential energy per unit mass required to raise matter from the lower to the higher station. Although substitution of Equation 2.5-2 in Equation 2.4-10 produces the appropriate vector form of the fundamental equation of fluid statics for a gravitational field, it is simpler to note that from Equations 2.4-6 and 2.5-1 we obtain $p = p(z)$ only and, therefore,

$$\frac{dp}{dz} = -\rho g, \tag{2.5-3}$$

which is an ordinary differential equation. Henceforth, unless other categories of field forces are involved, when the phrase *fundamental equation of fluid statics* is used, Equation 2.5-3 is meant. Most of our attention in this chapter will be directed toward applying Equation 2.5-3, or its integrated forms, to a variety of situations.

ISOBARIC SURFACES

Since $p = p(z)$ only, in a gravitational field, a surface $z = $ constant is a surface of constant pressure, called an *isobaric* surface.

THE PROBLEM OF INTEGRATING THE FUNDAMENTAL EQUATION OF STATICS

To integrate Equation 2.5-3 in the atmosphere, for example, an additional relation is needed in the form $\rho = \rho(p)$, which implies the existence of a *barotropic relation*. What actually is required is an energy balance relation that can be reduced to a barotropic relation. As a practical matter, however, the equation actually used is an empirical relation for the atmospheric temperature distribution, which bypasses the calculation of energy exchanges in the atmosphere; see Section 2.15 for the details of the analysis.

2.6 PRESSURE–HEIGHT RELATION IN A LIQUID

Variations in the density of a liquid due solely to pressure changes are relatively insignificant compared with those resulting from variations in temperature. For example, starting with water at 4°C and at standard atmospheric pressure, a pressure increase to 10 atm would cause a decrease in the specific volume of about 4%. On

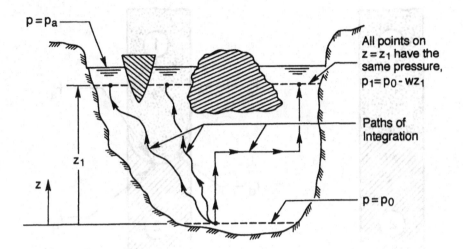

FIGURE 2.6-1 Isobaric surfaces in a fluid.

the other hand, while keeping the pressure constant, it would take only an increase of about 100°C to achieve the same percentage change (an increase) in specific volume. Naturally, heat addition to water at standard pressure and 100°C will cause boiling, but we do not consider this more-complicated situation.

For the reasons cited, in most problems involving liquids not subject to large temperature variations, it is an excellent approximation to treat the density as constant. Furthermore, with liquids it is customary to work with *weight density*, also called *specific weight*, rather than mass density. We shall use the symbol $w \equiv \rho g$ for specific weight. Equation 2.5-3 becomes, for a liquid,

$$dp = -w\,dz. \tag{2.6-1}$$

If we choose our origin so that at $z = 0$, the pressure is denoted as p_0, then

$$\int_{p_0}^{p} dp = -w \int_{0}^{z} dz$$

or

$$p = p_0 - wz, \tag{2.6-2}$$

which is henceforth called the *pressure–height* relation for a liquid, or simply the *hydrostatic* pressure variation.

Note that Equation 2.6-2 is independent of the path of integration as long as the path remains entirely within the same fluid. Furthermore, there may be several regions partially separated by a vertical barrier, which would be governed by Equation 2.6-2 as long as it is possible to construct a path between the two regions that remains entirely in the fluid. This point is illustrated in Figure 2.6-1.

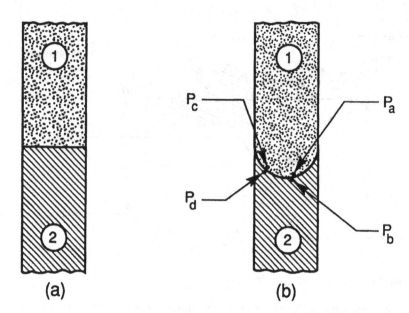

FIGURE 2.6-2 Fluids with interfaces.

In Equation 2.6-2, z is positive upward. Therefore, p decreases as z increases,* as expected. The pressure difference between any two levels z_1 and z_2 is obtained by applying Equation 2.6-2 twice to give $\Delta p = p_1 - p_2 = w(z_2 - z_1)$.

PRESSURE VARIATION AT AN INTERFACE

In dealing with liquids many situations occur when a *contact surface* or *interface* is present, separating a liquid from a gas, or two liquids from each other. If the interface is horizontal, as indicated in Figure 2.6-2a, experience has shown that the pressure must be continuous in the transition from fluid ① to ② even though $w_2 \neq w_1$. On the other hand, for a curved interface (i.e., a meniscus) the pressure difference between points b and d in the lower fluid must necessarily be different from the "same" points a and c in the upper, since the pressure differences are proportional to w_1 and w_2 in the lower and upper liquids, respectively, and to the height differential, which is the same for both fluids. This seeming anomaly can be explained only by the presence of another force. This turns out to be surface tension, which is always present in a meniscus and which results in a discontinuous pressure change as the interface is crossed. Surface tension forces are discussed in Section 2.21.

If fluid ② were a gas, then we could have neglected its pressure variations over small changes of height, such as occur near the meniscus. This can be seen by comparing the pressure gradient in a gas with that of a liquid. Selecting for comparison air (see Table 2.16-1), and water (see Table 1.13-1), at (nearly) the same conditions, we have

---

* For some purposes it is preferable to use the positive z-axis as vertically down. In this case the minus sign in Equation 2.6-2 would be replaced by a plus.

$$\frac{(dp/dz)_{H_2O}}{(dp/dz)_{air}} = \frac{\rho_{H_2O}g}{\rho_{air}g} = \frac{1}{v_{H_2O}\,\rho_{air}} = \frac{1}{(1.0010)(1.225\times10^{-3})} \approx \frac{816}{1}.$$

Therefore, for systems involving gases and liquids in contact, it is an excellent approximation to neglect pressure variations in the gas, providing that the only height changes involved are of the same order of magnitude as height changes in the liquid.

2.7 ELEMENTARY PRESSURE-MEASURING INSTRUMENTS

THE BAROMETER

The earliest devised and simplest pressure-measuring instruments are still those most widely used, and for many purposes they continue to provide the standards by which other instruments are calibrated. The first instrument employed to measure atmospheric pressure was the barometer, devised by one of Galileo's students, Evangelista Torricelli, in 1644.* It is sometimes overlooked that the barometer does not measure pressure directly; rather it is inferred from the fact that, by acting on a finite area, pressure creates a force that can support a column of liquid. The height of the column is related to the pressure through Equation 2.6-2.

The setup is shown in Figure 2.7-1. A closed-end tube filled originally with liquid of weight density w is inverted in a dish of the same liquid; if the tube is long enough, the level will fall to a height z_1 above the free surface of the liquid in the dish, which is designated $z = 0$. At the upper end of the tube the pressure depends on the vapor pressure of the barometer fluid. For a fluid to be useful as a barometer fluid, it must readily form a vapor at the ambient temperature, and the vapor pressure must be small compared with the pressure being read.

In Figure 2.7-1, assuming that p_1 is negligible compared with p_a, application of Equation 2.6-2 gives

$$p_1 \approx 0 = p_a - wz_1, \text{ or } p_a = wz_1. \qquad (2.7\text{-}1)$$

Thus, the measurement depends on the accuracy with which the liquid height and fluid density can be determined. A simple calculation shows that a liquid that permits construction of a barometer of modest dimensions must have a very high specific weight. At room temperature only mercury satisfies this requirement. Fortunately, the vapor pressure of mercury at 20°C (room temperature) is only 0.012 mm of mercury; in other words neglecting p_1 in the preceding analysis results in an error of about 1/63,000 of standard atmospheric pressure.**

---

* Middleton (1964) gives a thorough discussion of the origin and development of the barometer.
** At 20°C the density of mercury is $\rho = 13.546\ 2 \times 10^3$ kg/m³. Thus, for a pressure of 1 standard atmosphere $p_a = 1.013\ 25 \times 10^5$ Pa (see Table 2.16-1), $z_1 = 1.013\ 25/13.546\ 2(9.806\ 65) = 0.762\ 7$ m = 30.03 in. When corrected to the density of mercury at 0°C, for which $\rho = 13.595\ 5 \times 10^3$ kg/m³, the corresponding value is $z_1 = 0.760\ 0$ m = 29.92 in., values frequently cited in connection with 1 atmosphere of pressure.

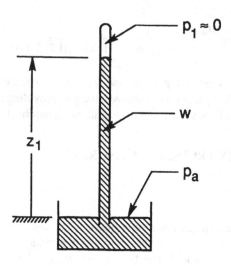

FIGURE 2.7-1 The barometer.

THE U-TUBE MANOMETER

The barometer is the standard instrument for measurement of atmospheric pressure. However, for pressures that vary only slightly from atmospheric pressure (or from any other fixed reference), the barometer is not convenient because its sensitivity to small pressure changes is low. Instead, use is made of the U-tube manometer, Figure 2.7-2. One leg of the manometer is maintained at a known reference pressure p_1. In the other leg, the unknown pressure p_2 is introduced (for the moment it is assumed that the fluids generating the pressures p_1 and p_2 are gases), which results in a height difference between the legs.

Since the pressure at the base level in both legs is the same, say, p_0, we can apply Equation 2.6-2 at heights z_1 and z_2, and subtract to obtain the pressure difference:

$$\Delta p \equiv p_2 - p_1 = w(z_1 - z_2) = w\Delta z. \qquad (2.7\text{-}2)$$

Equation 2.7-2 can be considered as an alternative form of Equation 2.6-2 that is particularly useful for manometers. Since $\Delta z = \Delta p/w$, we see that the sensitivity of a manometer (the deflection change for a given pressure difference) varies inversely with the manometer specific weight.

GAUGE PRESSURE

For some purposes, particularly in hydraulics work, it is convenient to work with *gauge pressure*, which is defined to be the pressure above absolute zero (also called static pressure) minus ambient atmospheric pressure. Gauge pressure is not a useful concept for compressible flows and is used but occasionally in this work.

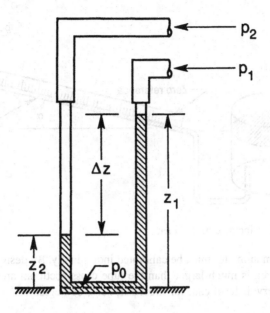

FIGURE 2.7-2 The U-tube manometer.

2.8 AN APPLICATION IN MANOMETRY

There are many special types of manometers which are designed to permit measurement of small pressure differences, some going as low as 10^{-6} in. of water. A rather elementary design is shown in Figure 2.8-1. To a reservoir of large capacity is attached a straight, glass tube inclined at an angle α to the horizontal. The manometer fluid has a specific gravity* of 0.827 at room temperature. The problem is to determine α so that the instrument sensitivity is 1 in. of slant deflection along the tube for a pressure difference of 1 lb ft^{-2}.

The 1-in. slant displacement corresponds to a vertical displacement of $(1)\sin \alpha$ in. of the meniscus. Therefore, from the manometer equation (Equation 2.7-2),

$$p_1 - p_2 = (1/12)w_{H_2O}(0.827)\sin \alpha = 1 \text{ lb ft}^{-2}$$

or

$$\sin \alpha = 12/(62.4)(0.827) = 0.232,$$

$$\alpha = 13.4°.$$

---

* *Specific gravity* (denoted herein as S.G.) is actually a misnomer for the ratio of the mass density of any substance at a specified temperature and pressure, divided by the mass density of water at 4°C and 1 atm of pressure, the latter value being $\rho_{H_2O} = 1.000\ 00 \times 10^3$ kg/m^3. This, of course, is the maximum density of liquid water at standard atmospheric pressure p_a. Since specific weight w equals the product of density by the gravitational constant g, then specific gravity is also equal to the ratio of the substance specific weight divided by that of water at 4°C and p_a, for which $w_{H_2O} = 9.806\ 65$ N/m^3 = 62.43 lb$_f$/ft^3.

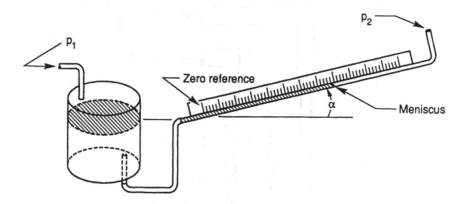

FIGURE 2.8-1 The inclined manometer.

In practice each manometer must be calibrated individually. By design, the reservoir cross-sectional area is much larger than the tube cross-sectional area. Hence, variations in the reservoir level can, for most purposes, be neglected.

2.9   FORCES AND MOMENTS ON SUBMERGED SURFACES

It is possible to derive certain formulas that enable the prediction of forces or moments on surfaces submerged in a liquid in terms of an "average" pressure and certain geometric properties of the submerged surface. However, if resort is made solely to these expressions, which involve a good deal of artificiality, it is possible that the more important idea of integrating a varying pressure distribution over a nonelementary surface would be slighted. In this section emphasis is given to the latter idea, proceeding from an elementary case to the more complicated.

FORCE AND MOMENT ON A FLAT VERTICAL WALL

Figure 2.9-1 illustrates a flat rectangular wall of area $A = Lh$ lying in the x,z-plane. It is required to compute the force exerted on the wall by a liquid of specific weight w. It is obvious that $F_x = F_z = 0$. The differential wall area is

$$dA = e_n dA = -e_y dA_y,$$ (2.9-1)

where the outer normal is, by convention, taken to be directed into the fluid. As for any situation in which the coordinate system is specified, it is preferable to shift over to the corresponding unit vectors; in this case we use $e_n = -e_y$.

The differential force on dA is

$$dF = -pdA = e_y pdA_y$$

or

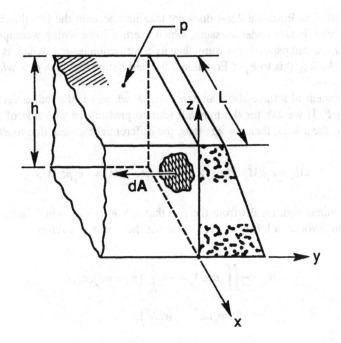

FIGURE 2.9-1 The geometry of a pressurized wall.

$$dF_y = pdA_y,$$

since a pressure force always acts opposed to the outer normal, and where dA_y lies in the x,z-plane. From Equation 2.6-2, for a coordinate system with origin fixed at the wall foot, and with atmospheric pressure p_a on the surface $z = h$,

$$p = p_o - wz = p_a + wh - wz; \qquad (2.9\text{-}2)$$

therefore,

$$F_y = \int dF_y = \iint_A pdA_y = \iint_A pdxdz. \qquad (2.9\text{-}3)$$

Since $p = p(z)$ only, and since $dA_y = Ldz$,

$$F_y = L\int_0^h pdz = L\int_0^h (p_o - wz)dz,$$

$$= p_oLh - \tfrac{1}{2}wLh^2 \qquad (2.9\text{-}4)$$

$$= (p_o - \tfrac{1}{2}wh)A.$$

The expression in Equation 2.9-4 does not take into account the fact that the other side of the wall is also under pressure, which exerts a force with a y-component. If the exterior is at atmospheric pressure, then its contribution is $-p_a A$, which is directed to the left. Adding this to F_y in Equation 2.9-4 and noting that $p_0 = p_a + wh$, the net force is $\frac{1}{2}whA$.

The moment of a force about an axis is $r \times F$ where r is the radius vector from the axis to F. If we ask for the moment (due to pressure on one side of the wall only) about the x-axis, then $r = e_z z$; thus, the differential moment due to dF is

$$dM_x = e_x dM_x = (e_z z) \times dF = ze_z \times e_y pdA_y = -e_x pzdA_y,$$

where the minus sign results from the fact that $e_z \times e_y = -e_x$, which indicates that the moment involves a left-hand rotation about the x-axis. Therefore,

$$M_x = -\iint_A pzdA_y = -L\int_0^h (p_0 - wz)zdz,$$

$$= -\left(\tfrac{1}{2} p_0 Lh^2 - \tfrac{1}{3} wLh^3\right). \tag{2.9-5}$$

It is the reasoning process lying behind Equations 2.9-4 and 2.9-5 that is important. However, for the record, it is pointed out that the first equation after Equation 2.9-3, which is valid for an area of arbitrary shape, can be written as

$$F_y = p_0 A - w \iint_A zdxdz,$$

$$= (p_0 - w\bar{z})A,$$

where the relation

$$\bar{z}A \equiv \iint_A zdxdz \tag{2.9-6}$$

defines a height $\bar{z}$ which is the distance from the x-axis to the center of area of A, and in which the integral is the first moment of area about the x-axis. Therefore,

$$F_y = (p_0 - w\bar{z})A \equiv \bar{p}A, \tag{2.9-7}$$

where $\bar{p}$ is the average uniform pressure that would have to act on A to produce the same horizontal force, and which corresponds to the actual fluid pressure at the center of area of A. Similarly, the moment about the x-axis can be put in the form

$$M_x = -\left[p_0 \iint_A z\, dA - w \iint_A z^2\, dA \right], \tag{2.9-8}$$

$$= -\left[p_0 A \bar{z} - w I_{xx} \right],$$

where

$$I_{xx} \equiv \iint_A z^2\, dA \tag{2.9-9}$$

is the second moment of area A about the x-axis.

The curious piece in the whole business, and the reason for downplaying this approach, is that the point of application $\bar{\bar{z}}$ of the force F_y that will produce the correct moment is *not* the same as $\bar{z}$, the coordinate of the point corresponding to the average pressure. The vertical coordinate of the point of application $\bar{\bar{z}}$ is given by

$$\bar{\bar{z}} \equiv -\frac{M_x}{F_y} \tag{2.9-10}$$

or, from Equations 2.9-4 and 2.9-8,

$$\bar{\bar{z}} = \frac{p_0 A \bar{z} - w I_{xx}}{F_y}. \tag{2.9-11}$$

Furthermore, from a theorem of plane statics,

$$I_{xx} = I_c + A\bar{z}^2 \tag{2.9-12}$$

where I_c is the second moment of area about an axis through the center of area of A parallel to the x-axis. Therefore, combining Equations 2.9-7, 2.9-11, and 2.9-12, we solve for

$$\bar{\bar{z}} - \bar{z} = -\frac{w I_c}{F_y}, \tag{2.9-13}$$

which states that the line of action of the force always lies below the center of area, since the right-hand side of Equation 2.9-13 is negative. For the rectangular area, Figure 2.9-2 shows for hydrostatic pressure variation why the point of application must lie below the center of area.

The relations 2.9-6 through 2.9-13 are valid for an arbitrary area A as long as it lies in a vertical plane. The ideas can readily be extended, if desired, to inclined areas and can be used if interpreted properly, even for nonplanar surfaces. However,

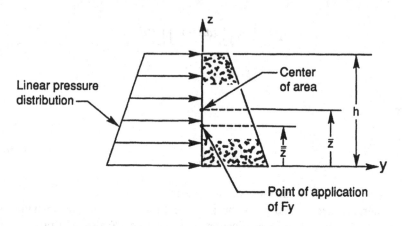

FIGURE 2.9-2 Relation of point of force application to center of area.

for curved areas we shall prefer the approach from first principles as employed in the next paragraph.

FORCES ON CURVED SURFACES

It is possible to indicate a method for evaluating forces on curved surfaces in a relatively simple way. However, the apparent simplicity is somewhat illusory since for a given shape the resulting integral may turn out to be troublesome to evaluate. For nonregular geometric shapes it would probably be necessary to use a numerical integration scheme.

Figure 2.9-3 shows a curved surface, bounded by the simple curve C, which is submerged in a liquid of specific weight w. The differential force on area dA is

$$dF = -e_n p dA. \tag{2.9-14}$$

In general, to integrate a vector function that has more than one component, the equation must be broken down into its component scalar relations prior to integrating. That is, in rectangular cartesian coordinates,

$$dF = e_x dF_x + e_y dF_y + e_z dF_z,$$

from which, for example,

$$dF_z = e_z \cdot dF = -e_z \cdot e_n p dA, \tag{2.9-15}$$

where, by definition, the right-hand dot product is expressible as a direction cosine of the angle between the vectors e_z and e_n, i.e.,

$$e_z \cdot e_n = \cos(e_z, e_n).$$

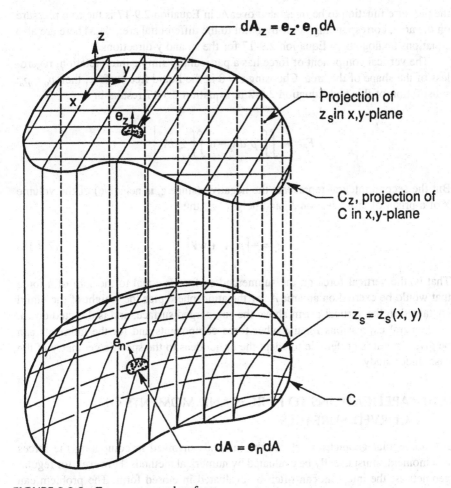

FIGURE 2.9-3 Forces on curved surfaces.

However, in Equation 2.9-15 we have the geometric interpretation that

$$dAe_z \cdot e_n = dA \cos(e_z, e_n) = dA_z, \qquad (2.9\text{-}16)$$

where dA_z is the projection of dA on a plane perpendicular to the z-axis, and where it is assumed that the angle (e_z, e_n) is acute. Otherwise, the right-hand side of Equation 2.9-16 requires a minus sign. Thus,

$$F_z = \iint_A e_z \cdot dF = -\iint_{A_z} p \, dA_z, \qquad (2.9\text{-}17)$$

where A_z is the projected area of A on a plane perpendicular to the z-axis, bounded by C_z and where $p = p(z)$ is the pressure on the actual surface $z_s = z_s(x, y)$. That is, the pressure function to be integrated over A_z in Equation 2.9-17 is the true pressure on the area, corresponding point-by-point to the differential area dA_z. There are also equations analogous to Equation 2.9-17 for the x- and y-directions.

The vertical component of force has a particularly simple interpretation, regardless of the shape of the area. Choosing $z = 0$ at the liquid free surface, then $p_0 = p_a$; substitution of this and Equation 2.6-2 in Equation 2.9-17 results in

$$F_z = -\left[\iint_{A_z} p_a dA_z - w \iint_{A_z} z_s dA_z\right].$$

But the second integral represents the negative (since z_s is negative) of the volume V of a prism of cross section A_z and height $-z_s$; thus,

$$F_z = -\left[p_a A_z + wV\right]. \tag{2.9-18}$$

That is, the vertical force on the submerged surface is equal to the sum of a force that would be exerted on an area A_z by the atmosphere plus the weight of the liquid contained in the vertical prism above the submerged surface, and is directed down.

General expressions for the other force components and for the moments can be given, but it is preferable to adapt the basic ideas to the specific geometry of the case under study.

2.10 APPLICATIONS TO FORCES AND MOMENTS ON CURVED SURFACES

For nonregular geometric surfaces the integrals involved in computing the forces and moments must usually be evaluated by numerical methods. However, for regular geometries, the integrals can often be evaluated in closed form. The problem can be reduced to the following procedure:

- Determine the pressure distribution as a function of the vertical coordinate z.
- Select a coordinate system that is suggested by the body geometry. This will define a differential area $dA = e_n dA$, where e_n is the outer normal to the area directed into the fluid.
- The differential force is $-\Delta p dA$, where Δp is the pressure difference across the differential area. Usually, $\Delta p = p - p_a$, where p_a is the atmospheric pressure. In other words, the difference Δp is actually the gauge pressure.
- Since the differential force is a vector whose direction varies from point to point, it is necessary to compute the x-, y-, and z-components of the differential force — for example, $dF_x = e_x \cdot dF$ — and integrate each component separately.
- For the moment, due to the distributed pressure, one must compute $dM = r \times dF$, where r is the moment arm about a specified axis, and then integrate.

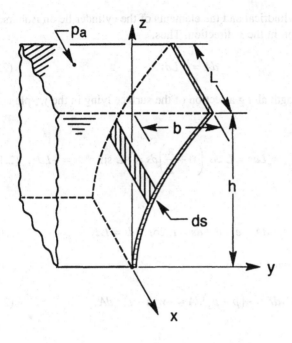

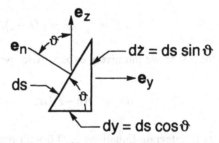

FIGURE 2.10-1 Forces on a parabolic surface.

EXAMPLE 1. THE FORCES ON A PARABOLIC SURFACE

Figure 2.10-1 shows a submerged parabolic, cylindrical surface of length L (in the x-direction) whose shape is specified by

$$\frac{y_{surf}}{b} = \left(\frac{z}{h}\right)^2. \tag{2.10-1}$$

The problem is to determine the y- and z-components of the hydrostatic force. Following the steps just specified we see that

$$p - p_a = w(h - z). \tag{2.10-2}$$

Since the surface is cylindrical and the elements of the cylinder lie on isobars, there is no pressure variation in the x-direction. Thus,

$$dA = e_n L ds, \qquad (2.10\text{-}3)$$

where ds is the arc length along a section of the surface lying in the y,z-plane. Note that

$$e_y \cdot dA = e_y \cdot e_n L ds = L \cos\left(\vartheta + \frac{\pi}{2}\right) ds = -L \sin \vartheta ds = -Ldz, \quad (2.10\text{-}4a)$$

and

$$e_z \cdot dA = e_z \cdot e_n L ds = L \cos \vartheta ds = Ldy. \qquad (2.10\text{-}4b)$$

The differential vector force is

$$dF = -(p - p_a)dA = -w(h - z)e_n dA, \qquad (2.10\text{-}5)$$

and, therefore,

$$dF_y = e_y \cdot dF = w(h - z)Ldz. \qquad (2.10\text{-}6a)$$

Note that $dF_y > 0$ as we intuitively expect. Also, we can put

$$dF_z = e_z \cdot dF = -w(h - z)Ldy = -wL(h - z)\frac{dy}{dz} dz. \qquad (2.10\text{-}6b)$$

It is simplest to integrate Equations 2.10-6 with respect to z. Thus, from Equation 2.10-1, on the parabolic surface, $dy/dz = (2b/h^2)z$. By substituting in Equation 2.10-6b,

$$dF_z = -(2wbL/h^2)(h - z)zdz. \qquad (2.10\text{-}6c)$$

Therefore,

$$F_z = -(2wbL/h^2) \int_{z=0}^{h}(h - z)zdz = -\frac{1}{3}wbLh, \qquad (2.10\text{-}7)$$

which is directed down, as we expect. From Equation 2.10-6a

$$F_y = wL \int_{z=0}^{h}(h - z)dz = \frac{1}{2} wLh^2. \qquad (2.10\text{-}8)$$

Note that F_y is identical to the result for the horizontal force on a vertical rectangular wall, obtained in Section 2.9.

The reader may find it instructive to compute the moment about the x-axis due to the pressure on the parabolic surface. The moment arm is $r = e_y y + e_z z$, so that

$$M = \int r \times dF = -e_x wL \int_{y=0,z=0}^{y=b,z=h} (h-z)(ydy+zdz), \qquad (2.10\text{-}9)$$

from which

$$M = -e_x \frac{wLh^3}{6} \left(1 + \frac{3b^2}{5h^2}\right). \qquad (2.10\text{-}10)$$

The moment is negative as we intuitively expect. Furthermore, for $b \to 0$ Equation 2.10-10 reduces to that of the rectangular wall in Section 2.9 if, in the latter case, we put $p_a \to 0$, thus $p_0 \to wh$.

It is worth remarking that to evaluate the integrals in Equations 2.10-6b and 2.10-9 we had to introduce the expression for the shape of the surface $y = y(z)$. This is a good illustration of how *line integrals*, discussed in most calculus texts, often appear in physical problems.

EXAMPLE 2. THE MOMENT ON A CIRCULAR ARC GATE

In the previous example it was convenient to reduce the integrations to a form dependent on the vertical coordinate z. However, in some cases it is simpler to utilize a curvilinear coordinate system suggested by the geometry of the surface.

Figure 2.10-2 shows a submerged gate, which is a quadrant of a cylinder, hinged along its upper edge. The problem is to compute the force in the actuator arm required to maintain the gate in place. If the actuator arm is parallel to the y-axis, and its line of application lies a distance of b units below the y-axis, then the requirement that the gate be in rotational equilibrium is expressed as

$$-be_z \times e_y F_a + M = 0,$$

or

$$e_x bF_a = -M, \qquad (2.10\text{-}11)$$

where F_a is the scalar force of the actuator arm on the gate, and M is the moment due to the water pressure on the gate. We proceed to compute M by the procedure outlined previously.

The pressure distribution is given by

$$p - p_a = w(h-z). \qquad (2.10\text{-}12a)$$

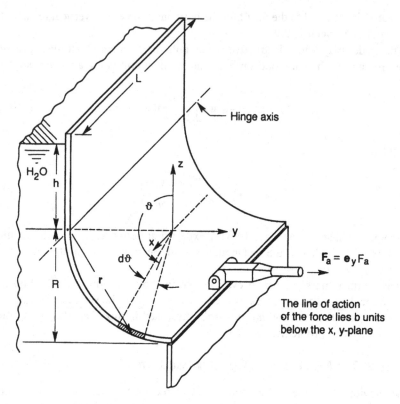

FIGURE 2.10-2   Moment on a cylindrical gate.

The differential area, in polar coordinates, is

$$dA = e_n LR d\vartheta.$$ (2.10-12b)

The differential force is

$$dF = -(p - p_a) dA = -e_n wRL(h - z) d\vartheta,$$ (2.10-12c)

and the moment arm of the differential force from the hinge line is

$$r = (R + y)e_y + ze_z.$$ (2.10-12d)

By noting that

$$e_y \times e_n = e_x \sin \vartheta, \quad e_z \times e_n = e_x \sin \left(\vartheta - \frac{\pi}{2} \right) = -e_x \cos \vartheta,$$

$$y = R \cos \vartheta, \quad z = R \sin \vartheta,$$ (2.10-12e)

the differential moment reduces to

$$dM = r \times dF,$$

$$= -e_x wR^2 L(h - R \sin \vartheta) \sin \vartheta d\vartheta.$$

(2.10-13)

Thus,

$$M = -wR^2 L \int_{\pi}^{3\pi/2} (h - R \sin \vartheta) \sin \vartheta d\vartheta,$$

$$= e_x wR^2 Lh \left(1 + \frac{\pi R}{4h}\right).$$

(2.10-14)

The limits in the preceding integral are determined by the fact that $d\vartheta$ is taken as positive in setting up the force relation. When Equation 2.10-14 is substituted into Equation 2.10-11, we find that the scalar part of the equation yields

$$F_a = -wR^2 L \frac{h}{b} \left(1 + \frac{\pi R}{4h}\right).$$

(2.10-15)

We see that F_a is negative (i.e., is directed to the left), confirming our intuition.

2.11 FORCES ON SUBMERGED BODIES

On a submerged body, hydrostatic pressure cannot exert a force with a horizontal component. This can be shown if, as illustrated in Figure 2.11-1a, we consider the intersections of a horizontal cylinder of arbitrarily small cross-sectional area with the body. There are two areas on the body surface delineated by the intersection, and their projections upon a plane perpendicular to the cylinder are equal except that the direction cosines between their outer normals and the cylinder axis are of opposite signs. Since the pressure is the same for both differential areas, their contributions to the horizontal force parallel to the cylinder cancel, and, since the net horizontal force can be considered as the summation over an infinity of such cylinders, it is zero.

Although Figure 2.11-1b for the vertical force is similar, the situation is different because pressure varies in the vertical direction. We have on the lower surface

$$dF_z = -p_\ell dA \cdot e_z = -p_\ell dA_\ell e_n \cdot e_z = p_\ell dA_z,$$

where dA_z is the cross-sectional area of the cylinder and where $e_n \cdot e_z$ is negative on the lower surface. Similarly, on the upper surface

$$dF_z = -p_u dA_z.$$

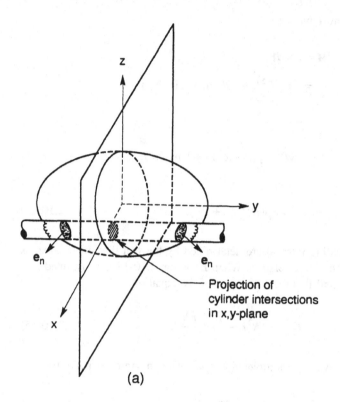

Projection of
cylinder intersections
in x,y-plane

(a)

FIGURE 2.11-1
Forces on submerged
bodies.

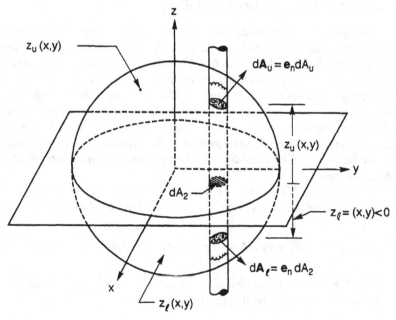

(b)

Hence, the net force on the body depends on the algebraic sum of the two preceding expressions; thus, the integrated pressure force over the body is given by

$$F_z = \iint_{A_z} (p_\ell - p_u)\,dA_z. \tag{2.11-1}$$

But, extending the result of Equation 2.7-2,

$$p_\ell - p_u = w(z_u - z_\ell), \tag{2.11-2}$$

where $z_u = z_u(x, y)$, $z_\ell = z_\ell(x, y)$ define the upper and lower portions of the body surface, and which together represent a double-valued function. Therefore,

$$F_z = w \iint_{A_z} (z_u - z_\ell)\,dA_z = wV \tag{2.11-3}$$

where V is the volume of the body.

ARCHIMEDES PRINCIPLE

Equation 2.11-3 gives only the pressure force on the body. For it to be in equilibrium with the gravity force $-w_B V$, where w_B is the weight density of the submerged body, there must be an external vertical force F_e acting such that

$$\sum F = F_e + F_z - w_B V = 0$$

or

$$F_e = (w_B - w)V. \tag{2.11-4}$$

That is, for a body whose specific weight exceeds that of the liquid, an external force must be supplied equal to the difference in the weight of the body and the weight of an equal volume of water. The principle is the same for a floating body except that the buoyancy force is equal to the submerged volume of the body times the specific weight of the liquid.

2.12 THE CONCEPT OF STATIC STABILITY

An object at rest with respect to an observer is said to be in *stable equilibrium* if, when it is displaced from its equilibrium position (without regard to how the forces causing the displacement arise), the remaining system of forces or moments acting on the object tends to return it to its original position.

There are various classifications of stability. The classical illustrative example is that of a right circular cone. If it were balanced on its vertex, it would be in *unstable equilibrium.* That is, any lateral displacement of the mass center, however small, would result in a toppling of the cone. Lying on its side on a flat, horizontal surface it would have *neutral stability.* Standing on its base on a horizontal surface, the cone would be in *stable equilibrium* providing that the displacement were not too large. For this reason it is said to be *stable in the small.*

The tendency of an object to return to its original position regardless of the displacement is called *stability in the large.* An example of this type is a floating sphere under a vertical displacement. On the other hand, for a lateral displacement, or for a rotation, a sphere is in neutral equilibrium.

Now everybody can be subjected to translations in three different and independent directions and three rotations about these axes. To cover all possible instabilities would take six criteria. Rather than write down the six general relations for stability, it is more instructive to consider a particular example from which the generalization readily follows if desired.

2.13 APPLICATION — STABILITY OF A BUOY

STATEMENT OF THE PROBLEM

A buoy is constructed in the form of a thin, uniform rod of length L, cross-sectional area A, and specific weight w_1. To ensure that it will float upright a weight W is attached to the bottom end. For the present analysis it is assumed that the volume of W is a negligible fraction of the rod volume and hence produces no buoyant forces of its own when submerged. It is required that the rod float upright in a liquid of specific volume w_2. A sketch of the setup appears in Figure 2.13-1.

It is worthwhile to point out that the nature of the problem limits the number of situations that must be examined. In the first place, the buoy must float with a portion above the liquid surface. In principle we ought to examine its stability for a small vertical displacement, but we know from experience that if it floats it is stable in this sense. Translations along either of the two horizontal axes are clearly of no interest (neutral stability), nor is a rotation about its vertical axis. This leaves only a rotation about a horizontal axis. By symmetry, any horizontal axis will do.

EQUILIBRIUM CONDITION

From the vertical component of the equation of equilibrium

$$\sum F_z = 0 = Ahw_2 - ALw_1 - W, \qquad (2.13\text{-}1)$$

or

$$h = L\left(\frac{w_1}{w_2} + \frac{W}{w_2 AL}\right) = L(\beta + \varepsilon), \qquad (2.13\text{-}2)$$

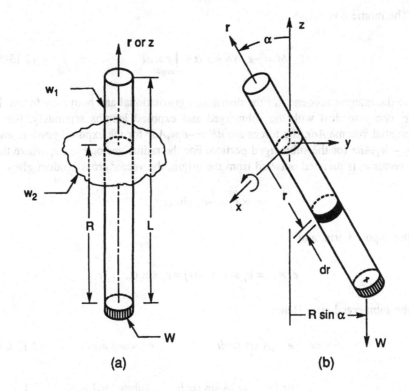

FIGURE 2.13-1 Geometry of weighted-rod buoy.

where, for convenience, we have introduced the dimensionless parameters $\beta \equiv w_1/w_2$ and $\varepsilon \equiv W/w_2AL$. The requirement that the rod float is given by

$$\frac{h}{L} = \beta + \varepsilon < 1, \qquad (2.13\text{-}3)$$

and, since ε can never be less than zero, the permissible range of ε is

$$0 \leq \varepsilon < 1 - \beta. \qquad (2.13\text{-}4)$$

MOMENT EQUATION FOR AN ARBITRARY ROTATION

It is now assumed that the rod is subjected to an arbitrarily small rotation (called a virtual displacement) about a horizontal axis of angle α measured with respect to the vertical, and such that the mean waterline remains unchanged. We choose the x-axis to coincide with the axis of rotation. It is not necessary to know the system of forces responsible for the displacement, nor are these forces considered in the stability analysis. When the angular displacement is achieved, we seek the moment on the rod due to the combined effects of gravity and the buoyancy forces. Note that we are *not* assuming that the rod is then in static equilibrium.

The moment is

$$M = -e_x Wh \sin \alpha + \int_{\text{length}} r \times dF, \qquad (2.13\text{-}5)$$

where the integral accounts for the distributed gravitational and buoyancy forces. It is necessary to deal with the submerged and exposed lengths separately. For a differential volume Adr, the forces are $dF = -e_z w_1 A dr$ for the exposed portion, and $e_z(w_2 - w_1)A dr$ for the submerged portion. For the radius vector $r = e_r r$, where the unit vector e_r is directed outward from the origin, the vector cross-product gives

$$e_r \times e_z = -e_x \sin \alpha,$$

for the exposed arm, and

$$e_r \times e_z = e_x \sin (\pi - \alpha) = e_x \sin \alpha,$$

for the submerged arm. Hence

$$r \times dF = e_x w_1 A \sin \alpha r dr, \qquad \text{exposed arm,} \qquad (2.13\text{-}6a)$$

$$= e_x(w_2 - w_1)A \sin \alpha r dr, \qquad \text{submerged arm.} \qquad (2.13\text{-}6b)$$

Therefore, combining Equations 2.13-5 and 2.13-6,

$$M = e_x M_x = e_x \left[-Wh \sin \alpha + w_1 A \sin \alpha \int_0^{L-h} r dr + (w_2 - w_1)A \sin \alpha \int_0^h r dr \right],$$

or

$$M_x = \left[-Wh + \tfrac{1}{2} w_1 A(L - h)^2 + \tfrac{1}{2}(w_2 - w_1)A h^2 \right] \sin \alpha. \qquad (2.13\text{-}7)$$

After substitution of the dimensionless parameters in the preceding relation, the following form is obtained:

$$M_x = -\tfrac{1}{2} w_2 A L^2 \left[\varepsilon^2 + 2\beta\varepsilon + \beta^2 - \beta \right] \sin \alpha. \qquad (2.13\text{-}8)$$

The first question is to determine the positions, if any, for which the rod is in rotational equilibrium. By definition, equilibrium occurs when $M_x = 0$. From Equation 2.13-8 it is evident that this point is when $\sin \alpha = 0$ or when $\alpha = 0$ or π. We do not need further analysis to convince ourselves that the second case, in which the weight is at the upper end of the rod, is unstable. However, in some problems there may be more than one stable solution, and each possibility should be examined before being rejected.

THE STABILITY CRITERION

Since equilibrium occurs at one specific point, we concern ourselves with stability under a small displacement only in the neighborhood $\alpha \to 0$. At the point itself there is no moment. *For a positive displacement, stability requires a negative restoring moment.* Thus, we are really asking how M_x changes under a small displacement. That is, the criterion is given by

$$\left. \frac{\partial M_x}{\partial \alpha} \right|_{\alpha \to 0} \begin{array}{l} < 0, \text{ stable equilibrium}; \\ = 0, \text{ neutral equilibrium}; \\ > 0, \text{ unstable equilibrium}. \end{array} \qquad (2.13\text{-}9)$$

In Equation 2.13-8, put

$$f(\varepsilon) \equiv \varepsilon^2 + 2\beta\varepsilon + \beta^2 - \beta. \qquad (2.13\text{-}10)$$

Then, applying the criterion Equation 2.13-9 to Equation 2.13-8 it is seen, for $\alpha \to 0$, that stability depends on the sign of the function $f(\varepsilon)$, i.e.,

$$f(\varepsilon) \begin{array}{l} < 0, \text{ stable equilibrium}; \\ = 0, \text{ neutral equilibrium}; \\ > 0, \text{ unstable equilibrium}. \end{array} \qquad (2.13\text{-}11)$$

Now, neutral equilibrium demarcates the boundary between the stable and unstable conditions. Denoting this value as ε_0, then, for neutral stability,

$$f(\varepsilon_0) = \varepsilon_0^2 + 2\beta\varepsilon_0 + \beta^2 - \beta = 0, \qquad (2.13\text{-}12)$$

from which

$$\varepsilon_0 = -\beta + \sqrt{\beta}, \qquad (2.13\text{-}13)$$

where the negative root has been discarded by virtue of Equation 2.13-4.

DISCUSSION

The preceding relations completely define the stability regime that is plotted in Figure 2.13-2. For a specified value of β the stable solutions for $\vartheta = 0$ lie below the curve $\varepsilon = 1 - \beta$ (at which the buoy is completely immersed) and above the curve

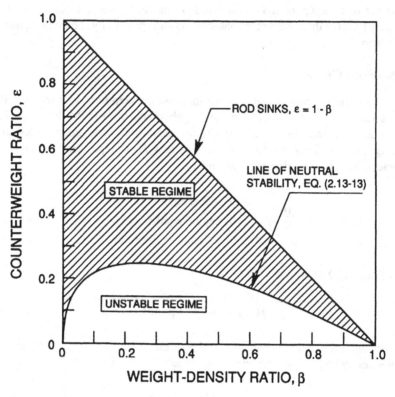

FIGURE 2.13-2 Stability plot for weighted-rod buoy.

$\varepsilon_0 = -\beta + \sqrt{\beta}$ on which the body is in neutral equilibrium. In the latter case, if the buoy is subjected to a small rotational displacement, it will tend to stay in that position.

For example, if $\beta = 0.49$, then $\varepsilon_0 = 0.21$; the rod will sink for $\varepsilon \geq 0.51$. Stable solutions are found in the range $0.21 < \varepsilon < 0.51$. As ε increases from ε_0, $f(\varepsilon)$ tends to become more negative and the buoy becomes more stable. It also sinks lower in the liquid, eventually becoming immersed at $\varepsilon = 0.51$. Thus, the designer must make a compromise that allows adequate exposure for the buoy and, at the same time, that provides a sufficient degree of stability. Presumably, the best compromise would be determined by testing.

2.14   ON THE STABILITY OF A FLOATING PARABOLIC SEGMENT

HISTORICAL NOTE

In the third century before the common era Archimedes (see Heath, 1897, pp. 252–300), who had earlier invented the sciences of statics and hydrostatics, undertook to perform the first analysis of stability of a floating body, which appears, in fact, to be the first stability analysis of any kind. Truesdell (1962, p. 36) writes:

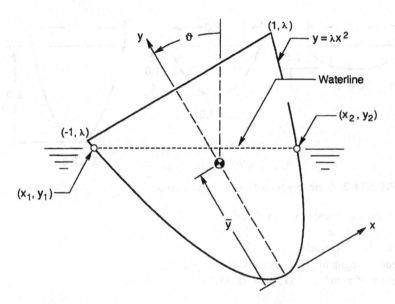

FIGURE 2.14-1   Effect of varying λ on shape of parabolic cross section.

Archimedes determined the positions of equilibrium, and their stability for any floating right segment of a paraboloid of revolution of arbitrary specific gravity. I doubt if a modern professor of mechanics, applying integral calculus and relying upon 2000 years' experience in hydrostatics, will solve this problem without at least a week of hard work.

There is a reason to believe that Truesdell's assessment of the difficulties for the axisymmetric geometry may be understated. Therefore, for the current purpose, we shall deal with an easier case — that of a two-dimensional segment.

DESCRIPTION

With reference to Figure 2.14-1 we consider a right, parabolic segment of unit (i.e., arbitrary) thickness normal to the plane, specified by $y' = ax'^2$, where the axes are fixed on the body, and where the vertex is at the origin. The constant a has the dimension of 1/length. The length of the base is taken to be $2b$ in each case; thus, the altitude is ab^2. Considerations are limited to a segment floating in a liquid of uniform density ρ. The density of the segment is also uniform and equal to $\alpha\rho$, where α is dimensionless and where the requirement that it not sink is specified by $0 < \alpha < 1$.

   Certain geometric or static properties of a parabola are used that are either well known or can be readily calculated. These are

Area of segment	$A_0' = 4ab^3/3$,
Center of area	$(\bar{x}', \bar{y}') = (0, 3ab^2/5)$.

It will be more convenient to work in nondimensional coordinates. Putting $x \equiv x'/b$, $y \equiv y'/b$, the corresponding nondimensional properties, with $\lambda \equiv ab$, are

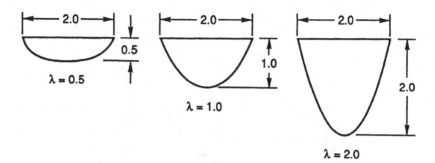

FIGURE 2.14-2 Geometry for a finite, positive rotation.

Equation of parabola	$y = \lambda x^2$,
Length of base	$= 2$,
Altitude of segment	$= \lambda$,
Area of segment	$A_0 \equiv A_0'/b^2 = 4\lambda/3$,
Center of area	$(\bar{x}, \bar{y}) = (0, 3\lambda/5)$.

Note that λ is essentially a slenderness ratio — the larger its value, the more slender the parabola. In this scheme, with a constant base of 2, we can vary the geometry simply by changing the value of λ. Sketches for typical values of λ are shown in Figure 2.14-2.

GEOMETRY OF THE TILTED SEGMENT

Working in body-fixed coordinates, Figure 2.14-2, we now consider the effect of rotating the segment through an angle ϑ about an axis normal to the plane through the center of mass which, in this case, corresponds to the center of the area. Note that the vertical location of the mass center with respect to the waterline depends on the angle ϑ.

Keeping in mind that Archimedes principle requires that the submerged volume (or area, in the two-dimensional case) remain constant, we can relate the angle of tilt to the coordinates (x_1, y_1), (x_2, y_2) of the waterline intersection with the parabola. If m is the slope, then

$$m = -\tan \vartheta = \frac{y_1 - y_2}{x_1 - x_2} = \frac{y_w - y_2}{x - x_2} < 0, \quad \text{if} \quad \vartheta > 0, \qquad (2.14\text{-}1)$$

where the equation of the waterline is obtained from Equation 2.14-1, i.e.,

$$y_w = m(x - x_2) + y_2 = mx + y_1, \qquad (2.14\text{-}2a)$$

and where $(0, y_i)$ the intersection of the waterline with the y-axis, fixes

$$y_i = y_2 - mx_2. \qquad (2.14\text{-}2b)$$

Putting $y_1 = \lambda x_1^2$, and similarly for y_2, then

$$m = \frac{\lambda\left(x_1^2 - x_2^2\right)}{x_1 - x_2} = \lambda\left(x_1 + x_2\right) = -\tan\vartheta, \qquad (2.14\text{-}3a)$$

$$y_i = y_2 - mx_2 = \lambda x_2^2 - \lambda\left(x_1 + x_2\right)x_2 = -\lambda x_1 x_2. \qquad (2.14\text{-}3b)$$

The area below the waterline is obtained by straightforward integration (with $y = \lambda x^2$):

$$A_s = \int_{x_1}^{x_2}\left(y_w - y\right)dx = \int_{x_1}^{x_2}\left(y_i + mx - \lambda x^2\right)dx.$$

After completing the integration, y_i and m are eliminated in favor of x_1 and x_2 by use of Equation 2.14-3, to produce

$$A_s = \frac{\lambda}{6}\left(x_2 - x_1\right)^3. \qquad (2.14\text{-}4)$$

Then, since the submerged area is α $(4\lambda/3)$, we can solve for

$$x_2 - x_1 = 2\alpha^{1/3}. \qquad (2.14\text{-}5)$$

It is readily seen that this relation is identically satisfied in the untilted position $\vartheta = 0$. If Equations 2.14-5 and 2.14-2a are solved simultaneously, then

$$x_1 = -\left(\alpha^{1/3} + \tan\vartheta/2\lambda\right), \qquad (2.14\text{-}6a)$$

$$x_2 = \alpha^{1/3} - \tan\vartheta/2\lambda. \qquad (2.14\text{-}6b)$$

LIMITING VALUE FOR THE TILT ANGLE

Although there is no practical limitation on the tilt angle ϑ, the current expressions are valid only for the case where the lower base junction lies on, or is above, the waterline. Thus, the maximum angle consistent with Equation 2.14-6a is obtained by putting $x_1 = -1$, from which the corresponding value is

$$\tan\vartheta_{max} = 2\lambda\left(1 - \alpha\right)^{1/3}. \qquad (2.14\text{-}7)$$

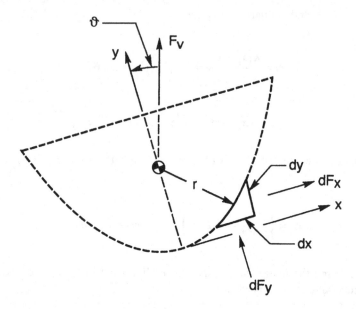

FIGURE 2.14-3   Geometry for calculating the differential force.

PRESSURE DISTRIBUTION ON SEGMENT SURFACE

If p_a is the ambient pressure on the waterline, then, in dimensional form, the equation of hydrostatics becomes

$$p - p_a = \rho g(y'_w - y')\cos \vartheta. \tag{2.14-8}$$

On the parabolic profile, therefore,

$$
\begin{aligned}
p = p_a &= \rho g\left(y'_w - ax'^2\right)\cos \vartheta, \\
&= \rho g b\left(y_w - \lambda x^2\right)\cos \vartheta,
\end{aligned}
\tag{2.14-9}
$$

in terms of the nondimensional variables.

THE DIFFERENTIAL BUOYANCY FORCE

Referring to Figure 2.14-3, and using the artifice that the interior of the parabola is maintained at pressure p_a, the vector differential force (per unit thickness) on an element of area is

$$dF = e_y\left(p - p_a\right)b\,dx - e_x\left(p - p_a\right)b\,dy. \tag{2.14-10}$$

If Equation 2.14-10 is integrated over the submerged area the scalar components of the force turn out to be

$$F_x = \tfrac{1}{3}\alpha\rho g b^2 \lambda \sin \vartheta, \quad F_y = \tfrac{1}{3}\alpha\rho g b^2 \lambda \cos \vartheta,$$

which can be resolved into the vertical force

$$F_v = \rho g b^2 \alpha (4\lambda/3),$$

as is required by Archimedes principle.

THE HYDROSTATIC MOMENT

Again referring to Figure 2.14-3, we note that the (dimensional) position vector of the differential force point of application is

$$r' = b\left[e_x x + e_y (y - y)\right]. \tag{2.14-11}$$

The differential moment (per unit thickness) is

$$dM = r' \times dF,$$

$$= b\left[e_x x + e_y (y - \bar{y})\right] \times \left[e_y (p - p_a) b\,dx - e_x (p - p_a) b\,dy\right],$$

or

$$dM = e_x b^2 \left[(p - p_a) x\,dx + (p - p_a)(y - \bar{y})\,dy\right]. \tag{2.14-12}$$

Thus,

$$M = e_z M_z = e_z \rho g b^3 \cos \vartheta \int_{x_1,y_1}^{x_2,y_2} \left[(y_w - \lambda x^2) x\,dx + (y_w - \lambda x^2)(y - \bar{y})\,dy\right]. \tag{2.14-13}$$

It is at this point the analysis becomes tedious. As before, we must perform a line integral along the curve $y = \lambda x^2$. After completing the integration, which involves six separate terms, Equations 2.14-2 through 2.14-6b are introduced, yielding an expression for M_z in terms of the constants ρ, g, and b, the parameters α and λ, and the angle of tilt ϑ. There results a surprising amount of algebra, which is omitted because of its length, and the expression for the moment reduces to

$$M_z = \rho g b^3 \sin \vartheta \left\{ \tfrac{1}{5}\lambda^2\alpha\left(1 - \alpha^{2/3}\right) - 2\alpha/3 \right] - (\alpha/3)\tan^2 \vartheta \right\}. \tag{2.14-14}$$

EQUILIBRIUM

At equilibrium, where the tilt angle (in nautical terms, the *list*) is designated as ϑ_e, the hydrostatic moment must be zero. We see that there are two possible solutions, each of which must be examined individually for stability. They are given by

$$\sin \vartheta_{e1} = 0, \quad \vartheta_{e1} = 0 \tag{2.14-15}$$

and

$$\tan^2 \vartheta_{e2} = (12/5)\,\lambda^2\left(1 - \alpha^{2/3}\right) - 2, \tag{2.14-16a}$$

from which the positive angle is

$$\vartheta_{e2} = \tan^{-1}\left[(12/5)\,\lambda^2\left(1 - \alpha^{2/3}\right) - 2\right]^{1/2}. \tag{2.14-16b}$$

Note, when the square bracket in Equation 2.14-16b is put equal to zero, that ϑ_{e2} $\to \vartheta_{e1} = 0$, i.e., the two solutions are identical. Solving, we obtain an expression for the corresponding slenderness ratio:

$$\lambda_{12} \equiv \left(\frac{5/6}{1 - \alpha^{2/3}}\right)^{1/2}. \tag{2.14-17}$$

We shall eventually show that the relation for λ_{12} demarcates the boundary between the stable equilibrium solutions ϑ_{e1} and ϑ_{e2}. That is, for $\lambda < \lambda_{12}$, only the solutions ϑ_{e1} are obtained; for $\lambda > \lambda_{12}$ solutions are of type ϑ_{e2}.

The upper bound to solutions of type ϑ_{e2} is obtained by equating Equation 2.14-16b to the expression for the angle of maximum tilt, Equation 2.14-7. This leads to

$$\lambda_{23} = \left(\frac{-5/4}{\left(1 - \alpha^{1/3}\right)\left(1 - 4\alpha^{1/3}\right)}\right), \tag{2.14-18}$$

where λ_{23} demarcates the boundary between ϑ_{e2} and type ϑ_{e3} (see Figure 2.14-6 later). Equation 2.14-18 is singular at $\alpha = 1/64 = 0.0156$ and at $\alpha = 1$. For the latter, the body is completely submerged, and is in neutral equilibrium. For $\alpha < 1/64$, values of λ are imaginary. Thus, for $\alpha < 1/64$ there are no equilibrium solutions. Figure 2.14-4 is a map of λ vs. α for the regime of equilibrium solutions ϑ_{e1} and ϑ_{e2} — all of which turn out to be stable (see following). It is probable, due to the generally narrow range of λ within which stable solutions for ϑ_{e2} are found, that experimental verification, particularly at the higher values of α, would be tenuous, even with precision-made models.

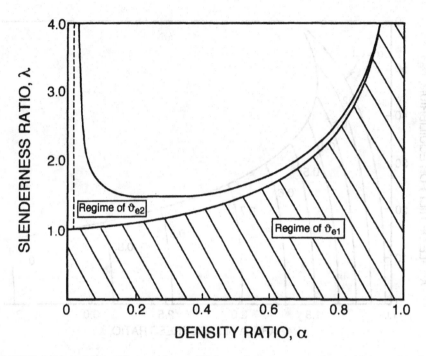

FIGURE 2.14-4   Map of λ vs. α for equilibrium solutions.

THE STABILITY CRITERION

Differentiating Equation 2.14-14 with respect to ϑ, we obtain

$$\frac{\partial M_x}{\partial \vartheta} = \rho g b^3 \alpha \left\{ \left[\frac{4}{5} \lambda^2 \left(1 - \alpha^{2/3} \right) - \frac{2}{3} \right] \cos \vartheta - \frac{\tan^2 \vartheta}{\cos \vartheta} \right\}. \qquad (2.14\text{-}19)$$

As indicated in Section 2.13, each equilibrium solution found must be tested to verify whether or not it satisfies the stability criterion that the derivative $\partial M_z/\partial \vartheta$ be negative.

For $\vartheta_{e1} = 0$, $\cos \vartheta_{e1} = 1$ and $\tan \vartheta_{e1} = 0$. Thus, for stability, the square bracket in Equation 2.14-19 must be negative, a requirement that reduces to

$$\lambda < \left(\frac{5/6}{1 - \alpha^{2/3}} \right)^{1/2} = \lambda_{12}, \qquad (2.14\text{-}20)$$

which condition is already fulfilled by the equilibrium criterion.

For the equilibrium solution ϑ_{e2}, as given in Equation 2.14-16b, to be stable, the wavy bracket in Equation 2.14-19 must be negative. Equivalently,

$$\frac{4}{5} \lambda^2 \left(1 - \alpha^{2/3} \right) - \frac{2}{3} > \frac{\tan^2 \vartheta_{e2}}{\cos^2 \vartheta_{e2}}. \qquad (2.14\text{-}21)$$

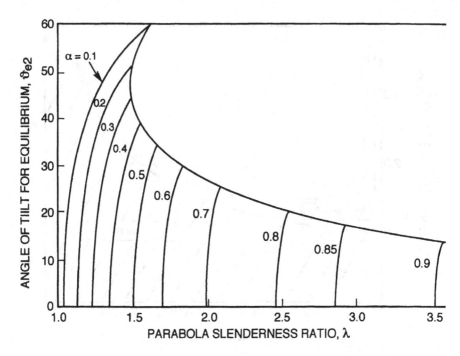

FIGURE 2.14-5 Equilibrium angle ϑ_{e2} plotted vs. λ, with α as parameter.

Now, $\tan^2\vartheta_{e2}$ is given by Equation 2.14-16a and is equal to three times the left-hand side of Equation 2.14-21. Also, we can write:

$$\left(\cos^2 \vartheta_{e2}\right)^{-1} = 1 + \tan^2 \vartheta_{e2} = \tfrac{12}{5}\lambda^2\left(1 - \alpha^{2/3}\right) - 1.$$

Substituting into Equation 2.14-21 and canceling the common factor, the following criterion for stability at ϑ_{e2} is obtained:

$$\lambda > \left(\frac{5/9}{1 - \alpha^{2/3}}\right)^{1/2}.\tag{2.14-22}$$

The condition for equilibrium, Equation 2.14-17, encompasses this requirement. We, therefore, conclude that all equilibrium solutions for both ϑ_{e1} and ϑ_{e2} are also stable.

Discussion

A map of the solutions of ϑ_{e2} as a function of λ, with α as a parameter, is found in Figure 2.14-5. It is noted again that the curves for large values of α involve a variation over a substantial range of ϑ_{e2}, with only a tiny corresponding change in λ. This lack of variation in λ is more emphatically illustrated in Figure 2.14-4.

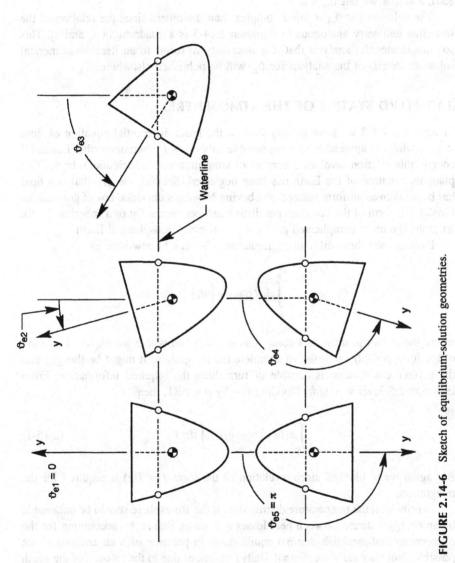

FIGURE 2.14-6 Sketch of equilibrium-solution geometries.

The solutions ϑ_{e1} and ϑ_{e2} represent only two of the five possible equilibrium configurations, all of which are sketched in Figure 2.14-6. It turns out that the solutions for ϑ_{e4} and ϑ_{e5} can be obtained from the preceding analysis merely by replacing α with $1 - \alpha$ in Equation 2.14-14 and those following, provided that ϑ_{e4} is taken in the second quadrant (i.e., the negative root is taken in Equation 2.14-15a), and that we put $\vartheta_{e5} = \pi$.

The solution for ϑ_{e3} is more complex than the others since the relation of the waterline geometry analogous to Equation 2.14-5 is a quadratic in x_1 and x_2. This so complicates the analysis that it is necessary to resort to an iterative numerical solution. Details of the analysis for ϑ_{e3} will be published elsewhere.

2.15 FLUID STATICS OF THE ATMOSPHERE

In Equation 2.5-3 we have already derived the basic differential equation of fluid statics, which is applicable to compressible as well as to incompressible fluids. Of course, this relation involves a number of simplifications (restrictions). In the first place the rotation of the Earth has been neglected. Second, the gravitational field has been taken as uniform rather than obeying Newton's universal law of gravitation. Finally, the form of the law does not distinguish between a flat or a spherical Earth, let alone the more complicated problem of a slightly nonspherical Earth.

Passing over these difficulties, Equation 2.5-3 can be rewritten as

$$\int_{p_a}^{p} dp / \rho = -g \int_{0}^{z} dz,$$

where the subscript a denotes conditions at $z = 0$ and where we require a relation of the form $\rho = \rho(p)$ in order to complete the integration. It might be thought that the perfect gas relation is capable of furnishing the required information. From Equation 2.5-3, $dp = -\rho g dz$. Dividing this by $p = \rho RT$, then

$$\int_{p_a}^{p} d(\ln p) = -(g/R) \int_{0}^{z} dz/T. \qquad (2.15\text{-}1)$$

But again we're blocked since a relation of the form $T = T(z)$ is required for the integration.

In principle, the temperature distribution in the atmosphere should be obtainable by an energy balance between heat losses and gains, that is, by accounting for the phenomena that establish thermal equilibrium. In practice such an analysis is not possible. Not only are there diurnal (daily) variations due to the rotation of the Earth on its axis, but also variations due to the rotation of the Earth about the sun, local effect of weather systems, moisture-content variation, and effects due to variation of the composition of the atmosphere for a variety of reasons, including dissociation, ionization, radiation, and chemical reactions. Furthermore, there are major variations of the temperature–height distributions with latitude as well as with season. In the following sections the general properties of the atmosphere are described, followed by a description of the *U.S. Standard Atmosphere, 1976* (Dublin et al., 1976). Most of this material has been abstracted from the publications *U.S. Standard Atmosphere, 1976* and the article by Cole et al. (1965). Atmosphere data as a function of latitude may be found in Chapter 14 of Jursa (1985).

DESCRIPTION OF THE ATMOSPHERE

The simplest way to classify the various atmospheric regions is by the temperature–height profile. The data for this classification has been gathered over a number of decades, especially in the Space Age, using soundings taken by aircraft, balloons, and rockets. A study of the profile indicates that there are four distinguishable spherical shells that surround the Earth and that blend together at their contiguous surfaces.

The *troposphere* is the shell adjacent to the Earth and is distinguished by an almost linearly decreasing temperature distribution. It should be kept in mind that we are talking about average behavior since, for example, local temperature inversions on a given day are common. The troposphere, the domain of weather, is in convective (i.e., depends on air circulation) equilibrium with the sun-warmed surface of the Earth. Near an altitude of 11 km, the temperature becomes essentially constant (about 227 K) over a finite height change. The altitude at which this first occurs is called the *tropopause*.

The second shell, the *stratosphere*, begins at the tropopause. The region of (almost) constant temperature is thicker at the poles and is almost nonexistent at the equator. The appearance of ozone occurs in the stratosphere. Eventually, the temperature starts to increase and continues to rise until a second (almost) constant-temperature region is encountered near 47 km. The beginning of the second constant-temperature (about 270 K) region is called the *stratopause*. As in all layers, the pressure and density decrease monotonically with height.

The stratopause marks the beginning of the third shell designated as the *mesosphere*. In the lower portion of the mesosphere the temperature is constant, but with increasing height the temperature again begins to fall, reaching a low of about 187 K near 86 km. The mesosphere is in radiation equilibrium between the ultraviolet ozone heating by the upper fringe of the ozone region, and the infrared ozone and carbon dioxide cooling by radiation to space. The upper limit of the mesosphere is designated as the *mesopause*.

The shell beyond the mesopause is the *thermosphere*, where the mean free path is so large that the concept of continuum is inapplicable and, in fact, where most aerodynamic effects (such as frictional heating and flow resistance) are negligible, even from a free-molecule-flow viewpoint. In this region the pressure and density asymptotically become zero, whereas the temperature (actually the *kinetic temperature*, which is defined by the mean free speed) asymptotically increases to a value the order of 1000 K.

SUPPLEMENTARY SHELLS

There are ways of classifying the atmosphere other than by temperature distribution. These alternative shells necessarily overlap the primary shells just described. In the *homosphere* (up to about 100 km) there is a continual mixing process that keeps the relative proportions of the constituents approximately constant. Surrounding the homosphere is the *heterosphere* in which the chemical process of diffusion dominates. Exterior to the heterosphere is the *exosphere* from which particles (especially those with the smallest atomic or molecular weights) may escape the Earth's gravitational field and be transported to outer space.

The *ionosphere* is the domain of charged particles and consists of a number of layers lying between 80 and 300 km. Ionization of the upper atmosphere depends primarily on the sun and its activity. The major part of the ionization is produced by solar ultraviolet- and x-radiation, and by corpuscular radiation from the sun. Cosmic rays may make a very minor contribution. As the Earth rotates with respect to the sun, ionization increases in the sunlit atmosphere, and decreases on the darkened side.*

2.16 U.S. STANDARD ATMOSPHERE, 1976

The *U.S. Standard Atmosphere, 1976* (Dublin et al., 1976) is the most recent in a series of standard atmospheres, the first of which was proposed by Diehl (1925). The 1976 version encompasses extensive data accumulated in the Space Age. It tabulates data in SI units primarily, and presents a vast tabulation of physical properties as a function of altitude.

THE COMPOSITION OF AIR

Air is a mixture of gases of differing molecular weights whose composition may vary even on the surface of the Earth as a result of changes primarily in moisture content and carbon dioxide. Based on careful measurements, Table 2.16-1 presents the normal composition of clean, dry air near sea level, which has been adopted as standard by international agreement. Molecular weights (which are not weights at all but dimensionless ratios) are based on the carbon isotrope $C^{12} = 12.000\ 0$ rather than on the older standard using oxygen as reference. These and other primary data are taken from *U.S. Standard Atmosphere, 1976*.

SEA-LEVEL REFERENCE CONDITIONS

Based on decades of measurements, the mean sea level reference conditions are specified in Table 2.16-2.

TEMPERATURE–HEIGHT RELATIONS

For the purpose of defining a temperature profile, up to 86 km the temperature function is represented by a series of straight-line segments with discontinuous slopes at the junctions. A complete tabulation of these values is given in Table 4 of *U.S. Standard Atmosphere, 1976*. Above 86 km, where the particles are not in thermal

---

* See Cormier et al. (1965).

TABLE 2.16-1
Fractional Volume Composition and Molecular Weights for Sea Level Dry Air

Gas	Fractional Volume	Molecular Weight $(\overline{m})$ (Dimensionless)
Nitrogen (N_2)	0.780 84	28.013 4
Oxygen (O_2)	0.209 476	31.998 8
Argon (A)	0.009 34	39.948
Carbon Dioxide (CO_2)	0.000 314	44.009 95
Total Fraction = 0.999 97	Mean Molecular Weight = 28.964 4	

Note: The remaining fraction of 0.000 03 is made up of neon, helium, krypton, xenon, methane, and hydrogen in extremely small proportions. There are also present, in varying low concentrations, a number of pollutants including nitrous oxide, nitric oxide, nitrogen dioxide, nitric acid vapor, hydrogen sulfide, ammonia, sulfur dioxide, and carbon monoxide. There is also an ozone concentration that varies with altitude. The most highly variable constituent of ordinary air is, of course, water vapor. For details see *U.S. Standard Atmosphere, 1976.*

Source: Dublin, M., et al., Eds., *U.S. Standard Atmosphere, 1976,* U.S. Government Printing Office, Washington, D.C., 1976.

equilibrium, the relations used to specify temperature variations are more complicated. A description is given in *U.S. Standard Atmosphere, 1976.*

TEMPERATURE–HEIGHT RELATION IN THE TROPOSPHERE

The tropopause* has arbitrarily been defined as $z^* = 11$ km, where the tropopause temperature is taken as $T^* = 216.65$ K $= 389.97$ °R. The *lapse rate a* in the troposphere is defined to be the negative of the mean temperature gradient between sea level and 11 km. Thus,

$$a \equiv \left(T_a - T^*\right)/z^* = 6.5 \times 10^{-3} \text{ K/m} = 0.003\,566\,2 \text{ °R/ft}. \qquad (2.16\text{-}1)$$

Therefore, in the troposphere, the temperature gradient for the standard atmosphere is

$$dT/dz = -a, \quad z < z^* \qquad (2.16\text{-}2)$$

or

$$T = T_a - az, \quad z < z^*. \qquad (2.16\text{-}3)$$

---

* All heights in these and succeeding relationships are given in *geopotential* rather than *geometric* units, a distinction that is explained in Section 2.17.

TABLE 2.16-2
Sea-Level Reference Conditions for Dry Air

Quantity	Symbol	SI (kg, m, s, K)	Engineering System (slug, ft, sec, °R)
Uniform gravitational constant	g	9.806 65 m/s²	32.174 ft/sec²
Universal gas constant	R^*	8.314 32 × 10³ N·m/kmol·K = 8.314 32 × 10³ m²·kg/kmol·s²·K	
Mean molecular weight	$\overline{m}$	28.964 4	28.9644
Specific gas constant	$R = R^*/\overline{m}k$	287.053 m²/s²·K	1716.6 ft²/sec² °R
Pressure[a]	p_a	1.013 250 × 10⁵ N/m²	2.116.2 lb/ft²
Temperature	T_a	288.150 K	518.67 °R
Density	ρ_a	1.225 0 kg/m³	0.0023769 slug/ft³
Speed of sound	a_a	340.294 m/s	1116.45 ft/sec
Dynamic viscosity	μ_a	1.789 4 × 10⁻⁵ kg/m·s	3.7372 × 10⁻⁷ lb sec/ft²
Coefficient of heat condition	k_a	2.532 6 × 10⁻² J/(s·m·K)	3.1631 × 10⁻³ ft lb/(sec ft °R) = 1.463 × 10⁻² Btu/(hr ft °F)

Note: For practical purposes the ratio of specific heats is taken as $\gamma = 1.400$.

[a] There are various units used to measure pressure. The *atmosphere* (atm) is 1.013 250 × 10⁵ N/m² = 1.013 250 bar = 0.101 | 325 | 0 MPa = 760 mm Hg. For the *millibar* (mb) 1 mb = 10⁻³ bar. For high-vacuum work there is the *torr*, where 1 torr = 1 mm Hg = 1/760 atm. There is also the *micron* of mercury where 1 μ Hg = 10⁻³/760 atm. Since the symbol μ is used as a multiplier prefix in SI, the use of the unit μ Hg is officially discouraged. The CGPM also recommends that the use of the torr unit be abandoned.

Source: Dublin, M., et al., Eds., *U.S. Standard Atmosphere, 1976*, U.S. Government Printing Office, Washington, D.C., 1976.

PRESSURE–HEIGHT RELATION IN THE TROPOSPHERE

With Equation 2.16-2 we can now integrate Equation 2.15-1:

$$\int_{P_a}^{p} d(\ln p) = -(g/R) \int_{0}^{z} \frac{dz}{dT}\frac{dT}{T} = (g/aR) \int_{T_a}^{T} d(\ln T)$$

or

$$p/p_a = \left(T/T_a\right)^{g/aR}. \tag{2.16-4}$$

But this is recognized to be a form of the polytropic relation, which, with a little algebra and use of the equation of state for a perfect gas, reduces to

$$pv^n = p_a v_a^n = \text{constant}, \tag{2.16-5}$$

where $n \equiv (1 - a\mathcal{R}/g)^{-1} = 1.233$. Combining Equations 2.16-4 and 2.16-3 gives the pressure–height relation in the troposphere

$$p/p_a = \left(1 - az/T_a\right)^{n/(n-1)}. \qquad (2.16\text{-}6)$$

A similar relation is obtainable for the density.

Similarly, the pressure–height relations can be obtained for the other regions of the atmosphere by utilizing the specified temperature distribution. For example, in the stratosphere over the range $z^* \leq z < z^{**} = 20$ km, $T = T^* = $ constant, and integration of Equation 2.15-1 produces

$$p/p^* = \exp\left[-g\left(z - z^*\right)/\mathcal{R}T^*\right], \quad z^* \leq z < z^{**}. \qquad (2.16\text{-}7)$$

In Table 2.16-3 a limited number of values are tabulated for these and other atmospheric variables, as selected from *U.S. Standard Atmosphere, 1976*. It is noted again, however, that the tabulated values represent averages over long periods of time and that specific conditions at a given day, location, and altitude may vary substantially from the standard value.

Figure 2.16-1 shows the range of systematic variability about the 1976 standard. The arrows show the lowest and highest mean monthly temperatures obtained for any location between the equator and pole. Estimates are given for the 1% maximum and minimum temperatures that occur during the warmest and coldest months, respectively.

OTHER STANDARD ATMOSPHERES

For the arctic and tropical regions there exist also the *Cold, Polar, Tropical*, and *Hot Atmospheres* prepared by U.S. Military organizations.

FOUR-DIMENSIONAL GLOBAL REFERENCE ATMOSPHERE MODEL

NASA has recently made available* to interested users a program that generates a four-dimensional atmosphere along any simulated trajectory from surface level to orbital altitudes. The fourth dimension refers, of course, to the wind profile along the trajectory in addition to the pressure, density, and temperature. Quoting from the COSMIC Program Abstract:

> The program was developed for design applications in the Space Shuttle program, such as the simulation of external tank re-entry trajectories. Other potential applications would be global circulation and diffusion studies, and generating profiles for comparison with other atmospheric measurement techniques, such as satellite measured temperature profiles and infrasonic measurement of wind profiles.

---

* For details readers should contact Computer Software Management and Information Center (COSMIC), The University of Georgia, 382 East Broad Street, Athens, Ga. 30602, telephone (404) 542-3265.

TABLE 2.16-3
Selected Values from U.S. Standard Atmosphere, 1976

Geometric Altitude z(m)	Geopotential Altitude H(m)	Temperature (K)	Pressure Ratio p/p_a	Density Ratio ρ/ρ_a	Particle Number Density $\nu(m^{-3})$	Particle Mean-Free-Path Length $\lambda(m)$	Particle Mean Free Speed C(m/s)	Speed of Sound a(m/s)	Dynamic Viscosity Ratio μ/μ_a	Thermal Conductivity Ratio k/k_a
-1 000	-1 000	294.651	1.124 4 +0	1.099 6 +0	2.800 7 +25	6.032 4 -8	464.09	344.11	1.017 4 +0	1.020 1 +0
0	0	288.150	1.000 0	1.000 0	2.547 0	6.633 2	458.94	340.29	1.000 0	1.000 0
300	300	286.200	9.649 4 -1	9.715 2 -1	2.474 4	6.827 7	457.39	339.14	9.947 3 -1	9.939 4 -1
600	600	284.250	9.308 8	9.436 6	2.403 5	7.029 3	455.83	337.98	9.894 5	9.878 7
1 000	1 000	281.651	8.870 0	9.074 8	2.311 3	7.309 5	453.74	336.43	9.823 7	9.797 6
2 000	1 999	275.154	7.846 1	8.216 8	2.092 8	8.072 8	448.48	332.53	9.645 6	9.594 0
3 000	2 999	268.659	6.920 4	7.422 5	1.890 5	8.936 7	443.15	328.58	9.465 6	9.389 1
4 000	3 997	262.166	6.085 4	6.688 5	1.703 6	9.917 3	437.76	324.59	9.283 6	9.183 0
5 000	4 996	255.676	5.334 1	6.011 7	1.531 2	1.103 4 -7	432.31	320.55	9.099 5	8.975 7
6 000	5 994	249.187	4.660 0	5.388 7	1.372 5	1.231 0	426.79	316.45	8.913 3	8.767 1
8 000	7 990	236.215	3.518 5	4.292 1	1.093 2	1.545 4	415.53	308.11	8.534 3	8.346 2
10 000	9 984	223.252	2.615 3	3.375 6	8.597 6 +24	1.965 1	403.97	299.53	8.146 1	7.920 3
12 000	11 977	216.650	1.914 5	2.546 4	6.485 7	2.604	397.95	295.07	7.944 7	7.701 4
15 000	14 965	216.650	1.195 3	1.589 8	4.049 3	4.172 3	397.95	295.07	7.944 7	7.701 4
17 500	17 452	216.650	8.075 2 -2	1.074 0	2.735 5	6.176 0	397.95	295.07	7.944 7	7.701 4
20 000	19 937	216.650	5.457 0	7.258 0 -2	1.848 6	9.139 3	397.07	295.07	7.944 7	7.701 4
25 000	24 902	221.552	2.515 8	3.272 2	8.334 1 +23	2.027 2 -6	402.43	298.39	8.094 5	7.864 1
30 000	29 859	226.509	1.181 3	1.502 9	3.827 8	4.413 7	406.91	301.71	8.244 6	8.027 8
40 000	39 750	250.350	2.833 8 -3	3.261 8 -3	8.307 7 +22	2.033 6 -5	427.78	317.19	8.946 8	8.804 6
50 000	49 610	270.650	7.873 5 -4	8.382 7 -4	2.135 1 +22	7.913 0	444.79	329.80	9.521 0	9.452 1
75 000	74 125	208.399	2.356 9 -5	3.258 9 -5	8.300 3 +20	2.035 4 -3	390.30	289.40	7.689 2	7.426 0

TABLE 2.16-3 (CONTINUED)
Selected Values from *U.S. Standard Atmosphere, 1976*

Geometric Altitude z(m)	Geopotential Altitude H(m)	Temperature (K)	Pressure Ratio p/p_a	Density Ratio ρ/ρ_a	Particle Number Density $v(m^{-3})$	Particle Mean-Free-Path Length $\lambda(m)$	Particle Mean Free Speed C(m/s)	Speed of Sound a(m/s)	Dynamic Viscosity Ratio μ/μ_a	Thermal Conductivity Ratio k/k_a
100 000[a]	98 451	195.08	3.159 3 −7	4.575 −7	1.189 +19	1.42 −1	381.4	—	—	—
200 000	193 899	854.56	8.362 8 −10	2.074 −10	7.182 +15	2.4 +2	921.6	—	—	—
300 000	286 480	976.01	8.655 7 −11	1.564 −11	6.509 +14	2.6 +3	1 079.7	—	—	—
500 000	463 540	999.24	2.984 0 −12	4.257 −13	2.192 +13	7.7 +4	1 215.0	—	—	—
750 000	670 850	999.99	2.230 3 −13	1.460 −14	1.637 +12	1.0 +6	1 793.9	—	—	—
1 000 000	864 071	1 000.00	7.415 5 −14	2.907 −15	5.422 +11	3.1 +6	2 318.1	—	—	—

[a] The speed of sound, viscosity, and thermal conduction coefficients are not tabulated above 86 km.

Source: Dublin, M., et al., Eds., *U.S. Standard Atmosphere, 1976*, U.S. Government Printing Office, Washington, D.C., 1976.

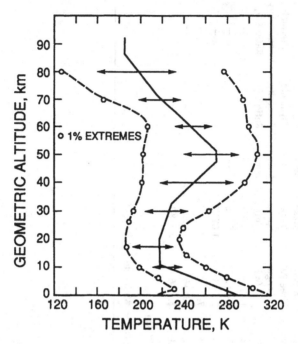

FIGURE 2.16-1   Range of systematic variability of temperature around the *U.S. Standard Atmosphere, 1976.*

2.17 CONCEPT OF GEOPOTENTIAL HEIGHT

The relative magnitude of errors contributed by the approximation of a uniform gravitational field is small as long as only small changes of height are involved. The simplicity afforded by this approximation justifies its retention, although the results can be converted to the inverse-square law by a simple artifice.

Assuming the Earth to be a sphere whose radius is equal to the mean radius of the actual ellipsoid, $R_o = 6\ 356.766$ km $= 3,949.427$ mi., then for a geometric height H above the Earth's surface, we know that on an arbitrary mass m the force exerted by the Earth's mass M is

$$F = -mg, \quad \text{or} \quad F = -GmM/\left(R_o + H\right)^2, \tag{2.17-1}$$

for the uniform and inverse-square gravitational laws, respectively, and where the value of the *universal gravitational constant* is $G = 6.672\ 59 \times 10^{-11}$ N·m²/kg². Since at the Earth's surface, where $H = 0$, both laws must yield the same force, we have

$$GM = gR_o^2. \tag{2.17-2}$$

The corresponding potential energy functions are obtained by integrating, respectively,

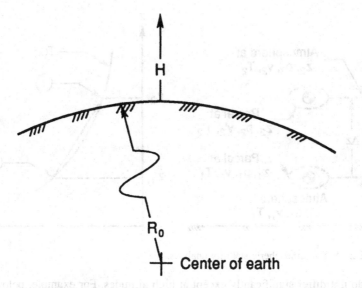

FIGURE 2.17-1   Geometric (H) vs. geopotential (h) height.

$$dW/dz = -F/m = g,$$

where z is the *geopotential height*; and in terms of the geometric height, by using Equation 2.17-2 in Equation 2.17-1,

$$dW/dH = gR_o^2 / (R_o + H)^2.$$

That is, depending on the model chosen for the gravitational field,

$$W = gz, \quad \text{or} \quad W = gR_o H / (R_o + H). \tag{2.17-3}$$

By the first relation of Equation 2.16-3 we have defined the geopotential height z as the fictitious height required to raise an arbitrary mass in a uniform gravitational field above the Earth's surface to produce the same change of potential energy as is required to raise the same mass to an actual height H in the real gravitational field. By equating the potential energy functions,

$$z/H = 1/(1 + H/R) \approx 1 - H/R_o, \tag{2.17-4a}$$

$$H/z = 1/(1 - z/R) \approx 1 + z/R_o, \tag{2.17-4b}$$

where the right-hand expressions are the resulting approximate forms if the denominators are expanded in series and only the first-order terms retained. Obviously, z

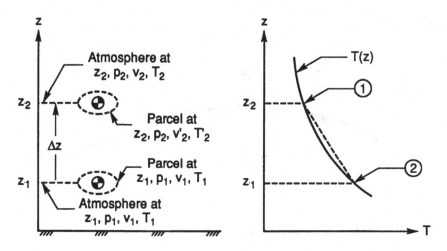

FIGURE 2.18-1 Atmospheric stability analysis.

and H do not differ significantly except at high altitudes. For example, below 63.6 km the difference is less than 1%, and a 10% difference is not reached until a height of 636 km.

Since all relations in Section 2.15 were derived on the basis of a uniform gravitational field, z must be interpreted therein as geopotential altitude. All quantities computed from these expressions are readily converted to a corresponding geometric altitude by use of Equation 2.17-4.

2.18 STABILITY OF THE ATMOSPHERE

The question of atmospheric stability is essentially unrelated to that of defining a standard atmosphere. Instead, for an arbitrary temperature (hence, pressure and density) distribution, the problem is concerned with determining whether or not a mass of air, displaced from its original equilibrium position, is subject to forces that tend to return it or further displace it from equilibrium.

We visualize the situation indicated in Figure 2.18-1. The air is assumed to be in equilibrium with some temperature distribution $T = T(z)$. At altitude z_1 conditions are p_1, v_1, T_1. We suppose that a certain parcel of air at z_1 — which can be thought of as being contained within a membrane incapable of exerting any stresses of its own — is displaced (for example, by an air current) to station z_2, where surrounding conditions are p_2, v_2, T_2. Within the membrane, conditions after displacement are denoted as $p_2' = p_2$, v_2', T_2'. The pressure adjusts to the local value, of course, since the membrane cannot support any pressure difference. From our knowledge of buoyancy, therefore, the stability criterion can be specified by

$$\frac{\rho_2'}{\rho_2} = \frac{v_2}{v_2'} \begin{array}{l} > 1, \ \text{stable}, \\ = 1, \ \text{neutral}, \\ < 1, \ \text{unstable}; \end{array}$$

or, equivalently, by

$$v_2 - v_2' \begin{cases} > 0, & \text{stable,} \\ = 0, & \text{neutral,} \\ < 0, & \text{unstable.} \end{cases} \qquad (2.18\text{-}1)$$

Now, since in an actual situation the temperature does not ordinarily decrease linearly, over the region between z_1 and z_2 the curve of $T = T(z)$ is approximated, Figure 2.18-1, by the chord connecting two points. Thus, from ① to ② we can imagine a linear temperature distribution such that the corresponding pressure-specific volume relation is $pv^m = $ constant, where m is related to the slope of the line from ① to ② as in Equation 2.16-5. By differentiating,

$$v^m \, dp/dz + mpv^{m-1} \, dv/dz = 0.$$

By substituting in the fundamental relation $dp/dz = -\rho g = - g/v$,

$$(dv/dz)_{\text{atmos.}} = g/mp \qquad (2.18\text{-}2)$$

is the relation governing the specific-volume gradient in the atmosphere corresponding to the approximate linear temperature distribution.

RELATION FOR AN ADIABATIC DISPLACEMENT

If the displacement is adiabatic,* that is, if there is no heat added to, or extracted from, the air within the membrane, then by applying the first law of thermodynamics,

$$d\bar{q} = de + pdv' = 0, ** \qquad (2.18\text{-}3)$$

we are led to the relation $p(v')^\gamma = $ constant for the displacement of a perfect gas, where γ is the ratio of specific heats. By manipulations identical to that leading to Equation 2.17-2

$$(dv'/dz)_{\text{displ.}} = g/\gamma p. \qquad (2.18\text{-}4)$$

For a small displacement, with the derivatives dv/dz and dv'/dz evaluated at ①,

---

* In meteorology it is the practice to refer to an *adiabatic* displacement as one in which there is neither heat added nor an exchange of energy with the environment due to friction forces on the boundary. In Chapter 4 we refer to such a situation as *isentropic* and employ adiabatic in a more-restricted sense. The first law of thermodynamics is reviewed in Section 3.15.
** The mathematical significance of the symbol d as in $d\bar{q}$ is discussed in Section 7.2.

$$v_2 - v_2' \approx \left[v_1 + \Delta z \left(\frac{dv}{dz} \right)_{\text{atmos.}} \right] - \left[v_1 + \Delta z \left(\frac{dv}{dz} \right)_{\text{displ.}} \right],$$

$$= \left(\frac{1}{m} - \frac{1}{\gamma} \right) g \Delta z / p_1 .$$

Obviously, a result equivalent to Equation 2.18-1 for a vanishingly small displacement is

$$\frac{1}{m} - \frac{1}{\gamma} \begin{array}{l} > 0, \ \text{stable}, \\ = 0, \ \text{neutral}, \\ < 0, \ \text{unstable}. \end{array} \qquad (2.18\text{-}5)$$

But it turns out to be more useful to rewrite the criterion of Equation 2.18-5 in terms of effective lapse rates (the lapse rate is the negative of the temperature gradient associated with a linearly varying temperature distribution) involving the exponents m and γ. By putting

$$a_m = (1 - 1/m) g/\mathcal{R}, \quad a_\gamma = (1 - 1/\gamma) g/\mathcal{R},$$

the criterion becomes, with a little algebra,

$$a_\gamma - a_m \begin{array}{l} > 0, \ \text{stable}, \\ = 0, \ \text{neutral}, \\ < 0, \ \text{unstable}. \end{array} \qquad (2.18\text{-}6)$$

For dry air the lapse rate associated with an adiabatic displacement is $a_\gamma = 0.031\ 64$ K/m = 0.005357 °R/ft. If the actual atmospheric temperature distribution corresponds to standard conditions, where $a_m \rightarrow a = 0.006\ 5$ K/m, we see that stable equilibrium prevails. In the case of a temperature inversion* a_m is negative, a condition that is extraordinarily stable.

THE EFFECT OF MOISTURE ON STABILITY

In the preceding analysis no pretense is maintained that the groundwork has been laid for a serious study** of meteorology. Nevertheless, an inkling of the essential features of atmospheric stability is obtained as long as the air contains negligible moisture. However, when this condition does not exist, a new variable, the moisture, is introduced, which leads to a greater complexity of analysis and which is primarily responsible for the most interesting and dramatic weather situations.

---

* Anyone who has ascended Mount Wilson, near Los Angeles, has probably noticed the temperature increase with height, an almost chronic condition.
** See Haltiner and Martin (1957) or Holton (1972).

When displaced upward, a parcel of unsaturated air initially behaves essentially the same as dry air; that is, the situation is stable. However, when the dew point is reached, resulting from expansion of the air, condensation occurs. If the moisture appears in the form of extremely fine droplets of liquid, or of ice, a cloud forms that remains with the displaced air and the process is adiabatic (but not isentropic). However, the heat released by condensation is absorbed by the air parcel so that the degree of cooling is less than it would be for dry air. If the condensation occurs such that the droplets are sufficiently large, precipitation results, which means that the original parcel loses mass in the form of precipitated moisture. Meteorologists call this a *pseudo-adiabatic* process.

To illustrate the main point, we can neglect mass changes of the parcel due to precipitation, but not the heat addition. Thus, in applying the first law only to the air of the parcel, the process appears to be *diabatic*, i.e., one with heat addition. We can represent a small displacement by a locally polytropic process, pv^k = constant, where $k > 1$. Then, for a perfect gas, the first law can be juggled into the form

$$\frac{d\bar{q}}{RT} = \left(\frac{\gamma}{\gamma - 1} \frac{k-1}{k} - 1 \right) \frac{dp}{p}. \qquad (2.18\text{-}7)$$

For heat addition $d\bar{q} > 0$, and for an upward displacement $dp < 0$, in which case the parenthesis in Equation 2.18-7 must be negative, a situation possible only if $k < \gamma$. By converting this into a lapse rate as before, $a_k < a$. The value of a_k is not trivial to obtain, and it varies with altitude in any case. Since $a_k < a_\gamma < a$, saturated air at a temperature distribution corresponding to U.S. Standard would be unstable. Of course, in a given case the degree of instability is related to the actual temperature–height distribution, which would usually different from the U.S. Standard.

In general, for a meteorological process involving displacement of saturated air, instabilities may always be possible. The most dramatic example is that of a cold front underrunning a region of warm, moist air. The violent upward displacements lead to enormous quantities of condensation accompanied by heat release, which further accentuates the updrafts. The process quickly (astonishingly so, considering the scale of the disturbance) leads to formation of thunderheads, one of the more spectacularly violent items in nature's catalog of phenomena.

A good example of how the various possibilities fit together is that of the *chinook** or *foehn* (wind), which is associated with a prevailing westerly breeze blowing over an ocean where it picks up moisture, toward a mountain range such as the Cascades. Figure 2.18-2 illustrates schematically the temperature history of a parcel of air as it passes over the mountain. A parcel originating near sea level, point ① is supposed not to be initially saturated. As it is forced upward by the presence of the mountain, its temperature falls off along line *ij* at the lapse rate a_r.

At ① the parcel reaches saturation and condensation occurs. The accompanying heat release turns the process unstable, resulting in strong updrafts and huge quantities of precipitation from the clouds formed on the upwind side of the mountain

---

* Derived from the Indian word meaning "snow-eater," which refers to its ability to sublimate snow.

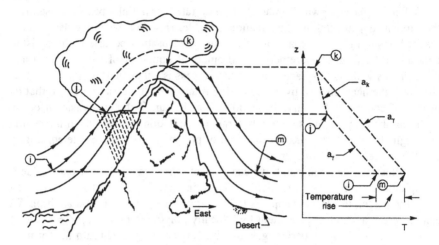

FIGURE 2.18-2 A highly idealized temperature history of a particle in a foehn.

range. The process continues (line jk with $a_k < a_\gamma$) until the air passes the summit at which point the air parcel has been "wrung" almost dry. On the downwind side of the mountain, the temperature follows line km with lapse rate a_γ resulting in a temperature substantially higher than its original value at the same altitude.

The foehn produces fantastic quantities of precipitation on the western slopes of the Cascades and almost none to the east. This, of course, explains why the western slopes are lush, verdant regions whereas to the east there exists a desert region with sparse vegetation at best.

2.19 BUOYANCY IN THE ATMOSPHERE

In a liquid in a gravitational field, the density variation over a vertical displacement is only a small fraction relative to the pressure variation over the same height. This, of course, is the justification for the incompressible assumption. However, in a gas the fractional change in density may be of the same relative order as for the pressure. The question then arises, how does one calculate the buoyancy provided by a gas?

Consider the case, Figure 2.19-1, of a balloon filled with a gas of density $\rho_i \neq$ constant, supported by an atmosphere of density $\rho_e \neq$ constant. Over a differential area dA, the net differential force is

$$dF = -e_n\left(p_e - p_i\right)dA,$$

and, therefore, the integrated pressure force over the bounding surface A is

$$F = -\iint_A \left(p_e - p_i\right)e_n dA. \qquad (2.19\text{-}1)$$

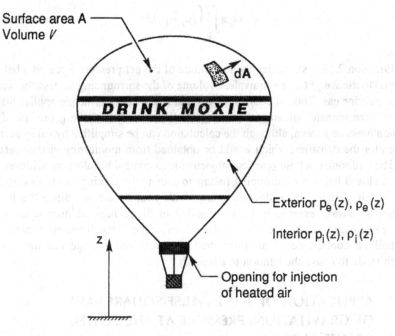

FIGURE 2.19-1 Balloon in a nonuniform atmosphere.

However, by a standard vector integral theorem,* the integral of a scalar point function over a closed surface can be transformed into an integral over the enclosed volume; i.e.,

$$\iint_A p e_n \, dA = \iiint_V \text{grad } p \, dV, \tag{2.19-2}$$

where V is the volume enclosed by A. Therefore,

$$F = -\iiint_V \text{grad}(p_e - p_i) \, dV. \tag{2.19-3}$$

But, from the fundamental equation of statics, Equation 2.5-3, we have, for a uniform gravitational field,

$$\text{grad}(p_e - p_i) = \left[(dp/dz)_e - (dp/dz)_i \right] e_z = -(\rho_e - \rho_i) g e_z, \tag{2.19-4}$$

where $\rho_e = \rho_e(z)$ and $\rho_i = \rho_i(z)$, only. Hence,

---

* For example, see Willis (1931, p. 96) or, more recently, Olmstead (1961, p. 632).

$$F = e_z g \iiint_V (\rho_e - \rho_i) dV. \qquad (2.19\text{-}5)$$

Equation 2.19-5 states that the magnitude of the net pressure force on a balloon is equal to the weight of an equivalent volume of the surrounding air less the weight of the interior gas. This, of course, proves that Archimedes principle applies also to non-constant-density situations. To integrate Equation 2.19-5 the geometry of the balloon must be known, although the calculation can be simplified by using average values for the densities, which would be obtained from monitoring instruments.

The balloonist, whose practical objective is to control his altitude, achieves this with a closed balloon by dropping ballast to ascend, or venting gas to descend. In this hot-air balloon the balloonist simply adds heat to ascend. Since the hot-air balloon is always essentially fully expanded in flight, heat addition reduces the average density by driving off some of the interior gas through the open mouth. As the balloon cools, exterior air infiltrates, increasing the average interior density, which tends to cause the balloon to descend.

2.20 APPLICATION OF THE INVERSE-SQUARE LAW OF GRAVITATION: PRESSURE AT THE CENTER OF THE EARTH

An interesting problem in "fluid" statics is the calculation of the pressure at the center of the Earth. Although it is possible to obtain the differential equation from the vector form of the fundamental equation (Equation 2.4-10) expressed in spherical coordinates, it is instructive to derive the applicable relation from first principles.

By choosing an elemental volume, Figure 2.20-1, consisting of a truncated conical plug of central angle α, we automatically account for symmetry in rotation about the R-axis. The element is centered on the surface of radius R and has an axial length ΔR; its volume is $\Delta V = \pi R^2 \alpha^2 \Delta R$. The equation of equilibrium puts the sum of the pressure forces and gravitational forces in the R-direction equal to zero:

$$\sum \left(F_{\text{press}} + F_{\text{grav.}} \right) = 0. \qquad (2.20\text{-}1)$$

For the two faces normal to the R-axis the combined pressure force is

$$\left[p\pi R^2 \alpha^2 - \frac{d}{dR}\left(p\pi R^2 \alpha^2 \right) \frac{\Delta R}{2} \right] - \left[p\pi R^2 \alpha^2 + \frac{d}{dR}\left(p\pi R^2 \alpha^2 \right) \frac{\Delta R}{2} \right]$$

$$\qquad (2.20\text{-}2)$$

$$= -\frac{d}{dR}\left(p\pi R^2 \alpha^2 \right) \Delta R.$$

For the conical side, assuming an effective mean pressure p, the R-component is

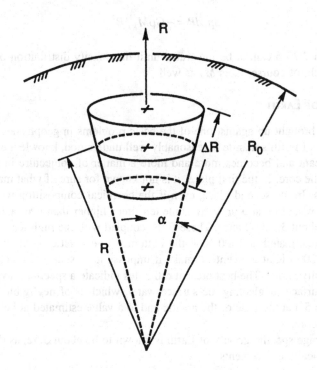

FIGURE 2.20-1 Elemental volume for Earth modeled as a sphere with Newton's universal law of gravitation.

$$p\left[\pi\left(R+\frac{\Delta R}{2}\right)^2\alpha^2 - \pi\left(R-\frac{\Delta R}{2}\right)^2\alpha^2\right] = p\left(2\pi R\alpha^2\right)\Delta R. \qquad (2.20\text{-}3)$$

By combining Equations 2.19-2 and 2.19-3, the resultant pressure force is

$$F_{press.} = -\left(\pi R^2\alpha^2\Delta R\right)\frac{dp}{dR} = -\Delta V\frac{dp}{dR}. \qquad (2.20\text{-}4)$$

The gravitational force on the element is determined by *Newton's universal law of gravitation.* It acts opposite to the +R-direction:

$$F_{grav.} = -G\frac{\rho\left(\pi R^2\alpha^2\Delta R\right)M_R}{R^2} = -G\frac{(\rho\Delta V)M_R}{R^2}, \qquad (2.20\text{-}5)$$

where the local density is ρ and where only that portion of the mass of the Earth within the sphere of radius R, denoted by M_R, contributes to the net gravitational force. By combining Equations 2.20-4 and 2.20-5 with Equation 2.20-1 and dividing out the elemental volume,

$$dp/dR = -G\rho M_R/R^2 .\qquad(2.20\text{-}6)$$

Equation 2.20-6 cannot be integrated until the density distribution $\rho = \rho(R)$ is known, which, of course, fixes M_R as well.

STRUCTURE OF EARTH

We are thus brought up against one of the basic problems in geophysics. Although the structure of Earth's mantle is reasonably well understood, knowledge of Earth's interior is scant and becomes more and more a matter of conjecture in the neighborhood of the core. In the first place, it is not known for sure of what materials the core consists. In the second place, even if the chemical composition were known, the material would be at a pressure some ten times higher than ever achieved in a laboratory (about 300,000 atm). These facts, coupled with the high temperatures in the interior (estimated at 1500 K at the bottom of the mantle — about 3000 mi. down — to 2000 K at the center), make it impossible to state accurately how the internal density varies. The best present estimates indicate a specific gravity* of 3.5 on Earth's surface (neglecting the surface water, which is of negligible thickness) increasing to 5.5 at the base of the mantle and to a value estimated at between 14.5 to 18 in the core.

The average specific gravity of Earth is known to be about 5.52, as determined by astronomical measurements.

CORE PRESSURE OF A CONSTANT-DENSITY EARTH

A simple first estimate of the core pressure can be obtained by assuming an average density $\overline{\rho}$, so that $M_R = (4\pi/3)\overline{\rho}R^3$. Under this assumption, Equation 2.20-6 becomes

$$dp/dR = -(4\pi/3)G\overline{\rho}^2 R.\qquad(2.20\text{-}7)$$

Integrating from the surface at a mean radius R_0, with atmospheric pressure p_a,

$$p - p_a = (2\pi/3)G\overline{\rho}^2 \left(R_0^2 - R^2\right).\qquad(2.20\text{-}8)$$

At Earth's center, where $R = 0$, we employ Equation 2.16-2 to eliminate the universal gravitational constant; thus,

$$p\big|_{R=0} \equiv p_0 = \overline{\rho}gR_o/2.\qquad(2.20\text{-}9)$$

For an average specific gravity of 5.52, this leads to a value for the core pressure of $p_o = 1.69 \times 10^6$ atm. The density variation of Earth is neither constant nor is it a well-behaved continuous function. Beneath Earth's crust there are three other layers,

---

* Specific gravity is defined in Section 2.8.

the mantle and the outer and inner cores. At the interface between the mantle and the outer core, which is liquid, there is a sharp, nearly discontinuous increase of density. Taking this nonlinear distribution of density into account yields estimates of pressures at Earth's center of 3.5 to 4.0×10^6 atm.

2.21 SURFACE TENSION

INTRODUCTION

Surface tension, or *capillarity*, is one of the oldest branches of physical science. The latter name comes from the Latin (*capilla*, a hair) because of the ability of water to rise in a very-small-bore tube when the lower end is immersed in a container of the liquid, a phenomenon that was first discovered by Leonardo da Vinci (1452–1519). The science was sufficiently well developed by the end of the 19th century as to generate a 27,000-word article in the *Encyclopedia Britannica* by James Clerk Maxwell (1911).

Few individuals are so unobservant as to be able to claim that he or she has never observed surface tension at work. Most of us were enthralled at some time in our childhood by blowing soap bubbles* and enjoying the thrill of "popping" them with an instrument of some kind. The more thoughtful observers might have concluded that the popping noise was due to the sudden release of air at a higher pressure inside the bubble than out. Other common examples of surface tension in action are raindrops, drops of agglomerated fat in chicken soup, carbon dioxide bubbles in a cola or, even better, in sparkling wine, and drops of mercury stationary on a flat surface. Another and related effect, that of capillary action, enables water to seep through a porous material, petroleum through porous rock, and liquid solder, with the aid of flux, to flow into a plumbing joint.

Apparently, the external surface of a bubble of liquid within a gas acts like a thin membrane in tension. This brings up the question: What is the source of the force which keeps the surface in tension? The answer is the intermolecular forces, of which there are two, one attractive — the long-range force (on a molecular scale) and one repulsive, which dominates in the short range. At an appropriate spacing, the two balance out such that the net force becomes zero. See Tabor (1991, p. 24) for a short discussion. The location and shape of the surface encasing the bubble would be determined, in principle, by summing the forces on each surface molecule induced by each of the other molecules in the bubble. In addition, since the bubble is in contact with another substance, such as surrounding air, the additional resulting intermolecular forces due to the unlike molecules play a role in the balance. Partington (1955, p. 178) indicates that there is evidence to show that only the exterior layer of the surface molecules in contact actually play a role in this matter. Nevertheless, we are left with a Herculean task of summation.

Thus, we resort to another approach characterized as *phenomenological*, which analyzes the phenomenon observed without dealing with the underlying causes that

---

* The classic book on soap bubbles is due to Boys (1959) with many excellent drawings of a wide variety of elegant soap bubble experiments.

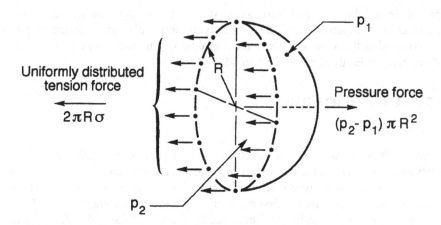

FIGURE 2.21-1 Hemispherical gas bubble in static equilibrium.

produce it — in this case the intermolecular forces. This section relies partly on the work of Bikerman (1958), the first chapter of which is an extensive monograph on surface tension.

Consider first the most elementary case, a gas bubble within a liquid that is presumed to be spherical in shape. We postulate that the "membrane" on the liquid surface of radius R is a shell in uniform tension. If we place a hemispherical segment in equilibrium, Figure 2.21-1, then on the "cut" edge of the segment we must provide for a distribution of tension force per unit length — the *surface tension coefficient* denoted* σ — on the segment periphery. Put p_2, p_1 for the interior, exterior pressure of the bubble, respectively. Then the equation of equilibrium in the direction parallel to the hemisphere axis is $(p_2 - p_1)(\pi R^2) - \sigma(2\pi R) = 0$, or putting $\Delta p \equiv p_2 - p_1$, we obtain

$$\Delta p = 2\sigma/R. \tag{2.21-1}$$

Equation 2.21-1 is the defining relation for the surface tension coefficient, which has the dimensions of force/length or in SI units N/m = kg/s². In a soap bubble, which has two distinct interfaces, the applicable equation is essentially identical except that the factor 2 is replaced by 4.

When the fluid surface is not spherical — such as within an asymmetrical droplet of fluid on a flat surface, where the upper surface is dragged down by gravity — then a surface with double curvature results. Placing a small, finite (near) rectangular surface element ΔA of the droplet in equilibrium, Figure 2.21-2a, we see that the pressure force on the surface element, directed out from the concavity, can be expressed as $\Delta p \Delta A = \Delta p (\Delta s_1 \Delta s_2)$. From Figure 2.21-2b, the surface tension force acting on a short edge of length Δs_2 is $2\sigma \Delta s_2$, directed normal to the edge and locally tangent to the lengthwise centerline.

---

* Most chemistry texts use the symbol γ.

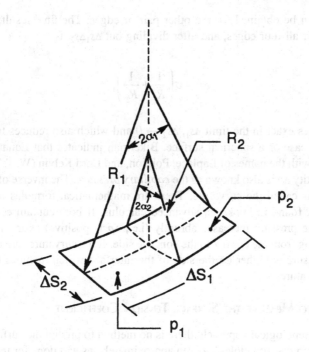

FIGURE 2.21-2a Geometry of a surface with double curvature.

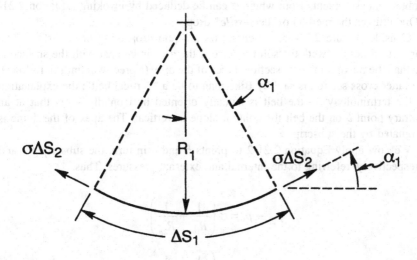

FIGURE 2.21-2b Surface tension on a strip of length Δs_1 and width Δs_2.

By taking into account that there is a pair of parallel edges, the combined component of these forces perpendicular to the center of area, directed inward, is $2\sigma\Delta s_2 \sin \alpha_1 = 2\sigma\Delta s_2 (\Delta s_1/2R_1) = \sigma\Delta s_1\Delta s_2/R_1$, where, since $\Delta s_1 = R_1(2\alpha_1)$, we have introduced the approximation $\sin \alpha_1 \approx \alpha_1 = (\Delta s_1)/2R_1$ for small angles. A similar expression can be obtained for the other pair of edges. The final result, therefore, accounting for all four edges, and after dividing out $\Delta s_1 \Delta s_2$, is

$$\Delta p = \sigma \left(\frac{1}{R_1} + \frac{1}{R_2} \right), \tag{2.21-2}$$

which becomes exact in the limit Δs_1, $\Delta s_2 \to 0$ and which also reduces to Equation 2.21-1 in the case of a spherical surface. Bikerman indicates that Equation 2.21-2 is associated with the names of Laplace, Poisson, and Lord Kelvin (W. T. Thomson).

The quantity Δp is also known as the *capillary pressure*. The inverse of the radius of curvature is called the *curvature*. There are mathematical formulas to calculate the curvature, found in books on advanced calculus. If both curvatures are of the same sign, the pressure increases abruptly (i.e., Δp is positive) as the interface is crossed moving from the convex to the concave side. If the curvatures are of opposite sign, the pressure is higher on the side of the interface which includes the shorter radius of curvature.

A METHOD TO MEASURE THE SURFACE TENSION COEFFICIENT

In the phenomenological approach, there is no method to predict the surface tension coefficient from first principles — even approximately, as was done for the viscosity coefficient in the kinetic theory of gases. Therefore, experiments must be devised to obtain measurements from which σ can be deduced by invoking Equation 2.21-2. One utilizes the method of the *sessile** drop.

Consider Figure 2.21-3a, which shows a liquid drop lying on a surface. The drop is said not to "wet" the solid surface if the area in contact with the surface is less than the maximum cross-sectional area of the drop (a precise definition follows). The latter cross section is said by Bikerman to be a "vertical belt"; the explanation for the terminology — the belt is actually oriented horizontally — is that at an arbitrary point b on the belt the contour slope is vertical. The apex of the dome is designated by the subscript a.

We now apply Equation 2.21-2 to points b and a in turn, the subscripts 2 and 1, respectively, referring to the internal and external pressures. Thus,

$$P_{2b} - P_1 = \sigma \left(\frac{1}{R_{1b}} + \frac{1}{R_{2b}} \right),$$

$$P_{2a} - P_1 = \sigma \left(\frac{1}{R_{1a}} + \frac{1}{R_{2a}} \right).$$

We next subtract the second of these from the first and apply the fundamental equation of fluid statics from b to a within the liquid to obtain

---

* "Sessile" — attached to a base.

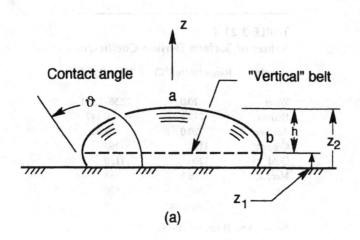

(a)

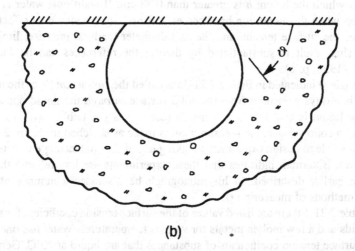

(b)

FIGURE 2.21-3 (a) Geometry of a sessile drop in surface tension. (b) Measurement of the contact angle in a gas bubble underlying a solid surface.

$$p_{2b} - p_{2a} = wh = \sigma \left(\frac{1}{R_{1b}} + \frac{1}{R_{2b}} - \frac{1}{R_{1a}} - \frac{1}{R_{2a}} \right), \qquad (2.21\text{-}3)$$

where $w = \rho g$ is the liquid specific weight. The pressure changes in the gas are treated as negligible. If the drop is much broader than high, the curvatures with subscript a may be neglected compared with the other two, and the formula simplifies.

The radii of curvature and the height can be measured by optical means, although the measurement is by no means trivial, so that other methods tend to be used in

TABLE 2.21-1
Values of Surface Tension Coefficient

Liquid	Temperature (°C)	$10^3 \, \sigma \, (kg/s^2)$
Water	20.0	72.58–72.91
Helium	–272.1	0.147
Acetone	20.0	23.32
Copper	1120	1269
Gold	1200	1120
Mercury	16.5	484
Tin	500	570
Tin	Melting point	610

Source: After Bikerman (1958).

preference. One fact of minor interest can be mentioned. Experimental results show that it is not possible to create a drop of water at 20°C — where $\sigma = 7.27 \times 10^{-2}$ kg/s — in which the height h is greater than 0.38 cm. If additional water is added to the drop when this condition has been established, the gravitational effect dominates over the surface tension and the belt diameter tends to increase, limiting h thereby. This result is substantiated by theory; the references can be found in Bikerman (1958, p. 6).

The angle ϑ indicated in Figure 2.21-3a is called the *contact angle* or the *wetting angle*. It is always measured from the solid surface *through* the liquid phase. If $\vartheta > \pi/2$, the liquid is said not to wet the surface. For a gas bubble within a liquid underlying a solid surface, the measurement is made as sketched in Figure 2.21-3b.

It is possible to design experiments making use of the contact angle to determine σ. However, Bikerman indicates that these experiments are less reliable than the procedure earlier described. In his monograph he discusses a number of other classical methods of measuring σ.

In Table 2.21-1 there are listed values of the surface tension coefficient for several pure liquids and a few molten metals for various temperatures. Water has one of the highest surface tension coefficients of substances that are liquid at 20°C. Generally, σ decreases with temperature (assuming that the pressure is increased to maintain the fluid in the liquid state) until the critical point is reached, where there is no longer any distinction between a liquid and a gas, and where $\sigma \rightarrow 0$.

Surface Tension at a Meniscus

It would be next to impossible to work in a chemistry laboratory without observing one of the most common examples of surface tension, namely, that of a meniscus (meaning crescent-shaped) within a glass tube containing a liquid. In most cases the meniscus is concave down, a description of the surface geometry of the liquid. If the liquid is mercury, however, the surface is convex up. In either case, there is an interface between the liquid and the surrounding gas. If the liquid is attracted to (i.e., wets) the solid, the meniscus is concave; if repulsed, then convex. A point where the meniscus line meets the solid surface is actually a triple intersection of interfaces: a gas/liquid, a liquid/solid, and a gas/solid interface.

Since there must be a surface tension force acting along each interface, there must be three surface tension coefficients, one for each, i.e., σ_{gl}, σ_{ls}, and σ_{gs}. It might seem reasonable to analyze the relationships between these three coefficients by putting in equilibrium, on a vertical plane through the axis of symmetry, a small region around the triple intersection analogous to what was done for a liquid drop surrounded by gas. Unfortunately, with two of the three coefficients undetermined, there are not enough independent relations to solve for the unknowns, including the contact angle, so the procedure fails.

CAPILLARY ACTION WITHIN A TUBE

The case of a small-bore tube one end of which is immersed in liquid within a reservoir might be termed *Leonardo's problem*. If the liquid wets the solid surface, the meniscus appears to be a hemispherical segment concave down. The diameter of the meniscus equals (approximately) the tube diameter. It turns out that the aspherical distortion due to gravity can be neglected in the first approximation.

Referring to Figure 2.21-4 — where the dimensions are greatly exaggerated — let the tube radius be r. The upper surface of the meniscus and of the reservoir are at the local ambient atmospheric pressure. If we apply the manometer equation within the fluid between the meniscus bottom and the top of the reservoir and equate it to the pressure drop of Equation 2.21-1, we obtain

$$\Delta p = wh = \sigma/2r. \qquad (2.21\text{-}4a)$$

Note that h, the height to which the liquid is raised in the tube, is inversely proportional to the radius of the meniscus.

The reader may object that there is no obvious basis for choosing the bottom of the meniscus in preference, say, to the location of the triple intersection and the reader is correct. The uncertainty is created by gravity and the unknown surface tension forces parallel to the wall that cause the meniscus to deviate slightly from the hemispherical. Maxwell suggests that a correction be made by adding to h an amount Δh based on the mean height of the fluid above the meniscus bottom, which depends on the contact angle. By assuming that ϑ is known (from experimental data), then, by neglecting the hemispherical deviation, the correction turns out to be

$$\frac{\Delta h}{r} = \frac{1}{6}\left(\frac{1-\sin\vartheta}{\cos\vartheta}\right)\left[3-\left(\frac{1-\sin\vartheta}{\cos\vartheta}\right)^2\right]. \qquad (2.21\text{-}4b)$$

Unlike the height h, the correction is directly proportional to the radius of the tube. Partington gives corrections of higher order in his work.

For a numerical example, consider a tube such that $r = 0.125$ cm, which corresponds approximately to a diameter of 0.1 in. Let the fluid be water at 20°C for

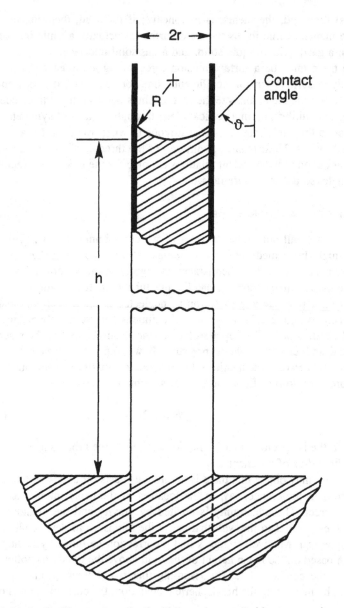

FIGURE 2.21-4 Capillary action of a liquid in a tube supplied by a reservoir.

which $w = \rho g = 9.807/1.0017 \times 10^{-3} = 9.790 \times 10^3$ N/m^3, using Table 1.13-1. Then, with data given above, $h = 2\sigma/rw = 2 \times 72.58 \times 10^{-3}/1.25 \times 10^{-2} \times 9.790 \times 10^3 = 0.001\ 19$ m $= 0.119$ cm, which is about 0.05 in.

For the correction we need the contact angle, which depends on the three substances at the meniscus — in this case air/water/glass. This must be determined as a separate experiment. In fact, the contact angle depends also on the purity of the water and the condition of the glass surface. At the time of the Maxwell (1911)

TABLE 2.21-2
Contact Angle for Air/Liquid/Solid
Interfaces at 20 to 25°C

Liquid	Solid	ϑ (°)
Mercury	PTFE[a]	150
	Glass	128–148
Water	Paraffin	110
	Human Skin	90
	Glass	Small ($\approx$0)

Source: After Adamson (1982).

[a] Polytetrafluoroethylene (Teflon).

article, the accepted value was ϑ = 25.5°, for which Equation 2.21-4b yields $\Delta h/r$ = 0.589. Thus, in this case Δh = 0.074 cm $\approx$ 0.03 in., which is a 6% correction. Based on the more recent data, as Bikerman suggests, one should use ϑ = 0, for which $\Delta h/r$ = $^1/_3$, reducing the correction to 3.5%. Presumably this correction should also be applied to manometer tubes, where h continues to be that corresponding to Leonardo's problem and not the manometer deflection.

As a practical matter in running the capillarity experiment, to reduce the time for the meniscus to reach its terminal value, it is quicker and more reliable to put suction on the top of the tube to raise the meniscus above its expected final value, then remove the suction and let the liquid descend to its correct level. Note also, if the fluid were mercury, the deflection h would be negative and the meniscus would be convex up.

Accurate measurement of the contact angle is difficult, so much so that results must be carefully verified to ensure that they are reliable. Tabulated values are difficult to find in the literature. One useful source is Adamson (1982), who references the original measurements. Several values from his text are listed in Table 2.21-2.

ON ENERGY METHODS

It is obvious that to raise a column of fluid above its reservoir level requires work and a source of energy. The energy source is the atmosphere, which depresses the reservoir level ever so slightly, and the energy is stored in the liquid column and the meniscus itself. The phenomenon of surface tension can be analyzed strictly from work/energy viewpoints, a method which Bikerman says is "safer." We shall not pursue the energy approach herein.

ON VIBRATIONS IN LIQUIDS UNDER SURFACE TENSION

When a water jet is emitted horizontally from an orifice that is circular in cross section, the jet cross section tends to maintain its circularity while falling under gravity. Eventually, irregularities on the surface become magnified by surface tension, distinct drops form, and the continuity of the jet is destroyed. On the other hand, if the exit of the orifice is elliptical in shape, the jet behavior is radically different. If the major axis of the exit ellipse is oriented vertically, the observer notes that a short distance downstream the long axis now occurs at the horizontal station. This alternating behavior* can be sustained through several cycles before the jet starts to break up into individual droplets.

This action can be explained by considering a cylinder of fluid made up of a thin, transverse slice of fluid as it is ejected from the orifice. Kelvin** showed that the fluid-slice contour would oscillate about a circle of radius a whose area is equivalent to that of the elliptical orifice. The circular frequency of oscillation is given by the relation

$$\omega^2 = 6\sigma/\rho a^3. \tag{2.21-5}$$

The frequency is independent of the aspect ratio of the ellipse. However, for a given value of a, the greater the aspect ratio, the greater the energy stored in the elliptical cross section, which at any instant is distributed between kinetic energy of the fluid particles and potential energy stored in membrane under tension. The jet, with its pattern of alternating vertical and horizontal major axes appears to be subject to a standing wave. A computer-graphics visualization of this phenomenon (omitting gravity) is shown in Figure 2.21-5.

Suppose that water at 20°C is ejected from an elliptical orifice whose equivalent area is a circle of radius $a = 0.2$ cm. From Equation 2.21-5 the circular frequency of oscillation is $\omega = 233.5$ rad/s, and thus the period of vibration is $\tau = 2\pi/\omega = 0.026$ 9 s. Now if the hydrostatic head is 15 cm, from Torricelli's law (Equation 4.6-3), the exit velocity is $u = \sqrt{2gh} = 1.715 \, \text{m/s}$. Consequently, the wavelength between successive peaks is $L = u\tau = 4.62$ cm. The pattern replicates itself until the jet begins to break up into separate drops. Measurement of the wavelength of an oscillating jet provides an alternative method for determining σ.

Similarly, a globule of liquid (think of a balloon filled with water) can oscillate around an imaginary spherical shell of equivalent volume of radius a. The frequency of the simplest mode of vibration is given by

$$\omega^2 = 8\sigma/\rho a^3. \tag{2.21-6}$$

The diameter of a globule with a period of $\tau = 1$ s is $2a = 2.45$ cm, approximately 1 in.

---

* This experiment is one which can be readily duplicated by at least half the potential readership of this text.
** The references can be found in Lamb (1932, pp. 471–475).

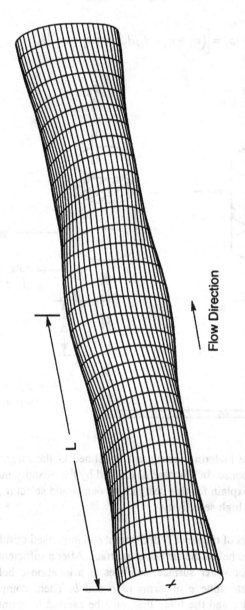

Flow Direction

FIGURE 2.21-5 Computer-generated model of a horizontal free jet emitted (neglecting gravitational effects) from an elliptical orifice. Each vertical slice is subject to surface tension effects that cause the slice to oscillate in space about its mean station, which moves from left to right, resulting in a standing wave on the jet surface. (Image generated by Thomas Citrini of Rensselaer Polytechnic Institute.)

PROBLEMS

2.1. The manometer shown employs two different fluids, a design that permits the measurement of small pressure differences. Show that

$$\Delta p = p_1 - p_2 = \left[\left(w_1 + w_2\right)\left(d_1/d_2\right)^2 + \left(w_1 - w_2\right)\right]h.$$

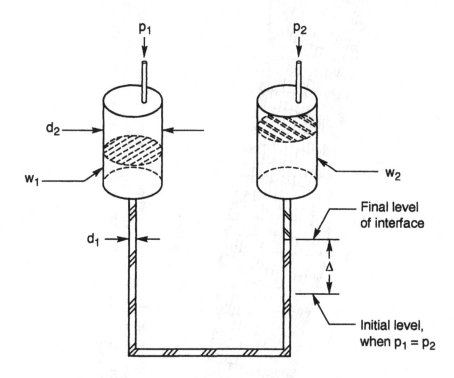

The *sensitivity* of a metering instrument is defined as the magnitude of the instrument response (in this case h) divided by the quantity measured (in this case Δp). Explain for this design how one would select d_1, d_2, w_1, and w_2 to obtain a high sensitivity.

2.2. A cylindrical bucket of diameter D and height b is immersed upside down in water to a depth a below the air/water interface. After a sufficient period of time, the interior water surface stabilizes at a location c below the bucket bottom. Determine c in terms of a and b. Then, compute the numerical value of c and the force that must be exerted to maintain the bucket in equilibrium if its weight is 20 ton, and if $a = 20$ ft, $b = 30$ ft, $D = 10$ ft. The main water surface pressure is standard atmospheric.

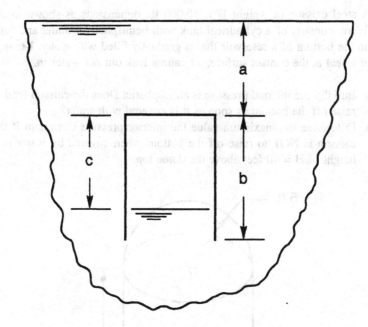

2.3. A gate is immersed in water and inclined at 30° to the vertical. The gate is designed so that its width increases parabolically from the bottom up according to the relation $b = s^2/10$ measured in feet. If the water height is $h = 10$ ft, determine the force on the gate due to the water pressure. Compute the moment exerted by the water on the gate about an axis along the top of the gate.

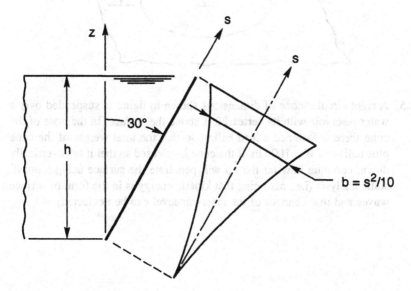

2.4. A steel caisson of weight W = 10,000 lb, dimensions as shown in the figure, consists of a cylindrical tank with hemispherical dome and rests on the bottom of a reservoir that is gradually filled with water. Because of a seal at the contact surface, air cannot leak out nor water in.

    a. Initially the internal pressure is atmospheric. Does the caisson tend to raise off the bottom as soon as it is covered with water?

    b. Determine the maximum value the interior pressure can attain if the caisson is NOT to raise off the bottom when covered by water to a height of H = 40 feet above the dome top.

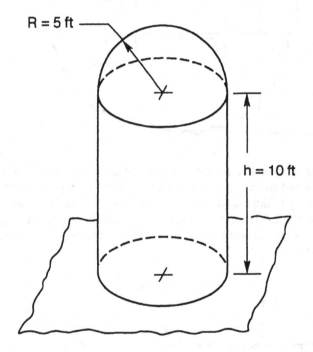

2.5. A right circular cone of dimensions shown in figure is suspended over a water reservoir with its vertex 20 ft above the surface. In the nose of the cone there is fastened a lead ballast so that the total weight of the cone plus ballast is W = 1000 lb. If the cone is released so that it falls vertically down, compute how far the tip will penetrate the surface using a quasi-static analysis (i.e., assuming that kinetic energy is in the form of surface waves and that changes of the reservoir level can be neglected).

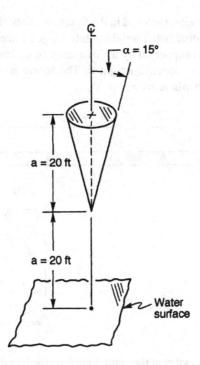

2.6. Water of depth 10 ft lies on a bed of soluble material. By diffusion the solid is absorbed throughout the liquid but in varying proportion; i.e., there is a greater concentration of suspended material near the bottom than at the top. Samplings indicate that the mixture density varies from that of pure water near the top to 1.5 times standard water density near the bottom.

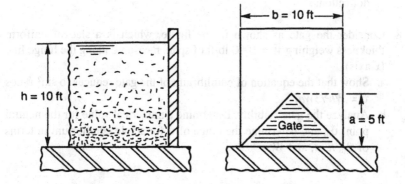

  a. Assuming that the mixture density varies linearly with depth, determine the pressure–height relation in the mixture and compute the pressure at the bottom.
  b. Determine the force on the triangular gate shown in the diagram.

2.7. An indicating device, sketched in the figure, consists of a rigid arm, length
  L, of uniformly distributed weight q lb/ft, hinged at the left end. The other
  end of the arm is supported by the buoyancy force developed on a sphere
  of radius R and of specific weight w. The sphere is immersed in a two-
  layer liquid bath where $w_2 > w > w_1$.

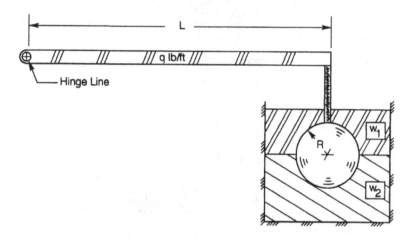

  a. Determine w so that at the neutral position the liquid interface coincides
     with the sphere equator. Neglect the weight and buoyancy force con-
     tribution of the rigid support connecting the arm and the sphere.
  b. If $\vartheta = 0$ denotes the neutral position of the system, determine the
     stability derivative $dM/d\vartheta$ of the system at the neutral point. *Hint*:
     Determine the restoring moment associated with an arbitrarily small
     displacement $d\vartheta$ of the system.
  c. Determine the undamped natural frequency of the system for small
     deflections.

2.8. Consider the gate as shown in the figure, which is a slab of uniform
  thickness weighing $\overline{W} = 1000$ lb/ft of span, measured along the hinge line
  (x-axis).
  a. Show that the equation of equilibrium of the gate reduces to $\sin^2 \vartheta \cos$
     $\vartheta = Wh^3/3Wa$.
  b. Analyze the gate stability. Determine a numerical value for the neutral
     point ϑ_0, and determine the range of ϑ for stable equilibrium in terms
     of ϑ_0, h, a, and $\overline{W}$.

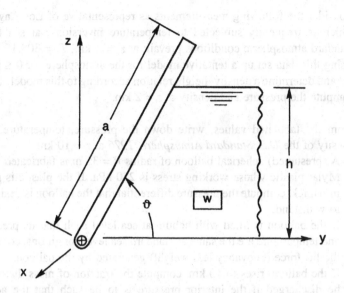

2.9. A concentrated weight of $W = 1000$ lb is supported by four identical, hollow conical floats. For the current purpose the weight of the floats and supporting arms is considered to be negligible.

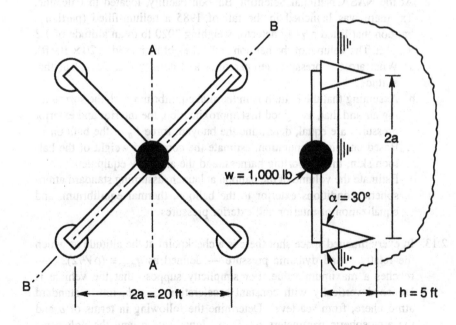

a. Determine the equilibrium position in floating.
b. Determine the pitching stability derivative $dM/d\vartheta$ for axis AA.
c. Determine the stability derivative for axis BB. Show that the result can be explained from a theorem of moments of area from plane mechanics.

2.10. Consider the following measurements as representative of Los Angeles, which is frequently subjected to temperature inversions: at sea level, standard atmospheric conditions prevail; at $z_1 = 5$ km, $T_1 = 308.15$ K. Using this data set up a tentative model for the atmosphere for $0 \le z \le 5$ km and determine a density–height relation according to this model. Then, compute the pressure and density at $z = 2$ km.

2.11. From the tabulated values, write down the pressure, temperature, and density of the *U.S. Standard Atmosphere, 1976* at $z = 10$ km.

a. A (presumed) spherical balloon of radius $R = 35$ m is fabricated from Mylar plastic whose working stress is 250 kPa. If the plastic is $t = 5$ mm thick, compute the pressure differential that the balloon is designed to withstand.

b. If the balloon is filled with helium at sea level such that its pressure and temperature are the same as standard sea level conditions, compute the lift force (buoyancy less weight) generated by the balloon.

c. If the balloon rises to 10 km, compute the fraction of mass that must be discharged if the interior pressure is to be such that the actual pressure differential is half the allowable value. Assume thermal equilibrium.

2.12. At the NASA National Scientific Balloon Facility, located in Palestine, Tx, there was launched in the fall of 1985 a helium-filled (partially) balloon that lifted a γ-ray detector weighing 3020 lb to an altitude of 1.3×10^5 ft. The volume of the balloon at that height increased to 20×10^6 ft^3.

a. What are the pressure, temperature, and density (p_h, T_h, ρ_h) at that altitude?

b. Assuming that the helium is in thermal equilibrium with the surrounding air and that, as a good first approximation, the interior and exterior pressures are equal, determine the buoyant force F_B of the balloon.

c. Based on this computation, estimate the combined weight of the balloon skin, the supporting harness, and the auxiliary equipment.

d. Estimate the volume of the balloon at launch, assuming standard atmospheric conditions exterior to the balloon, thermal equilibrium, and equalization of interior and exterior pressures.

2.13. In every manned space shot there is a checkpoint at the altitude at which the relative wind dynamic pressure — defined as $q_{max} \equiv (\rho V^2/2)_{max}$ — reaches a maximum value. For simplicity suppose that the vehicle is launched vertically with constant acceleration $\ddot{z} \equiv b$, into a standard atmosphere, from sea level. Determine the following in terms of b and the atmospheric parameters p_a, T_a, ρ_a, *lapse rate* a, and the *polytropic exponent n*:

a. The relation for the density variation in the troposphere ρ/ρ_a;

b. The time t_m and altitude z_m at which the vehicle attains q_{max}. Verify that the latter is independent of b;

   c. The vehicle velocity $\dot{z}_m$ and q_{max};

   d. At q_{max}, the numerical values of z_m, t_m, $\dot{z}_m$, if $b = 2g$;

   e. The vehicle Mach number $M_m \equiv \dot{z}_m/a_m$ at q_{max}, where a_m is the local speed of sound.

2.14. Refer to Section 2.19. In an attempt to obtain a more realistic value of the pressure at the center of Earth, the following linear density-distribution function is proposed:

$$\rho/\rho_H = A + B(R/R_0),$$

where $\rho = \rho(R)$ is the local density, ρ_H is the standard density of water, and R_0 is the mean radius of Earth.

   a. Determine A and B if the specific gravity on Earth's surface is S.G. = 3.5 and is S.G. = 16 at the center.

   b. Show, corresponding to the prescribed density distribution, that the mass M enclosed within a sphere of radius R, divided by the total mass M_0, is given by

$$\frac{M}{M_0} = \frac{4A\left(R/R_0\right)^3 + 3B\left(R/R_0\right)^4}{4A + 3B}.$$

   c. Next determine an expression for $p(R) - p_a = f(R/R_0)$. Then, compute $p_c - p_a$ and compare with the value from Equation 2.19-9.

3 The Equations of Fluid Dynamics

I cannot but feel well concerning those physical principles with which I became strongly involved, since indeed they led me by the hand to exposing many new properties concerning both the equilibrium and the motion of fluids, which, unless the love of the undertaken work deceives me, will some day promote Hydrodynamics significantly, if they are refined more than I was allowed (by circumstances) to do.

Daniel Bernoulli (1738), Translated from the Latin by T. Carmody and H. Kobus

3.1 INTRODUCTION

Although the ancients had developed a number of remarkable practical water-distribution systems — the Roman water aqueducts can be particularly mentioned — there is no evidence that there existed prior to the time of Galileo (1564–1642) any real understanding of the underlying principles that enable one to calculate flows. Starting in the 17th century, as the science of mechanics developed, these principles were applied to fluids, eventually giving rise to a tremendous body of literature with which some of the most illustrious names in engineering and science are associated. Rouse and Ince (1957) have furnished us with a most interesting study of the development of hydraulics through the early part of the 20th century. In this they touch upon most of the names of those who laid the groundwork out of which grew the diverse specialties of today, including meteorology, hydrodynamics, aerodynamics, viscous flow, and compressible flow.

In the following the equations of fluid dynamics are developed in a general way under the restriction of one space variable. For many problems of engineering importance, at least for the purposes of preliminary design, this may be sufficient. The exposition of the three-dimensional (3D) equations is best left to specialized courses. In many cases the solution of the 3D equations requires a numerical analysis for which there are various programs well suited to the powerful desktop computers developed in the last two decades of the 20th century.

3.2 CONCEPTS FROM KINEMATICS

Consider first the motion of a single particle through space; its trajectory is known in fluid dynamics as a *particle pathline*. In Figure 3.2-1, let P, P' denote its position along the pathline at times t, $t + \Delta t$, respectively. The position of the particle can be specified in terms of the radius vector

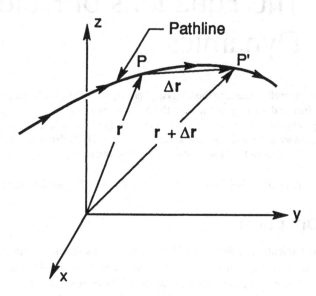

FIGURE 3.2-1 Particle pathline.

$$r = e_x x + e_y y + e_z z, \qquad (3.2\text{-}1)$$

measured with respect to an observer whose location, by convention, is taken to coincide with the origin. As yet no restriction is placed on the motion of the observer. At a given instant of time we know the location of a particle if we know its coordinates $r = r(t)$ or, equivalently, $x = x(t)$, $y = y(t)$, $z = z(t)$ which, as indicated, depend on time alone.

VELOCITY OF A PARTICLE

We define *particle velocity* as follows:

$$V \equiv \lim_{\Delta t \to 0} \frac{r(t + \Delta t) - r(t)}{\Delta t} = \frac{dr}{dt}. \qquad (3.2\text{-}2)$$

Substitution of Equation 3.2-1 in 3.2-2 gives

$$V = e_x \frac{dx}{dt} + e_y \frac{dy}{dt} + e_z \frac{dz}{dt},$$
$$= e_x u + e_y v + e_z w, \qquad (3.2\text{-}3)$$

where the scalar components of the velocity in rectangular cartesian coordinates are

$$u \equiv dx/dt, \quad v \equiv dy/dt, \quad w \equiv dz/dt. \qquad (3.2\text{-}4)$$

The vector V and the differential dr, in the limit as $\Delta t \to 0$, are evidently tangent to the pathline.*

It turns out that, instead of using the definition of velocity in terms of the radius vector r, it is often more convenient to work with a curvilinear coordinate s defined such that $(ds)^2 \equiv dr \cdot dr = (dx)^2 + (dy)^2 + (dz)^2$. We also introduce a unit vector $e_s \equiv dr/|dr|$, tangent to the particle pathline, such that e_s always points in the direction in which s increases. Consequently, e_s depends on s, which, for a particle, is a function only of time. Therefore, $dr = e_s ds$, and

$$V = \frac{dr}{dt} = e_s \frac{ds}{dt}.$$ (3.2-5)

The scalar part of the vector V is called the particle *speed* $V \equiv |V|$. Since the direction of V is e_s,

$$V = e_s V = e_s \, ds/dt, \quad V = ds/dt.$$ (3.2-6)

In this case the direction of increasing s was defined by the motion itself. In other cases, e.g., motion along the x-axis in rectangular cartesian coordinates, the velocity can be directed either in the plus or minus directions, in which case the scalar, velocity components u, v, w can be either positive or negative.

ACCELERATION OF A PARTICLE

Acceleration of a particle is defined as the time derivative of its velocity. Thus, applying the product rule of differentiation,

$$a \equiv \frac{dV}{dt} = \frac{d}{dt}(e_s V),$$ (3.2-7)

$$= e_s \frac{dV}{dt} + V \frac{de_s}{dt} = e_s \frac{dV}{dt} + V \frac{ds}{dt} \frac{de_s}{ds},$$

$$= e_s \frac{dV}{dt} + V^2 \frac{de_s}{ds}.$$ (3.2-8)

It is a standard demonstration in most mechanics texts that the derivative of the tangential unit vector, in the direction along the pathline, is given by

$$de_s/ds = e_n/R,$$ (3.2-9)

---

* In the present case the mathematical operations indicated by dr/dt, or by dx/dt, etc., are unambiguous but, in a flow composed of an infinity of particles, it is *not* necessarily true that the mathematical symbol dr/dt stands for particle velocity. An attempt to clarify this point will be made when the concept of a flow field is introduced.

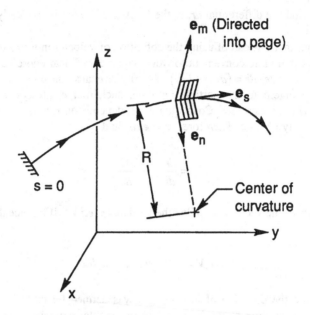

FIGURE 3.2-2   Natural coordinates.

where e_n is a vector normal to e_s, lying in the osculating plane to the pathline, directed toward the center of curvature, and where R is the radius of curvature defined so as to be always positive. The osculating plane is defined by the particle position and two points on the pathline, lying on either side of it, in the limit as the three points merge. From Equations 3.2-8 and 3.2-9, therefore,

$$a = e_s \, dV/dt + e_n \, V^2/R. \tag{3.2-10}$$

Thus all particles moving in a curved trajectory have a component of acceleration normal to the path. The pair of vectors in Equation 3.2-10 allows definition of a third unit vector, $e_m \equiv e_s \times e_n$, which is called the *binormal*, and which is orthogonal to the other two, as shown in Figure 3.2-2. In order to maintain the right-handedness of the coordinates, whose cyclic order is s, n, m, s, ..., the vector e_m, for a path of the curvature shown, is directed into the page. The coordinates s, n, m are variously called *natural*, or *path*, or *intrinsic* coordinates.

CONCEPT OF AN INERTIAL OBSERVER

It should be kept in mind that when we speak of velocity or acceleration of a particle, we mean the velocity or acceleration with respect to a particular observer. In principle, the choice of an observer is completely arbitrary, but, in fact, for any given problem there is usually a preferred observer (or, equivalently, a preferred frame of reference) with respect to whom the equations of motion take the simplest form. Usually, although not always, the simplest form results for an inertial observer, and the phenomenon is said to be viewed in an *inertial frame of reference*.

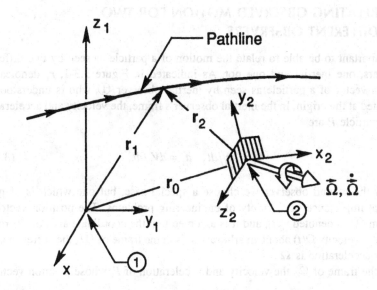

FIGURE 3.3-1   Relating motion observed by an inertial and noninertial observer.

Galileo pointed out the impossibility of detecting "absolute motion." This idea is inherent in *Newton's first law*, which states that every body continues in a state of rest, or of uniform motion in a straight line, in the absence of external forces. This gives a criterion* for determining whether or not an observer is an inertial observer. If the observer observes the motion of a body in the absence of external forces to be other than a straight line, or motionless, then the observer is not an inertial observer.

On the other hand, for purpose of discussion it is often convenient to speak of an inertial observer as someone who moves with uniform velocity with respect to some (undefined) fixed point. More precisely, it means simply that the observer is not accelerating with respect to any other inertial observer, as defined just above.

For some purposes it is possible to neglect the nonuniform motion of the observer if the observer's acceleration is sufficiently small compared with that of the phenomenon being studied. For example, for most engineering problems not involving space flight it is possible to neglect the Earth's rotation on its axis and treat an earthbound observer as an inertial. However, an earthbound observer, when calculating motion of the atmosphere, or of an ocean, or a space vehicle trajectory, must take into account that he or she is not an inertial observer.

---

* Such a criterion employs the concept of force without defining it. Every individual has an instinctive sense of force, particularly as a muscular sensation. We consider that *force* is understood if we can assign a magnitude and direction to it and fix its point of application. The concept is fundamental to mechanics (some writers call it a primitive concept) but is indefinable in terms of more elemental concepts. It is a building block upon which the postulates of mechanics are based. This view is in flat opposition to that of some writers who maintain that Newton's second law is a definition of force. For a historical view of the concept of force see Jammer (1962); unfortunately even Jammer's historical researches do not lead to a final clear, and unambiguous, explanation of force.

3.3   RELATING OBSERVED MOTION FOR TWO DIFFERENT OBSERVERS

It is important to be able to relate the motion of a particle as seen by two different observers, one inertial and one not. As indicated in Figure 3.3-1, r_1 denotes the position vector of a particle as seen by inertial observer ① who is understood to be located at the origin. In the inertial observer's frame, the velocity and acceleration of the particle P are

$$V_1 = dr_1/dt, \quad a_1 = dV_1/dt. \tag{3.3-1}$$

For the second observer we choose a special case, but one which is of great practical importance in a variety of engineering problems. The position vector of ② from ① is denoted $r_0(t)$, and it is assumed that the coordinate axes of ② rotate at angular velocity $\Omega(t)$ about an arbitrary axis in the frame of ①; the corresponding angular acceleration is $\dot{\Omega}$.

In the frame of ② the velocity and acceleration of P, whose position vector is r_2, are

$$V_2 = dr_2/dt, \quad a_2 = dV_2/dt. \tag{3.3-2}$$

It can be shown* that the velocities and accelerations of P, as seen by the two observers, are related by

$$V_1 = dr_0/dt + V_2 + \Omega \times r_2, \tag{3.3-3}$$

$$dV_1/dt = d^2r_0/dt^2 + dV_2/dt + 2\Omega \times V_2 + \Omega \times (\Omega \times r_2) + \dot{\Omega} \times r_2. \tag{3.3-4}$$

In Equation 3.3-4 d^2r_0/dt^2 is the acceleration of ② with respect to ①. The term $2\Omega \times V_2$ is the *Coriolis acceleration*, and the term $\Omega \times (\Omega \times r_2)$ is called the *centripetal* acceleration. As will be noted further along, in fluid mechanical applications we replace the terms dV_1/dt, dV_2/dt by substantial derivatives DV_1/Dt, DV_2/Dt, for reasons to be explained.

EXAMPLE

Determine the relation between the accelerations of a particle as determined in the frame of ①, an inertial observer, and of ②, an observer coincident with ① but whose frame rotates with respect to that of ① at angular velocity $\Omega = e_z\Omega$ about the common z-axis, as indicated in Figure 3.3-2. Express the result in polar coordinates (r_2, ϑ_2, z) with the particle velocity in the frame of ② given by $V_2 = e_{r2}u_{r2} + e_{\vartheta2}u_{\vartheta2}$, where u_{r2} and $u_{\vartheta2}$ are completely arbitrary.

---

* For example, see Meriam and Kraige (1986, pp. 346–351).

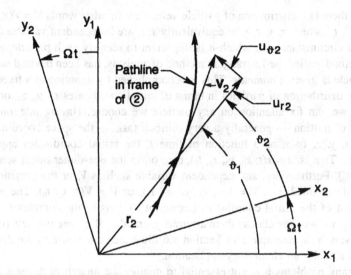

FIGURE 3.3-2 Relationship between accelerations in different frames.

The necessary relation is provided by Equation 3.3-4. Since $d^2r_0/dt^2 = 0$, $\dot{\Omega} = 0$, Equation 3.3-4 reduces to

$$\frac{dV_1}{dt} = \frac{dV_2}{dt} + 2\Omega \times V_2 + \Omega \times (\Omega \times r_2). \qquad (3.3-5)$$

However,

$$\Omega \times V_2 = \Omega e_z \times (e_{r2}u_{r2} + e_{\theta2}u_{\theta2}) = e_{\theta2}\Omega u_{r2} - e_{r2}\Omega u_{\theta2},$$

and

$$\Omega \times r_2 = \Omega e_z \times e_{r2}r_2 = e_{\theta2}\Omega r_2,$$

$$\Omega \times (\Omega \times r_2) = \Omega e_z \times e_{\theta2}\Omega r_2 = -e_{r2}\Omega^2 r_2.$$

Therefore,

$$\frac{dV_1}{dt} = \frac{dV_2}{dt} - e_{r2}(2\Omega u_{\theta2} + \Omega^2 r_2) + e_{\theta2}2\Omega u_{r2}. \qquad (3.3-6)$$

3.4 KINEMATICS IN A FLOW FIELD

For a flow, in which there is always a continuous distribution of particles, the space variables x, y, z, or, equivalently, the pathline coordinate s, are no longer associated with a unique particle. This gives rise to the concept of a velocity field in which, at any time, there is a distribution of particle velocities; in other words $V = V(x, y, z, t)$ or $V = V(s, t)$, where x, y, z, t, or equivalently s, t, are independent variables.

Under circumstances in which it is important to identify each particle, an analytical method, called the Lagrangian method of analysis, has been devised whereby each particle is given a *nametag*. This is accomplished by denoting at a fixed time, say, t_0, the distribution of particles in terms of initial coordinates x_0, y_0, z_0, or s_0 our nametag; we can fix attention on any particle we choose. Having integrated the equations of motion — generally a very difficult task — the space coordinates of particles x, y, z, or s, as a function of time t, the initial coordinates appear as parameters. That is, $x = x(t; x_0, y_0, z_0, t_0)$, etc., or, in the one-dimensional scheme, $s = s(t; s_0, t_0)$. Furthermore, any dependent variable such as V, or the pressure p, is similarly determined, i.e., $V = V(t; x_0, y_0, z_0, t_0)$, or $V = V(t; s_0, t_0)$. The explicit involvement of the initial coordinates (subscript 0) adds to the complexity of the equations since we may choose them arbitrarily, and thus there are two sets of space and time variables. Example 2 of Section 3.6 illustrates the nature of the difficulty in what is actually an elementary application.

For many problems it is not essential to display the analytical dependence on the initial coordinates. That is, except for certain important cases, we bypass the analytical problems in determining the time history of a given particle. This implies that — barring the exceptions — we do not usually care which particles occupy a given point as long as we can determine the velocity, pressure, density, etc. at that point as a function of time. The latter viewpoint, called the *Eulerian method*, does not sacrifice the ability to do the computation for a particular (i.e., arbitrary) particle; it simply bypasses the analysis involved, which may actually be quite difficult. In the Eulerian scheme the independent coordinates (x_0, y_0, z_0, t_0) or (s_0, t_0) are usually dispensed with completely and we are left with one of the sets (x, y, z, t) or (s, t). It should be kept in mind, therefore, that (x, y, z) or s are not necessarily associated with an identifiable particle. This fact has implications in computing certain derivatives, as we will attempt to clarify in the following section.

One of the cases where the Lagrangian representation may have to be invoked is in one-dimensional, nonsteady flows where knowledge of the particle path is needed. An elementary example can be found in Section 8.8.

3.5 TIME DERIVATIVES AND THE SUBSTANTIAL DERIVATIVE

In a flow, any variable, such as pressure, temperature, or velocity, will likely depend on all of the independent variables. This can be expressed in a one-dimensional flow as $F = F(s, t)$, where F may represent the temperature, or pressure, or velocity. The differential of such a function is given by

$$dF = \frac{\partial F}{\partial s}\, ds + \frac{\partial F}{\partial t}\, dt. \tag{3.5-1}$$

By dividing by dt, it follows that the total time derivative is

$$\frac{dF}{dt} = \frac{\partial F}{\partial t} + \frac{ds}{dt}\frac{\partial F}{\partial s}. \tag{3.5-2}$$

In fluid mechanics, it is understood that the independent variables are the space coordinate s and time t, unless otherwise specified. Thus, it is customary to write the partial derivatives as

$$\frac{\partial}{\partial t} \equiv \left(\frac{\partial}{\partial t}\right)_s, \quad \frac{\partial}{\partial s} \equiv \left(\frac{\partial}{\partial s}\right)_t, \tag{3.5-3}$$

where the subscripts can be omitted for convenience. Furthermore, the differentials ds and dt in Equation 3.5-2 may be chosen arbitrarily, and thus the ratio ds/dt is the velocity of a point moving in the s,t-plane whose total time derivative operator is

$$\frac{d}{dt} = \frac{\partial}{\partial t} + \frac{ds}{dt}\frac{\partial}{\partial s}. \tag{3.5-4}$$

In other words, the application of Equation 3.5-4 to some function $F = F(s, t)$ allows one arbitrary condition to be specified, namely, the value of ds/dt.

THE SUBSTANTIAL DERIVATIVE

For a particle in motion with respect to an observer there is often a need in fluid mechanics to compute the time derivative of some function for the particle as it moves along its path. We could indicate such by using the temporary subscript notation

$$\left(\frac{d}{dt}\right)_P = \frac{\partial}{\partial t} + \left(\frac{ds}{dt}\right)_P \frac{\partial}{\partial s},$$

where, in consequence of specifying the derivative for a fixed particle,

$$\left(\frac{ds}{dt}\right)_P = V.$$

The subscript P is used to denote the fact that the differentiation is to be computed for a fixed (though arbitrary) particle. In fact, the desirability for a special notation to denote a time derivative for a particle was recognized first by Stokes* whose notation is now universally accepted. Accordingly, we put

---

* Sir George Stokes (1819–1903), British scientist, especially renowned for his derivation of the equations of 3D viscous flow. The equations are called the Navier–Stokes equations in honor of Stokes and the French scientist L. Navier who discovered them independently.

$$\left(\frac{d}{dt}\right)_P \equiv \frac{D}{Dt} \equiv \frac{\partial}{\partial t} + V\frac{\partial}{\partial s}, \tag{3.5-5}$$

from which it follows that

$$\left(\frac{ds}{dt}\right)_P \equiv \frac{Ds}{Dt} \equiv V. \tag{3.5-6}$$

The operator D/Dt enables the time derivative of an arbitrary quantity to be computed for a point moving with a particle of fixed identity in terms of partial space, and time, derivatives and the speed V, as computed in the frame of the observer.

Thus, the analog of Equation 3.2-5, which is valid only for a discrete particle, becomes, in Eulerian notation,

$$V(s,\ t) = e_s\, Ds/Dt. \tag{3.5-7}$$

Similarly, the expression for particle acceleration in natural coordinates is obtained from Equation 3.2-10:

$$a = DV/Dt = e_s\, DV/Dt + e_n\, V^2/R. \tag{3.5-8}$$

The reason for these gymnastics, we repeat, is the fact that s serves both as a space (independent) variable *and* as the particle coordinate. In Equations 3.5-7 and 3.5-8 $a = a(s,\ t),\ e_s = e_s(s,\ t)$, etc., where d/dt has been replaced by the substantial derivative operator D/Dt, in consequence of the auxiliary restriction on how the time derivative is to be computed. It should be noted that, whereas $Ds/Dt = V$, $DV/Dt = \partial V/\partial t + V\partial V/\partial s$.

3.6 ILLUSTRATIVE EXAMPLES

EXAMPLE 1

Suppose, in a horizontal duct ($V \to u,\ s \to x$), that the velocity distribution $u = u(x, t)$ is given by

$$u = \left(Ax + Bx^2 + Cx^3\right)\cos \omega t. \tag{3.6-1}$$

The constants A, B, and C, must have the appropriate dimensions to ensure that the right-hand side has the dimension of velocity. Determine the general expression for $a_x(x,\ t)$, and for $\omega = 1$ Hz determine $a_x(1,\pi/4)$, $a_x(2,0)$. From the velocity expression we differentiate to obtain

$$\frac{\partial u}{\partial x} = \left(A + 2Bx + 3Cx^2\right)\cos \omega t, \quad \frac{\partial u}{\partial t} = -\omega\left(Ax + Bx^2 + Cx^3\right)\sin \omega t.$$

Therefore,

$$a_x = \frac{Du}{Dt} = \frac{\partial u}{\partial t} + u\frac{\partial u}{\partial x},$$

$$= -\omega\left(Ax + Bx^2 + Cx^3\right)\sin \omega t + \left(A + 2Bx + 3Cx^2\right)\left(Ax + Bx^2 + Cx^3\right)\cos^2 \omega t. \tag{3.6-2}$$

From this last relation, we then find, for $\omega = 1$ Hz,

$$a_x(1, \pi/4) = -\left(1/\sqrt{2}\right)(A + B + C) + \tfrac{1}{2}(A + 2B + 3C)(A + B + C), \tag{3.6-3}$$

$$a_x(2, 0) = (A + 4B + 12C)(2A + 4B + 8C). \tag{3.6-4}$$

EXAMPLE 2

We suppose that we know the Eulerian — i.e., $u = u(x, t)$ — expression for the velocity distribution in a horizontal duct (of unspecified shape):

$$u(x,t) = (2 - x)t. \tag{3.6-5}$$

The objective is to determine the following:

- The particle acceleration in the Eulerian scheme;
- The flow speed variation $u(0,t)$ and $u(1,t)$;
- The flow speed variation $u(x,0.5)$ and $u(x,1)$;
- In the Lagrangian scheme, for a particle whose initial coordinates are (x_0, t_0), the particle position $x = x(t;x_0,t_0)$, velocity $u = u(t;x_0,t_0)$, and acceleration $a_x(t;x_0,t_0)$. Then, plot u vs. t, and u vs. x for the particles $(x_0,t_0) = (0,0)$ and $(0,1)$.

To answer these questions we proceed as follows:

- The particle acceleration requires the calculation of the substantial derivative of the velocity. Thus,

$$a_x = a_x(x,t) = Du/Dt = \partial u/\partial t + u\partial u/\partial x,$$

$$= (2 - x) + (2 - x)t(-t) = (2 - x)\left(1 - t^2\right). \tag{3.6-6}$$

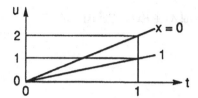
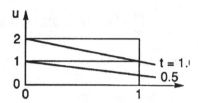

FIGURE 3.6-1   Example of plotting velocity as a function of time, or as a function of space, for a nonsteady flow. (a) u vs. t for fixed x; (b) u vs. x for fixed t.

- From Equation 3.6-5, putting $x = 0,1$ in turn,

$$u(0,t) = 2t, \quad u(1,t) = t. \tag{3.6-7}$$

Plots of these two distributions in time are shown in Figure 3.6-1a.
- From Equation 3.6-5, putting $t = 0.5, 1.0$ in turn,

$$u(x,0.5) = 0.5(2 - x), \quad u(x,1) = 2 - x. \tag{3.6-8}$$

Plots of the two space distributions are given in Figure 3.6-1b.
- In the Lagrangian scheme, as noted, a particle is identified by its initial coordinates, in this case by the pair (x_0,t_0). If the Eulerian velocity is known, the Lagrangian expressions can always be determined, at least in principle. With $u = Dx/Dt$, we can separate the variables in Equation 3.6-5 and integrate as follows:

$$\int_{x_0}^{x} \frac{Dx}{2 - x} = \int_{t_0}^{t} tDt. \tag{3.6-9}$$

By setting the lower limits as the initial coordinate x_0 of an arbitrary particle at time t_0 (also arbitrary), we have then established a procedure to identify *any* particle simply by selecting values for the pair (x_0,t_0), the nametag of a particle in one-dimensional flow. Integration of Equation 3.6-9 produces

$$x = x(t; x_0, t_0) = 2 - (2 - x_0)e^{-(t^2 - t_0^2)/2}. \tag{3.6-10}$$

The semicolon in the preceding equation calls attention to the fact that the pair (x_0,t_0) is treated as parameters, used for the purpose of identifying the particle.

We can obtain the expression for the velocity function for a fixed particle by differentiation of Equation 3.6-10, keeping (x_0,t_0) constant. Thus,

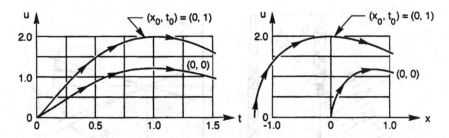

FIGURE 3.6-2 Example of plotting velocity as a function of time or space for a fixed particle. (a) u vs. t for a fixed particle; (b) u vs. x for a fixed particle.

$$u(t;x_0,t_0) \equiv \left.\frac{\partial x}{\partial t}\right|_{x_0,t_0} = (2-x_0)te^{-(t^2-t_0^2)/2}. \qquad (3.6-11)$$

The subscript pair on Equations 3.6-11 indicates that during the differentiation we are keeping the quantities x_0, t_0 constant, thereby specifying a derivative for a fixed particle. After completing the differentiation, x_0 and t_0 may be chosen arbitrarily.

Similarly, the acceleration of a fixed particle in the Lagrangian scheme is obtained from

$$a_x(t;x_0,t_0) = \left.\frac{\partial u}{\partial t^2}\right|_{x_0,t_0} = \left.\frac{\partial^2 x}{\partial t^2}\right|_{x_0,t_0} = (2-x_0)(1-t^2)e^{-(t^2-t_0^2)/2}. \qquad (3.6-12)$$

It is worth emphasizing that although Equations 3.6-2 and 3.6-12 both give the acceleration of a particle, the former does not permit continuous identification of a specific particle as it traverses the duct.

In Figure 3.6-2 we show the plots of velocity both as a function of the space and time coordinates for a pair of specific particles.

3.7 EXAMPLE INVOLVING OTHER TIME DERIVATIVES

Two radio transmitters, one a part of the instrumentation of a sounding rocket, the other located at the top of a tall tower, transmit continuous recordings of the instantaneous atmospheric temperatures as sensed by equipment located on the transmitters. The air mass, which has both horizontal and vertical temperature gradients, is moving over the ground; hence, the temperature recorded on the tower is a function of time, i.e., $T_2 = T_2(t)$. Because the rocket moves through a temperature gradient, its recording $T_1 = T_1(t)$ also varies with time. The recordings are plotted as a function of time in Figure 3.7-1b. The rocket passes the tower apex at time t_0 with speed of W ft/sec. The problem is to determine for the atmosphere the vertical temperature gradient $\partial T/\partial z$ at t_0, from the data.

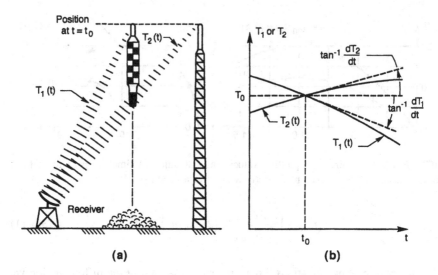

FIGURE 3.7-1   Setup for simultaneous temperature recording.

For the rocket, which is assumed to rise vertically, the temperature varies with its altitude, which itself is a function of time. According to Equation 3.5-4, in terms of the local time and space partial derivatives, for the rocket-mounted instrument,

$$\left(\frac{dT}{dt}\right)_1 = \left(\frac{\partial T}{\partial t}\right)_z + \left(\frac{dz}{dt}\right)_1\left(\frac{\partial T}{\partial z}\right)_t,$$

$$= \partial T/\partial t + W \partial T/\partial z.$$

For the tower-based instrument,

$$\left(\frac{dT}{dt}\right)_2 = \left(\frac{\partial T}{\partial t}\right)_z + \left(\frac{dz}{dt}\right)_2\left(\frac{\partial T}{\partial z}\right)_t = \frac{\partial T}{\partial t},$$

where $(dz/dt)_2 = 0$, since the second instrument is fixed with respect to the ground. Eliminating $\partial T/\partial t$ from those two relations gives us the vertical temperature gradient at the top of the tower at the instant when the rocket passes by, in terms of the time derivatives actually recorded by the two instruments, i.e., at $t = t_0$,

$$\left.\frac{\partial T}{\partial z}\right|_{t=t_0} = W^{-1}\left[\left(\frac{dT}{dt}\right)_1 - \left(\frac{dT}{dt}\right)_2\right]_{t=t_0}.$$

The principles involved in this somewhat artificial example can be extended to other situations where ds/dt is not necessarily a flow velocity, particularly in nonsteady flow situations.

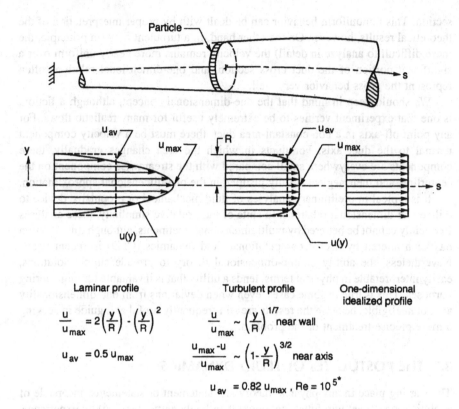

FIGURE 3.8-1 Velocity profiles in duct flow. The significance of the Reynolds number Re is developed in Chapters 4, 5, and 6.

3.8 CONCEPT OF A ONE-DIMENSIONAL FLOW*

In the following we will deal with internal flow through ducts in which it is assumed that the duct cross-sectional area A and the average flow variables p, T, V, etc. depend at most on only one space variable s (measured along the duct axis) and time t; i.e., $p = p(s, t)$, $T = T(s, t)$, $V = V(s, t)$, etc.; this is illustrated in Figure 3.8-1. As developed herein, the equations are valid only when effects of nonuniformities at a cross section are negligible.** Thus, we treat one-dimensional flow as though the pressure, temperature, velocity, etc. are uniform in the transverse direction.

However, as noted in Section 1.8, viscous effects in a fluid flow require that its relative velocity at a solid boundary be zero. Hence, the velocity relative to the wall actually varies from zero to a maximum value at the duct axis. In a fully developed laminar flow*** large deviations from the average occur over most of the cross

---

* The originator of the one-dimensional concept is one of the founders of fluid mechanics, Daniel Bernoulli (1738). His term for the model was the "method of thin slices."
** It is possible to handle non-negligible nonuniformities by one-dimensional theory if they are properly averaged. See Foa (1960, pp. 160–166).
*** Laminar and turbulent flows are discussed in Chapter 6.

section. This nonuniform behavior can be dealt with by proper interpretation of the theoretical results, however. On the other hand, in a turbulent flow (in principle, the more difficult to analyze in detail) the velocity remains more nearly uniform over a significant portion of the duct cross section, and one-dimensional relations often represent the gross behavior very well.

We should keep in mind that the one-dimensional concept, although a fiction, is one that experiment verifies to be extremely useful for many realistic flows. For any point off-axis in a non-constant-area duct, there must be a velocity component normal to the duct axis. For ducts in which the area changes gradually, these components are everywhere small compared with the streamwise component on the centerline, and their neglect is fully justified in the one-dimensional approximation.

It is typical in preliminary analyses of fluid mechanical or propulsive device to utilize one-dimensional relations. In spite of their relative simplicity, these analyses frequently cannot be bettered by multidimensional treatments, although this situation has been altered by use of computational fluid dynamics (CFD) in recent years. Nevertheless, the ability of one-dimensional theory to provide simple solutions, easily interpretable in physical terms, lends a utility that is invaluable for engineering purposes. This is true in some cases, even when deviations from one-dimensionality are not negligible, because the results are still frequently useful as a guide in seeking a more precise treatment of the problem.

3.9 THE POSTULATES OF FLUID DYNAMICS

The starting place in any physical theory is a statement or statements, incapable of analytical proof, yet unrefuted (and more than likely amply verified) by experience. This statement is translated into mathematical language, which then becomes a working tool of that science. It is sometimes overlooked that the massive structure of fluid mechanics is built upon a foundation of only a few independent statements, each yielding one independent relation. These are

CONSERVATION OF MASS

In the present usage we shall employ conservation in an extended sense to include the possibility of mass "creation." The resulting equation is termed the *continuity equation.*

NEWTON'S SECOND LAW

When applied to a fluid particle, Newton's second law yields a relation usually called the *dynamic equation of motion.* Some writers refer to it as the momentum equation, but herein we shall reserve the latter term for a relation in which continuity and the dynamic equation are combined.

FIRST LAW OF THERMODYNAMICS

In classical thermodynamics the *first law* represents the expression of conservation of energy. The law is written for an observer who is motionless with respect to the center of mass of the thermodynamic system. To analyze the details of a flow, each particle is conceived of as a system that interacts with its neighbors. However, it is rarely convenient to analyze a flow in the frame of an observer attached to an arbitrary particle. Thus, the first law is rewritten for some specific observer who sees each particle in motion. In this form the first law is combined with the dynamic equation of motion, written (usually) for an inertial observer; the resulting relation we refer to as the *energy equation*.

SECOND LAW OF THERMODYNAMICS

The second law of thermodynamics deals with the way that thermal processes may behave. One of the important consequences of the second law leads to the postulation of a scalar, point function, called the *specific entropy*, which may be taken to depend on any two independent, thermodynamic state functions, e.g., pressure and temperature. In certain flows, in particular the flow through a shock wave, the so-called conservation equations (*mass, momentum, energy*) permit two solutions. By appeal to the second law, and consideration of the behavior of the change of entropy between two points in the flow, one of these solutions can be discarded and the solution turns out to be unique.

EQUATIONS OF STATE

It is postulated that there exists a function relating any three of the independent, thermodynamic state variables. One such relation is expressed as $f(p, \rho, T) = 0$. We refer to this functional relationship as the *thermal equation of state* to distinguish it from the *caloric equation of state*, which relates the *internal energy function* e to a pair of independent state variables, e.g., $e = e(p, T)$. This is discussed more fully in Chapter 7.

Thus, in the analytical formulation of the equations of one-dimensional fluid flow, the problem is to determine the five dependent variables p, ρ, T, e, and V in terms of the independent variables s and t. There are five working relations: the continuity, dynamic, and energy equations (sometimes called the conservation relations), and the thermal and the caloric equations of state. The second law must sometimes be invoked to eliminate the possibility of multiple solutions. With the initial and boundary conditions specified, the problem becomes mathematically determinate, although the solution may still be difficult to obtain.

3.10 THE CONTINUITY EQUATION

Figure 3.10-1 illustrates the general setup for a one-dimensional flow. For an arbitrary observer (located at an origin that is not necessarily motionless with respect to the duct wall), a duct is described in terms of a single, curvilinear coordinate s, and time t. The duct cross-sectional area* $A = A(s, t)$ is arbitrary in shape and, as

---

* Recall that in mechanics the direction of a flat area is given by a unit vector normal to the surface.

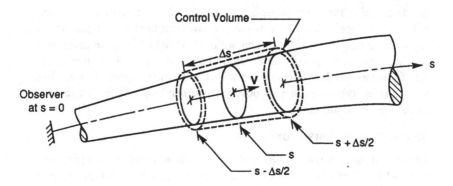

FIGURE 3.10-1   Control volume in duct flow.

indicated, may vary with both s and t. For a fixed-geometry, nonuniform duct, an observer motionless with respect to the duct wall sees $A = A(s)$ only. On the other hand, for a duct of time-dependent geometry even an observer fixed along the duct axis sees $A = A(s, t)$. An example of the latter is that of arterial blood flow, where the pump (the heart) generates pulses that result in a time-dependent arterial area distribution.

A *control volume* can be thought of as a closed mathematical surface, fixed with respect to a specified observer. In principle, its dimensions may be vanishingly small, but, in fact, it is subject to the restriction that its smallest dimension shall be large enough to guarantee that the continuum approximation (see Chapter 1) shall hold. A control-volume surface may coincide with a material boundary, or it may be thought of as an infinitely permeable membrane through which fluid may pass without obstruction. In the current model the centroid coordinate is s, and the boundaries consist of cross sections of the duct at stations $s + \Delta s/2$ and $s - \Delta s/2$ and the duct wall between the two end stations. By the principle of mass conservation* one ordinarily means that mass is neither created nor destroyed. However, the concept of a source in which mass appears at a point, or within a region, is so useful that we extend the term to include mass "creation" (or its negative, destruction). It must be acknowledged that mass creation is only a convenient mathematical way of treating mass addition to, or extraction from, the flow. In the case of one-dimensional flow, mass creation can be envisioned as fluid injection through the duct walls, for example. Another possibility is change of phase, for example, as occurs in two-phase flow where there may be a transfer of matter from one phase (say, saturated liquid water) to another (saturated steam). It would be necessary to write an equation for each phase including a term that allows for the apparent creation of matter.

---

* Leonardo da Vinci (1452–1519), on the basis of astute observations, was apparently the first to formulate mass conservation in a quantitative sense. However, his work was not known to his contemporaries and, in fact, remained unknown for several centuries after his death. Rouse and Ince (1957, p. 49) suggest that the principle should bear his name. They also point out that since Leonardo recorded his observation primarily for his own benefit, it was really his countryman Benedetto Castelli (c1577–c1644) who published in 1628 the first unambiguous statement of the mass-conservation principle.

The word statement of the extended principle of mass conservation, therefore, is that, for the control volume during the interval $t - \Delta t/2$ to $t + \Delta t/2$,

$$\left\{\begin{array}{c} \text{Net mass flow} \\ \text{in through} \\ \text{end boundaries} \end{array}\right\} + \left\{\begin{array}{c} \text{Mass} \\ \text{created} \\ \text{in volume} \end{array}\right\} = \left\{\begin{array}{c} \text{Mass} \\ \text{accumulated} \\ \text{in volume} \end{array}\right\}. \qquad (3.10\text{-}1)$$

By recalling that the tangential unit vector $e_s = e_s(s, t)$, then, at the centroid station, the vector cross-sectional area is $A = e_s A(s, t)$ and the flow density and velocity are $\rho = \rho(s, t)$, $V = e_s V(s, t)$. The mass rate of flow through the centroid station is a scalar denoted by $\dot{m}(s, t)$, such that

$$\dot{m}(s, t) \equiv \rho A \cdot V = \rho A V. \qquad (3.10\text{-}2)$$

The dimensions of mass rate of flow are *mass/time*. At station $s + \Delta s/2$ the average mass rate of flow during Δt is obtained from Taylor's expansion where only the first two terms are retained:

$$\dot{m}(s + \Delta s/2, t) = \rho A V + \frac{\partial}{\partial s}(\rho A V)\frac{\Delta s}{2}.$$

Were the higher-order terms retained, it could be shown by a limiting process similar to that used for the fundamental equation of fluid statics that their contribution is negligible. A similar expression is obtained at $s - \Delta s/2$, t, and thus the first bracket in Equation 3.10-1, for the time increment Δt, is

$$\left[\dot{m}(s - \Delta s/2, t) - \dot{m}(s + \Delta s/2, t)\right]\Delta t = -\frac{\partial}{\partial s}(\rho A V)\Delta s\Delta t. \qquad (3.10\text{-}3)$$

To account for mass creation we introduce a *growth factor* term $\kappa \equiv \kappa(s, t)$, whose definition is given in the following relation:

$$\left\{\begin{array}{c} \text{Mass created} \\ \text{in volume} \\ \text{during } \Delta t \end{array}\right\} \equiv \kappa \left\{\begin{array}{c} \text{Mass in} \\ \text{control volume} \\ \text{at time } t \end{array}\right\}\Delta t = \kappa \rho A \Delta s\Delta t. \qquad (3.10\text{-}4)$$

The growth factor is thus a fraction with dimension of 1/time.

At time t, the instantaneous mass in the control volume is $\rho A\Delta s$. The accumulated mass is obtained by subtracting the mass at time $t - \Delta t/2$, computed by Taylor's expansion holding s constant, from the final mass at $t + \Delta t/2$. Thus, the right-hand side of Equation 3.10-1 becomes*

---

* The error made in taking the average density to be equal to the density at the center of the control volume vanishes as $\Delta s \to 0$, $\Delta t \to 0$.

$$\left[\rho A\Delta s+\frac{\partial}{\partial t}(\rho A\Delta s)\frac{\Delta t}{2}\right]-\left[\rho A\Delta s-\frac{\partial}{\partial t}(\rho A\Delta s)\frac{\Delta t}{2}\right]=\frac{\partial}{\partial t}(\rho A)\Delta s\Delta t, \quad (3.10\text{-}5)$$

where it is permissible to take the quantity Δs outside the partial time derivative since it is not a function of time. Equations 3.10-3 through 3.10-5 are now substituted into Equation 3.10-1. The common factor $\Delta s\Delta t$ is divided out and the process of going to the limit, as $\Delta s \to 0$, $\Delta t \to 0$, has been reduced to a formality since none of the remaining terms involves either. We finally obtain, after rearranging, the *equation of continuity*:

$$\frac{\partial}{\partial t}(\rho A)+\frac{\partial}{\partial s}(\rho A V)=\kappa\rho A. \quad (3.10\text{-}6)$$

Equation 3.10-6 is valid for a very general category of flows: steady or nonsteady, viscous or nonviscous, constant or time-dependent duct geometry. Since no statement was necessary about the motion of the observer, Equation 3.10-6 is also valid for an arbitrary observer, including even a noninertial observer. It is important to keep in mind that s denotes the space coordinate measured from the origin, which is coincident with the observer. At a particular point in the flow, and at an arbitary time, all observers would see the same duct cross-sectional area A, and the same state variables p, ρ, T, etc., which, by definition, are independent of the observer. However, the velocity V is measured with respect to the observer, and consequently any pair of observers in relative motion will record different values for V everywhere in the flow.

An alternative expression to Equation 3.10-6 can be obtained by expanding the second term on the left by the product rule,

$$\frac{\partial}{\partial s}(\rho A V)=V\frac{\partial}{\partial s}(\rho A)+\rho A\frac{\partial V}{\partial s}.$$

Then, dividing both sides by ρA, and employing the logarithmic differential and the definition of the substantial derivative (Equation 3.5-5), Equation 3.10-6 can be rewritten as

$$\frac{D}{Dt}(\ell n\rho A)+\frac{\partial V}{\partial s}=\kappa. \quad (3.10\text{-}7)$$

Discussion

Equations 3.10-6 and 3.10-7 are equivalent forms of the continuity equation. Each is a nonlinear, nonhomogeneous, partial differential equation. As such they cannot be integrated in general. However, if — as is usually the case — certain restrictions apply, the equations simplify.

For example, the duct may be like a blood vessel, consisting of a flexible wall that deforms under pressure, and thus the cross-sectional area depends on both time and space coordinates. On the other hand, most engineering problems involve a *duct of fixed geometry*, which we specify by putting $A = A(s)$ only. In such case, we can put

$$\frac{\partial}{\partial t}(\rho A) = A\frac{\partial \rho}{\partial t}.$$

(3.10-8)

Another common example of a simplifying restriction is a *steady flow*. This restriction is denoted by

$$\frac{\partial}{\partial t} \equiv O; \quad F(s,t) \rightarrow F(s) \text{ only}, \quad \frac{\partial}{\partial s} \rightarrow \frac{d}{ds}.$$

(3.10-9)

Equation 3.10-9 — which effectively states that conditions at *any* fixed but arbitrary point do not vary with time — represents an enormous simplification. The function F, which stands for any flow variable and may represent area, or pressure, or velocity, or entropy, etc., then becomes a function of the space variable only, and the equation is reduced to an ordinary differential equation.

Applying Equation 3.10-9 to Equation 3.10-6, we obtain

$$\frac{d(\rho A V)}{ds} \equiv \frac{d\dot{m}}{ds} = \kappa \rho A,$$

(3.10-10)

where

$$\dot{m} \equiv \rho A V$$

(3.10-11)

is called the *mass rate of flow*. We see that Equation 3.10-10 can be integrated to obtain

$$\dot{m} = \dot{m}(s) = \int_{s=0}^{s} \kappa \rho A \, ds.$$

(3.10-12)

The growth factor, $\kappa = \kappa(s, t)$ in the general case, is used to account for mass addition or removal from the flow, for example, when the boundary layer is removed by suction, or cool fluid injected, as in certain gas turbine setups.

More commonly, however, mass is neither removed nor added to most flows, a condition that is indicated by the restriction

$$\kappa \equiv O.$$

(3.10-13)

In the current work we refer to such a flow as *mass conserving*. Thus, for a steady, mass-conserving flow, Equation 3.10-10 reduces to

$$\frac{d\dot{m}}{ds} = O, \quad \text{or} \quad \dot{m} \equiv \rho A V = \text{constant}.$$

(3.10-14)

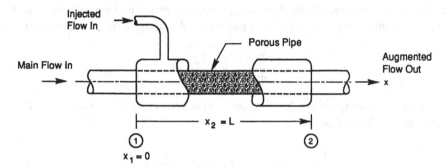

FIGURE 3.10-2   Flow with mass addition in constant area pipe.

One of the most important restrictions is that of a *constant-density flow*, a restriction that can be applied to the flow of a liquid, although it is also applicable to the flow of gases at low speeds. The appropriate equation is, simply,

$$\rho \equiv \text{constant}. \qquad (3.10\text{-}15)$$

Another commonly encountered restriction is that of *horizontal flow* in which we put

$$\frac{\partial}{\partial s} \to \frac{\partial}{\partial x}, \quad F(s,t) \to F(x,t), \quad V \to u. \qquad (3.10\text{-}16)$$

A comprehensive list of possible restrictions and the associated equations that we use to simplify the governing equations can be found in Section 4.2.

EXAMPLE 1

Determine the appropriate form of the continuity equation for a constant-area, horizontal, mass-conserving flow. The answer is

$$\frac{\partial \rho}{\partial t} + \frac{\partial}{\partial x}(\rho u) = \frac{\partial \rho}{\partial t} + u\frac{\partial \rho}{\partial x} + \rho\frac{\partial u}{\partial x} = \frac{D\rho}{Dt} + \rho\frac{\partial u}{\partial x} = 0. \qquad (3.10\text{-}17)$$

If we divide Equation 3.10-17 by ??, an alternative form results:

$$\frac{D}{Dt}(\ln \rho) + \frac{\partial u}{\partial x} = 0. \qquad (3.10\text{-}18)$$

EXAMPLE 2

Consider a constant-density, steady flow in a horizontal, constant-area duct into which fluid is injected from the surrounding jacket through a very large number of small holes in the duct wall, such that the growth factor is constant, i.e., for $0 \leq x \leq L$, $\kappa (x, t) \equiv \kappa = $ constant. The flow velocity ahead of the jacket is such that $x \leq 0$, $u = u_1$ for all time. Determine the velocity in the other two segments of the pipe.

Taking into account the other restrictions cited, then $A = $ constant, $\rho = $ constant, $\partial/\partial t \equiv 0$, $\partial/\partial x \rightarrow d/dx$, $V \rightarrow u$. Hence, Equation 3.10-10, the continuity equation for the region of flow injection, reduces to

$$\frac{du}{dx} = \kappa, \text{ or } \int_{u_1}^{u} du = \kappa \int_{0}^{x} dx, \quad 0 \leq x \leq L. \tag{3.10-19}$$

Therefore, the complete velocity distribution is specified by

$$x = \begin{cases} u_1, & x < 0, \\ u_1 + \kappa x, & 0 \leq x \leq L, \\ u_1 + \kappa L, & x > L. \end{cases} \tag{3.10-20}$$

Note that between stations ① and ② there is an increase in the axial momentum of $(\dot{m} u)_2 - (\dot{m} u)_1 = \rho A(2u_1 \kappa L + \kappa^2 L^2)$. This implies that the fluid from the jacket is injected at any station with the local axial velocity $u(x)$, an unlikely possibility.

More realistically, the fluid would be injected with zero axial velocity generating a region of continuous mixing as the flow moves through the pipe along the jacket length. The treatment of such a flow would require a significantly more complex analysis.

EXAMPLE 3

For a steady, constant-density, horizontal, mass-conserving flow, design a duct so that the velocity distribution increases linearly from station $x_1 = 0$, where the velocity is u_1 and the area is A_1, to station x_2, where $u_2 = 4u_1$. Sketch the duct shape assuming a circular cross section.

Under the restrictions cited, the continuity equation reduces to

$$uA = u_1 A_1 = \text{constant}. \tag{3.10-21a}$$

Thus, at any station

$$A/A_1 = u_1/u. \tag{3.10-21b}$$

The expression for the specified velocity distribution is

$$u/u_1 = 1 + 3x/L, \tag{3.10-22}$$

Therefore, combining Equations 3.10-22 and 3.10-21b the duct shape is given by

$$A/A_1 = (1+3x/L)^{-1}. \tag{3.10-23}$$

If the duct is circular in cross section, then the diameter variation is given by

$$D/D_1 = (A/A_1)^{1/2} = (1+3x/L)^{-1/2}. \tag{3.10-24}$$

An illustration of the duct is shown in Figure 3.10-3.

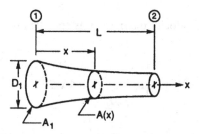

FIGURE 3.10-3 Duct shape for linear velocity distribution.

3.11 INVARIANCE OF CONTINUITY EQUATION UNDER CHANGE OF OBSERVER

The ability to transform the governing relations from one frame of reference to another is one to be cultivated since the equations that apply may be greatly simplified if the proper frame of reference can be determined. In Figure 3.11-1 a horizontal duct is shown. For observer ①, who is motionless with respect to the duct, the continuity equation is given by

$$\frac{D}{Dt_1}(\ln \rho A) + \frac{\partial u_1}{\partial x_1} = \kappa, \tag{3.11-1}$$

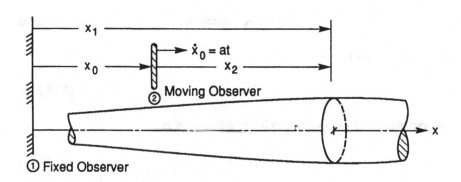

FIGURE 3.11-1 Change of reference.

where x_1 is the horizontal coordinate replacing s, u_1 is the flow speed with respect to ① and where D/Dt_1 is the substantial derivative operator in the frame of ①. Observer ② is moving to the right with respect to ① such that the horizontal acceleration of ② is uniform at a. Their positions are assumed to coincide at $t = 0$. It is shown, if we express all quantities and derivatives in the frame of ① in terms of quantities in the frame of ②, that Equation 3.11-1 is transformed into an equation of identical form.

If $x_0 = \frac{1}{2}at_1^2$ denotes the distance that ② has traveled in time t_1 from ①, then the coordinates of an arbitrary station in the duct with respect to the two observers at time t_1 are related by

$$x_2 = x_1 - x_0 = x_1 - \tfrac{1}{2}at_1^2. \tag{3.11-2}$$

This represents one of our transforming relations. The second is simply that both measure time by the same clock, i.e.,

$$t_2 = t_1. \tag{3.11-3}$$

Equations 3.11-2 and 3.11-3 are of the form $x_2 = x_2(x_1, t_1)$, $t_2 = t_2(x_1, t_1)$. We now transform

$$\frac{D}{Dt_1} = \frac{\partial}{\partial t_1} + u_1 \frac{\partial}{\partial x_1}$$

into terms of quantities measured in the frame of ②. From the chain rule

$$\frac{\partial}{\partial x_1} = \frac{\partial x_2}{\partial x_1}\frac{\partial}{\partial x_2} + \frac{\partial t_2}{\partial x_1}\frac{\partial}{\partial t_2} = \frac{\partial}{\partial x_2}, \tag{3.11-4a}$$

$$\frac{\partial}{\partial t_1} = \frac{\partial x_2}{\partial t_1}\frac{\partial}{\partial x_2} + \frac{\partial t_2}{\partial t_1}\frac{\partial}{\partial t_2} = -at_1\frac{\partial}{\partial x_2} + \frac{\partial}{\partial t_2}. \tag{3.11-4b}$$

Furthermore, the velocity of a particle in the duct with respect to ① is equal to the velocity with respect to ② plus the relative velocity of the observers, i.e., $u_1 = u_2 + at_2$. Therefore,

$$\frac{D}{Dt_1} = -at_2\frac{\partial}{\partial x_2} + \frac{\partial}{\partial t_2} + (u_2 + at_2)\frac{\partial}{\partial x_2} = \frac{\partial}{\partial t_2} + u_2\frac{\partial}{\partial x_2} \equiv \frac{D}{Dt_2}.$$

Also,

$$\frac{\partial u_1}{\partial x_1} = \frac{\partial}{\partial x_2}(u_2 + at_2) = \frac{\partial u_2}{\partial x_2}.$$

Combining these relations, we have

$$\frac{D}{Dt_1}(\ell n\rho A) + \frac{\partial u_1}{\partial x_1} = \frac{D}{Dt_2}(\ell n\rho A) + \frac{\partial u_2}{\partial x_2} = \kappa, \qquad (3.11\text{-}5)$$

which is of the same form as Equation 3.11-1, and where it is now understood that ρ and A are functions of the new independent variables (x_2, t_2). As shown in Equation 3.11-4, the spatial partial derivative operator is invariant, but the time partial is not. This is left to the reader to explain.

Of course, the final result was implicit in the derivation of the continuity equation since it holds for a completely arbitrary observer and thus needs no further justification. Nevertheless, the procedure used in the transformation has utility beyond the current verification.

3.12 THE DYNAMIC EQUATION OF MOTION

We now postulate the applicability of Newton's second law,*

$$ma = m\frac{dV}{dt} = \Sigma F, \qquad (3.12\text{-}1)$$

to a mass of fluid lying within a control volume that is fixed with respect to the observer. Since fluids are compressible, the mass within a control volume may vary with time; hence, we are postulating that Equation 3.12-1 is valid for variable-mass systems. Furthermore, it is understood that Equation 3.12-1 is to be used only for inertial observers. For a noninertial observer, it would be necessary to derive the applicable expression by means of Equation 3.3-4. The quantity $m = m(t)$ is the instantaneous mass within the control volume, and dV/dt is the acceleration of the instantaneous center of mass. On the right, ΣF stands for the vector sum of all applied forces.

We shall obtain appropriate expressions for the forces acting on the control volume and then, by letting the volume become arbitrarily small, obtain a differential equation known as the *dynamic equation of motion*. Referring to Figure 3.12-1, since there is a continuous distribution of matter in the duct, the acceleration of the mass center of the control volume must be expressed as a substantial derivative $a = DV/Dt$. Therefore, the left-hand side of Equation 3.12-1 becomes

$$ma = \rho A \Delta s \frac{DV}{Dt}, \qquad (3.12\text{-}2)$$

---

* See Section 3.13 for a discussion of the form of Newton's second law. Truesdell (1967) notes that although Newton had formulated a word statement of the second law and published it in the *Principia* in 1686, he never formulated it as an equation. It was the great mathematician L. Euler who first published (the equivalent of) Equation 3.12-1 in 1747.

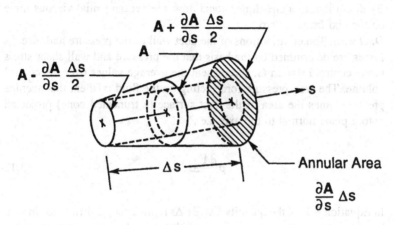

FIGURE 3.12-1 Projected area of duct wall.

where the usual assumption is employed that average conditions for the fluid within the control volume, and over a small time increment, are those at the central point (s, t).

In macroscopic mechanics there are only two recognized mechanisms by which forces can be exerted on a fluid, specifically, surface forces and field forces (of which body forces, the only ones treated herein, are a special case).

SURFACE FORCES

There are two possible contributions to surface forces. These are *pressure forces* that act in a direction opposed to the outer normal of a surface, and *viscous forces** that generally may have an arbitrary orientation. The viscous force on any surface can be resolved into a component normal to the surface and two orthogonal components tangential to the surface. The end faces of the control volume are treated separately from the duct wall.

- *End faces.* On any surface, a pressure force is defined to be positive such that it acts in the direction opposed to the outer normal. On the other hand, a positive, normal, viscous force acts in the same direction as the outer normal. By denoting the normal viscous stress on a surface perpendicular to the duct axis as τ_s, then the combined pressure and normal viscous force acting on the centroid of area A is $(-p + \tau_s)A$. By use of Taylor's expansion, the net contribution of the pressure and normal viscous forces acting on the ends is computed as

$$\frac{\partial}{\partial s}\left[(-p + \tau_s)A\right]\Delta s. \tag{3.12-3}$$

---

* A glimpse of how viscous stresses are treated in two-dimensional flows is given in Chapter 6.

By definition, in a one-dimensional flow, the net tangential viscous force on the end faces is zero.

- *Duct wall.* The contributions of the duct wall to the pressure and viscous forces are determined on the basis that the pressure and wall shear stress at the centroid station (s, t) represent the average values over the control volume. The wall pressure force acting on the fluid is, then, this average pressure times the area of the duct surface (a truncated cone) projected onto a plane normal to the axis, i.e.,

$$p \frac{\partial A}{\partial s} \Delta s. \tag{3.12-4}$$

In Equation 3.12-4 the quantity $(\partial A/\partial s)\Delta s$ represents the difference in area between the control volume ends of the duct. For ducts whose cross-sectional area is sufficiently slowly varying, the wall viscous force can be assumed to act parallel to the duct axis, and is given by the product of the wall shear stress τ_w times the sidewall area of the duct, and is directed upstream, i.e.,

$$-e_s \tau_w P \Delta s, \tag{3.12-5}$$

where P is the perimeter of the duct at station s. In the one-dimensional treatment this force is considered to be distributed uniformly over the entire slug of fluid in the control volume. The convention is adopted that τ_w is always positive, and that the direction of the viscous force is opposite to that of the particle velocity relative to the wall, i.e., opposite to $V - V_w$, where V_w is the wall velocity relative to the observer. If V_w is directed in the e_s-direction such that $V_w - V > 0$, then the minus sign in Equation 3.12-5 would be omitted.

By adding terms in Equations 3.12-3 through 3.12-5, we have the combined surface forces:

$$\frac{\partial}{\partial s}\left[(-p + \tau_s)A\right]\Delta s + p \frac{\partial A}{\partial s} \Delta s - e_s \tau_w P \Delta s = \frac{\partial}{\partial s}(\tau_s A)\Delta s - \left(A \frac{\partial p}{\partial s} + e_s \tau_w P\right)\Delta s. \tag{3.12-6}$$

BODY FORCES

Assuming that the only field forces under consideration contribute a force proportional to the mass of the control volume (hence the name for this special class of forces), the *body forces* can be written

$$\rho A \Delta s f, \tag{3.12-7}$$

where f is the force per unit mass contributed by the action of the field. The sum of the expressions in Equations 3.12-6 and 3.12-7 represents the sum of the applied forces.

THE DYNAMIC EQUATION

Combining Equations 3.12-2, 3.12-6, and 3.12-7, dividing out Δs, and going to the limit as $\Delta s \to 0$, we obtain

$$\rho A \frac{DV}{Dt} = \frac{\partial}{\partial s}\left[\left(-p + \tau_s\right)A\right] + p \frac{\partial A}{\partial s} - e_s \tau_w P + \rho A f \qquad (3.12\text{-}8)$$

On the other hand, since it turns out that the normal viscous stresses are almost always negligible (a prominent exception is in a shock wave), we shall put $\tau_s \equiv 0$, except when specifically stated otherwise. Then, Equation 3.12-8 becomes

$$\rho A \frac{DV}{Dt} = -\frac{\partial}{\partial s}\left(pA\right) + p \frac{\partial A}{\partial s} - e_s \tau_w P + \rho A f, \qquad (3.12\text{-}9a)$$

$$= -A \frac{\partial p}{\partial s} - e_s \tau_w P + \rho A f. \qquad (3.12\text{-}9b)$$

In general, a vector relation can be decomposed into three scalar component relations. Introducing Equation 3.5-8, and taking the dot product with e_s, the following relation results for the s-direction:

$$\frac{DV}{Dt} = -\frac{1}{\rho} \frac{\partial p}{\partial s} - \frac{\tau_w P}{\rho A} + e_s \cdot f, \qquad (3.12\text{-}10)$$

where $e_s \cdot f$ is the component of the body force in the s-direction. If the only body force present is due to a uniform gravitational field, then $f = -e_z g$, and

$$e_s \cdot f = -g \frac{\partial z}{\partial s}, \qquad (3.12\text{-}11)$$

where $z = z(s, t)$ denotes the vertical coordinate of the duct centerline above reference, Figure 3.12-2.

INTEGRATING THE DYNAMIC EQUATION — EXAMPLES

Equation 3.12-10, with $DV/Dt = \partial V/\partial t + V \partial V/\partial s$, is a nonlinear, nonhomogeneous first-order partial differential equation with independent variables s, t and dependent variables V, p, and ρ. Depending on the nature of the problem, the quantities τ_w, P, A, and f may depend on either the independent or dependent variables. In the simplest cases they are specified in advance. For example, in a constant-area duct, A and P are constants. Or, in a uniform gravitational field, $f = -e_z g$.

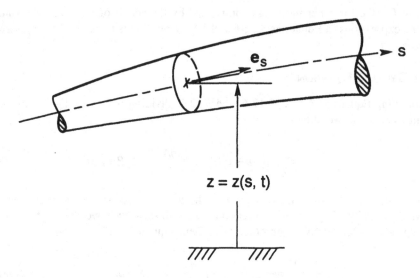

FIGURE 3.12-2 Defining the duct centerline coordinate.

Since Equation 3.12-10 has no general solution, we must seek to integrate it under special conditions, i.e., under restrictions that are applicable to several of the more common cases encountered in engineering applications and that were involved in the integration of the continuity equation. Thus, suppose that we restrict considerations to steady flow in a horizontal, constant-area duct in a uniform, gravitational field. Consequently, if D is the duct diameter, we put

$$V \to u, \quad s \to x; \quad \frac{DV}{Dt} \to u\frac{du}{dx}; \quad \frac{P}{A} = \frac{4}{D}; \quad e_s \cdot f = 0$$

and Equation 3.12-10 becomes

$$u\frac{du}{dx} = -\frac{1}{\rho}\frac{dp}{dx} - \frac{4\tau_w}{\rho D}. \tag{3.12-12}$$

By multiplying by dx and integrating from station $x_1 = 0$, with initial conditions u_1, p_1, ρ_1, to an arbitrary station denoted by x, u, p, ρ,

$$\int_{u_1}^{u} d\left(\frac{1}{2}u^2\right) + \int_{p_1,\rho_1}^{p,\rho} \frac{dp}{\rho} = -\frac{4}{D}\int_{p_1,x_1=0}^{p,x} \frac{\tau_w}{\rho}\,dx,$$

or, simply,

$$\frac{1}{2}\left(u^2 - u_1^2\right) + \int_1 \frac{dp}{\rho} = -\frac{4}{D}\int_1 \frac{\tau_w}{\rho}\,dx. \tag{3.12-13}$$

Additional information is required to evaluate the two remaining integrals. In the first we need a relation of the form $\rho = \rho(p)$. A thermodynamic relation between the two states that has this form is said to be *barotropic*. We will consider three cases for which the applicable relations are given:

- Constant-density flow — for liquids at a fixed temperature, or for gases at very low speeds:

$$\rho = \text{constant}; \tag{3.12-14a}$$

- Isothermal flow of a perfect gas. Such a flow involves heat transfer from, or to, a particle of gas:

$$T = \text{constant}, \quad \text{or} \quad \rho/\rho_1 = p/p_1; \tag{3.12-14b}$$

- Isentropic flow of a perfect gas, where γ is the ratio of specific heats. Such a flow can be achieved in two ways. In one the heat transfer to, or from, a particle is zero. Such a flow is said to be *adiabatic*. Furthermore, the dissipation due to viscous effects must simultaneously be zero, that is, the flow is *inviscid*. In the other, the heat transfer out from a particle must exactly balance the viscous dissipation. The latter case should generally be regarded as a theoretical curiosity except in a shockwave. Then,

$$\rho/\rho_1 = \left(p/p_1\right)^{1/\gamma}. \tag{3.12-14c}$$

To evaluate the second integral in Equation 3.12-13, a relation is needed that links the wall shear stress to the flow dynamics (i.e., the fluid velocity, or the velocity gradients) as well as to the fluid properties. The reader can verify the integrated expressions for the three following cases, under the cited restrictions.

- Constant-density flow with constant wall stress ($\tau_w = \text{constant}$):

$$\frac{p}{\rho} + \frac{1}{2}u^2 = \frac{p_1}{\rho} + \frac{1}{2}u_1^2 - \frac{4\tau_w}{\rho}\frac{\left(x - x_1\right)}{D}; \tag{3.12-15a}$$

- Isothermal flow of a perfect gas:

$$RT_1 \ln\frac{p}{p_1} + \frac{1}{2}\left(u^2 - u_1^2\right) = -\frac{4}{D}\int_1\frac{\tau_w dx}{\rho}; \tag{3.12-15b}$$

Details of a systematic treatment of isothermal flow, including the viscous term are given in Section 7.17.
- Isentropic (i.e., adiabatic and inviscid) flow of a perfect gas:

$$\frac{\gamma}{\gamma-1}\frac{p}{\rho}+\frac{1}{2}u^2 = \frac{\gamma}{\gamma-1}\frac{p_1}{\rho_1}+\frac{1}{2}u_1^2, \qquad (3.12\text{-}15c)$$

or, equivalently, where the speed of sound is given by $a^2 = \gamma RT = \gamma p/\rho$,

$$\frac{a^2}{\gamma-1}+\frac{1}{2}u^2 = a_1^2+\frac{1}{2}u_1^2. \qquad (3.12\text{-}15d)$$

Since the viscous term has been neglected in deriving the last two relations the explicit dependence on the duct geometry has simultaneously been eliminated. Thus, these two relations apply to *any* isentropic, steady flow of a perfect gas and are forms of the energy equation.

THE CENTRIFUGAL FORCE EQUATION

For the n-direction, the equation corresponding to Equation 3.12-10 is

$$\frac{V^2}{R} = e_n \cdot f.$$

The preceding relation is meaningless, however, since there are many situations where the field force is such that $e_n \cdot f = 0$ and where neither $V = 0$ nor $R \rightarrow \infty$, e.g., curved horizontal flow with non-negligible gravitational forces.

This simply indicates one of the deficiencies of one-dimensional theory. A more-refined analysis would have taken into account the pressure gradient normal to the duct axis, yielding (for inviscid flow)

$$\frac{V^2}{R} = -\frac{1}{\rho}\frac{\partial p}{\partial n}+e_n \cdot f. \qquad (3.12\text{-}16)$$

This is sometimes called the *centrifugal force equation*; it has important applications in treating flows in more than one space dimension. For the current purpose, we shall ignore Equation 3.12-16.

3.13 ON NEWTON'S SECOND LAW

One occasionally encounters the statement, usually in physics and mechanics texts, that the most general form of Newton's second law is

$$\frac{d}{dt}(mV) = F, \qquad (3.13\text{-}1)$$

rather than the form given in Equation 3.12-2, which is employed in the present work. It is frequently claimed* that the latter form must be used when dealing with systems of variable mass. (We specifically omit from the present discussion the possibility of relativistic effects.) Evidently, except for the case of constant mass, this is not equivalent to Equation 3.12-2, so that one form or the other must be wrong.

Experience has shown that it is fruitless to attempt resolution of the question by appeal either to authority or illustrative example. Scientific truths do not lie on the reputation of an individual, even the greatest of whom has on occasion been in error; on the other hand, examples that have been cited to bear on the current question have frequently been found fallacious, although the fallacy usually hinges on a subtle point that is difficult to detect or that is not readily grasped.

We reproduce here a simple but unassailable argument of Foa (1960, p. 24), who argues from first principles. We have already noted that every inertial observer would record the same acceleration of a moving body; i.e., acceleration is an invariant. It is postulated that force and mass (for nonrelativistic speeds) are also invariants. Thus, for a time-dependent mass, every observer would record the same derivative $\dot{m} \equiv dm/dt$. Velocity, on the other hand, depends on the relative motion of the observer and is therefore not an invariant.

Expanding the preceding relation and solving for V, we obtain

$$V = \frac{F - mdV/dt}{\dot{m}},$$
(3.13-2)

which states that the velocity, a noninvariant, is equal to a quantity made up solely of invariants. This contradiction invalidates the alternative formulation and leaves Equation 3.12-1 as the only possibility.

EXAMPLE: WATER DROPLET FALLING IN A VACUUM

In spite of the caveat regarding illustrative examples, the following counterexample has a certain utility in demonstrating the invalidity of Equation 3.13-1. Suppose that at $t = 0$ a water droplet of initial radius R_0 is released from rest. Due to the environment of zero pressure, evaporation must occur, and it seems evident that the water vapor thus generated cannot interfere with the trajectory of the remaining liquid, which, because of surface tension, will retain its spherical shape.

We can deal with the evaporation question independently of the motion. Assuming that the rate of evaporation is proportional to the surface area, i.e., if $V = V(t)$ is the instantaneous volume, then

---

* For example, in the book by the eminent physicist A. Sommerfeld, he states unequivocally, on pages 3–5 of Sommerfeld (1952), that the formulation $F = ma$ holds only for constant mass and that the alternative formulation is correct for variable mass. On page 28, however, in an application to a variable-mass system, it turns out that it is necessary to interpret mV as something other than the instantaneous momentum of the variable mass. Having made this "interpretation," the correct result is given as Equation 4 on page 29. It takes only a little reorganization to show that his Equation 4 is in fact equivalent to Equation 3.12-2 above. The reason for the success of this interpretation is that, by including the ejected mass as part of his system, Sommerfeld actually deals with a constant-mass system, in which case either formulation is permissible. To some, at least, the idea that the meaning of certain symbols must be interpreted according to the problem raises a doubt of whether or not we are really dealing with a science. The seeds of confusion, however, are so widely sown through the literature, even by acknowledged authorities, that it is not likely that unanimity of viewpoint will ever be achieved.

$$\frac{dV}{dt} = \frac{d}{dt}\left(\frac{4}{3}\pi R^3\right) = 4\pi R^2 \dot{R} = -k\left(4\pi R^2\right), \qquad (3.13\text{-}3)$$

where k is a constant to be determined by experiment. Thus, $\dot{R} = -k$, or

$$R(t) = R_0 - kt. \qquad (3.13\text{-}4)$$

Then, the droplet will have completely vaporized at

$$t_f = R_0/k. \qquad (3.13\text{-}5)$$

Now, we employ the (incorrect) form of Newton's second law:

$$\frac{d}{dt}(mV) = -mg. \qquad (3.13\text{-}6)$$

Since $m = m(t) = 4\pi\rho R^2/3$, Equation 3.13-6, with Equation 3.13-4, becomes

$$\frac{dV}{dt} - \frac{3k}{R_0 - kt}\,V = -g. \qquad (3.13\text{-}7)$$

The solution of Equation 3.13-7, under the stated initial condition, is

$$V = \frac{g}{4k}\left[R_0 - kt - \frac{R_0^4}{\left(R_0 - kt\right)^4}\right], \qquad (3.13\text{-}8)$$

which indicates that the droplet velocity becomes infinite when $t = t_f$, a result that is manifestly erroneous. On the other hand, if Equation 3.12-1 is used in place of Equation 3.13-6, then the usual result that $V = -gt$ is obtained, as expected, over the interval $0 \le t \le t_f$.

3.14 ON CONCEPTS FROM THERMODYNAMICS

The ability to analyze energy transformations is one of the principal objectives of classical thermodynamics. This objective is no less important in fluid dynamics, but some of the thermodynamic concepts cannot be taken over into fluid dynamics unless they are extended or revised to adapt to the needs of fluid dynamic systems. Those concepts of immediate interest are described in this section with the appropriate extensions.

SYSTEMS

As used in thermodynamics there are several categories of systems. An *isolated system* is one in which the material within a control volume (of macroscopic dimensions) exchanges neither matter nor energy with its environment. A *closed system* may exchange only energy, and an *open system* may exchange either, or both.

TEMPERATURE

On a macroscopic level we can no more define *temperature* than we can *force*, although, like force, we have an instinctive sense for it through sensations of hot and cold. From the molecular point of view, temperature is defined in terms of the mean kinetic energy of random motion of an ensemble of molecules. The ensemble must be sufficient in number that, within it, the statistical variation due to the random motion is negligible. Equivalently, this means that the characteristic dimension of the region to which it is desired to assign a temperature shall be of the order of, or greater than, some length. This length may be chosen, for example, as ten mean-free-path lengths of the ensemble. For gases at ordinary (i.e., at standard atmospheric) densities, this is a very small length indeed, about 6.6×10^{-8} m. Within a cube of ten mean-free-path lengths on a side there would be about 7.4×10^{6} molecules under these conditions, which would be large enough a sample to reduce statistical fluctuations to a negligible value. For analytical reasons we shall speak of temperature of a particle, or at a point. This implies the existence of a continuous temperature distribution in a fluid continuum, which is essential for analytical purposes. But, in so doing, it is desirable to keep in mind that the concept breaks down just as does the concept of pressure when the Knudsen number* approaches unity.

On the macroscopic level it is the practice to introduce an absolute-temperature scale, based on the experimental approach of Lord Kelvin, which requires no statement about its molecular interpretation whatever. Both approaches, of course, lead to the same practical result for out purpose.

EQUILIBRIUM

If two bodies at different temperatures are brought into contact with each other, there results a transfer of heat until the two temperatures become equal. In classical thermodynamics these bodies are said to be in *thermal equilibrium*. Also, two systems *A* and *B* are said to be in equilibrium with each other if each is in equilibrium with system *C*. This is usually called the *zeroth law of thermodynamics*.

However, in a typical flow (e.g., in the expulsion of air at high pressure from a reservoir) there are substantial gradients in flow variables, in particular of temperature, and, according to the preceding definition, the flow is not in equilibrium. Thus, in fluid dynamics, we are forced to modify our point of view. In the first place we choose as our system an arbitrarily small quantity of mass — a particle. The flow is then analyzed as a continuous distribution of systems. We then employ the following definition: Equilibrium is a thermodynamic condition that does not change with time as long as the environment remains unchanged. Therefore, equilibrium may exist even in the presence of temperature gradients, whereas, according to the first definition, fluid particles that lie on opposite sides of an intermediate particle subject to a temperature gradient would not be in equilibrium with each other.

---

* See Chapter 1, Section 1.7.

EXTENSIVE AND INTENSIVE VARIABLES

In running a laboratory chemical reaction the chemist is most often working with known quantities of reactants, as well as with a known volume within which the reaction takes place. Typical variables that the chemist may desire to measure or calculate are the internal energy E or the entropy S of the system, denoted by uppercase letters. If the chemist were to double the quantity of each reactant, E or S would double. Any quantity that depends on the mass of the system (or, more precisely, whose value is the sum of the contributions of the constituents) is said to be an *extensive* variable. On the other hand, the chemist would probably compute the internal energy per mole of the system, a quantity that is independent of the total mass. Or, as is the practice in fluid mechanics, the chemist would compute the internal energy per unit mass, which is called a *specific* variable (and which is independent of the total mass present). Specific variables of state, which are denoted by lowercase letters in the present work, are members of a class called *intensive* variables, of which such variables as temperature and pressure are also members.

In fluid dynamics, where we are dealing with flows involving gradients, and where the extensive properties of the system are usually of little interest, it is convenient — and, in fact, essential — in most cases to employ intensive variables for analytical purposes.

3.15 THE FIRST LAW OF THERMODYNAMICS*

For reasons stated in the preceding section we shall write the first law of thermodynamics for an infinitesimal system, i.e., a particle, and in intensive units. The observer in the formulation of the first law is assumed to be at rest with respect to the center of mass of the particle. We postulate the existence of a state function, the internal energy** e, which can be expressed in terms of any two independent state variables, e.g., $e = e(p, T)$. The first law relates changes in this function to the processes by which energy is added to, or subtracted from, our closed system. The word statement of the first law is

$$\begin{Bmatrix} \text{Change of} \\ \text{internal} \\ \text{energy per} \\ \text{unit mass} \end{Bmatrix} = \begin{Bmatrix} \text{Compressive work} \\ \text{per unit mass} \\ \text{performed on} \\ \text{boundaries} \end{Bmatrix} + \begin{Bmatrix} \text{Energy per unit mass} \\ \text{exchanged with environment} \\ \text{in forms other than} \\ \text{compressive work} \end{Bmatrix}$$

or

$$de = -pdv + d\bar{q}. \tag{3.15-1}$$

---

* For an excellent introduction to classical thermodynamics see Fermi (1956).

** Recall that in Section 1.9 the internal energy was identified as the molecular kinetic energy of random motion. In the macroscopic, or thermodynamic, view it is not necessary to provide a physical model to explain the existence of internal energy.

According to the present scheme of bookkeeping $d\bar{q}$ is a catchall differential that includes the following possible energy transformations:

a. Energy added in the form of heat, e.g., by conduction, radiation, combustion, it is denoted $d\bar{q}$;
b. Energy added due to the action of viscous forces on the particle boundary and denoted $d\psi$. This is defined in terms of an observer fixed with respect to the particle. This contribution of the viscous forces is converted entirely to heat, and the process is called *dissipation*.

Therefore, $d\bar{q} = dq + d\psi$, and the first law, rewritten,* is

$$d\bar{q} = d\bar{q} + d\psi = de + pdv. \qquad (3.15\text{-}2)$$

Differentials written dq and $d\psi$ (and, of course, $d\bar{q}$) have a bar through the vertical to emphasize that these differentials are not exact, i.e., that there do *not* exist variables q or ψ that are functions of two independent state variables; therefore, dq and $d\psi$ are not exact differentials. Hence, any integrals involving them must be handled with special care.

To illustrate the fact that $d\bar{q}$ (and thus dq and $d\psi$) cannot be exact, we start with the equation formed by omitting the center part of Equation 3.15-2. Since the internal energy is a state function, we can write $e = e(p, v)$; hence, $de = (\partial e/\partial p)_v\, dp + (\partial e/\partial v)_p\, dv$. Therefore, we must have

$$d\bar{q} = de + pdv = \left(\frac{\partial e}{\partial p}\right)_v dp + \left[p + \left(\frac{\partial e}{\partial v}\right)_p dv\right].$$

If we assume (erroneously) that $\bar{q} = \bar{q}(p, v)$, then $d\bar{q}$ must also be an exact differential. In any advanced calculus text it is proved that the necessary condition for this to be valid is

$$\frac{\partial}{\partial v}\left(\frac{\partial e}{\partial p}\right)_v = \frac{\partial}{\partial p}\left[p + \left(\frac{\partial e}{\partial v}\right)_p\right],$$

$$\frac{\partial^2 e}{\partial p \partial v} = 1 + \frac{\partial^2 e}{\partial v \partial p}.$$

Since we also know that the internal energy function must be analytic — i.e., that all derivatives exist and that the order of differentiation is immaterial — then

---

* As written, Equation 3.15-2 is valid only for pure substances. If it is desired to consider phase changes or mixtures, Equation 3.15-2 must be appropriately modified.

the two second-order derivatives are equal. Thus, we are left with the contradiction that $0 = 1$. Consequently, the assumption that $\bar{q}$ is a state function is invalid. Therefore, $d\bar{q}$ is a path function, which means, in effect, that the process by which heat is added to a particle must be specified before it is possible to evaluate the integral $\int d\bar{q}$.

ENTHALPY

In fluid mechanics it is convenient to eliminate the internal energy in favor of a defined function, the *specific enthalpy h*, where

$$h \equiv e + pv. \tag{3.15-3}$$

In fluid mechanics, as shown in Section 3.19, the term pv has a special significance and is called the *flow work*. Enthalpy is a function of state although it does not have an elementary physical interpretation as do its component functions. Introducing Equation 3.15-3 in Equation 3.15-2 produces

$$dq + d\psi = dh - vdp = dh - dp/\rho. \tag{3.15-4}$$

Because each of the functions refers to a system that has no motion relative to the observer, each one must also be a function only of the independent variable *time*. Therefore, dividing by dt, we obtain the relation

$$\frac{dq}{dt} + \frac{d\psi}{dt} = \frac{dh}{dt} - v\frac{dp}{dt}, \tag{3.15-5}$$

where no ambiguity in the meaning of d/dt arises.

FIRST LAW FOR AN ARBITRARY OBSERVER

If we now shift to an observer who sees a distribution of particles in relative motion, we have a situation where there are two independent variables (s, t) and where, as in Section 3.5, we must replace total time derivatives of functions computed for a particle by substantial derivatives. This holds, however, only for functions of state. Although dq and $d\psi$ are written for a particle, they are not expressible as functions of state and they cannot be expressed as a function of s and t either, until the processes by which heat is added to the particle, or by which heat is dissipated to the particle, are specified. That is, dq and $d\psi$ are *path functions*. Therefore, put*

$$dq/dt \equiv \dot{q}, \quad d\psi/dt \equiv \dot{\psi}, \tag{3.15-6}$$

---

* In other words it is not valid to write $dq / dt = \partial q / \partial t + V \partial q / \partial s$ and similarly for $d\psi / dt$.

to distinguish these derivatives from the substantial derivatives such as Dh/Dt. Consequently, the first law, in an arbitrary frame of reference, becomes*

$$\dot{q} + \dot{\psi} = \frac{Dh}{Dt} - v \frac{Dp}{Dt}. \qquad (3.15\text{-}7)$$

For emphasis, it is noted again that all terms in Equation 3.15-7 refer to a particle and, also, that $\dot{\psi}$, the rate of viscous dissipation, must be computed in the frame of an observer moving with the particle.

THE ENTROPY FUNCTION**

Truesdell (1962) raises the fundamental question: "What is entropy? Heads have split for a century trying to define entropy in terms of other things. I tell you, entropy, like force, is an undefined object, and if you try to define it, you will suffer the same fate as the force-definers of the seventeenth and eighteenth centuries: either you will get something too special, or you will run around in a circle."

It is straightforward to show for a thermally perfect gas that if Equation 3.15-2 is divided by the absolute temperature T, the right-hand side can be replaced by the total differential of another state variable s, called the *specific entropy*, which can be expressed as a function of any two independent state variables, e.g., $s = s(p, T)$, such that

$$T ds = dq + d\psi = de + pdv = dh - vdp. \qquad (3.15\text{-}8)$$

It is hereafter assumed without proof, that Equation 3.15-8 applies to any substance, regardless of the applicable equations of state.

If the particle is in relative motion with respect to the observer, the appropriate form of Equation 3.15-8, after dividing by dt, and replacing all time derivatives by substantial derivatives, then

$$T \frac{Ds}{Dt} = \frac{De}{Dt} + p \frac{Dv}{Dt} = \frac{Dh}{Dt} - v \frac{Dp}{Dt}. \qquad (3.15\text{-}9)$$

Therefore, combining Equations 3.15-7 and 3.15-9, we have the alternative forms for the first law of thermodynamics:

---

* Foa noted, in a private comment to the author: "what is so useful [in viewing the first law as the fundamental statement of energy conservation] is that, being entirely made up of terms that are invariant with respect to changes of reference, it is itself invariant under a change of observer."

** Of the familiar thermodynamic functions, such as entropy, enthalpy, internal energy, etc., it is probable that to most students a correct understanding of entropy, above all others, appears to be elusive, in spite of its importance and utility in practical problems. It may be a small consolation to observe that even some of the greatest minds of an earlier day struggled with the same question. For a view of their debate the reader is directed to "The Entropy Polemics" in *Electrician*, quoted by Stodola and Loewenstein (1945, p. 1319). It is remarkable for the most part how unenlightening these remarks are, as Stodola good-humoredly points out.

$$T \frac{D\phi}{Dt} = \dot{q} + \dot{\psi} = \frac{De}{Dt} + p \frac{Dv}{Dt} = \frac{Dh}{Dt} - v \frac{Dp}{Dt}. \qquad (3.15\text{-}10)$$

Isentropic Flow

Very important categories of flows can be identified in terms of how the entropy function varies. A flow is called *particle isentropic* or *multi-isentropic* if the entropy of every particle remains constant with time, i.e.,

$$\frac{D\phi}{Dt} = \frac{\partial \phi}{\partial t} + V \frac{\partial \phi}{\partial s} = 0, \qquad (3.15\text{-}11)$$

which does not necessarily imply that the entropy of all particles is the same. If, in addition to Equation 3.15-11, there are no entropy gradients in the flow field $\partial\phi/\partial s = 0$, the flow is said to be *homentropic* (i.e., homogeneous entropy).

3.16 THE SECOND LAW OF THERMODYNAMICS

A proper treatment* of the second law of thermodynamics is one of the major objectives of any course in thermodynamics and is, therefore, beyond the scope of this text. There are various formulations of the second law, notably those of Clausius and of Kelvin, which have far-reaching consequences. To develop their ideas requires the use of *reversible* vs. *irreversible* processes. However, since all real processes are irreversible, it is proposed to bypass these arguments. Instead, we shall be satisfied to state the consequences of the second law in a form useful for our purpose, specifically, a statement about the way the entropy function must behave in any thermodynamic transformation.

As Planck (1926, p. 83) says:

> There is but one way of clearly showing the significance of the second law, and that is to base it on facts by formulating propositions which may be proved or disproved by experiment. The following proposition is of this character: It is in no way possible to completely reverse any process in which heat has been produced by friction...

In Equation 3.15-8 the entropy function was introduced such that $Td\phi = dq + d\psi$, where $d\psi$ is the frictional work on the boundary dissipated to heat, which the fluid senses as an input. For our purpose the second law states that this is always positive or zero, i.e., that

$$d\psi \geq 0. \qquad (3.16\text{-}1)$$

All known measurements point to the truth of Equation 3.16-1 thereby giving it a sound experimental basis. If Equation 3.16-1 is employed in the entropy relation,

---

* See Fermi (1956).

then it follows, for *any* thermodynamic transformation between two states, designated by the subscripts 1 and 2, that

$$s_2 - s_1 \equiv \int_1^2 ds = \int_1^2 \frac{dq + d\psi}{T} ,$$

$$\geq \int_1^2 \frac{dq}{T} .$$

(3.16-2)

In a thermodynamic transformation from T_1 to T_2 we can replace the inequality of Equation 3.16-2 by

$$s_2 - s_1 \geq \frac{1}{T_{max}} \int_1^2 dq,$$

(3.16-3)

where T_{max} is the highest temperature encountered by the fluid particle during the transformation.

Now, if $dq \equiv 0$, the transformation is said to be adiabatic and we see that the entropy change cannot be negative. However, if the rate of heat transfer is scheduled just such that $dq \neq 0$, but so that $\Delta q = \int dq = 0$, this is referred to herein as a *globally adiabatic* process for which, from Equation 3.16-3,

$$s_2 - s_1 \gtreqqless 0.$$

(3.16-4)

That is, in a globally adiabatic process as well, the entropy may not decrease.

As shown in Section 7.14 this situation is encountered when a particle in motion passes through a standing shock wave. In the same section the viscous dissipation term, the heat-conduction term, and the entropy-production term are all given in detail for a viscous, heat-conducting, perfect gas, as well as an expression for the particle velocity history as it traverses the shock.

3.17 THE ENERGY EQUATION

The continuity equation (Equation 3.10-6), the dynamic equation (Equation 3.12-10),* the first law of thermodynamics (Equation 3.15-10), plus a thermal and caloric equation of state, constitute a complete set of independent equations governing one-dimensional flow. However, for analytical purposes it has been found convenient to combine the dynamic equation and the first law of thermodynamics into what we call the energy equation, and which will be considered one of our principal working relations. It must be kept in mind that of the dynamic equation, first law, and energy equation, only two are independent.

---

* If normal viscous stresses are not negligible, then Equation 3.12-8 would be used.

Combining the equations is accomplished as follows: Equation 3.5-8 is substituted on the left-hand side of Equation 3.12-9b, and then the dot product of this with V is formed. Dividing through by ρA yields*

$$\frac{D}{Dt}\left(\frac{1}{2}V^2\right) = -\frac{1}{\rho}\left(V\frac{\partial p}{\partial s}\right) - V\frac{\tau_w P}{\rho A} + V \cdot f. \tag{3.17-1}$$

The expression within the parentheses on the first term on the right of Equation 3.17-1 is converted to a substantial derivative by adding and subtracting the quantity $\rho^{-1}\partial p/\partial t$. After rearranging terms, we have

$$\frac{1}{\rho}\frac{Dp}{Dt} + \frac{D}{Dt}\left(\frac{1}{2}V^2\right) = \frac{1}{\rho}\frac{\partial p}{\partial t} - V\frac{\tau_w P}{\rho A} + V \cdot f. \tag{3.17-2}$$

The first law is next introduced by solving for Dp/Dt in Equation 3.15-10 and substituting in Equation 3.17-2. This leads to the alternative forms

$$\frac{Dh^o}{Dt} = \frac{1}{\rho}\frac{\partial p}{\partial t} + T\frac{Ds}{Dt} - V\frac{\tau_w P}{\rho A} + V \cdot f \tag{3.17-3a}$$

$$= \frac{1}{\rho}\frac{\partial p}{\partial t} + \dot{q} + \dot{\psi} - V\frac{\tau_w P}{\rho A} + V \cdot f, \tag{3.17-3b}$$

where we have made the substitution

$$h^o \equiv h + \tfrac{1}{2}V^2 = e + pv + \tfrac{1}{2}V^2 \tag{3.17-4}$$

on the left-hand side of Equation 3.17-3. The quantity h^0 is called the *total enthalpy* (or, equivalently, *stagnation enthalpy* by some writers). Equation 3.17-3 is called herein the *energy equation*, although some objections can be made to this terminology for reasons noted in Section 3.19.

3.18 DISSIPATION AND THE ROLE OF FRICTION

Equation 3.17-3b exhibits dependence upon a dissipation term that requires further discussion. Prior to dealing with viscous dissipation, we introduce a simpler example to help clarify the role of friction in energy transformations.

---

* Since Equation 3.17-1 is an expression for the time rate of change of kinetic energy per unit mass of any fixed particle, it is sometimes called the differential form of the mechanical energy equation.

DISSIPATION DUE TO SOLID FRICTION

Solid friction, also called *Coulomb friction*, arises when two surfaces in contact with each other are in relative motion. The friction force exerted by one on the other side is proportional to the normal force that presses the two surfaces together.

Suppose that an observer were standing on a long flat surface that we assume, for the present argument, to be non-heat-conducting. The observer propels a block away from the observer so that it slides on the surface. For this observer the energy of the block is initially all kinetic. Because of the action of the surface friction force on the block, its speed decreases continuously until it reaches a state of zero velocity relative to the surface.

For the surface-fixed observer we have two questions: What has happened to the initial kinetic energy of the block? Does the frictional force do any work?

The answer to the second question is *no*; for the observer standing on the surface, the force does no work. Work requires the displacement of the mechanism that exerts the force under consideration — in this case the floor relative to the observer. We see, therefore, that work is uniquely defined for each observer who is calculating it and is thus *not* an invariant under a change of observer. Only for two observers not in relative motion does work have the same value.

Returning to the case of a sliding block we are in a position to answer the first question. In the frame of a second observer moving with the sliding block, the wall that exerts a friction force on the block is doing work. This work is immediately converted to heat, which, barring any heat transfer from the block to its surroundings, is converted entirely into kinetic energy of random motion of the block molecules, i.e., is converted entirely into internal (thermal) energy. Such a process is called dissipation.

NONDISSIPATIVE WORK PERFORMED BY FRICTION FORCES

Suppose, instead, that an observer were to drop the same block onto a surface (e.g., a belt) that moves at uniform velocity with respect to the observer. Assuming that the belt moves rapidly enough, there is a relative motion between block and belt generating a friction force that gradually accelerates the block until it acquires the same speed as the belt. In this case the mechanism (the belt) exerting the force on the block moves with respect to the observer and therefore does work in the observer's frame. This shows up as kinetic energy of the block. In addition, since, during the acceleration, the belt also moves with respect to an observer *on* the block, dissipative work is also being performed. The final energy of the block has increased and, in the frame of the original observer, shows up in the forms of both thermal energy (kinetic energy of random motion) and kinetic energy of directed motion.

DISSIPATION IN A FLUID FLOW — FIXED WALL

In a duct the friction force is due to viscous action exerted by the walls. On a fluid element of length Δs the force is $-e_s \tau_w P \Delta s$. If the observer for whom Equation 3.17-3 is written is fixed with respect to the duct wall, then the velocity of the wall with respect to the observer fixed on a fluid particle is $-V$. In this case, the rate of work dissipated* to heat, per unit mass, is

$$\frac{\left(-e_s \tau_w P \Delta s\right) \cdot (-V)}{\rho A \Delta s} = V \frac{\tau_w P}{\rho A} \equiv \dot{\psi}. \qquad (3.18\text{-}1)$$

Thus, in Equation 3.17-3b the sum of the third and fourth terms on the right is zero, reducing the energy equation** to

$$\frac{Dh^o}{Dt} = \frac{1}{\rho} \frac{\partial p}{\partial t} + \dot{q} + V \cdot f. \qquad (3.18\text{-}2)$$

It is interesting to note that although Equation 3.18-2 accounts properly for any viscous stresses imposed by the wall on the fluid, the equation does not explicitly exhibit any dependence on friction.

DISSIPATION IN A FLUID FLOW — MOVING WALL

Consider the case where the wall moves at velocity $V_w = e_s V_w$ with $V_w < V$. Then the wall viscous force still tends to decelerate the flow and is given by $-e_s \tau_w P \Delta s$. The velocity of the wall with respect to an observer moving with a particle is $-e_s(V - V_w)$, which results in a dissipation function of

$$\dot{\psi} = \left(V - V_w\right) \frac{\tau_w P}{\rho A}, \quad V_w < V. \qquad (3.18\text{-}3)$$

Then, when Equation 3.18-3 is combined with Equation 3.17-3b, the energy equation takes the form

$$\frac{Dh^o}{Dt} = \frac{1}{\rho} \frac{\partial p}{\partial t} + \dot{q} - V_w \frac{\tau_w P}{\rho A} + V \cdot f, \quad V_w < V. \qquad (3.18\text{-}4)$$

The term on the right of Equation 3.18-4 involving the wall velocity is precisely the rate of energy per unit mass extracted from an arbitrary particle as a result of the tendency of the flow to do work on the wall by moving it to the right.

For the case of $V_w > V$ the direction of the viscous stress of the wall is to the right, and tends to accelerate the flow. The signs of the viscous terms in the dynamic equations, Equation 3.12-9, would have to be reversed to account for this, and

---

* If the normal viscous stresses are of significance, for example, as is the case if the details of the flow through a shock wave are to be examined, the normal viscous stress term must be accounted for in the dissipation function.

** The energy equation in a more restricted version of Equation 3.18-2 was first derived by Eck (1957, p. 50). Dean (1959) made use of it to point out that flow machines, such as turbines, can produce work only by processes that are nonsteady in the frame of an inertial observer.

similarly in the energy equations (Equations 3.17-1 through 3.17-3). The dissipation function, which must always be positive, becomes

$$\dot{\psi} = \left(V_w - V\right)\frac{\tau_w P}{\rho A}, \quad V_w > V, \tag{3.18-5}$$

and the final form of the energy equation becomes

$$\frac{Dh^o}{Dt} = \frac{1}{\rho}\frac{\partial p}{\partial t} + \dot{q} + V_w \frac{\tau_w P}{\rho A} + V \cdot f, \quad V_w > V. \tag{3.18-6}$$

Clearly, the viscous term in Equation 3.18-6 represents an energy input (mechanical work) to the fluid from the wall.

CONCLUSION

From Equation 3.18-2 we conclude that the only ways to change the total enthalpy of a particle (in the frame of an inertial observer) are by the following mechanisms: by an unsteady pressure field (e.g., created by a compressor rotor); by viscous forces in which the wall moves with respect to the observer; by field forces (e.g., in a gravitational field by changes of height); and by heat addition (i.e., by conduction, radiation, or combustion).

3.19 WORK AND THE INTEGRATED FORM OF THE ENERGY EQUATION

MECHANICAL WORK

It is customary to define *mechanical work* (sometimes called *useful work*) in fluid mechanics as positive when it represents an energy input to the flow. Furthermore, it is defined uniquely for the observer in whose frame the dynamic equation is written. The rate of work per unit mass done on a fluid particle moving at velocity V by the *field force f* is $V \cdot f$. If the field force is due solely to a uniform gravitational field, then

$$V \cdot f = Ve_s \cdot \left(-e_z g\right) = -Vg\partial z/\partial s, \tag{3.19-1}$$

where $\partial z/\partial s$ is the instantaneous cosine of the angle between the velocity vector and the vertical. Thus, if the velocity has an upward component, the rate of work done *on* the particle is negative. Obviously, this means that energy is extracted from the flow in order to raise the particle above the horizontal.

Another possibility of doing mechanical work is illustrated in Section 3.17 using viscous forces due to a moving wall. The associated rate of work per unit mass done on a fluid particle, in the frame of an inertial observer, is a product of the viscous force per unit mass times the velocity of the mechanism exerting the force — in this case the wall — which is

$$\pm \frac{\tau_w P}{\rho A} \, e_s \cdot e_s V_w = \pm V_w \, \frac{\tau_w P}{\rho A}, \quad V_w \gtrless V. \tag{3.19-2}$$

The third possibility of doing mechanical work on a fluid particle, as can be deduced from Equation 3.18-2, is by an unsteady pressure field. However, there is no elementary way to treat this term in one-dimensional theory. We thus put

$$\frac{dw}{dt} \equiv \dot{w} = \pm V_w \, \frac{\tau_w P}{\rho A} + V \cdot f, \quad V_w \gtrless V, \tag{3.19-3}$$

as the rate of mechanical work input exclusive of nonsteady pressure forces. Then, between stations ① and ② the work (per unit mass) performed on a particle is

$$\Delta w \equiv \int_{1 \text{ path}}^{2} \dot{w} Dt, \tag{3.19-4}$$

where a path integral is involved.

As a practical matter, the mechanical work done on a fluid particle is more readily calculated from measurements of the flow properties (p, T, and V) at stations ① and ② than by actual integration of Equation 3.19-4. Furthermore, many practical devices involve a work input by local nonsteady pressure terms, but where the flow at ① and ② is steady to a close approximation. Thus, we can enlarge the meaning of Δw in Equation 3.19-4 to include contributions of nonsteady pressure forces, as long as the flow is steady at the monitoring stations ① and ②, upstream and downstream of the input region, respectively. Again, the value of Δw is computed from measurements of appropriate flow quantities at these stations. Examples of work calculation or, equivalently, power calculation, can be found in Chapters 4, 6, and 7. A flow in which $dw/dt \equiv 0$, or $\Delta w \equiv 0$, is said to be *workless*.

HEAT ADDED

Energy added in the form of heat is also obtained as a path integral, i.e.,

$$\Delta q \equiv \int_{1 \text{ path}}^{2} \dot{q} Dt. \tag{3.19-5}$$

The rate of heat input must be specified, or, as for the mechanical work term, it is determined from appropriate measurements at the stations ① and ②.

THE INTEGRATED ENERGY EQUATION

Since the total enthalpy is not a path function, the following integral depends only on the end points:

$$\int_1^2 \frac{Dh^o}{Dt} \, Dt = h_2^o - h_1^o. \tag{3.19-6}$$

Consequently, Equation 3.19-6 can be equated to the sum of the inputs calculated in Equations 3.19-4 and 3.19-5, to produce

$$h_2^o = h_1^o + \Delta w + \Delta q, \tag{3.19-7}$$

where Δw now stands for the mechanical work input between ① and ② in any form. It has the restrictions, however, that it is valid only for inertial observers, and that the flow at stations ① and ② be steady. We shall refer to Equation 3.19-7 as the *integrated energy equation for steady flow.*

FLOW WORK

In the definition of total enthalpy $h^0 \equiv e + pv + \frac{1}{2}V^2$, the first and third terms are the familiar internal energy and kinetic energy per unit mass, respectively. In a steady flow the pv term is called *flow work* per unit mass. This can be visualized at an arbitrary station as follows: for the area A with pressure p, the pressure force exerted on the face by the fluid just upstream is pA. The rate at which this fluid does work in the frame of the observer who sees a steady flow is $pA \cdot V = pAV$. Since the mass rate of flow at this station is $\dot{m} = \rho AV$, then the flow work per unit mass is $pAV/\rho AV = p/\rho = pv$. Thus, in a steady flow, *total enthalpy* is the sum of the internal energy, flow work, and kinetic energy terms.

DISCUSSION

Equations 3.19-3, 3.19-4, and 3.19-7 are valid for inertial observers. In some rotating machines the flow is steady at the inlet to, and exit from, the rotating part, but is nonsteady between, for an observer on the case of the machine. However, for an observer whose frame rotates with the moving device, the flow appears to be steady between inlet and exit. Since the latter is not an inertial observer, the definition of work (Equation 3.19-3), and thus the energy equation, has to be modified appropriately. The resulting equation reflects the fact that the mechanical work Δw, the flow work term, and the kinetic energy are all defined uniquely for the particular observer, and all differ for the two observers. For this reason, total enthalpy loses its significance as a measure of the *energy level* of a flow except in situations where Equation 3.19-7 applies, or where it is possible to transform to a frame of reference, i.e., to a noninertial observer, in which the flow appears to be steady.

   A flow that becomes steady under a change of observer is said to be *crypto-steady*. Foa (1960, Chap. 5) gives the relation between the work computed in the

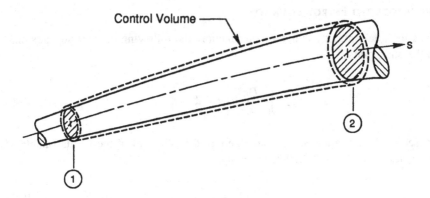

FIGURE 3.20-1 Control volume for momentum equation.

two frames under such a transformation. He also explains (Chapters 5 and 9) how to compute the work for a noninertial observer.

In situations where total enthalpy loses its utility as a working variable — for example, in flows which are not crypto-steady — instead of Equation 3.19-7, it is more convenient to return to the original form of the first law of thermodynamics, Equation 3.15-10, to develop the governing relations. This is demonstrated in Chapter 8.

3.20 THE LINEAR MOMENTUM EQUATION

In many practical and important situations in fluid mechanics it is required only to determine the integrated force on a body, or on a body of fluid, without regard to the relative contribution of viscous and pressure effects, and without concern for the details of how the force may be distributed. The *linear momentum equation* enables one to determine this integrated force in terms of quantities that are relatively easily obtained by computation or measurement. The advantage that this equation has over other equivalent relations is its great simplicity and utility in situations where prospects of a detailed analysis may be unlikely.

The momentum equation is obtained by combining the continuity and dynamic equations. If the substantial derivative in Equation 3.12-9a is expanded by Equation 3.5-5, we then form the sum

$$V[\text{Equation } 3.10\text{-}6] + [\text{Equation } 3.12\text{-}9a]$$

and combine terms according to the product rule. Thus,

$$\frac{\partial}{\partial t}(\rho A V) + \frac{\partial}{\partial s}(\rho A V V) = \kappa \rho A V - \frac{\partial}{\partial s}(pA) + \rho A f + p\frac{\partial A}{\partial s} - e_s \tau_w P. \quad (3.20\text{-}1)$$

We regard Equation 3.20-1 as the differential form of the momentum equation. The sum of the last two terms on the right is the vector force per unit length of duct exerted by the duct wall (or boundary) on the fluid by the pressure and viscous forces combined. Solving for this sum, we can then do a partial integration (holding time constant) to obtain the integrated force, denoted $F_{1,2}$, that the boundary exerts *on* the fluid within the control volume of Figure 3.20-1. Thus, noting again that $\dot{m} = \rho AV$,

$$F_{1,2} \equiv \int_1^2 \left(p \frac{\partial A}{\partial s} - e_s \tau_w P \right) \partial s,$$

$$= \int_1^2 \frac{\partial}{\partial t}(\rho AV)\partial s - \int_1^2 \kappa \rho AV \partial s - \int_1^2 \rho Af \partial s + \int_1^2 \frac{\partial}{\partial s}(pA + mV)\partial s.$$

(3.20-2)

Since time is held constant during the integration, the operator $\partial/\partial t$ can be taken through the integral sign, where it becomes d/dt, because the space variable s is integrated out of the expression. Thus, Equation 3.20-2 becomes

$$F_{1,2} = \frac{d}{dt} \int_1^2 \rho AV \partial s - \int_1^2 \kappa \rho AV \partial s - \int_1^2 \rho Af \partial s + p_2 A_2 + \dot{m}_2 V_2 - \left(p_1 A_1 + \dot{m}_1 V_1 \right).$$ (3.20-3)

The quantity $pA + \dot{m}V$ is called the *stream force* or *impulse*.* It should not be confused with the quantity $F_{1,2}$ to which it is related by Equation 3.20-3.

If the only field force acting is the gravitational term, then $f = -e_z g$, and the third term on the right in Equation 3.20-3 can be replaced by $-e_z g m_{1,2}$, where the mass in the duct contained between stations ① and ② is denoted as

$$m_{1,2} = \int_1^2 \rho A \partial s.$$ (3.20-4)

Equation 3.20-3 can be stated in the following way: The instantaneous force $F_{1,2}$ exerted by a duct wall (or by a stream tube whose boundaries are partially or wholly fluid surfaces) on a body of fluid contained between two stations is given by the vector sum:

---

* If the terms

$$\int_1^2 \rho A f \partial s + p_1 A_1 - p_2 A_2$$

were added to both sides of Equation 3.20-3, the resulting equation would be in the form found in some texts. The left-hand side

$$F_{1,2} + \int_1^2 \rho A f \partial s + p_1 A_1 - p_2 A_2$$

would represent the sum of all applied forces on the control volume, including the gravitational force and the pressure forces on the end faces. Remaining on the right-hand side would be only the momentum terms including the term representing the momentum of the injected ($\kappa > 0$) fluid.

a. Of the rate at which the momentum of the fluid contained within the control volume is changing;
b. Minus the rate at which the momentum is "created," associated with the fluid "created" within the control volume;
c. Plus a force equal and opposite to the field force on the fluid within the control volume;
d. Plus the change in stream force between stations ① and ②.

In the vast majority of cases some of these terms will be zero. In particular, few problems require either the unsteady term (a), or the mass creation term (b). Furthermore, the gravitational term is likely to be negligibly small in gas flows, or, if the duct is horizontal, its only effect is to create a pressure gradient that can be handled by the relations of fluid statics. In either case, it can be ignored in the dynamic problem.

MOMENTUM EQUATION FOR A DUCT WITH BOTH INTERNAL AND EXTERNAL PRESSURE FIELDS

In some practical problems we seek to determine the net force on a duct with an interior flow, or whose exterior is also exposed to a flow, or at least a constant-pressure field. In either case, the internal force exerted by the fluid on the duct is $-F_{1,2}$. To determine the exterior pressure and/or viscous forces, for example, on a jet engine hung below a wing, the problem is very complicated and may require a detailed aerodynamic study. For the special case of a duct at rest with (constant) atmospheric pressure on the boundary, it is easy to account for the exterior pressure forces.

In Figure 3.20-2, if the exterior pressure is constant at p_a and if the inlet and exit ducts were not present, the surface would be closed and there would be no net exterior force. With inlet and exit areas specified, and if the exterior pressure is constant at p_a, static equilibrium (for a fixed duct) requires

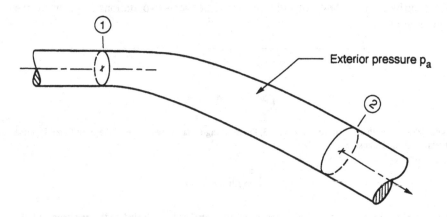

FIGURE 3.20-2 Duct with exterior pressure field.

$$R - F_{1,2} + p_a(A_2 - A_1) = 0,$$

or

$$R = F_{1,2} - p_a(A_2 - A_1),$$ (3.20-5)

where R is the reaction force supplied by external supports acting on the duct and where $p_a(A_2 - A_1)$ represents the difference in external pressure force between the closed-surface configuration and that with the duct openings.

3.21 SUMMARY

The equations derived in Chapter 3 are very general, and, as long as a one-dimensional representation is meaningful, they can handle an extremely wide range of problems. In the succeeding chapters an attempt is made to treat some branches of fluid mechanics in detail. Then there are many specialized aspects — for example, incompressible nonsteady, inviscid flow — that are illustrated by only one or two examples.

In the following, the principal relations are repeated for convenient reference.

1. Relation between fluid particle velocity and acceleration as seen by two observers: ① an inertial observer and ② an observer accelerating with respect to ① and rotating about an axis through origin of ② and adapted to a fluid particle from equations (3.3-3) and (3.3-4)

$$V_1 = \frac{Dr_0}{Dt} + V_2 + \Omega \times r_2,$$

$$\frac{DV_1}{Dt} = \frac{d^2 r_0}{dt^2} + \frac{DV_2}{Dt} + 2\Omega \times V_2 + \Omega \times (\Omega \times r_2) + \dot{\Omega} \times r_2.$$

2. Continuity equation:

$$\frac{\partial}{\partial t}(\rho A) + \frac{\partial}{\partial s}(\rho A V) = \kappa \rho A.$$ (3.10-6)

3. Streamwise component of dynamic equation of motion in natural coordinates, inertial observer:

$$\frac{DV}{Dt} = \frac{\partial V}{\partial t} + V \frac{\partial V}{\partial s} = -\frac{1}{\rho} \frac{\partial p}{\partial s} - \frac{\tau_w P}{\rho A} + e_s \cdot f.$$ (3.12-10)

4. Body force term for uniform gravitational field:

$$e_s \cdot f = -g \frac{\partial z}{\partial s}. \tag{3.12-11}$$

5. First law of thermodynamics, arbitrary observer, and relation to entropy:

$$T \frac{D\delta}{Dt} = \dot{q} + \dot{\psi} = \frac{Dh}{Dt} - v \frac{Dp}{Dt}. \tag{3.15-9}$$

6. Energy equation, inertial observer, to account for moving with

$$h^\circ \equiv h + \tfrac{1}{2} V^2 \equiv e + pv + \tfrac{1}{2} V^2, \tag{3.17-4}$$

$$\frac{Dh^\circ}{Dt} = \frac{1}{\rho} \frac{\partial p}{\partial t} + \dot{q} \mp V_w \frac{\tau_w P}{\rho A} + V \cdot f, \quad \left\{ \begin{matrix} 0 < V_w < V \\ V_w > V \end{matrix} \right\}.$$

7. Integrated energy equation; flow must be steady at stations ① and ②:

$$h_2^\circ = h_1^\circ + \Delta w + \Delta q. \tag{3.19-7}$$

8. Linear momentum equation for inertial observer; $F_{1,2}$ is the force that the boundary exerts on fluid between stations ① and ② (i.e., $F_{1,2}$ does not include forces on end faces):

$$F_{1,2} = \frac{d}{dt} \int_1^2 \rho A V \partial s - \int_1^2 \kappa \rho A V \partial s - \int_1^2 \rho A f \partial s + p_2 A_2 + \dot{m}_2 V_2 - (p_1 A_1 + \dot{m}_1 V_1). \tag{3.20-3}$$

9. Reaction force exerted on duct with exterior at ambient pressure p_a:

$$R = F_{1,2} - p_a(A_2 - A_1). \tag{3.20-5}$$

PROBLEMS

3.1. The velocity of a one-dimensional flow in a horizontal duct is given by

$$u = u_0 (1 + \sin \alpha x) e^{-at},$$

where $\alpha = 2$ ft^{-1}, $a = 3$ sec^{-1}. Calculate the acceleration for a particle at $x = 5$ ft, $t = 0.1$ sec, if $u_0 = 10$ ft/sec.

3.2. Suppose that in a railroad flatcar the coal were piled such that its height can be approximated by $h = h_0 \sin (\pi x_2/L)$, where h_0 is the maximum height of the pile and x_2 is the horizontal coordinate measured from the left end of the car. The car is now set in motion moving to the right at uniform speed U with respect to the ground.

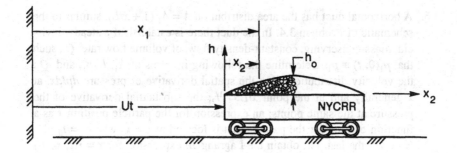

a. Determine the expression for the height of the coal $h_1 = h_1(x_1, t)$ as a function of the coordinate x_1 measured from a fixed point on the ground, and time t.
b. Another train moves past the first at velocity $-V$ with respect to the ground. An instrument attached to the second train can detect the coal height at the point immediately opposite the instrument. Compute the time rate of change of the coal pile height that the instrument would detect. Assume that at $t = 0$ the station of the instrument is at the right end of the coal car.

3.3. Show that for steady, incompressible, mass-conserving, one-dimensional flow in a circular, horizontal, conical duct the acceleration of a particle is inversely proportional to the fifth power of the duct diameter.

3.4. It is desired to design a horizontal duct such that in a steady, incompressible, mass-conserving flow the particle acceleration is constant in the interval $0 \le x \le L$. Determine the area distribution $A(x)$ required to accomplish this if $A(0) \equiv A_0$, $u(0) \equiv u_0$, and the acceleration is a_0. Make a sketch that shows the general configuration of the duct shape.

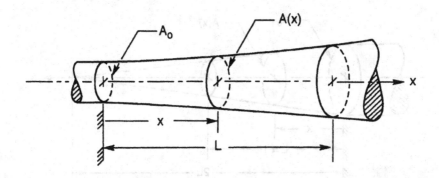

3.5. Write the continuity equation applicable to the restrictions of constant density, horizontal, mass-conserving flow. Determine the required area distribution $A(x, t)$, such that $u = (1 + x)e^{-t}$ under the initial condition that $\dot{Q}(x, 0) \equiv \dot{Q}_0 =$ constant.

3.6. A horizontal duct has the area distribution $A = A_1 (1 + x/L)$, similar to the schematic of Problem 3.4. In the duct there is established a steady, inviscid, mass-conserving, constant-density flow, of volume flow rate $\dot{Q}$, such that $p_1(0, t) \equiv p_1$. Determine the following in terms of A_1, L, p_1, and $\dot{Q}$: the velocity distribution u/u_1; the spatial derivative of pressure dp/dx, at x generally and at the point $x/L = \frac{1}{2}$; the substantial derivative of the pressure at the same points; an expression for the particle position x as a function of time for the particle that is located at $x = x_0$ when $t = t_0$. *Hint*: for the last part obtain the Lagrangian expression for $x = x(t; x_0, t_0)$ by integrating

$$\int_{t_0}^{t} Dt = \int_{x_0}^{x} \frac{Dx}{u(x)}.$$

3.7. Design a duct for steady, horizontal, inviscid, incompressible, mass-conserving flow such that the pressure increases linearly with the duct coordinate, i.e., such that $p - p_1 = kx$. If at $x_2 = L$, $p_2 - p_1 = 2p_a$, determine k and an expression for u/u_1. Find $A/A_1 = F(x/L)$. Suppose that $q_1 = 4p_a$; compute and plot A/A_1 and $(p - p_1)/q_1$ over the interval $0 \leq x/L \leq 1$.

3.8. The duct shown in the figure is a truncated, right, circular cone. Within it is established a nonsteady, horizontal water flow, such that $\dot{Q} = \dot{Q}_0 (1 + \sin \omega t)$. At the exit $x = L$ the flow leaves as a horizontal (approximately) uniform jet. Determine the area distribution $A(x)$ and the velocity distribution $u(x, t)$. If the exit pressure is p_a, determine the pressure distribution $p(x, t)$ required to maintain the flow neglecting viscous effects. For $\dot{Q} = 0.2$ ft³/sec, $\omega = 1$ rad/sec, $A_0 = 1$ ft², determine the time (i.e., ωt_0 in the first quadrant) at which $p(0, t) - p_a$ is a maximum, and compute the value $p(0, \omega t_0) - p_a$.

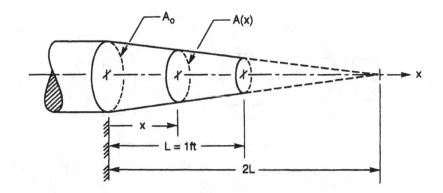

3.9. The figure shows a straight, horizontal, constant-area duct in which a steady, inviscid flow of water is maintained between the stations $x = 0$ and $x = L$, and where mass is injected (i.e., "created" within the duct), such that the growth factor $\kappa =$ constant. At $x = 0$, the velocity and pressure are maintained constant at U_0 and $p_0 = 2p_a$, respectively. Determine the duct velocity and pressure distributions downstream of $x = 0$. If $L = 10$ ft, $p_1 = p_a$, $U_0 = 10$ ft/sec, determine κ.

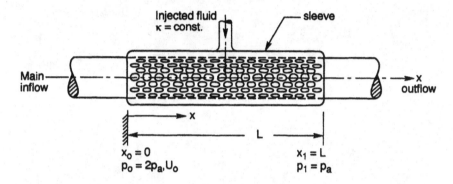

Note: Even neglecting the fact that viscous effects would be important in such a flow, the model is unrealistic. In the first place, the use of a one-dimensional approximation precludes consideration of the mechanics of the mixing process, which are multidimensional. Furthermore, the exit flow has kinetic energy and momentum that exceed that of the entry fluid. These increases could be provided only by a jacket designed so that the entry flow would transport its own momentum and energy into the main flow, as well as mass.

3.10. A rectangular metal block of weight Mg is placed on a belt moving horizontally at velocity $u_B \equiv U_0$ with respect to an inertial observer fixed with respect to the belt-driving mechanism. The coefficient of solid dynamic friction is μ.

   a. Determine the time t_f and the distance x_f required for the block to attain zero speed relative to the belt.
   b. Let W be the useful work performed by the belt on the block and ψ the dissipative work. Compute W and ψ in terms of Mg, t_f, and μ. Determine the values of t_f, x_f, W, and ψ for $Mg = 5$ lb, $\mu = 0.8$, and $U_0 = 100$ ft/sec.
   c. Repeat parts (a) and (b) if, instead, the belt moves such that $u_B = U_0 (2 - e^{-kt})$, where $k = 0.391$ sec^{-1} and the placement is made at $t = 0$.

4 Applications in Constant-Density Flow

One knows that until now our understanding of hydraulics has been extremely limited; for however many great geniuses have applied themselves to it at different times, we are still, after so many centuries, in almost absolute ignorance of the true laws to which the movement of water is subject; after one hundred fifty years one has barely discovered, with the aid of experiment, the duration, the quantity, and the velocity of the flow of water from a given orifice. All that concerns the uniform course of the waters of the surface of the earth is unknown to us; and to obtain an idea of how little we do know, it will suffice to cast a glance over what we do not.

To estimate the velocity of a river of which one knows the width, the depth, and the slope; to determine to what height it will rise if it received another river in its bed; to predict how much it will fall if one diverts water from it; to establish the proper slope of an aqueduct to maintain a given velocity, or the proper capacity of the bed to deliver to a city at a given slope the quantity of water which will satisfy its needs; to lay out the contours of a river in such a manner that it will not work to change the bed in which one has confined it; to calculate the yield of a pipe of which the length, the diameter, and the head are given; to determine how much a bridge, a dam, or a gate will raise the level of a river; to indicate to what distance backwater will be appreciable, and to foretell whether the country will be subject to inundation; to calculate the length and the dimensions of a canal intended to drain marshes long lost to agriculture; to assign the most effective form to the entrances of canals, and to the confluences or mouths of rivers; to determine the most advantageous shape to give to boats or ships to cut the water with the least effort; to calculate in particular the force necessary to move a body which floats on the water. All these questions, and infinitely many others of the same sort, are still unsolvable: who would believe it?...

Written in 1786 by P.L.G. du Buat; translated by Rouse and Ince (1957, p. 130)

4.1 INTRODUCTION

An effort has been undertaken in Chapter 3 to derive the equations of fluid mechanics for one-dimensional flow (i.e., for one space dimension) in quite general form. The equations involving three space dimensions introduce complexities — particularly those involving viscous effects — beyond the objectives of this text. In recent years, however, computers have been programmed to solve these more difficult equations for applications once considered almost inaccessible to analysis.

This fact has not rendered the one-dimensional approach obsolete, however. In the first place, an adequate grasp of the one-dimensional techniques serves as a good introduction to the more complex task. Further, for some engineering problems, it often provides an invaluable, and inexpensive, method of preliminary analysis that can be used to guide the more complex approach.

Even the one-dimensional equations are nonlinear. This fact militates against the possibility of obtaining general solutions applicable to all cases. However, in almost every case there are physical restrictions on a flow that can be translated into mathematical relations that make life — or, in this case the mathematics — simpler.

Some of the more frequently encountered restrictions are cataloged in the following section. One of them — that of *constant-density flow* — is so important that the remainder of Chapter 4 is devoted to engineering applications under this restriction.

4.2 A CATALOG OF RESTRICTIONS

The existence of a certain physical restriction applicable to a flow should stimulate in the analyst a mental picture of a corresponding relation that is employed to simplify the mathematical model of the flow. The following list, although not exhaustive, includes those most useful for the current purpose. Although the most common and useful restrictions are cataloged, not all of them apply simultaneously to the equations of Chapter 4, or to later chapters. In most cases the applicable restrictions are either stated or will be evident to the informed reader.

INCOMPRESSIBLE FLOW

In the most general case an *incompressible* flow is one in which the density of a particle of fixed identity is constant, i.e.,

$$\frac{D\rho}{Dt} \equiv \frac{\partial\rho}{\partial t} + V\frac{\partial\rho}{\partial s} \equiv 0. \qquad (4.2\text{-}1)$$

Equivalently, using the specific volume V, instead of the density ρ, $Dv/Dt \equiv 0$. Equation 4.2-1 does not say that every particle has the same density, but only that density of an arbitrary particle does not vary with time. It incorporates the possibility of spatial density gradients in a fluid. Flows in which this effect is present are said to be *stratified*. They are not further considered in this work.

CONSTANT-DENSITY FLOW

More common by far is the case of uniform and constant density

$$\rho \equiv \text{constant}. \qquad (4.2\text{-}2)$$

Unfortunately, it is a widespread practice to refer to flows in which Equation 4.2-2 applies as *incompressible*. Therefore, though technically incorrect, in what follows the terms *constant density* and *incompressible* are used interchangeably. An equivalent term is *isochoric* flow, which actually refers to the constancy of *specific volume* $v \equiv 1/\rho$.

INVISCID FLOW

Flows in the presence of solid boundaries *always* involve some viscous effects. And, except for the gas clouds of astrophysics, all flows involve solid boundaries somewhere. However, in many situations, viscous effects can be ignored compared with other effects, at least in the first approximation to the solution. This leads to the definition of *inviscid* flow:

$$\tau_w \equiv 0. \tag{4.2-3}$$

STEADY FLOW

In general, in a one-dimensional flow, any function F depends on the space variable s, and time t, i.e., $F = F(s, t)$. Thus, for the pressure, $p = p(s, t)$, and the velocity, $V = V(s, t)$, etc. If a flow is *steady*, then, at every point,

$$\frac{\partial}{\partial t} \equiv 0, \tag{4.2-4a}$$

and, consequently,

$$F(s, t) \rightarrow F(s), \quad \text{only.} \tag{4.2-4b}$$

The implications of Equation 4.2-4 are important since they imply that, for a steady, one-dimensional flow, there is only the one independent variable s, and the differential equations involved reduce to ordinary differential equations. This can be symbolized by

$$\frac{\partial}{\partial s} \rightarrow \frac{d}{ds}. \tag{4.2-4c}$$

UNIFORM FLOW

The restriction of *uniform flow* — though infrequently encountered — has governing relations analogous to those for steady flow, i.e.,

$$\frac{\partial}{\partial s} \equiv 0, \quad F(s, t) \rightarrow F(t) \ \text{only}, \quad \frac{\partial}{\partial t} \rightarrow \frac{d}{dt}. \tag{4.2-5}$$

An example of uniform flow is found in Section 8.2.

SOURCE-FREE FLOW

This expresses the absence of mass "creation," or mass injection, and is specified by

$$\kappa = \kappa(s, t) \equiv 0. \tag{4.2-6}$$

A flow in which Equation 4.2-6 applies is known as *mass conserving*, a situation that may be assumed to apply in most situations.

ADIABATIC FLOW

This is a flow in which no energy is received by a particle from an external source; the defining relation is

$$dq \equiv 0, \text{ or, equivalently, } \dot{q} \equiv 0. \tag{4.2-7}$$

This states that there is no combustion, radiation, or heat conduction from the walls or from the adjacent particles.

PARTICLE-ISENTROPIC FLOW

A *particle-isentropic* flow is one in which the specific entropy of a fixed, but arbitrary, particle remains constant, and is expressed as

$$\frac{D\sigma}{Dt} = 0. \tag{4.2-8}$$

ISOENERGETIC FLOW

A steady flow is *isoenergetic* if both $\dot{q} \equiv 0$ (hence, $\Delta q \equiv 0$) and $dw/dt \equiv 0$ (hence, $\Delta w \equiv 0$). It thus follows from Equation 3.19-7 that

$$h^0 \equiv \text{constant}. \tag{4.2-9}$$

We would also have an isoenergetic flow if the quantities $\dot{q}$ and dw/dt were such that the sum $\dot{q} + dw/dt$ is identically zero in a steady flow.

HORIZONTAL FLOW

For a duct where the centerline equation is expressed as $z = z(s, t)$, the horizontal restriction results in

$$\frac{\partial z}{\partial s} \equiv 0, \ z(s, t) \rightarrow z(t) \text{ only}. \tag{4.2-10a}$$

If, in addition, the flow is steady, this reduces simply to

$$z = \text{constant}. \tag{4.2-10b}$$

GRAVITATIONAL FIELD FORCE

We repeat a result previously stated for a uniform gravitational field:

$$e \cdot f = f_s = -g \frac{\partial z}{\partial s}. \qquad (4.2\text{-}11)$$

Since, for most engineering purposes, the uniform gravitational approximation is sufficiently accurate, the inverse-square law is not treated in this chapter.

4.3 THE EQUATIONS OF INCOMPRESSIBLE (CONSTANT-DENSITY) FLOW

Steady, incompressible flows in a uniform gravitational field without mass creation, i.e., flows involving restrictions (Equations 4.2-2, 4.2-4, 4.2-6, and 4.2-11) are historically and practically so important that it is desirable to give special emphasis to the relations that result under these restrictions.

CONTINUITY EQUATION

The expression for the mass rate of flow for steady, mass-conserving flow was previously obtained as Equation 3.10-14, i.e.,

$$\dot{m} = \rho A V = \text{constant}. \qquad (4.3\text{-}1a)$$

That is, the mass rate of flow is the same at every station and for all time. If the flow is also incompressible, then the *volume rate of flow* is given by

$$\dot{Q} \equiv \dot{m}/\rho = AV = \text{constant}. \qquad (4.3\text{-}1b)$$

DYNAMIC EQUATION

For steady flow in a uniform gravitational field, the dynamic equation (Equation 3.12-10) becomes

$$V \frac{dV}{ds} = -\frac{1}{\rho} \frac{dp}{ds} - \frac{\tau_w P}{\rho A} - g \frac{dz}{ds}. \qquad (4.3\text{-}2)$$

Since all the terms in Equation 4.3-2, except for the viscous term, are already in exact-derivative form, it can be rearranged to produce

$$\frac{d}{ds}\left(\frac{p}{\rho} + \frac{1}{2} V^2 + gz\right) = -\frac{\tau_w P}{\rho A}, \qquad (4.3\text{-}3)$$

which can be integrated from a fixed station s_1 along the duct, where the height, pressure, and speed are z_1, p_1, and V_1, respectively, to a second point denoted by (s_2, z_2, p_2, V_2). Thus,

$$\frac{p_2}{\rho} + \frac{1}{2} V_2^2 + g z_2 = \frac{p_1}{\rho} + \frac{1}{2} V_1^2 + g z_1 - \int_{s_1}^{s_2} \frac{\tau_w P}{\rho A} \, ds. \tag{4.3-4}$$

FIRST LAW OF THERMODYNAMICS

From Equation 3.15-10, the expression for entropy variation for constant-density flow ($Dv/Dt \equiv 0$) reduces to

$$T \frac{D\phi}{Dt} = \dot{q} + \dot{\psi} = \frac{De}{Dt}. \tag{4.3-5}$$

It is interesting to note from Equation 4.3-5 that in a particle-isentropic, incompressible flow the internal energy is constant. Also, since $\dot{\psi} \geq 0$, a viscous flow can be isentropic, providing heat is removed exactly at the right rate, i.e., if $\dot{q} = -\dot{\psi}$. This last result is valid for compressible flows as well.

The internal energy change for a fixed particle, between two points in the flow, can be obtained by integrating Equation 4.3-5, i.e.,

$$\int_1^2 De \equiv \Delta e = \int_1^2 TD\phi = \int_{1\,\text{path}}^2 (\dot{q} + \dot{\psi}) Dt = \Delta q + \int_{1\,\text{path}}^2 \dot{\psi} Dt, \tag{4.3-6}$$

where Δq was defined in Equation 3.19-5. Making use of Equation 3.18-1 for the case of a nonmoving wall, we have the alternative expressions for the internal energy increment:

$$\Delta e \equiv e_2 - e_1 = \int_1^2 TD\phi = \Delta q + \int_{1\,\text{path}}^2 V \frac{\tau_w P}{\rho A} Dt \tag{4.3-7}$$

The differentials $D\phi$ and Dt in Equations 4.3-6 and 4.3-7 indicate that the integration is to be performed following a fixed particle.

ENERGY EQUATION

The steady-flow energy equation was previously obtained as Equation 3.19-7:

$$h_2^0 = h_1^0 + \Delta q + \Delta w, \tag{4.3-8}$$

where, from Equation 3.19-7, the rate of mechanical work on the fluid is given by

$$\dot{w} = \pm V_w \frac{\tau_w P}{\rho A} + V - f, \quad V_w \gtrless V, \tag{4.3-9}$$

hence,

$$\Delta w = \int_1^2 \left[\pm V_w \frac{\tau_w P}{\rho A} + v \cdot f \right] Dt, \quad V_w \gtrless V. \tag{4.3-10}$$

Now, the particle velocity is $V = e_s Ds/Dt$. Therefore, it follows for a uniform gravitational field, that

$$V \cdot f \, Dt = (Ds/Dt) e_s \cdot e_z (-g) \, Dt = -g Ds \, \cos(e_s, e_z) = -g Dz. \tag{4.3-11}$$

Hence, Equation 4.3-10 becomes

$$\Delta w = \pm \int_1^2 V_w (\tau_w P/\rho A) \, Dt + g(z_1 - z_2), \quad V_w \gtrless V. \tag{4.3-12}$$

In Equations 4.3-8 through 4.3-12 the incompressible restriction has not been used. Consequently, for a steady flow in a uniform gravitational field, and for a duct with a fixed wall ($V_w \equiv 0$), Equation 4.3-12 reduces to $\Delta w = g(z_1 - z_2)$, and the energy equation becomes

$$h_2^0 = h_1^0 + g(z_1 - z_2) + \Delta q. \tag{4.3-13}$$

Thus, we see that if $z_2 < z_1$, the gravitational field does work on the particle, whereas, if $z_2 > z_1$, work is done by the flow to increase the particle height.

LINEAR MOMENTUM EQUATION

For steady, source-free flow, in a uniform gravitational field, the momentum equation is obtained by simplifying Equation 3.20-3. Thus, the boundary force $F_{1,2}$, which is the vector sum of the pressure and viscous forces exerted on the fluid between stations ① and ② by the boundaries, excluding the end faces, reduces to

$$F_{1,2} = \dot{m}(V_2 - V_1) + p_2 A_2 - p_1 A_1 + e_z g m_{1,2}. \tag{4.3-14}$$

Consequently, from Equation 3.20-5, under the same restrictions, the reaction force that must be exerted on a duct, with exterior at ambient pressure p_a, to maintain the duct in place, is

$$R = \dot{m}(V_2 - V_1) + (p_2 - p_a)A_2 - (p_1 - p_a)A_1 + e_z g m_{1,2}. \tag{4.3-15}$$

We can combine Equations 4.3-14 and 4.3-15 to reproduce the relation between R and $F_{1,2}$, previously obtained as Equation 3.20-5:

$$R = F_{1,2} - p_a(A_2 - A_1).$$ (4.3-16)

An oft-used alternative to Equation 4.3-15 is to define

$$\Sigma F \equiv F_{1,2} + p_1 A_1 - p_2 A_2 - e_z g m_{1,2}.$$ (4.3-17)

The quantity ΣF is the vector sum of all applied forces, pressure, viscous, and gravitational, acting on the fluid within the control volume between ① and ②. Therefore,

$$\Sigma F = \dot{m}(V_2 - V_1).$$ (4.3-18)

Equation 4.3-18 states that in a steady flow the vector sum of the forces acting on the fluid contained within the control volume is equal to the rate of momentum outflow minus the rate of momentum inflow over the control volume.

Equation 4.3-18 is the steady-state version of the more complicated general relation for nonsteady flow referred to in the footnote following Equation 3.20-3.

Applications involving both forms Equations 4.3-15 and 4.3-17 will be given in the following sections.

4.4 BERNOULLI'S EQUATION

If, in addition to the restrictions applied in Section 4.3, we require the flow to be inviscid, then Equation 4.3-4 becomes

$$\frac{p_1}{\rho} + \frac{1}{2} V_1^2 + g z_1 = \frac{p_2}{\rho} + \frac{1}{2} V_2^2 + g z_2 = \text{constant}.$$ (4.4-1a)

This is the relation named for Daniel Bernoulli (1700–1782).

Rouse and Ince (1957, pp. 95–105) point out that although Bernoulli was able to make correct predictions for a variety of flow situations, the result expressed in the form of Equation 4.4-1a is essentially due to his contemporary, the great mathematician Leonhard Euler (1707–1783). Dividing Equation 4.4-1a by the gravitational constant g produces the alternative form:

$$\frac{p}{\rho g} + \frac{V^2}{2g} + z = \text{constant},$$ (4.4-1b)

in which each term has the dimension of *length*, a form particularly convenient for purposes of hydraulics engineers. The term z is called the *head* and $p/\rho g$, $V^2/2g$, the *pressure head* and *velocity head*, respectively.

4.5 TOTAL PRESSURE

In Equation 3.17-4 the total enthalpy was defined as $h^0 = e + p/\rho + V^2/2$. For a steady flow the relation between the total enthalpy at a specified station ① and that at any downstream station, designated h^0, is given by

$$h^0 = h_1^0 + \Delta q + \Delta w. \tag{4.5-1}$$

If the flow is also isoenergetic, then $\Delta q \equiv \Delta w \equiv 0$. For the case of constant-density, isoenergetic flow, therefore,

$$h^0 = e + p/\rho + V^2/2 = e_1 + p_1/\rho + V_1^2/2. \tag{4.5-2}$$

We now define *total pressure** p^0 as the hypothetical pressure a particle would acquire if brought to rest (in the frame of an observer who sees the flow speed as V) in a process that is both isentropic and isoenergetic. Denote the end state of this hypothetical process by ①′. However, we know from Equation 4.3-5, in an isentropic, incompressible flow process, that $e =$ constant. Therefore, from Equation 4.5-2, the total pressure associated with conditions at ① is

$$p_{1'}^0 = p_1 + \tfrac{1}{2}\rho V_1^2.$$

In fact, dropping the subscripts in the preceding relation we have

$$p^0 = p + \rho V^2/2, \tag{4.5-3}$$

for the total pressure associated with a point at which the static pressure is p and the flow speed is V. The product $\rho V^2/2$ is called the *dynamic pressure*** at that point. The reader is cautioned that Equation 4.5-3 serves as the definition of total pressure for constant-density flow only.

The total-pressure relation between any two points in a constant-density, steady flow is obtained by substituting Equation 4.5-3 into the dynamic equation (Equation 4.3-4); then between point ① and any downstream point ②:

$$p_2^0 = p_1^0 - \int_1^2 \frac{\tau_w P}{A}\, ds - \rho g(z_2 - z_1), \tag{4.5-4}$$

---

* Some writers call p^0 the stagnation pressure. We prefer to think of *total* pressure as a hypothetical quantity calculable from Equation 4.5-2, even in a flow in which the speed is everywhere nonzero. The term *stagnation* pressure is reserved in this book for situations in which the flow is actually stagnated, i.e., in which the speed is actually zero at some point. We further note that without heat transfer, the isentropic condition requires that the hypothetical deceleration process be inviscid. However, the flow (in contrast to the hypothetical deceleration) need not be inviscid. In fact the variation of total pressure from point to point in a flow is used as a measure of flow losses, i.e., of viscous effects.

** The term *pressure* is used interchangeably with *static pressure*. If dynamic pressure, or total pressure, is intended, this will always be explicitly indicated.

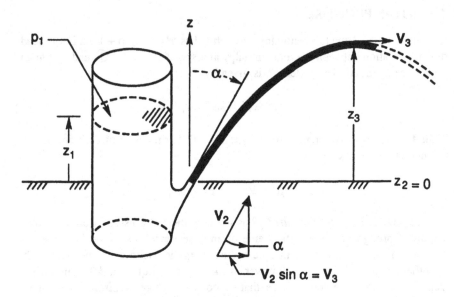

FIGURE 4.6-1   Flow out of a pressurized reservoir.

which shows — excepting the possibility of a flow with a moving wall, see Section 3.12 — that the effect of viscosity is to reduce the total pressure. In a horizontal flow over a fixed wall the total pressure always decreases in the direction of the flow. Equation 4.5-4 also brings out the hypothetical nature of total pressure, since it is always possible to calculate it (and sometimes to measure it), even though the flow is neither isentropic nor isoenergetic.

4.6 FRICTIONLESS FLOW OUT OF A PRESSURIZED RESERVOIR

Many flows can be treated — at least in the first approximation — as inviscid. Even in cases where viscous effects are not negligible, an inviscid analysis is helpful in establishing certain theoretical limits for the more complicated viscous situation.

In Figure 4.6-1 a tank of liquid of density ρ is pressurized at p_1 with a head z_1 above the outlet, $z_2 \equiv 0$. A nozzle, indicated with a streamlined contour to reduce viscous effects, emits a jet into the atmosphere at speed V_2 inclined locally at angle a to the vertical. The problem is to determine the jet exit speed, and the maximum height z_3 to which the jet will rise, assuming that viscous effects are negligible.

Although this is not truly a steady-flow situation, if we consider the case when the nozzle exit area is much smaller than the tank cross section, i.e., when $A_2 \ll A_1$, then, by the continuity equation, $V_1 \ll V_2$. The meaning of the latter inequality is that it is permissible to ignore V_1^2 compared with V_2^2. Furthermore, intuition suggests that the nonsteady effects should also be negligible (see Section 4.7), i.e., that the Bernoulli equation, derived on the basis of restriction (Equation 4.2-4a), may be applied between stations ① and ②, and ② and ③. Thus,

$$\frac{p_1}{\rho}+\frac{1}{2}V_1^2+gz_1 = \frac{p_2}{\rho}+\frac{1}{2}V_2^2+gz_2 = \frac{p_3}{\rho}+\frac{1}{2}V_3^2+gz_3. \qquad (4.6\text{-}1)$$

Therefore, neglecting $V_1^2/2$ in Equation 4.6-1, and with $z_2 = 0$, $p_2 = p_a$,

$$V_2 = \left[2gz_1 + 2(p_1 - p_a)/\rho\right]^{1/2}. \qquad (4.6\text{-}2)$$

If the tank is open to the atmosphere, then, $p_1 = p_a$, and

$$V_2 = \left(2gz_1\right)^{1/2}, \qquad (4.6\text{-}3)$$

a result named in honor of its discoverer, Evangelista Torricelli* (1608–1647). It is noteworthy, according to Equation 4.6-3, that a particle attains the same exit speed as in a free fall from height z_1, a value that is independent of the nozzle angle.

To determine the maximum jet height we note, since there is no horizontal force acting on a particle once it has left the nozzle (assuming negligible air resistance), that the horizontal component of velocity is everywhere the same; i.e., $V_3 = V_2 \sin \alpha$. Hence, from the second and third relations of Equation 4.6-1,

$$V_2^2\left(1 - \sin^2 \alpha\right) = 2gz_3$$

or

$$z_3 = \left(V_2^2/2g\right)\cos^2 \alpha = z_1 \cos^2 \alpha, \quad 0 \le \alpha \le \pi/2. \qquad (4.6\text{-}4)$$

According to the inviscid theory, when $\alpha = 0$ the jet height is the same as the liquid free-surface height in the tank.

4.7 ON QUASI-STEADY FLOW

A flow in which we put $\partial/\partial t \approx 0$ is said to be *quasi-steady*. This was the case of the example of Section 4.6. It is possible to obtain an approximate expression that brings out the magnitude of error involved by applying this restriction to the flow of this example, which is truly nonsteady.

From Equation 3.12-10, the dynamic equation for an inviscid, nonsteady flow in a gravitational field is

$$\frac{\partial V}{\partial t} + V\frac{\partial V}{\partial s} = -\frac{1}{\rho}\frac{\partial p}{\partial s} - g\frac{\partial z}{\partial s}. \qquad (4.7\text{-}1)$$

---

* Rouse and Ince (1957) note that Torricelli (1643) actually had established only that the velocity was proportional to $z^{1/2}$. It was the father of Daniel, Johann Bernoulli (1744), who provided the factor $2g$.

If the term $\partial V/\partial t$ is neglected in Equation 4.7-1, then the differential form leading to the Bernoulli equation is obtained, which is identically satisfied by Equation 4.6-1.

Now, combining the continuity equation $\dot{Q} = VA$, which also holds exactly in the incompressible, nonsteady case, with Equation 4.6-3 for the exit speed, for the case of a tank vented to the atmosphere, the free-surface speed is obtained:

$$V_1 = \dot{Q}/A_1 = (A_2/A_1)V_2 = \left(\sqrt{2g}\ A_2/A_1\right)z_1^{1/2}, \qquad (4.7\text{-}2)$$

where, in an unsteady flow, $z_1 = z_1(t)$. Therefore,

$$\partial V_1/\partial t = \left(\sqrt{2g}\ A_2/A_1\right)\tfrac{1}{2}z_1^{-1/2}\ \dot{z}_1 = \left(\sqrt{2g}\ A_2/A_1\right)\tfrac{1}{2}V_1/z_1^{1/2}, \qquad (4.7\text{-}3)$$

since $\dot{z}_1 = V_1$ denotes the free-surface speed. Again, eliminating V_1 in Equation 4.7-3,

$$\partial V_1/\partial t = \left(A_2/A_1\right)^2 g. \qquad (4.7\text{-}4)$$

That is, the unsteady acceleration of the free-surface is equal to the gravitational constant multiplied by the square of the area ratio. Now, at the free surface, since $\partial/\partial s = -\partial/\partial z$, the last term in Equation 4.7-1 is

$$-g(\partial z/\partial s)_1 = g, \qquad (4.7\text{-}5)$$

which means that to neglect the expression in Equation 4.7-4, by comparison, is a very good approximation indeed.

This does not tell the whole story. If the procedure is repeated for the exit station, then

$$\partial V_2/\partial t = \left(A_2/A_1\right)g. \qquad (4.7\text{-}6)$$

However, if $A_2/A_1 \ll 1$, this is still very small, and its neglect involves minor errors, at worst.

Since, at other points in the flow, we expect the nonsteady accelerations to be bracketed by the expressions of Equations 4.7-4 and 4.7-6, we have confirmed the validity of the decision to ignore $\partial V/\partial t$ as a first approximation.*

This proof is only approximate since we have used the quasi-steady-flow solution to calculate terms in the more exact nonsteady relation. Nevertheless, it confirms our intuition, and we can employ this approximation in similar situations with confidence. If the reservoir were pressurized, the free-surface speed, as well as the

---

* If the tank pressure is maintained at some constant value above atmospheric pressure, it turns out that, although the exit speed increases, the quasi-steady condition is achieved even more rapidly, and nonsteady effects become even less important.

acceleration, would be greater than that given by Equation 4.7-4 and further analysis would be required to determine under what conditions the same approximation could be employed. Full treatments of three truly nonsteady flows are presented in Sections 8.2, 8.3, and 8.4.

4.8 FLOW LOSSES IN INTERNAL INLETS

An *internal inlet* is a device that taps fluid from a plenum, or chamber, in which the fluid is essentially stagnant prior to entering the inlet, and conducts it to a conduit, or ejects it from a nozzle. In the flow of Figure 4.6-1 the inlet is formed by modifying the tank wall shape into a nozzle. Not only is such an inlet difficult to fabricate — although one can be found in most any teapot — but, since the fluid in the neighborhood of the walls is subject to viscous retarding forces, an inlet designed without regard to sound fluid mechanical principles may have large flow losses associated with it.

It is possible to define the *loss* term between any two points in an incompressible flow as the difference

$$p_1^0 + \rho g z_1 - \left(p_2^0 + \rho g z_2\right).$$

Clearly, from Equation 4.5-4, this is dependent solely on the frictional contributions of the wall, i.e.,

$$p_1^0 + \rho g z_1 - \left(p_2^0 + \rho g z_2\right) = \int_1^2 \frac{\tau_w P}{A} \, ds. \qquad (4.8\text{-}1)$$

Further, if, in addition to the restrictions on Equation 4.3-13, we require the flow to be adiabatic ($\Delta q \equiv 0$), the energy equation takes the form

$$h_2^0 = \frac{p_2^0}{\rho} + \frac{1}{2} V_2^2 + e_2 = \frac{p_1^0}{\rho} + \frac{1}{2} V_1^2 + g z_1 + e_1 + g\left(z_1 - z_2\right). \qquad (4.8\text{-}2)$$

Consequently, Equations 4.8-1 and 4.8-2 combine to produce the following expression for adiabatic, workless (except height changes), incompressible flow in a gravitational field:

$$\left(p_1^0 + \rho g z_1\right) - \left(p_2^0 + \rho g z_2\right) = \rho \Delta e = \int_1^2 \frac{\tau_w P}{A} \, ds, \qquad (4.8\text{-}3)$$

where $\Delta e \equiv e_2 - e_1$. Again, we see, for the restrictions enumerated, that the effect of viscosity is to increase the (specific) internal energy of a fluid particle, hence its temperature, and consequently, its entropy. This represents a flow loss.

We shall specify the loss in terms of a dimensionless coefficient* k, which is the ratio of the quantity given in Equation 4.8-3 divided by the dynamic pressure of the flow at ②, i.e.,

$$k \equiv \frac{p_1^0 + \rho g z_1 - \left(p_2^0 + \rho g z_2\right)}{\frac{1}{2}\rho V_2^2} = \frac{\Delta e}{V_2^2/2},$$ (4.8-4a)

which, for horizontal flow, reduces to

$$k \equiv \frac{p_1^0 - p_2^0}{\frac{1}{2}\rho V_2^2}.$$ (4.8-4b)

For an inlet, station ② is taken as close as possible to the point where the cross section first becomes uniform, but not in a region where the flow is separated from the wall. The second relation in Equation 4.8-4a relates the flow losses to the increase of internal energy as the particle moves downstream.

Figure 4.8-1 shows how important the design of the inlet can be if the loss is to be minimized. In general, the inlet should not be located flush on the reservoir wall, since this allows a buildup of viscous effects on the bounding stream tube. Ideally, the shape of the inlet, as the area decreases, should maintain a monotonically decreasing curvature in the flow direction, joining smoothly to the duct with a continuous radius of curvature. This helps maintain a favorable pressure distribution (i.e., continuously decreasing pressure) on the wall, which tends to prevent separation of the viscous boundary layer.

It is generally very difficult to obtain theoretical expressions for k, and the values given** are experimental. One standard mathematical shape that produces a low loss is the lemniscate, Figure 4.8-1a. Only the solid portion of the curve follows the mathematical expression. The rest of the shape (dashed line) is not critical, and is obtained rather arbitrarily by joining a circular arc to the leminscate at the point of 90° surface inclination and to a straight-line segment to complete the contour.

Sharp corners, as in the shapes of Figures 4.8-1c, d, and f, are highly undesirable since the flow will *always* separate at a sharp corner and, if the flow is laminar ahead of the inlet, transition to turbulence will likely result. Both conditions tend to increase losses drastically.

Figure 4.8-1d shows a device known as a Borda mouthpiece, after the French investigator Jean Charles Borda (1733–1799). He found that if the tube is kept short, so that the flow empties into a second plenum on the downstream side, a free jet (no solid boundaries downstream of separation) is formed, Figure 4.8-2a, such that the jet area tends to one half the tube area. In such a flow, no losses are involved.

---

* The flow loss coefficient is related both to the internal energy change of Equation 4.8-3, and to the *head loss*, h_ℓ, to be discussed more fully in Chapter 6, where $\Delta e \equiv g h_\ell = k(V_2^2/2)$ is the defining relation for h_ℓ.

** The values are from Henry (1944). The report lays down useful and practical procedures for the design of internal ducting so as to minimize flow losses.

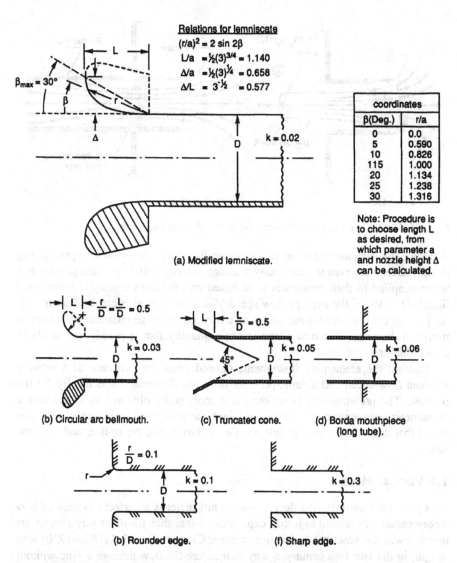

Relations for lemniscate

$$(r/a)^2 = 2 \sin 2\beta$$
$$L/a = \tfrac{1}{2}(3)^{3/4} = 1.140$$
$$\Delta/a = \tfrac{1}{2}(3)^{1/4} = 0.658$$
$$\Delta/L = 3^{-\tfrac{1}{2}} = 0.577$$

$\beta_{max} = 30°$

$k = 0.02$

coordinates	
β(Deg.)	r/a
0	0.0
5	0.590
10	0.826
115	1.000
20	1.134
25	1.238
30	1.316

Note: Procedure is to choose length L as desired, from which parameter a and nozzle height Δ can be calculated.

(a) Modified lemniscate.

$\dfrac{L}{D} = \dfrac{r}{D} = 0.5$ $k = 0.03$

$\dfrac{L}{D} = 0.5$ 45° $k = 0.05$

$k = 0.06$

(b) Circular arc bellmouth.   (c) Truncated cone.   (d) Borda mouthpiece (long tube).

$\dfrac{r}{D} = 0.1$ $k = 0.1$

$k = 0.3$

(b) Rounded edge.   (f) Sharp edge.

FIGURE 4.8-1 Losses in internal inlets. (Abstracted by Henry, J.R., NACA ARR No. L4F26, from a Russian report.)

However, if the tube is long, the jet reattaches downstream of a region of recirculating flow as sketched in Figure 4.8-2b. It is to this case that the flow loss coefficient shown in Figure 4.8-1d applies. The theoretical result for the short tube is derived in Section 4.14.

4.9 ELEMENTARY FLOW-METERING DEVICES

Every supplier dispensing arbitrary quantities of a product, and every customer who foots the bill, wants to be assured that the prescribed quantity is delivered or received.

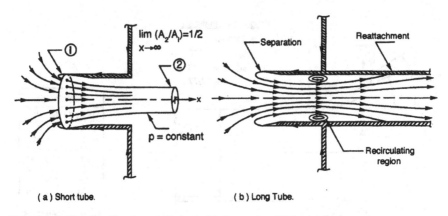

FIGURE 4.8-2 The Borda mouthpiece: (a) short tube; (b) long tube.

Although the Romans had developed* a magnificent water-distribution system, they did not understand even the continuity relation; consequently, their attempts to meter water supplied to their customers were based on erroneous concepts. Since for a liquid $\dot{Q} = VA$, if the average flow speed V at a given station can be determined, $\dot{Q}$ follows readily because the area is an easily measured geometrical parameter. It turns out, however, that flow speed is not a quantity that is readily accessible to direct measurement.

Instead, the elementary instruments depend upon the existence of a relation between flow speed and a static-pressure drop usually created specifically for the purpose. The pressure-drop measurement is most easily obtained by connecting a manometer between the two stations at which the pressure drop is required. Instruments that depend on this principle are the venturi meter, the orifice, and the flow nozzle.

The Venturi Meter

As a practical flow-metering device, the venturi meter is a perfect example of how theory sometimes is used to justify experiment rather than the other way around. Its inventor was the American hydraulics engineer Clemens Herschel (1842–1930) who sought, in the late 19th century, a way to measure the flow through a pipe without creating significant additional resistance for the flow to overcome. Giovanni Venturi (1746–1822) had noted a century earlier** that, in the flow of a liquid through a contraction, the pressure dropped and was minimum at the section of least area, but this observation did not lead to the design of a metering instrument. Herschel reasoned, for a given contraction, that the pressure drop should be related to the flow rate. By calibrating the pressure drops vs. a series of known flow rates, he

---

* As recounted by Rouse and Ince (1957, pp. 25–32), from the writings of the great hydraulics engineer Sextus Julius Frontinus (A.D. 40–103). The limited historical material in this chapter is based mostly upon Rouse and Ince's exposition.
** Venturi's work had actually been foreshadowed by the experiments of D. Bernoulli and others.

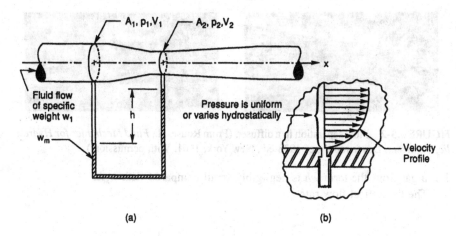

FIGURE 4.9-1 The venturi meter.

concluded that the device now known as a venturi meter would serve satisfactorily for his purpose.

If the flow is assumed to be steady, inviscid and horizontal, combination of the continuity equation (Equation 4.3-1b) and the Bernoulli equation (Equation 4.4-1a) leads to the expression

$$V_2^2 = \frac{2(p_1 - p_2)/\rho}{1 - (A_2/A_1)^2} = \frac{2\Delta p/\rho}{1 - (A_2/A_1)^2},$$ (4.9-1)

where $\Delta p \equiv p_1 - p_2$. Since the flow density ρ (or, equivalently, the specific weight w) and the area ratio A_2/A_1 are presumed known, the problem remains to measure the pressure difference Δp where it is understood that p_1 and p_2 are the values on the instrument centerline.

It is an extraordinary fact that although viscous effects on the flow through a venturi meter are usually small, it is viscosity that makes the measurement possible. Static pressures are defined in terms of a hypothetical instrument moving with the local center of mass of a fluid element. However, in the neighborhood of a straight, material boundary, the no-slip condition on the wall results in a velocity profile (see Figure 4.9-1) over which the pressure is essentially constant for a gas or, for a liquid, where only the hydrostatic effect results in a variation of pressure normal to the wall. This has been amply verified by a variety of experimental and theoretical research. Thus, by suitably designed pressure taps (also called *piezometers*), it is possible to measure the pressure difference with a manometer fixed with respect to the wall. If the manometer fluid has a specific weight w_m, then

$$\Delta p = (w_m - w)h.$$ (4.9-2)

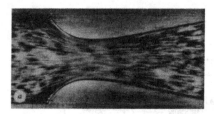 

FIGURE 4.9-2 Flow separation in a diffuser. (From Rouse, H., *Fluid Mechanics for Hydraulic Engineers*, Dover reprint of 1938 ed., New York, 1961. With permission.)

For a gas flow, the term *wh* is negligibly small compared with $w_m h$.

The theoretical flow rate is

$$\dot{Q}_{th} = A_2 V_2 = A_2 \left[\frac{2\Delta p/\rho}{1-\left(A_2/A_1\right)^2} \right]^{1/2} \qquad (4.9\text{-}3)$$

However, careful experimentation shows that the actual flow rate is somewhat less than inviscid theory predicts, due to viscous effects that show up in two ways. In the first place, maintenance of a viscous flow requires a total-pressure drop in the direction of the flow, which, for a venturi meter, corresponds to a greater static-pressure drop Δp for a given flow rate than if viscous effects were not present. Expressed differently, a given Δp corresponds to a lower flow rate than predicted by inviscid theory. Second, in the portion of the venturi meter downstream of the throat, where the area is increasing, the pressure is also increasing. In the neighborhood of the wall the fluid, which is retarded by viscosity, has inadequate kinetic energy to allow it to move against the increasing pressure. If such a configuration, called a diffuser, is not well designed, the boundary layer may separate from the walls resulting in a flow completely different in character than predicted by one-dimensional theory.

Figure 4.9-2 shows* the flow through a venturi, first as the flow has just started and when viscous effects are still small, and, second, after a period of time when the boundary layer has built up and the flow has separated from the diffuser walls to form a jet surrounded by regions of backflow. In well-designed and carefully machined venturis the tendency of the flow to separate is minimized.

The actual flow rate is determined by multiplying the theoretical value by a correction factor C_v, i.e.,

$$Q = C_v \dot{Q}_{th} = C_v A_2 \left[\frac{2\Delta p/\rho}{1-\left(A_2/A_1\right)^2} \right]^{1/2}. \qquad (4.9\text{-}4)$$

* The photographs in Figure 4.9-2 were made by taking a short time exposure of the light reflected from fine aluminum particles on the surface of a flow of water.

Equation 4.9-4* serves as a definition of the *velocity coefficient* C_v, an empirical factor that can be determined only by calibration. In principle every venturi requires its own calibration curve. However, the Research Committee on Fluid Meters of the American Society of Mechanical Engineers (ASME) has, after careful experimentation, provided standardized design procedures and calibration curves. Venturis made and installed to the recommended specifications may be used directly without further calibration; see Bean (1971).

For reasons that are detailed in Chapter 5, data are best presented in dimensionless form. For the venturi tube the velocity coefficient is plotted vs. the approach Reynolds number, $Re_1 = \rho V_1 D_1/\mu = V_1 D_1/\nu$, where $\nu \equiv \mu/\rho$ is the kinematic viscosity of the fluid. Figure 4.9-3a shows the main features of the Herschel-type venturi tube, and Figure 4.9-3b gives a plot of C_v vs. Re_1, which represents a faired curve through data obtained from the calibration of hundreds of venturi tubes.

It should be kept in mind that, if, for a gas flow, speeds in the throat become sufficiently high, compressibility effects must be taken into account, whereas, for a liquid, the throat pressure must remain sufficiently higher than the liquid vapor pressure as to avoid cavitation. Both of these phenomena are discussed in the ASME report.

THE PLATE ORIFICE

There are two major drawbacks to the venturi meter, its high cost and its excessive length, which have led to the introduction of the plate orifice and the flow nozzle, both of which are much shorter, much less expensive to machine, and simpler to install. However, this simplicity is achieved at the cost of higher flow losses than in a venturi meter. More than any other device, the plate orifice is widely used for metering flows — compressible as well as incompressible.

The orifice depends upon the creation of a pressure difference resulting from an area change but one in which the flow is deliberately caused to separate. Figure 4.9-4a is a photograph of a two-dimensional flow through a sharp-edged orifice. The

---

* This relation was obtained only after Herschel (1887) had developed the venturi meter and learned how to calibrate it. It is included in his paper almost as an afterthought. His disdain for theoretical research points up the antagonism that existed between hydraulics engineers and some of the 19th-century theoretical hydrodynamicists, who had actually discovered many remarkable results mostly for flows in which viscosity played a negligible role. As Herschel wrote:

> I know well the argument that groping, even aimless, research must precede applied science; but I have found that practical inventions more often precede their scientific explanation...while aimless research only results in nothing better than having its printed record encumber the shelves of libraries...the study of theoretical hydraulics having proved a wholly barren field for several centuries....

However, Herschel's point of view had its own serious deficiencies. Much of hydraulics in the last half of the 19th century was sheer empiricism backed by only the faintest understanding of the basic principles governing flows of real fluids. It was not until the beginning of the 20th century that theoretical methods were developed to treat in detail the effects of viscosity, which ultimately enabled researches — theoretical and experimental — to occupy a common ground of understanding. It is hard to argue with his observation that libraries are encumbered with journals laden with unread articles, however.

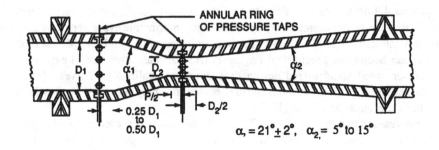

(a)

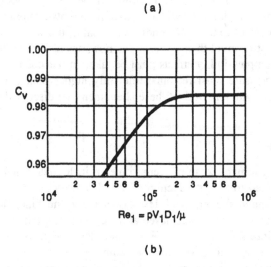

(b)

FIGURE 4.9-3 The ASME venturi meter and calibration curve.

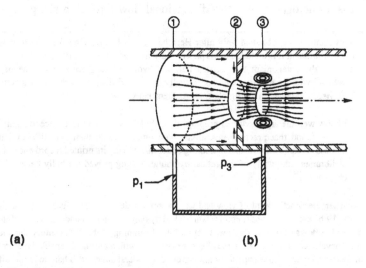

(a)   (b)

FIGURE 4.9-4 The plate orifice.

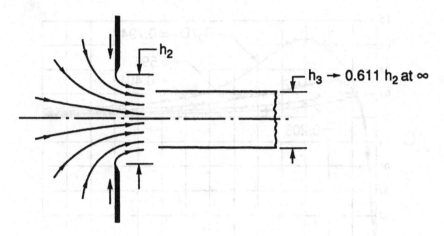

FIGURE 4.9-5 Flow through a two-dimensional slot.

essential features are separation of the jet from the sharp edge followed by contraction of the jet, which is bounded by regions of recirculating fluid called eddies. It is an experimental fact that pressure variations within the eddies are small, and that the wall pressure at station ③ is approximately the same as that of the *vena contracta*, the jet minimum area downstream of the orifice.

An idealized theoretical treatment of the orifice is based on Figure 4.9-4b. For an axisymmetric configuration, Bernoulli's equation, based on an assumed inviscid flow of the jet from station ① to station ③, gives

$$V_3^2 = \frac{2\Delta p/\rho}{1-\left(A_3/A_1\right)^2}.$$

(4.9-5)

where $\Delta p = p_1 - p_3$. However, since there is no general theory to predict A_3, an empirical *contraction coefficient* C_c is introduced, such that $A_3 \equiv C_c A_2$.

That a jet emitted from an orifice tends to contract to a minimum value was first recognized by Newton, who conducted experiments to measure the contraction coefficient. For the case of an inviscid, two-dimensional jet, in the absence of gravity and emitted from an orifice connecting an infinite upstream plenum with an unbounded downstream region, (Figure 4.9-5), G. R. Kirchhoff (1824–1887) showed,* by the method of conformal mapping, that the theoretical value of the contraction coefficient is $C_c \equiv h_3/h_2 = \pi/(\pi + 2) = 0.611$.

For the case of an axisymmetric jet emanating from an infinite plenum, Garabedian (1956), by a numerical analysis, determined that $C_c = 0.5793$. Using a more sophisticated approach, also numerical, Bloch (1968) found that $C_c = 0.59135 \pm 0.00004$. For a jet emanating from an upstream channel of area A_1, the contraction coefficient can be expected to depend on the ratio A_2/A_1.

---

* The original work is found in Kirchhoff (1869). The analysis can also be found in Milne-Thomson (1968, p. 318).

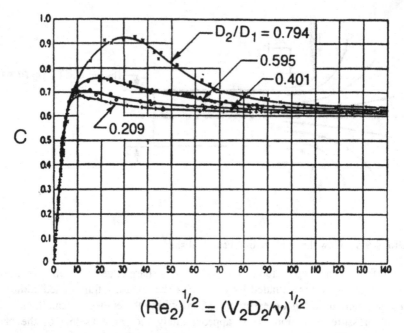

$$(\text{Re}_2)^{1/2} = (V_2 D_2/\nu)^{1/2}$$

FIGURE 4.9-6 Discharge coefficient for a plate orifice. (From Johansen, F.C., *Proc R. Soc. A*, 126(801), 231, 1930. With permission.)

The theoretical flow rate is $\dot{Q}_{th} = A_3 V_3$ and, as was done for the venturi meter, the actual flow *for an orifice* can be expressed as

$$\dot{Q} = C_v \dot{Q}_{th} = C_v C_c A_2 \left[\frac{2\Delta p/\rho}{1 - \left(C_c A_2/A_1\right)^2} \right]^{1/2}. \tag{4.9-6}$$

The coefficients C_v and C_c are obviously interrelated and are not determinable from theory. To get around this fact Johansen (1930) expressed the flow rate in terms of a *discharge coefficient* C_d, determined by measurement and defined in the following relation:

$$\dot{Q} = \frac{C_d A_2}{\left[1 - \left(A_2/A_1\right)^2\right]^{1/2}} \left(2\Delta p/\rho\right)^{1/2}. \tag{4.9-7}$$

In other words $C_d \equiv \dot{Q}/\dot{Q}_{th}$.

A summary of his data is presented in Figure 4.9-6. It shows, in the low-speed regime, that the viscous effects result in a Reynolds number dependence radically different from that at high speeds. In fact, photographs printed in the original article indicate a flow pattern quite unlike that in Figure 4.9-4a. At low speeds, the flow is

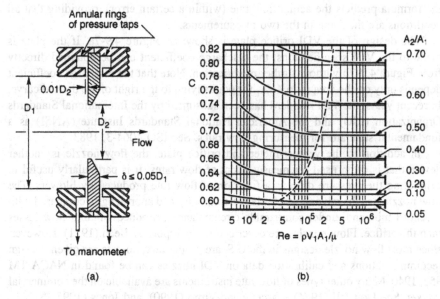

FIGURE 4.9-7 The VDI orifice plate.

laminar and nearly symmetrical about the plane of the orifice. Variations in C_d with changes in the diameter ratio D_2/D_1 are important only in an intermediate range of Re_2. In the highest range of Reynolds number, C_d approaches a constant, asymptotic value. Johansen points out that to compare other data with his, both sets of equipment must have been fabricated to the same standards of relative roughness (i.e., smoothness).

For engineering purposes it was decided about a century ago by the Verein Deutscher Ingenieuere (VDI) in Germany to follow a different course, one which is still widely used. See NACA TM 972, 1940, for a translation of the original report. It happens that the mean pressure on the back face of an orifice plate is essentially equal to that of the downstream region bounding the jet. Thus, Equation 4.9-7 was replaced by the following, simplified expression:

$$\dot{Q} = CA_2\sqrt{2\Delta p/\rho}, \tag{4.9-8}$$

where Δp is the static-pressure drop from the front face of the orifice to the back. The multiplying factor C is called the *discharge coefficient*.

It is remarkable that Equation 4.9-8 has only a marginal theoretical justification. The quantity $\sqrt{2\Delta p/\rho}$ is a fictitious velocity. When multiplied by the orifice area A_2 we have a quantity that has the dimensions of volume flow rate and that, by definition, is proportional to the actual flow rate. The formula works because the discharge coefficient is ultimately determined by correlation with experimentally determined flow rates. It is expressed as a function of the pipe Reynolds number, with the area ratio A_2/A_1 as a parameter. It is a virtue of the correlation process that the formula predicts the actual flow rate (within a certain error), providing that all conditions are the same in the two measurements.

The design of the VDI orifice plate is shown in Figure 4.9-7a. If the plate is made to the VDI specifications, the discharge coefficient can be obtained directly from Figure 4.9-7b without further modification. Note that the discharge coefficient depends only on the area ratio A_2/A_1 for points lying to the right of the dashed curve. In recent years, the VDI standard has been subsumed by the International Standards Organization (ISO) and the American National Standards Institute (ANSI) as a fundamental standard for measuring flow rates. See ISO 4064-3, 1983.

In addition to the venturi meter and orifice plate, the flow nozzle is another device that is often used for measurement of flow rates. It is particularly useful in calibrating the pressure drop as a function of flow rate produced by blowers. The flow nozzle has the advantage of relative simplicity and short length, compared with a venturi tube and, under comparable circumstances, produces smaller flow losses than the orifice. Flow nozzles are described in the report by Bean (1971). However, since most flow nozzle designs in the U.S. are proprietary, no data are given. Design recommendations and calibration data on VDI nozzles can be found in NACA TM 952, 1940. Many other types of flow rate instruments are available on the commercial market. See Dowdell (1974), Cascetle and Virgo (1990), and Jones (1995).

A NEW APPROACH TO ORIFICE METERING

The awkwardness of the standard formula for orifice metering caused the author* to rethink the situation. Up to the time that he became aware of the work of Garabedian and, later, of Bloch, there had been no theoretical expression for the contraction coefficient $C_c \equiv A_3/A_2$ in the axisymmetric version of the flow as it appears in Figure 4.9-5. Further, the appearance of the velocity-of-approach factor $[1 - (A_3/A_1)^2]^{1/2}$ had always seemed to be an inapposite ornament.

The alternative is to replace the static-pressure drop in the VDI formula by a pressure difference involving the local total pressure just upstream of the orifice station — sensed by a Pitot tube — and the pressure on the jet boundary which is linked to the velocity at the asymptotic downstream station through Bernoulli's equation. Thus, referring to Figure 4.9-4b a Pitot tube would be located at ①. Then, treating the flow as steady and one-dimensional between stations ① and ③, the latter the station where the jet approaches its asymptotic uniform value, Bernoulli's equation produces

$$p^0 = p_1 + \tfrac{1}{2}\rho u_1^2 = p_3 + \tfrac{1}{2}\rho u_3^2 = \text{constant}.$$

As noted, a variety of experiments has shown that the plenum pressure p_3, which bounds the jet, is essentially equal to that on the back face of the orifice.

Thus, we put $\Delta P \equiv p^0 - p_3 = \tfrac{1}{2}\rho u_3^2$, or $u_3 = \sqrt{2\Delta P/\rho}$. Then, the theoretical volume flow rate is $\dot{Q}_t = u_3 A_3 = C_c A_2 u_3$, where we have introduced Bloch's contraction

---

* See Brower, W.B., Jr., U.S. Patent No. 5,365,795. Improved Method for Determining Flow Rates in venturis, Orifices, and Flow Nozzles Involving Total Pressure and Static Pressure Measurements, 1994.

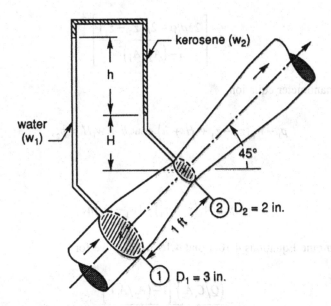

FIGURE 4.10-1 Venturi meter.

coefficient $C_c \equiv A_3/A_2 = 0.5914$. Finally, introducing a discharge coefficient C, then $\dot{Q} = C \dot{Q}_t$ and our expression for the actual flow rate becomes

$$\dot{Q} = CC_c A_2 \sqrt{2\Delta P/\rho}. \qquad (4.9\text{-}9)$$

In Equation 4.9-9 ΔP is understood to be the pressure difference between the Pitot tube value and the static pressure on the back face of the orifice plate. The discharge coefficient is introduced to deal with the inevitable problem of viscous effects entering, particularly in the transition range. This, of course, means that a calibration is required, and where the discharge coefficient may depend on the flow Reynolds number. By introduction of the Pitot tube, dependence of the discharge coefficient on the orifice-to-pipe-area ratio is bypassed. A further advantage is that the velocity-of-approach factor does not enter either. Preliminary measurements indicate that C is close to unity as long as the pipe Reynolds number does not fall too close to the transition value of $Re = 2 \times 10^5$.

4.10 TWO FLOW RATE EXAMPLES

VENTURI METER

Water (w_1), at 68°F, $\nu = 1.080 \times 10^{-5}$ ft²/sec, flows through the ASME venturi meter of Figure 4.10-1 such that $V_1 = 10$ ft/sec. Determine the manometer reading if kerosene $(w_2$, specific gravity = 0.82) is used as the manometer fluid.

For a venturi installed in an inclined pipe, Equation 4.9-4 must be modified to account for the variation in height, i.e.,

$$\dot{Q} = C_v A_2 \left[\frac{2\Delta p/\rho - 2g(z_2 - z_1)}{1 - (A_2/A_1)^2} \right]^{1/2}. \tag{4.10-1}$$

From the manometer equation

$$p_1 - w_1(z_2 - z_1 + H + h) + w_2 h + w_1 H = p_2,$$

or

$$p_1 - p_2 - w_1(z_2 - z_1) = (w_1 - w_2)h. \tag{4.10-2}$$

We can combine Equations 4.10-1 and 4.10-2 and solve for

$$h = \frac{(\dot{Q}/C_v A_2)^2 \left[1 - (A_2/A_1)^2 \right]}{2g(1 - w_2/w_1)},$$

$$= \frac{V_1^2 \left[(A_1/A_2)^2 - 1 \right]}{2g(1 - w_2/w_1)C_v^2}.$$

$$\tag{4.10-3}$$

A final expression restates the functional relation from experiment:

$$C_v = C_v(\text{Re}_1). \tag{4.10-4}$$

In this case, since $\text{Re}_1 = V_1 D_1/\nu = 2.3 \times 10^5$, we find from Figure 4.9-3 that $C_v = 0.983$; therefore,

$$h = \frac{100(81/16 - 1)}{2 \times 32.17 \times 0.18 \times (0.983)^2} = 36.4 \text{ ft.}$$

This value of h, considering the flow speeds involved, is very large, corresponding to about 1 atm of pressure. Since the use of a 36-ft-high manometer would not be satisfactory for most laboratories, the experimenter would almost certainly shift to a denser manometer fluid, probably mercury (S.G. = 13.6). The experimenter would also remember to put the manometer *below* the venturi, for reasons that should not be necessary to state. Since the manometer reading is inversely proportional to the difference of the fluid specific weights $w_1 - w_2$ ($w_2 - w_1$ for a manometer below the pipe), we see that, if w_1 and w_2 are nearly equal in magnitude, even low flow speeds may generate large manometer readings. On the other hand, in a gas flow the density is relatively negligible compared with that of the manometer fluid. Consequently, to obtain a reasonably large manometer deflection, either high flow speeds or a small contraction ratio A_2/A_1 may be required.

ORIFICE

Compute the volume flow rate for the flow of air ($\rho = 0.00234$ slug/ft^3, $\mu = 0.375 \times 10^{-6}$ lb ft/sec^2) through a VDI orifice, if $D_1 = 6$ in., $A_2/A_1 = 0.5$ and $\Delta p = 20$ in. H$_2$O.

From Equation 4.9-8 we can solve for the (unknown) coefficient

$$C = \frac{\dot{Q}/A_2}{(2\Delta p/\rho)^{1/2}} = \frac{V_1 A_1/A_2}{(2\Delta p/\rho)^{1/2}}. \tag{4.10-5}$$

Also, from the definition of the Reynolds number,

$$V_1 = (\mu/\rho D_1)\mathrm{Re}_1. \tag{4.10-6}$$

Eliminating V_1 from these two relations yields

$$C = \frac{\mu/\rho D_1}{(A_2/A_1)(2\Delta p/\rho)^{1/2}} \, \mathrm{Re}_1.$$

Substituting in the numerical values,

$$\left(\frac{2\Delta p}{\rho}\right)^{1/2} = \left(\frac{2w_2 h}{\rho}\right) = \left(\frac{2 \times 62.4 \times 20/12}{2.34 \times 10^{-3}}\right) = 298.1 \text{ ft/sec,}$$

and

$$C = \frac{0.375 \times 10^{-6}}{(2.34 \times 10^{-3})(0.5)^2 (298.1)} = 2.150 \times 10^{-6} \, \mathrm{Re}_1. \tag{4.10-7}$$

This provides one relation in the form $C = C(\mathrm{Re}_1)$. Figure 4.9-7b furnishes a second in the form of

$$C = C(\mathrm{Re}_1, A_2/A_1), \tag{4.10-8}$$

where, for our case, $A_2/A_1 = 0.5$. A simultaneous trial-and-error solution of Equations 4.10-7 and 4.10-8 results in

$$C = 0.697 \text{ @ } \mathrm{Re}_1 = 3.25 \times 10^5. \tag{4.10-9}$$

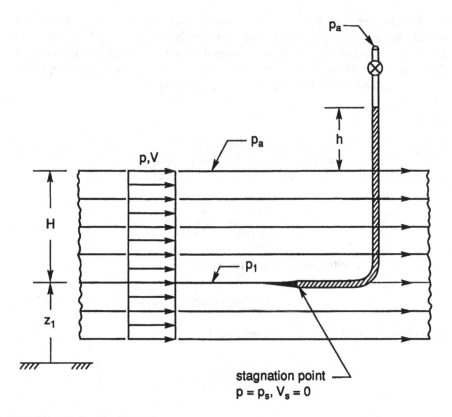

FIGURE 4.11-1 The Pitot tube.

Therefore, the flow rate is, from Equation 4.9-8,

$$\dot{Q} = 0.697 \ (\pi/64) \ (298.1) = 10.20 \ \text{ft}^3/\text{sec}. \qquad (4.10\text{-}10)$$

Note that the solution lies on the portion of the curve of C vs. Re which is horizontal, and thus, locally, C does not vary with Reynolds number. An alternative solution method would be, for a first approximation, to select C for the specified A_2/A_1 to the right of the dashed line and to iterate, varying C, until both Equations 4.10-5 and 4.10-6 yield the same value of V_1.

4.11 PITOT AND PITOT-STATIC TUBES

For situations in which it is desired to measure the flow speed *at a point*, the device invented by Henri de Pitot (1695–1771) is remarkably simple and accurate. The *Pitot tube*, also known as the total-head tube, is simply a uniform tube bent so that the open end can be positioned to oppose the oncoming flow. Figure 4.11-1 shows its use in a uniform flow of liquid.

The tube is immersed to depth H in a flow of speed U and density ρ. The liquid rises in the tube under the stagnation process to a height h above the free surface. The streamline on which the tube lies can be thought of as having a finite speed $V_1 = U$ and area $A_1 = 0$ upstream, and a finite area A_s and speed $V_s = 0$ at stagnation; thus, the continuity equation is "satisfied" in a sense along this streamline. For inviscid flow,

$$\frac{p_1}{\rho} + \frac{1}{2} U^2 + gz_1 = \frac{p_s}{\rho} + \frac{1}{2} V_s^2 + gz_1.$$

Thus,

$$U^2 = 2(p_s - p_1)/\rho. \tag{4.11-1}$$

But the pressure upstream varies hydrostatically. This can be seen if we transfer mentally to an observer embedded in the flow (with the Pitot tube removed). Since the fluid is at rest in this frame, the equations of fluid statics apply. That is, $p_1 = p_a + \rho g H$.

In the original reference frame, the equation of statics applies within the Pitot tube, so that $p_s = p_a + \rho g(h + H)$. Therefore, $p_s - p_1 = \rho g h$, and

$$U = (2gh)^{1/2}. \tag{4.11-2}$$

Thus, the flow speed is given in terms of the height h which is obviously an example of Torricelli's relation, Equation (4.6-3). Although this result is correct, experience has shown that certain precautions must be taken if reliable measurements for the flow of liquids in open channels are to be obtained by use of a Pitot tube. See King and Brater (1954) for a description of this and other instruments used in measuring water flows.

For the flow of gases, or of liquids in closed ducts or tunnels, atmospheric pressure is not necessarily a convenient reference, particularly for pressurized devices. An instrument that bypasses this difficulty is the Pitot-static tube designed by L. Prandtl (1875–1953), Figure 4.11-2a. Surrounding the total-head tube is a larger, round-nose tube that has about eight pressure taps drilled in it three to six diameters from the nose. Theory and experiment confirm that the surface pressure distribution varies as shown in Figure 4.11-2b. From the stagnation value on the centerline, it falls rapidly to well below free-stream value and then increases along the afterbody, asymptotically approaching the free-stream value. The direction of the arrow indicates whether the local surface pressure is above or below the free-stream value. Further details are given by Prandtl and Tietjens (1934, Vol. II, p. 120).

The stagnation pressure is piped through the central tube. A number of small holes are drilled on a circumference of the outer tube. These holes permit the interior of the large tube to adjust to an "average" cylinder static pressure, which is very

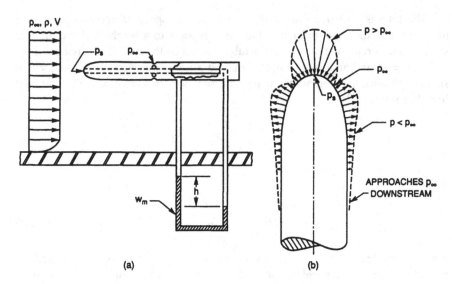

FIGURE 4.11-2 Schematic of the Pitot-static tube. (After Prandtl, L., and Tietjens, O.G., *Applied Hydro- and Aeromechanics*; Vol. 2, reprinted by Dover, New York, 1957. With permission.)

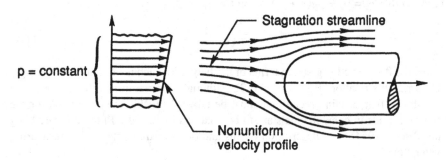

FIGURE 4.11-3 Pitot-static tube in a shear flow.

close to the free-stream value p_∞, if the tube is reasonably parallel to the flow. Connecting the two pressure sources across a manometer allows the flow speed to be calculated by Equation 4.11-1 and the manometer equation. The Pitot-static tube of the type shown does not require calibration for moderate flow speeds (i.e., low, subsonic Mach numbers).

There is another widely used style of Pitot-static tube that employs a sharp-edged nose but that is slightly more sensitive to errors in case of flow misalignment. A review of the literature on Pitot tubes has been given by Folsom (1956).

PITOT-STATIC TUBE IN NONUNIFORM FLOW

If a rounded-nose Pitot-static tube is used in a flow where the velocity profile is not uniform (such as in a boundary layer), the flow pattern is asymmetrical, Figure 4.11-3, resulting in an erroneous reading. It turns out that the stagnation pressure corre-

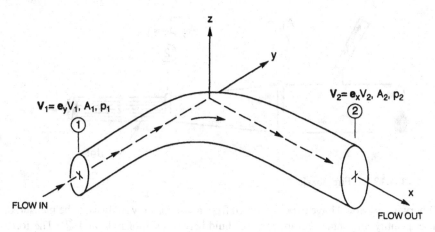

FIGURE 4.12-1 Flow through a pipe bend.

sponds to a streamline displaced (hence, the term *displacement effect*) from the Pitot-static tube centerline toward the higher velocities. It is possible, by fairly formidable mathematics, to correct the error. In fact, it is possible to use the peculiar properties of the flow to turn a properly designed Pitot-static tube into a flow-direction measuring device as well. See Cousins (1969) who gives the original reference.

4.12 ELEMENTARY APPLICATIONS OF THE MOMENTUM EQUATION

STEADY FLOW OF A LIQUID THROUGH A 90° PIPE BEND

Figure 4.12-1 illustrates a pipe bend of 90° in which there is also a change of cross section. The problem is to determine the reaction forces necessary to maintain the pipe in place solely due to the bend, i.e., assuming that the pipe is incapable of transmitting axial forces. Given date in this flow problem are ρ, $\dot{Q}$, p_1, A_1, and A_2. The ambient pressure is p_a.

From Equation 4.3-14, the reaction that must be exerted by an external structure to maintain the pipe in place is

$$R = \rho\dot{Q}\left(e_x V_2 - e_y V_1\right) + e_x\left(p_2 - p_a\right)A_2 - e_y\left(p_1 - p_a\right)A_1 + e_z g m_{1,2}, \quad (4.12\text{-}1)$$

where, for steady liquid flow, we have made use of the continuity relation $\dot{m} = \rho\dot{Q}$. The flow speeds, of course, are $V_1 = \dot{Q}_1/A_1$, $V_2 = \dot{Q}/A_2$. This leaves only p_2 as unknown.

If the flow were capable of being treated as inviscid, as may be possible for short pipe lengths, then p_2 would be obtained from Bernoulli's equation:

$$p_2 = p_1 + \tfrac{1}{2}\rho\left(V_1^2 = V_2^2\right) = p_1 + \tfrac{1}{2}\rho V_1^2\left[1 - \left(A_1/A_2\right)^2\right]. \quad (4.12\text{-}2)$$

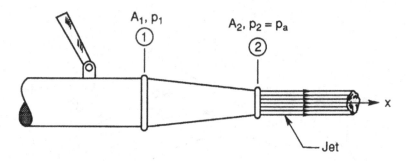

FIGURE 4.12-2   Fire nozzle.

Otherwise, p_2 would have to be computed from a viscous flow analysis, or be measured. The quantity $m_{1,2}$ stands for the mass of fluid between stations ① and ②. The force required to oppose the weight of the pipe must be accounted for separately.

FIRE NOZZLE

As shown in Figure 4.12-2, this is a familiar problem. From the data, and with $p_2 = p_a$ at the exit, then for inviscid flow the continuity and Bernoulli relations combine to yield

$$V_2 = \left[\frac{2(p_1 - p_a)/\rho}{1 - (A_2/A_1)^2} \right]^{1/2}. \tag{4.12-3}$$

Therefore, we have $\dot{m} = \rho \dot{Q} = \rho A_2/V_2$ and $V_1 = \dot{Q}/A_1$, in terms of the specified quantities. Consequently, the horizontal reaction force is

$$R_x = \dot{m}(V_2 - V_1) - (p_1 - p_a)A_1. \tag{4.12-4}$$

Since the momentum and pressure contributions in Equation 4.12-4 are of opposite sign, it is not possible to tell, therefrom, whether the force is directed left or right. It is left as an exercise to show, by elimination of V_1 and V_2, that

$$R_x = -(p_1 - p_a)A_1 \frac{1 - A_2/A_1}{1 + A_2/A_1}. \tag{4.12-5}$$

The form of Equation 4.12-5 indicates that the reaction force must be applied so as to oppose the jet.

THE WATER ROCKET

We wish to calculate the reaction force necessary to be exerted on a water rocket (of the type available as a child's toy), as the flow is initiated, to maintain the rocket

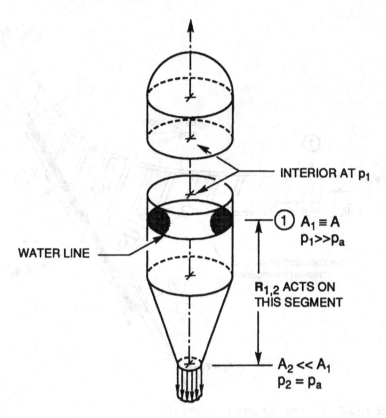

FIGURE 4.12-3 Water rocket.

stationary. The device is essentially a nozzle, Figure 4.12-3, for which the upper end has been capped off, and the interior pressurized at p_1. The z-axis is taken to be positive down; therefore $A_1 = e_z A_1$, $A_2 = e_z A_2$, and $V_2 = e_z V_2$, where A_1, A_2, and V_2 are all positive quantities. We apply Equation 4.3-14 between the free surface, station ①, and the exit, station ②. To justify use of the quasi-steady relations, we must put $V_1 \approx 0$, an approximation that will be valid if A_2/A_1 is sufficiently small.

For the control volume between ① and ②, then

$$R_{1,2} = \dot{m} V_2 - (p_1 - p_a) A_1 - e_z g (m_{1,2} + M), \qquad (4.12\text{-}6)$$

where the last term is due to the gravitational force on rocket (mass M) and liquid (mass $m_{1,2}$). Thus, the external force required on the casing is the sum of $R_{1,2}$ and the negative of the pressure force on the cap, i.e.,

$$R = R_{1,2} + (p_1 - p_a) A_1,$$

$$= e_z \left[\dot{m} V_2 - g (m_{1,2} + M) \right]. \qquad (4.12\text{-}7)$$

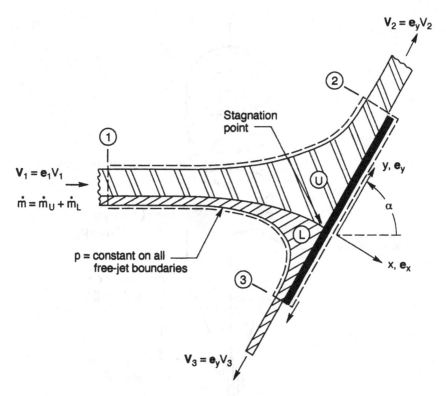

FIGURE 4.12-4   Jet impinging upon an inclined plate.

The component $\dot{m}\,V_2$ of the reaction force is numerically equal to, and is directed opposite to, the rocket thrust. This expression for the thrust applies for compressible as well as viscous flows. For incompressible, inviscid flows, Bernoulli's equation can be used to compute $\dot{m}$ and V_2. The analogous relation for a compressible flow is developed in Section 7.15.

If the flow is inviscid, the only mechanism contributing to the thrust is the nonuniform pressure distribution acting on the casing. Its evaluation as the term $\dot{m}\,V_2$ is simply a convenient way of computing it. If viscous effects are of significance, they show up as shear stresses on the casing, which tend to reduce the momentum outflow and, hence, the thrust.

STEADY TWO-DIMENSIONAL JET IMPINGING UPON AN INCLINED PLATE

It is an experimental fact that a jet that is directed against an inclined plate splits into two outlet flows, which leave the plate (approximately) tangentially, if gravitational and viscous effects are small. The geometry is illustrated in Figure 4.12-4.

Now the flow shown in the figure is clearly *not* one dimensional. However, by dividing the flow into two parts, designated as upper (U) and lower (L), we can demonstrate the power of one-dimensional techniques by applying the appropriate relations at stations where each flow behaves as though it were one dimensional.

The line that separates the two flows is called the stagnation streamline because its intersection S with the plate is a stagnation point. Along the stagnation streamline, the pressure varies and the force exerted by the upper flow on the lower must be equal and opposite to the force exerted by the lower on the upper.

For simplicity, we restrict considerations to steady, constant-density, inviscid, and mass-conserving flows. We also assume that the region shown within the control volume (the dashed line) is sufficiently small that the change of height is everywhere negligible.

Under the restrictions cited and by taking into account that on the surface of a free jet the pressure is everywhere the same, then, by Bernoulli's equation applied at stations ①, ②, and ③, where the flow is assumed to be uniform, $p_1 = p_2 = p_3 = p_a$; hence,

$$V_1 = V_2 = V_3 \equiv V = \text{constant}. \tag{4.12-8}$$

Therefore, by applying Equation 3.20-5 to the upper flow from ① to ②, and the lower flow from ① to ③,

$$R_U = \dot{m}_U V\left(e_y - e_1\right), \quad R_L = \dot{m}_L V\left(-e_y - e_1\right), \tag{4.12-9}$$

where it is convenient to select the cartesian axes normal and tangent, respectively, to the plate. It is recalled that R is the external force that must be exerted on a solid boundary to maintain the boundary fixed. Thus, the net y-component required due to the action of both jets is

$$R_y = e_y \cdot \left(R_U + R_L\right),$$
$$= \dot{m}_U V(1 - \cos \alpha) - \dot{m}_L V(1 + \cos \alpha), \tag{4.12-10}$$
$$= \rho A_2 V^2 (1 - \cos \alpha) - \rho A_3 V^2 (1 + \cos \alpha),$$

where A_2 and A_3 are the outlet areas of the two flows and where

$$e_1 \cdot e_y = \cos \alpha, \quad e_1 \cdot e_x = \sin \alpha. \tag{4.12-11}$$

However, since only a viscous flow can exert a force tangent to a flat surface, $R_y = 0$; hence,

$$\frac{A_2}{A_3} = \frac{1 + \cos \alpha}{1 - \cos \alpha}. \tag{4.12-12}$$

Further, by the continuity relation, $A_2 + A_3 = A_1$; thus,

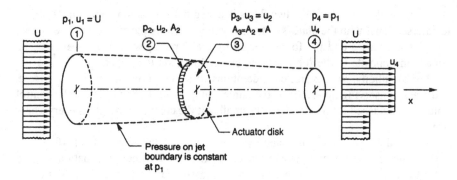

FIGURE 4.12-5 The propeller as an actuator disk.

$$A_2 = \tfrac{1}{2}(1+\cos\alpha)A_1, \quad A_3 = \tfrac{1}{2}(1-\cos\alpha)A_1. \tag{4.12-13}$$

The normal force on the plate is given by

$$R_x = e_x \cdot (R_U + R_L),$$

$$= -(\dot{m}_U + \dot{m}_L)V\sin\alpha, \tag{4.12-14}$$

$$= -\dot{m}V\sin\alpha.$$

Note that when $\alpha = \pi/2$, $R_x = -\dot{m}V$, which corresponds to the case for which the magnitude of the reaction force is maximum. On the other hand, when $\alpha = 0$ or π, $R_x = 0$, as one would expect.

ELEMENTARY THEORY OF A PROPELLER TREATED AS AN ACTUATOR DISK

One of the classical theories of fluid machinery is the analysis of propeller performance on the basis of one-dimensional flow relations. A propeller is a multibladed device, driven by a power plant, which moves the power plant and the vehicle to which it is attached at speed U through the medium (water or air), which is otherwise at rest. With respect to an observer embedded in the distant medium, the flow is unsteady.

However, with respect to an observer moving with the propeller, the flow can be treated as steady (actually quasi-steady). A slender stream tube of fluid upstream is sucked into the propeller disk, pressurized by the action of the blades, and is discharged downstream as a gradually narrowing jet within the overall body of fluid, as shown in Figure 4.12-5.

To formulate a model within the one-dimensional framework, it is necessary to make certain approximations. As always, certain of these approximations, although plausible, are not everywhere mutually consistent. In the first place, we assume that the flow is one-dimensional (at least at certain stations), steady, incompressible (constant density), and inviscid. However, anyone who has ever observed the action of a fan knows that the flow cannot be steady. In fact, for reasons discussed in Section 3.18, the propeller, a machine that puts energy into the flow, is an inherently nonsteady device. To avoid dealing with these details the propeller is replaced in the model by an actuator disk,* which is visualized as a porous disk through which the flow velocity is assumed to be constant, but across which the pressure increases discontinuously. This change in pressure generates a force on the disk; there is an equal and opposite force of the disk on the fluid.

In the frame of the observer at the propeller station, the flow upstream (the free stream) is at pressure p_1 and flow speed U. It is assumed that the curvature of the jet is everywhere sufficiently small that the exterior flow pressure remains at p_1, including the stations corresponding to the upstream and downstream locations of the actuator disk. (This is one of our inconsistencies.)

Work is done by the actuator disk as reflected by the increased (average) pressure at ③. This is exemplified by the tendency of the slipstream to accelerate downstream as it attempts to adjust to the uniform pressure $p_4 = p_1$ far downstream, imposed on it by the surrounding fluid. The downstream velocity u_4 is substantially greater than that of the free stream. For the current purpose the jet boundary is treated as a discontinuity in velocity, but across which the pressure remains constant at p_1.

The applicable equations are the steady-state versions of the continuity equation, Bernoulli's equation, and the momentum equation. Thus, we have the following relations:

From ① to ②:

$$p_1 + \tfrac{1}{2}\rho u_1^2 = p_1 + \tfrac{1}{2}\rho U^2 = p_2 + \tfrac{1}{2}o u_2^2$$

or

$$p_2 = p_1 + \tfrac{1}{2}\rho U^2 - \tfrac{1}{2}\rho u_2^2. \tag{4.12-15}$$

From ② to ③:

$$u_2 = u_3, \quad p_3 = p_2 + \Delta p,$$

which becomes, with Equation 4.12-15,

$$p_3 = p_1 + \tfrac{1}{2}\rho U^2 - \tfrac{1}{2}\rho U_3^2 + \Delta p. \tag{4.12-16}$$

From ③ to ④:

$$p_3 + \tfrac{1}{2}\rho u_3^2 = p_4 + \tfrac{1}{2}\rho u_4^2, \quad p_4 = p_1.$$

---

* A concept due to the great 19th-century engineer Rankine (1865), who devised the first analysis for marine propellers.

Combining this with Equation 4.12-16, we have

$$p_1 + \tfrac{1}{2}\rho U^2 - \tfrac{1}{2}\rho u_3^2 + \Delta p + \tfrac{1}{2}\rho u_3^2 = p_1 + \tfrac{1}{2}\rho u_4^2,$$

which simplifies to

$$\Delta p = \tfrac{1}{2}\rho\left(u_4^2 - U^2\right). \tag{4.12-17}$$

We now apply the momentum equation to Equation 4.3-14 to the jet from ① to ④. Thus, with $u_1 = U$, $p_4 = p_1$,

$$F_{1,4} = \dot{m}\left(u_4 - U\right) + p_1\left(A_4 - A_1\right), \tag{4.12-18a}$$

which is a scalar equation since the flows involves only velocities in the x-direction. However, since the jet boundary pressure is constant at p_1, the axial component of the pressure force must also be given by

$$F_{1,4} = -p_1\left(A_1 - A_4\right) + \Delta pA, \tag{4.12-18b}$$

where the actuator disk area is $A_2 = A_3 \equiv A$ and where the force of the actuator disk on the fluid is ΔpA to the right. Equating these two expressions we find that

$$\Delta pA = \tfrac{1}{2}\rho A\left(u_4^2 - U^2\right) = \dot{m}\left(u_4 - U\right). \tag{4.12-19}$$

In fact, the force ΔpA is numerically equal to the thrust (which is defined to be positive to the left) of the flow on the propeller.

If we divide out the common factor in the second and third expressions, then the mass rate of flow is

$$\dot{m} = \tfrac{1}{2}\rho A\left(u_4 + U\right), \tag{4.12-20a}$$

and the volume rate of flow is

$$\dot{Q} \equiv \dot{m}/\rho = \tfrac{1}{2}\left(U + u_4\right)A, \tag{4.12-20b}$$

so that the mean flow speed V through the actuator disk is

$$V = u_2 = u_3 = \dot{Q}/A = \tfrac{1}{2}\left(U + u_4\right). \tag{4.12-21}$$

OVERALL EFFICIENCY*

The overall efficiency η_0 of a propulsive flow is defined as the ratio of the useful power output of the flow to the rate of energy input to it. The useful power output in this case must be calculated in the frame of an observer embedded in the fluid far upstream (for practical purposes an earthbound observer). Thus, the force on the propeller is $-\Delta pA$ and its velocity relative to the observer is $-U$. The useful power output is thus ΔpAU. Hence,

$$\eta_0 = \frac{\Delta pAU}{P}, \qquad (4.12\text{-}22)$$

where P is the power input to the shaft (and must be measured on a dynamometer in a laboratory).

PROPULSIVE EFFICIENCY

Propulsive efficiency is defined as the ratio of the useful power output to the total mechanical energy output resulting from the transformations of the propulsive flow. The latter energy output is the sum of the useful work and the kinetic energy increment of the exhaust, also measured with respect to the earthbound observer. The velocity increment of the exhaust is $u_4 - U$. Thus, the rate of mechanical energy output is $1/2\, \dot{m}\, (u_4 - U)^2$. Hence, the propulsive efficiency is

$$\eta_P = \frac{\Delta pAU}{\Delta pAU + \frac{1}{2}\dot{m}\left(u_4 - U\right)^2} = \frac{U}{U + \frac{1}{2}\dot{m}\left(u_4 - U\right)^2 \big/ \Delta pA}.$$

Substituting Equation 4.12-19 into the denominator leaves

$$\eta_P = \frac{2U}{U + u_4} = \frac{2}{1 + u_4/U}. \qquad (4.12\text{-}23)$$

Thus, in order to optimize the propulsive efficiency, the ratio of u_4/U, which exceeds unity, should be as close to unity as possible.

FLOW THROUGH A DUCTED FAN

Although the flow through a ducted fan,** Figure 4.12-6, has certain similarities to that of a propeller, there are certain crucial differences. A fan cannot be analyzed by a stream-tube approach, since on the upstream side the fan draws fluid from the entire region of available fluid at rest exterior to the duct, including the fluid

---

* See Foa (1960, chap. 13) for an extensive discussion of efficiencies in propulsive devices, which goes beyond this elementary example.
** A fan is a device that creates motion of a fluid while remaining itself at rest, whereas a propeller provides propulsive power to a vehicle that moves relative to the main body of a fluid.

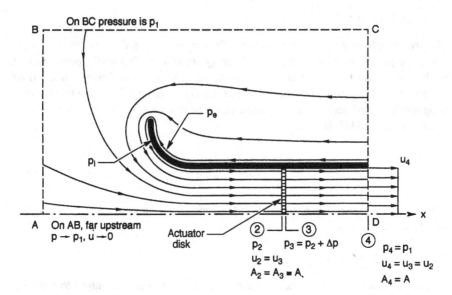

FIGURE 4.12-6 The ducted fan problem.

arbitrarily far downstream. For this reason it is more convenient to use the form of the momentum equation first stated as Equation 4.3-18. Thus, with reference to Figure 4.12-6,

$$\Sigma F_x = \dot{m}(V_2 - V_1),\qquad (4.12\text{-}24)$$

where ΣF_x is the sum of all forces exerted by the control-volume surfaces on the fluid it bounds. The right-hand side of Equation 4.12-24 is the net momentum outflow through the control volume.

For simplicity, only a meridional section through the duct axis is shown. Station AB is chosen sufficiently far upstream, and BC sufficiently far off-axis, that the flow speed is sensibly zero and the pressure is the ambient at p_1. The downstream control surface is chosen to coincide with the duct exit, although any parallel surface downstream will do equally well.

The action of the fan draws fluid into the duct where the pressure everywhere upstream of the disc is less than p_1. The fan restores the flow to pressure p_1 across the actuator disk without changing its velocity, thereby creating a uniform jet which, theoretically, extends infinitely far downstream.

Note the streamline on the upper side of the duct wall. It originates in the downstream reservoir, and the flow speed thereon (until it reaches the curved wall) is so low that the static pressure remains at p_1. However, to get around the inlet flange end, it accelerates. Correspondingly, the exterior pressure, denoted $p_e(x)$ starts to drop. On the inside of the inlet flange* the streamlines start to bunch up, and the pressure $p_i(x)$ continues to decrease accordingly. In other words, the flow is nonuniform in the inlet region. The x-component of the force caused by the pressure difference across an annulus of the duct of area dA (projected on a vertical plane) is $[p_i(x) - p_e(x)]dA$, which produces an integrated pressure force on the flange

$$F_p = \int_{x=0}^{x=L} [p_i(x) - p_e(x)] dA, \tag{4.12-25}$$

where at $x = L$, the duct cross-sectional area becomes constant at A. The origin is taken at the x-station of the outermost location of the curved inlet. The duct, of course, exerts a force of $- F_p$ on the fluid.

There is one more force within the duct, namely, the force that the actuator disk exerts on the fluid, which is

$$(p_3 - p_2)A = \Delta pA. \tag{4.12-26}$$

Therefore, since every particle originates in a region of zero velocity, we can put $V_1 = 0$ in Equation 4.12-24, $V_2 = u_4$, and substitute in Equations 4.12-25 and 4.12-26. Thus,

$$-F_p + \Delta pA = \dot{m}u_4 = \rho A u_4^2. \tag{4.12-27}$$

However, from Bernoulli's equation applied between ① and ④

$$p_1 = p_2 + \tfrac{1}{2}\rho u_2^2 = p_3 - \Delta p + \tfrac{1}{2}\rho u_4^2 = p_1 - \Delta p + \tfrac{1}{2}\rho u_4^2, \tag{4.12-28a}$$

since both the pressure and the flow speed must be constant from ③ to the control surface CD; i.e., $p_3 = p_4 = p_1$, $u_3 = u_4$. Therefore,

$$\Delta p = \tfrac{1}{2}\rho u_4^2, \tag{4.12-28b}$$

then, from Equation 4.12-27,

$$F_p = \Delta pA - \rho u_4^2 A = -\tfrac{1}{2}\rho u_4^2 A. \tag{4.12-29}$$

Therefore, we have determined the force imposed on the duct by the nonuniform pressure distribution (which is unknown) without having to integrate Equation 4.12-25.

Finally, the force exerted on the actuator disk by the fluid is

$$-\Delta pA = -\tfrac{1}{2}\rho u_4^2 A, \tag{4.12-30}$$

---

* Theoretically, on the flange edge the velocity must become infinite and the pressure negatively infinite. We will ignore this singularity since it does not affect the inlet flange, integrated pressure force.

also directed left. This force is transmitted from the actuator disk (the fan) through the structure that attaches the motor driving the fan to the duct itself. To maintain the duct in place requires an external reaction, such that

$$R_x = e_x \rho u_4^2 A,$$
(4.12-31)

which just balances the two forces computed in Equations 4.12-29 and 4.12-30.

4.13 THE BORDA–CARNOT RELATION FOR A SUDDEN ENLARGEMENT

It has been previously noted that in most cases it is difficult, if not impossible, to obtain an exact expression for the loss coefficient of a specific flow geometry. However, there is one case, that of the flow through a sudden enlargement, which admits of a theoretical solution. Along with that of Borda, its name is associated with Lazare N. M. Carnot (1753–1823), an eminent military engineer, one of the founders of the metric system, and father of the thermodynamicist Sadi Carnot (1796–1832).

THEORY

In the case of the flow through a sudden enlargement, Figure 4.13-1a, a jet is formed at the opening, bounded by eddies of recirculating fluid in a complex and irregular pattern. The region downstream is gradually filled until the flow occupies the entire channel. In the following application of the momentum equation, station ② is assumed to be taken sufficiently far downstream of the enlargement for this condition to be achieved. The area of the jet at station ① is taken to be that of the duct immediately upstream of the discontinuity.

On the surface of the control boundary extending from ① to ② there is a varying pressure distribution and there are viscous stresses exerted by the bounding, recirculating flow. The x-component of the momentum equation, Equation 4.3-13, gives the combined force due to both effects, i.e.,

$$F_{1,2} = e_x \left[\dot{m}(V_2 - V_1) + p_2 A_2 - p_1 A_1 \right].$$
(4.13-1)

Now the force exerted by the recirculating fluid on the main flow must be equal to the force exerted by the boundary of the duct on the recirculating fluid between ① and ②. If the viscous forces on the constant-area portion are assumed to be negligible, then $F_{1,2}$ is directly calculable in terms of the shoulder pressure, which is constant and equal to p_1. Thus,

$$F_{1,2} = e_x p_1 (A_2 - A_1).$$
(4.13-2)

Equating these two expressions produces

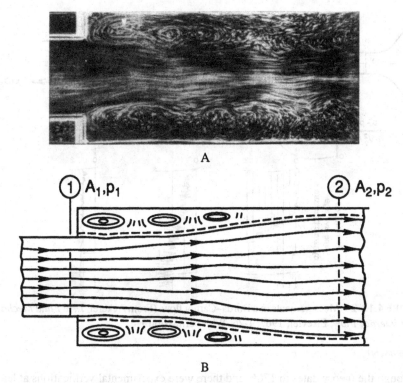

FIGURE 4.13-1   Borda–Carnot flow: (a) flow through a sudden enlargement (from Kaufman, *Fluid Mechanics*, McGraw-Hill, New York, 1963, with permission); (b) control volume.

$$\Delta p \equiv p_2 - p_1 = \rho V_2(V_1 - V_2). \tag{4.13-3}$$

From Equation 4.8-5, for horizontal flow, the loss coefficient becomes

$$k \equiv \left(p_1^o - p_2^o\right)/\tfrac{1}{2}\rho V_2^2. \tag{4.13-4}$$

Therefore, when Equation 4.13-3 and the definition of total pressure are combined with Equation 4.13-4

$$k = \left(V_1/V_2 - 1\right)^2 = \left(A_2/A_1 - 1\right)^2. \tag{4.13-5}$$

A dimensionless expression for the Borda–Carnot pressure rise can also be obtained from Equation 4.13-2:

$$\Delta p/\left(\rho V_2^2/2\right) = 2\left(A_2/A_1 - 1\right). \tag{4.13-6}$$

That these expressions are dependent solely upon the area ratio, and are independent of Reynolds number, is a consequence of the inviscid flow assumption.

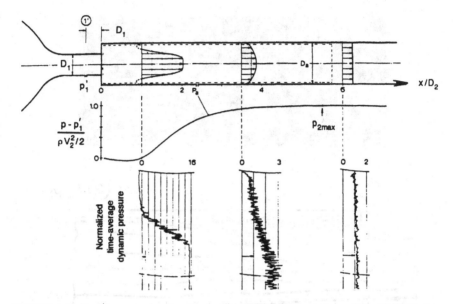

FIGURE 4.13-2 Measurements in a Borda–Carnot flow. (From Ackeret, J., in *Fluid Mechanics of Internal Flow*, Elsevier, 1967, 1-26. With permission.)

EXPERIMENT

Although the theory dates to 1766, and there were experimental verifications at least as early as the 19th century, the Borda–Carnot theory is still a subject of practical engineering interest. Ackeret (1967) has published some experimental measurements of H. Sprenger on the Borda–Carnot flow. For a circular tube with area ratio A_2/A_1 = 4.0, Figure 4.13-2 shows the pressure rise along the wall $\Delta p' = p(x) - p_1'$, normalized by the overall value Δp, given by Equation 4.13-6. Since the pressure on the downstream face of the shoulder is difficult to measure and, in any case, is slightly unsteady due to the eddies, Sprenger used, instead of p_1, the perssure p_1' one diameter D_1 upstream of the discontinuity. It is interesting to note the decrease in pressure immediately after the area change, which is associated with the reverse flow of the eddy and of the necking-down of the jet as it enters the large tube. This effect is also clearly shown in Figure 4.13-1a.

The illustrated velocity profiles show the reversed flow due to the eddies just after the area change. This profile eventually becomes nearly uniform, in this case about six diameters ($6D_2$) downstream. The lower plots are for the local dynamic pressure, made by traversing the station with a Pitot-static tube. The waviness indicates an unsteadiness associated with the turbulent eddies that are triggered by the jet as it emerges from the upstream section.

Figure 4.13-3 shows the normalized pressure rise along the wall of the larger tube for a number of different area ratios. It is remarkable that in every case the pressure ratio reaches 95 to 97% of the Borda–Carnot value within five diameters (D_2).

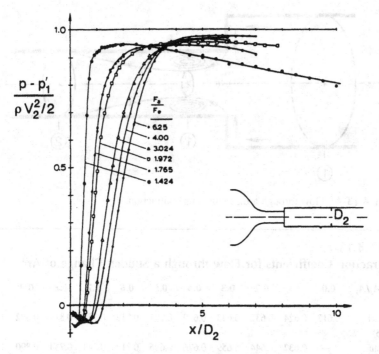

FIGURE 4.13-3 Borda–Carnot pressure variation for different area ratios. (From Ackeret, J., in *Fluid Mechanics of Internal Flow*, Elsevier, 1967, 1-26. With permission.)

APPLICATION

Ackeret cites an interesting and highly practical application of the Borda–Carnot principle. In automobile tunnels, it is essential to ventilate but there is little space in which to locate the fans. But, by locating a small, powerful, electric turbofan in the upper corner, the flow can be readily maintained because of the spreading effect on the downstream side of the fan for, as Sprenger also found, the main character of the flow is also maintained even when the upstream "channel" is off-center. As an extra fillip, by using reversible fans, the ventilation direction can be reversed along with the traffic flow, if desired.

ON FLOW THROUGH A CONTRACTION

For a sudden contraction, Figure 4.13-4, the flow separates at the corner, forms a *vena contracta*, and then fills the downstream channel much in the manner of the Borda–Carnot flow. If we treat the flow as essentially inviscid up to the minimum area, then the total pressure is constant from ① to ①'. From ①' to ②, we can use the Borda–Carnot results for horizontal flow, directly. Thus,

$$k \equiv \left(p_{1'}^o - p_2^o\right)\big/ \tfrac{1}{2}\rho V_2^2 = \left(A_{1'}/A_2 - 1\right)^2.$$

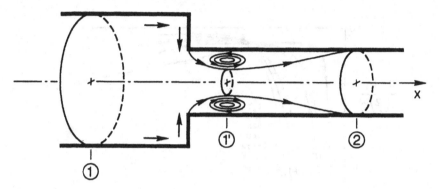

FIGURE 4.13-4   Flow through a discontinuous contraction.

TABLE 4.13-1
Contraction Coefficients for Flow through a Sudden Change of Area

C_c A_2/A_1	0.0	0.1	0.2	0.3	0.4	0.5	0.6	0.7	0.8	0.9	1.0
Weisbach (1855)	0.617	0.624	0.632	0.643	0.659	0.681	0.712	0.755	0.813	0.892	1.00
Freeman (1888)	—	0.632	0.644	0.659	0.676	0.696	0.717	0.744	0.784	0.890	1.00

But

$$A_{1'}/A_2 \equiv C_c ;$$

therefore,

$$k = \left(1/C_c - 1\right)^2 , \tag{4.13-7}$$

where $C_c = A_{1'}/A_2$. Unfortunately, there is no theoretical way of computing C_c, so that we must resort to experimental values. In Table 4.13-1 we list two independent sets of data made by the hydraulics engineers Julius Weisbach (1806–1870) and John R. Freeman (1855–1932), respectively. The deviations between the two sets of data are, fortunately, quite small, and there is no basis for preferring one set over the other.

4.14 THE BORDA MOUTHPIECE

THEORY

The Borda mouthpiece (the case previously designated as the short tube) provides an interesting illustration of how the one-dimensional equations can be applied to a flow that involves regions that are locally not one dimensional. For the purpose

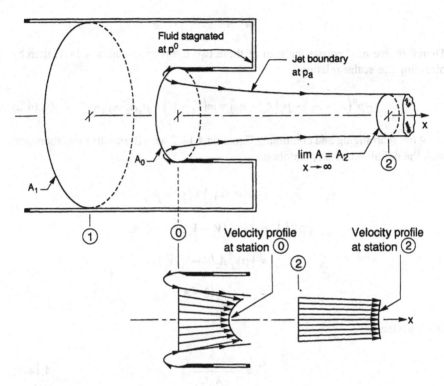

FIGURE 4.14-1   The Borda mouthpiece — short-tube analysis.

we initially ignore gravitational effects and assume viscous effects to be negligible. Instead of the infinite upstream region considered in Figure 4.8-2, we consider a liquid flow in a uniform channel, Figure 4.14-1, terminating in a vertical wall in which the mouthpiece is embedded. On the vertical wall the flow is essentially stagnated, and therefore the pressure is constant at p^0 (neglecting the hydrostatic term). On the mouthpiece interior surface the pressure decreases gradually from p^0 at the base, to the exit where it separates at pressure p_a, imposed by ambient conditions. The pressure remains constant at p_a over the entire jet surface.

Applying the momentum equation for the x-direction between ① and a downstream station ②, where the flow is assumed to have become uniform, the net force exerted on the boundaries of the fluid is

$$F_{1,2} = e_x \left[\dot{m}(V_2 - V_1) + p_2 A_2 - p_1 A_1 \right]. \tag{4.14-1}$$

But the x-component of the force exerted by the channel interior is

$$-p^0 (A_1 - A_0).$$

Also, the force exerted on the jet boundary by the surrounding atmosphere is

$$-p_a(A_0 - A_2).$$

Therefore, we may equate the sum of these two terms to Equation 4.14-1, thereby obtaining the scalar relation

$$-p^0(A_1 - A_0) - p_a(A_0 - A_2) = \dot{m}(V_2 - V_1) + p_a A_2 - p_1 A_1. \qquad (4.14\text{-}2)$$

After rearranging and combining Equation 4.14-2 with Bernoulli's equation, and with the definition of $\dot{m}$, we obtain

$$(p^0 - p_a)A_0 = \dot{m}(V_2 - V_1) + (p^o - p_1)A_1,$$

$$\tfrac{1}{2}\rho V_2^2 A_0 = \rho A_2 V_2 (V_2 - V_1) + \tfrac{1}{2}\rho V_1^2 A_1,$$

$$= \tfrac{1}{2}\rho V_2^2 A_2 (2 - V_1/V_2),$$

$$= \tfrac{1}{2}\rho V_2^2 A_2 (2 - A_2/A_1),$$

or, equivalently,

$$\frac{A_2}{A_0} = \frac{1}{2 - A_2/A_1}. \qquad (4.14\text{-}3)$$

To secure the classical expression for the infinite upstream reservoir, we let $A_2/A_1 \to 0$ and obtain $A_2/A_0 = C_c = {}^1/_2$. We note that an analogous derivation is not possible for the sharp-edged orifice, Figure 4.9-5, since the pressure on the upstream side of the orifice plate is not constant and cannot be determined by any elementary theory.

DISCUSSION

According to two-dimensional theory (there is no exact axisymmetric solution), A_2 is attained only asymptotically at infinity. Practically speaking, however, almost all of the jet contraction occurs within one or two widths (D_0) downstream of the exit. The jet shape does not depend on the flow speed, and changes in $p_0 - p_a$ merely alter the mass flow rate. The nonuniform nature of the flow is indicated by the illustrations of the jet velocity profile near the mouthpiece, and far downstream, in Figure 4.14-1. A straight, horizontal jet can be realized only in a zero-gravity environment. If gravity is present, the jet curves downward, the mean flow speed continually increases, and the area approaches zero asymptotically.

It should also be noted that a well-defined jet is obtained from the mouthpiece aperture (actually from any kind of aperture) only for the case of a liquid jet flowing into a gas-filled region. This suggests an additional necessary criterion for all free-jet problems, specifically, that $\rho/\rho_a \gg 1$. In the case of water flowing into standard air, $\rho/\rho_a \approx 800$.

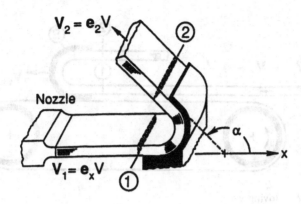

FIGURE 4.15-1 Fixed vane.

4.15 FORCES ON VANES, POWER PRODUCTION

FIXED VANE

We consider a steady, incompressible flow from a nozzle such that, at the outlet, conditions are assumed known at p_1, ρ, $V_1 = e_x V$ and $A_1 = e_x A$. The flow, Figure 4.15-1, is directed to a vane that turns it through the angle α measured with respect to the x-axis.

For the present purpose viscous effects and changes of height are assumed negligible. Since Bernoulli's equation applies and since the pressure is fixed by the environment at $p_1 = p_a$, the flow speed is the same at every point during the turn. The reaction force required to hold the vane in place is simply

$$R = \dot{m}(V_2 - V_1) = \dot{m}V(e_2 - e_x),\qquad(4.15\text{-}1)$$

where e_2 is the direction of the flow at the vane exit.

One of the peculiarities inherent in the one-dimensional approach is evinced by Equation 4.15-1. If, as the boundary condition states, the jet pressure is everywhere constant at p_a during the turning process, what produces the force on the vane? The answer is that the jet pressure is atmospheric only on its free boundary and that the pressure of the jet in contact with the vane is higher, as is required by the centrifugal force equation (Equation 3.12-12) previously mentioned. In spite of this apparent deficiency, Equation 4.15-1 predicts the integrated reaction force correctly. This is another demonstration of the great versatility of the momentum equation and the one-dimensional approach.

In the simple configuration of Figure 4.15-1, it is clear that the maximum reaction force must be furnished when the jet turns through $\alpha = 180°$, in which case $e_2 = -e_x$ and

$$R = -e_x 2\dot{m}V.\qquad(4.15\text{-}2)$$

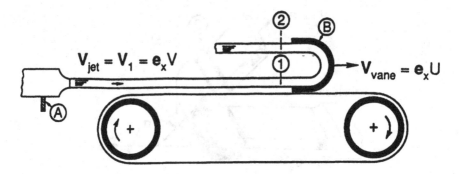

FIGURE 4.15-2   Moving vane.

Since the vane is fixed with respect to the observer on the nozzle, the reaction force is not displaced; hence, no work is done on, or by, the jet.

MOVING VANE

We next consider the rather artificial and idealized setup of Figure 4.15-2 in which the nozzle remains fixed with jet speed $V_1 = e_x V$, but where a vane with a 180° turning angle is now attached to a belt that allows it to move uniformly to the right at speed U with respect to the nozzle. If there is a reaction force furnished by the belt, its point of application is moving with respect to observer Ⓐ on the nozzle; hence, energy is being extracted from, or put into, the flow by the belt–pulley system.

The analysis of the fixed vane does not apply directly since, for a moving belt, the flow is unsteady in the frame of Ⓐ. This can be visualized if the "thought experiment" is conducted by fixing an instrument, say, a Pitot tube, with respect to the nozzle, in the space ahead of the vane. When the vane "overtakes" the instrument, the reading will increase from zero to some value and then drop back to zero after the vane has gone by. It would be possible to employ the unsteady momentum equation to compute R, but it is simpler if we note that in the frame* of observer Ⓑ, moving with the vane, the flow is steady. In the frame of observer Ⓑ the jet velocity components, denoted by primes, are

$$V_1' = V_1 - e_x U = e_x (V - U), \quad V_2' = -V_1' = -e_x (V - U).$$

Also, from the momentum equation, the reaction force, for observer Ⓑ, is

$$R = \dot{m}' (V_2' - V_1') = -e_x 2\dot{m}' (V - U), \tag{4.15-3a}$$

where, in the frame of Ⓑ, the mass rate of flow is $\dot{m}' = \rho A (V - U)$ and, therefore,

---

* A flow that appears to be steady over certain regions of space and intervals of time in a particular frame of reference is said to be *crypto-steady*. See a further note in Section 3.19.

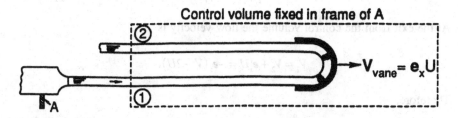

FIGURE 4.15-3   Energy considerations for a moving vane.

$$R = -e_x 2\rho A (V - U)^2. \tag{4.15-3b}$$

Now the power P developed by the moving vane, computed *in the frame of the observer on the nozzle*, is the product of the force $-R$ that the jet exerts on the belt, and the belt velocity relative to the observer on the nozzle, i.e.,

$$P = (-R) \cdot e_x U = 2\rho A U (V - U)^2. \tag{4.15-4}$$

Obviously, there is no power extracted from the jet when the vane speed is $U = 0$, or when it moves at the same speed as the jet, i.e., when $U = V$. Since we expect the power to be a continuous function of the vane speed U, there ought to be an extreme value, presumably a maximum, between these two cases. By the standard methods of calculus, this maximum occurs when $U = V/3$, from which

$$P_{max} = \frac{8}{27} \rho A V^3. \tag{4.15-5}$$

Thus, the maximum power of this configuration does not occur when the force on the vane is a maximum, nor is it theoretically possible to extract all the kinetic energy (in the frame of the nozzle) from the jet since, for maximum power, $V_2 = -e_x V/3$. The reason for this surprising result is explained in the next section.

THE MOVING VANE CONSIDERED FROM AN ENERGY VIEWPOINT

The preceding analysis can be approached from another viewpoint that illuminates the process by which power is generated by extraction from a fluid jet. In Figure 4.15-3, in the frame of observer Ⓐ, there is drawn a control volume which encloses the moving vane.

From Equation 4.3-6, for a particle in inviscid, adiabatic flow $\Delta e = 0$. Further, for an incompressible fluid, and with $p_1 = p_2$, the increment of total enthalpy is $\Delta h^0 = h_2^0 - h_1^0 = (V_2^2 - V_1^2)/2$. That is, under the applicable restrictions, the only energy transformations possible in the present case are changes of kinetic energy, i.e., by Equation 4.3-8, $\Delta w = \Delta h^0$. By denoting the rate of kinetic energy flow through a control surface by $\dot{K}$,

$$\dot{K}_1 = \tfrac{1}{2}\dot{m}_1 V^2 = \tfrac{1}{2}\rho A V^3. \tag{4.15-6}$$

At the exit from the control volume the flow velocity is

$$V_2 = V_2' + e_x U = -e_x(V - 2U).$$

Therefore,

$$\dot{m}_2 = \rho A(V - 2U),$$

and the kinetic energy outflow is

$$\dot{K}_2 = \tfrac{1}{2}\dot{m}_2(V - 2U)^2 = \tfrac{1}{2}\rho A(V - 2U)^3. \tag{4.15-7}$$

It might be thought that to maximize the energy extracted from the flow the belt speed U should be adjusted so that the difference

$$\dot{K}_1 - \dot{K}_2 = \tfrac{1}{2}\rho A\left[V^3 - (V - 2U)^3\right] \tag{4.15-8}$$

is maximized, a situation that occurs when the vane speed is $U = V/2$. But this reasoning is erroneous since it assumes that all of the energy difference is extracted by the vane. However, because the vane is moving, the portions of the jet segments within the control volume ahead of the vane grow ever longer, which means that the associated kinetic energy of the jet segments (which both increase in length at the rate of U) increases at the rate

$$\tfrac{1}{2}\rho A U\left[V^2 + (V - 2U)^2\right].$$

Therefore, the kinetic energy extracted by the vane is

$$\dot{K}_{\text{vane}} = \tfrac{1}{2}\rho A\left[V^3 - (V - 2U)^3 - UV^2 - U(V - 2U)^2\right]. \tag{4.15-9}$$

Differentiating Equation 4.15-9 with respect to U and setting the result equal to zero leads to two values at which $\dot{K}_{\text{vane}}$ is extreme: $U = V$, which is discarded since this is a minimum of zero, and $U = V/3$, a result already obtained by another means. Thus, the theoretical maximum power extractable from a jet is $(8/27)/(1/2) = 0.596$ of the rate of kinetic energy outflow from the nozzle. We see that this fraction is exactly the overall efficiency η_0 defined in Equation 4.12-22.

MULTIVANE DEVICES

The configuration of a belt and vane, or vanes, is not a practicable mechanical design. Rather than an endless belt, the Pelton wheel, Figure 4.15-4 directs the jet at a series of vanes mounted on a rotating wheel. Since a 180° turning angle does not permit satisfactory clearance for the deflected jet, the inlet angle is offset from the tangent to the wheel and the deflected flow is removed through a space in the wheel housing. To eliminate side thrust, a split vane is employed. Construction details of a typical wheel can be seen in Figure 4.15-4b.

4.16 THE ROCKET SLED WATER BRAKE

DESCRIPTION

The highest speed ever attained by a ground-supported vehicle was achieved on a rocket-propelled sled. In 1982, at Holloman Air Force Base in New Mexico, a rocket sled reached a top speed of 8972 ft/sec. Anyone who has seen a color motion picture of one of these sleds in action, fire spewing from its rocket exhaust, and brilliant cascades of sparks trailing from its rail contact points, will agree that it is a *Chariot of Fire*.

The rocket propulsion unit is supported on a rugged, welded-steel frame by *slippers*, metal clamplike devices that surround heavy steel rails, very accurately aligned, which hold the sled on the desired trajectory. In the deceleration phase, braking is achieved by scooping water from a trough under the sled and ejecting it forward. This action is capable of generating fantastic braking forces by fairly elementary means.

In the following we shall analyze a simplified model of a water brake design, by the quasi-steady momentum equation, which will then be compared with the full, nonsteady solution to bring out what a quasi-steady approach fails to include. We shall also point out the differences between the model employed and an actual sled construction.

QUASI-STEADY SOLUTION

For the current purpose we treat the model of Figure 4.16-1. The brake is supposed to consist of a steel, constant-area tube of semicircular shape terminating in a pair of straight pipes and mounted near the rocket nose. The sled has a mass M, and its speed with respect to the ground is u. At the instant the brake is actuated, its speed is u_i.

From Newton's second law the equation for the sled motion is

$$M \frac{du}{dt} = \Sigma F, \qquad (4.16\text{-}1)$$

where ΣF is the sum of the horizontal forces acting on the sled. This includes the aerodynamic drag, the rail friction force, and the water brake force. In the braking phase, of course, the rocket force is assumed to be zero. The equation of the sled motion for the vertical direction is of no interest as long as the sled stays attached to the rails.

Since for a ground-fixed observer ① the flow through the U-tube is unsteady, the analysis will be undertaken in the frame of ②, who rides with the sled. Although

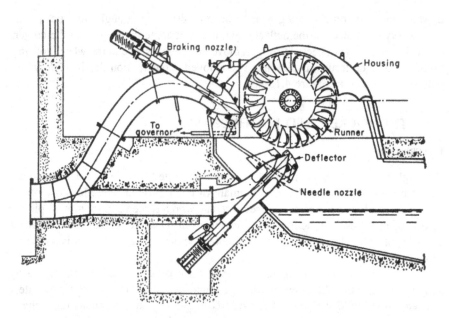

(a) Pelton wheel schematic with a dual supply, from Baumeister (1961).

(b) Finish grinding on a Pelton wheel, courtesy of Allis-Chelmers, Inc.

FIGURE 4.15-4   The Pelton wheel.

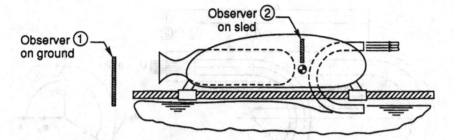

FIGURE 4.16-1 The water brake model.

flow conditions vary with time even for ②, they vary "slowly," and thus the quasi-steady relations can be employed. By taking the ambient pressure* to be $p_a = p_1 = p_2$, then the braking force is obtained from Equation 4.3-15:

$$F_b = e_x \cdot (-R), \qquad (4.16\text{-}2)$$

with

$$R = \dot{m}(V_2 - V_1) + e_z g m_{1,2}, \qquad (4.16\text{-}3)$$

where $\dot{m}$, V_2, V_1 are measured in the frame of ②, and where $m_{1,2}$ is the mass of water within the brake duct. Thus, since $A_2 = -A_1 = e_x A$, $V_2 = -V_1 = e_x u$, $\dot{m} = \rho A u$, then

$$R = e_x \left(2\rho A u^2 \right) + e_z g m_{1,2}. \qquad (4.16\text{-}4)$$

If included, the aerodynamic drag would be proportional to u_1^2, whereas the rail friction force would be proportional to the vertical force with which the sled presses down on the rails (i.e., proportional to the sum of the sled weight plus any vertical components of the aerodynamic and braking forces). By neglecting braking contributions other than the water brake force,

$$M \frac{du}{dt} = -2\rho A u^2, \qquad (4.16\text{-}5)$$

which can readily be integrated, by separation of variables, to yield

$$\frac{u}{u_i} = \frac{1}{1 + \dfrac{2\rho A u_i}{M} t}, \qquad (4.16\text{-}6)$$

---

* In fact, the calculation of the aerodynamic exterior force is beyond the scope of the current analysis.

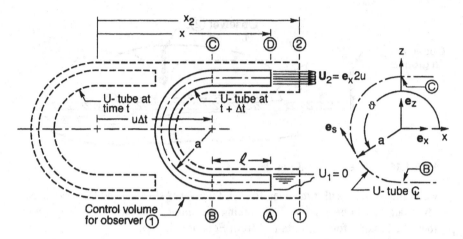

FIGURE 4.16-2 Control volume for nonsteady analysis of water brake.

where u_i is the sled horizontal velocity upon initiation of braking. Also, since for the sled $u = dx/dt$, the sled position can be obtained as a function of time by a second integration:

$$x = \frac{M}{2\rho A} \ln\left[1 + \frac{2\rho A u_i}{M} t\right]. \qquad (4.16\text{-}7)$$

According to Equations 4.16-6 and 4.16-7, the time and distance to bring the sled to a halt are both infinite. This anomaly occurs because the analysis has neglected the rail friction force, which actually dominates the braking process at low sled speeds.

THE UNSTEADY SOLUTION

One of the problems with the preceding quasi-steady analysis is that the observer is decelerating with respect to the ground and is therefore not an inertial observer. Consequently, use of the steady-state momentum equation (Equation 4.3-14), can be expected to contribute some error. We will therefore rework the same problem in the frame of ①, Figure 4.16-2, using the full, nonsteady momentum equation. To emphasize that we are in a different frame of reference we designate flow velocities in the frame of ① as U. Velocities in the two frames are related by $U = V + e_x u$, i.e., $U_1 = V_1 + e_x u = 0$, $U_2 = V_2 + e_x u = e_x 2u$, etc.

Therefore, from Equations 3.20-3 and 3.20-5, for a mass-conserving flow in a uniform gravitational field,

$$R = \frac{d}{dt}\int_1^2 \rho A U \partial s + e_z \int_1^2 \rho A g \partial s + \dot{m}_2 U_2 - \dot{m}_1 U_1. \qquad (4.16\text{-}8)$$

The pressure term drops out as before. It takes some stretching of the mind, however, to accept the fact* that $\dot{m}_1 = 0$, $\dot{m}_2 = 2\rho Au$. Thus, the sum of the last three terms in Equation 4.16-8 is

$$e_z g m_{1,2} + e_x 4\rho A U^2. \tag{4.16-9}$$

To evaluate the first integral in Equation 4.16-8, we note that $U = V + e_x u = e_s u + e_x u$, where $e_s u$ is the flow velocity within the U-tube in the frame of ① in natural coordinates, and where $e_s = e_s(s)$ is directed along the duct axis. Thus, choosing the control volume indicated, where the integration is performed holding t constant and where, by definition, ∂s is everywhere positive,

$$\int_1^2 \rho A U \partial s = e_x \rho A u \int_1^2 \partial s + \rho A u \int_1^2 e_s \partial s. \tag{4.16-10}$$

Now, in the first integral on the right of Equation 4.16-10,

$$\int_1^2 \partial s = \int_1^A \partial s + \int_A^D \partial s + \int_D^2 \partial s = L + 2(x_2 - x),$$

where $L \equiv \pi a + 2\ell$ is the axial length of the duct. In this expression, also, $x = x(t)$ is the coordinate** of the sled measured from observer ① at the instant during which the integration is performed, and x_2 is the coordinate (a constant) of the control boundary with respect to observer ①. Therefore,

$$e_x \rho A u \int_1^2 \partial s = e_x \rho A u \left[L + 2(x_2 - x) \right]. \tag{4.16-11}$$

In the second integral on the right of Equation 4.16-10,

$$\int_1^2 e_s \partial s = \int_1^B (-e_x) \partial s + \int_B^C (e_z \sin \vartheta - e_x \cos \vartheta) a \partial \vartheta + \int_C^2 e_x \partial s.$$

Since the sum of the first and the third integrals on the right is zero,

$$\int_1^2 e_s \partial s = \int_{3\pi/2}^{\pi/2} (e_z \sin \vartheta - e_x \cos \vartheta) a \partial \vartheta = -e_x 2a$$

---

* This seemingly incongruous result is a consequence of the fact that the control volume selected continuously experiences a loss of mass in the jet expelled from the U-tube exit.

** It will be seen that the physical point chosen to represent the sled coordinate is a matter of indifference. In this case, for convenience, the exit station of the duct was selected.

and

$$\rho A u \int_1^2 e_s \partial s = -e_x 2\rho A u a. \qquad (4.16\text{-}12)$$

In evaluating the preceding integrals, use has been made of the fact that $e_s = \pm e_x$ on the lower, and upper straight portions, respectively and that $e_s = e_z \sin \vartheta - e_x \cos \vartheta$, $\partial s = a \partial \vartheta$, on the curved portion of the duct. Therefore, combining Equations 4.16-10 through 4.16-12,

$$\frac{d}{dt} \int_1^2 \rho A V \partial s = \frac{d}{dt} \left\{ e_x \rho A u \left[L + 2(x_2 - x) - e_x 2\rho a A u \right] \right\},$$

$$= e_x \rho A \frac{d}{dt} \left\{ u \left[(\pi - 2)a + 2\ell + 2(x_2 - x) \right] \right\}, \qquad (4.16\text{-}13)$$

$$= e_x \rho A \left\{ \frac{du}{dt} \left[(\pi - 2)A + 2\ell + 2(x_2 - x) \right] - 2u \frac{dx}{dt} \right\}.$$

Finally, we note that in the frame of ① $dx/dt = u$ and, also, since we are interested in the force on the fluid *only* within the U-tube, we choose the position of the duct so that its exit coincides with boundary ② of the control volume, i.e., so that $x = x_2$. Therefore, Equation 4.16-13 becomes

$$\frac{d}{dt} \int_1^2 \rho A U \partial s = -e_x 2\rho A u^2 + e_x \rho A \left[(\pi - 2)a + 2\ell \right] \frac{du}{dt}. \qquad (4.16\text{-}14)$$

Substitution of Equations 4.16-9 and 4.16-14 into Equation 4.16-8 produces the final expression for the force exerted by the sled on the duct:

$$R = e_x 2\rho A u^2 + e_x m_e \frac{du}{dt} + e_z g m_{1,2}, \qquad (4.16\text{-}15a)$$

where it is convenient to define a term

$$m_e \equiv \rho A \left[(\pi - 2)a + 2\ell \right], \qquad (4.16\text{-}15b)$$

which is evidently the mass of a cylinder of water of cross-sectional area A and length $(\pi - 2)a + 2\ell$. Substitution of Equation 4.16-15b in Equation 4.16-1 gives the modified expression for the sled motion:

$$(M + m_e) \frac{du}{dt} = -2\rho A u^2. \qquad (4.16\text{-}16)$$

FIGURE 4.16-3   Sled using brake model of Figure 4.16-1.

By comparing Equation 4.16-16 with Equation 4.16-5 we see that, in the current problem, the effect of using quasi-steady theory is to neglect the portion of m_e of the "effective" mass being decelerated. Since m_e is small compared with the sled mass, it appears that the accuracy of the quasi-steady approach is entirely adequate in this instance, not to mention its greater simplicity.

Discussion

The model of Figure 4.16-1 corresponds to one actual sled configuration, shown in the photograph of Figure 4.16-3, which was tested at Holloman Air Force Base in 1965. A pair of U-tube brakes was mounted symmetrically with respect to the vertical plane of symmetry, to avoid asymmetrical braking loads.

Figure 4.16-4 shows another configuration employing a 180° change of the water flow direction. The piping is on centerline but is partially concealed by the pair of rocket engines. Also visible in the same photograph are the slippers clamped around the metal track. The wedgelike surfaces just above the slippers are intended to reduce the aerodynamic drag of the horizontal steel frame structure.

Figure 4.16-5 shows a configuration mounted on a sled that is powered by three rockets. Space limitations require that the flow deflection be limited to 90°, thereby reducing the total braking to about one half the maximum possible for the flow rate. The outlet ducting is split about the central welded steel support.

Rasmussen (1976) has given a very interesting survey of the general problems of designing brakes to meet a variety of objectives, for example, providing prepro-

FIGURE 4.16-4   Sled employing a 180° total deflection angle for braking.

FIGURE 4.16-5   Sled that employs a 90° total deflection angle for braking.

grammed deceleration time histories as are required to test inertial guidance equipment.

He points out that entry speeds (to the brake inlet) as high as 4000 ft/sec have been attained. This corresponds to an entry Reynolds number of 8.2×10^7, for water at 60°F, and an inlet diameter of 3 in. There is no other known technology that involves such high water speeds, so that there is no body of experimental data on duct liquid flows at these Reynolds numbers to which the designer has access.

From a practical design viewpoint, to avoid the appearance of cavitation,* flow separation, buildup of viscous effects, etc., experience has shown that it is essential to provide an exit area of the brake duct that is greater than the inlet area.

The water brake is the most effective way of achieving deceleration at high speeds. Using an example cited by Rasmussen, suppose that it is necessary to achieve $100g$ continuous deceleration for an interval of 1 sec. Since the maximum rate of deceleration is proportional to the square of the instantaneous sled speed, there is some minimum speed required, say, 750 ft/sec, below which the available braking force is insufficient to furnish the required deceleration. Thus, neglecting air drag, the brake must begin picking up water at the initial speed of $3220 + 750 = 3970$ ft/sec. The track length required for this part of the run, alone, is 2360 ft.** To achieve uniform deceleration with a fixed-geometry duct, the brake inlet must be only partly submerged initially and, as the run progresses, the scoop depth of the inlet relative to the water level must be gradually increased according to a precise schedule.

Deceleration levels of up to $200g$ have been attained (as of 1976) with the expectation of $300g$ in the period since then. Many other details on the general problem of braking are given in the Rasmussen paper.

4.17 THE WATER SPRINKLER PROBLEM

We have not previously considered any situation involving an observer (noninertial) whose frame rotates at a constant angular velocity with respect to an inertial observer. There are á number of important flows that are most conveniently analyzed in the frame of such an observer, e.g., turbines and compressors. However, the elementary case of a garden-variety water sprinkler provides a simpler though still instructive example. Not only does it involve a noninertial observer, but the theoretical results can be used to illuminate the principle of the centrifugal compressor.

THE MODEL

The model schematic is shown in Figure 4.17-1. A sprinkler arm, free to rotate about a vertical axis, is supported by a stem through which the water moves upward until it branches at the inlet, station ①. To observer ①, who is motionless with respect to the stem, the water is supplied at total pressure p_i^0, fixed by a pump or supply line. Due to the reaction generated in the curved sprinkler arm, the arm rotates at a constant angular velocity $\Omega = e_z \Omega$.

In each arm the vane turns the flow through an angle α_e with respect to a radial line between the axis and the exit station e. The radial distance of the exit from the axis is denoted as a.

---

* *Cavitation* involves the formation of vapor bubbles in a liquid flow, which may occur when the static pressure becomes equal to, or less than, the vapor pressure. Then, in regions of higher pressure, these bubbles may collapse very suddenly. The ensuing nonsteady pressure forces associated with collapse are capable of creating extremely high stresses in the surrounding walls, resulting in serious erosion. Designers of flow conduits must take extreme precautions to avoid conditions that result in cavitation.

** The longest track at Holloman Air Force Base (as of 1974) is 50,788 ft. See Holloman Air Force Base (1974) for a survey report on its facilities.

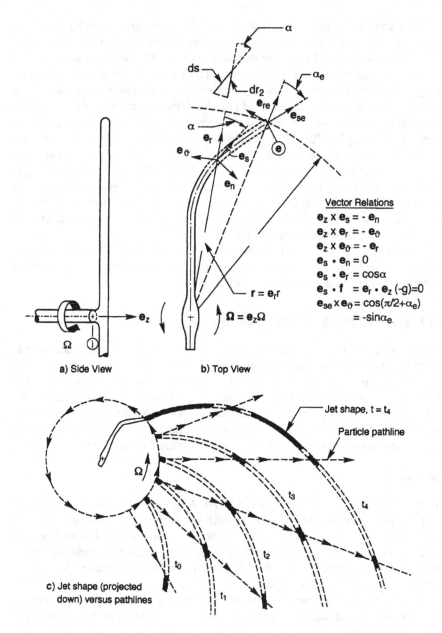

FIGURE 4.17-1 The water-sprinkler problem: (a) side view; (b) top view; (c) jet shape C projected down vs. pathlines.

Since observer ① sees a rotating arm, the flow appears to ① to be an unsteady, cyclical motion. To avoid the complexities of a nonsteady analysis it is preferable to shift to ②, an observer who rotates with the sprinkler arm. If Ω = constant, then we expect that ② will observe a steady flow. However, relations — such as Bernoulli's equation — derived from the form of the dynamic equation valid for an inertial observer (i.e., Equation 3.12-9) are not applicable in the frame of ②. Thus, we must instead introduce Equation 3.3-4, which relates the acceleration of a particle, as observed by a noninertial observer, to that of an inertial observer, and combine with Equation 3.12-9b to obtain the appropriate generalization of the dynamic equation.

In what follows the subscripts 1 and 2 denote quantities in the frame of ① and ②, respectively. Subscripts i and e denote the inlet and exit stations.

THEORY

From Equation 3.3-4, for the case $r_0 = 0$, $\Omega = 0$, when applied to a fluid particle,

$$\frac{DV_1}{Dt} = \frac{DV_2}{Dt} + 2\Omega \times V_2 + \Omega \times (\Omega \times r_2). \qquad (4.17\text{-}1)$$

Now, introducing natural coordinates at an arbitrary point $r_2 = e_r r_2$ along the arm, with $V_2 = e_s V_2$, $\Omega = e_z \Omega$ and using the vector relations tabulated on Figure 4.17-1, Equation 4.17-1 becomes

$$\frac{DV_1}{Dt} = \frac{DV_2}{Dt} + e_n 2\Omega V_2 - e_r \Omega^2 r_2. \qquad (4.17\text{-}2)$$

The second and third terms on the right of Equation 4.17-2* are the Coriolis, and centripetal accelerations, respectively.

Equation 4.17-2 is now introduced into Equation 3.12-9b. Noting that $A = e_s A$, we divide both sides by ρA and rearrange slightly. Thus,

$$\frac{DV_2}{Dt} = -e_s \frac{1}{\rho} \frac{\partial p}{\partial s} - e_n 2\Omega V_2 + e_r \Omega^2 r_2. \qquad (4.17\text{-}3)$$

Further, since $DV_2/Dt = e_s DV_2/Dt + e_n V_2^2/R$, we take the dot product of e_s with Equation 4.17-3 to obtain the dynamic equation along the path of the particle. If the flow is steady** and inviscid and if the sprinkler arm is everywhere horizontal, it reduces to

$$\frac{d}{ds}\left(\frac{1}{2}V_2^2\right) = -\frac{1}{\rho}\frac{dp}{ds} + \Omega^2 r_2 e_s \cdot e_r. \qquad (4.17\text{-}4)$$

---

* Equation 4.17-2 can be obtained from Equation 3.3-6 if, in the latter, ordinary derivatives are replaced by the corresponding substantial derivatives, and if we note that $e_{r2} = e_r$, $e_{\theta2} = e_n$, $u_{r2} = V_2$, and in the sprinkler problem, $u_{\theta2} = 0$.
** This is an excellent example of how a change of reference transforms the mathematical problem of solving a set of inherently nonsteady partial-differential equations to one which involves only the solution of an ordinary differential equation.

Note that Equation 4.17-4 is similar to the differential form of Bernoulli's equation except that an additional term appears on the right.

Now, from the geometry, $dse_s \cdot e_r = ds \cos \alpha = dr_2$; Equation 4.17-4 can therefore be rewritten, for constant-density flow, as

$$d\left(\frac{p}{\rho} + \frac{1}{2} V_2^2 - \frac{1}{2}\Omega^2 r_2^2\right) = 0. \qquad (4.17\text{-}5)$$

If Equation 4.17-5 is integrated from station ①, where $r_2 = 0$, $V_{2i} = V_i$, $p = p_i$, then

$$\frac{p}{\rho} + \frac{1}{2} V_2^2 - \frac{1}{2} \Omega^2 r_2^2 = \frac{p_i}{\rho} + \frac{1}{2} V_i^2 = \frac{p_i^0}{\rho}. \qquad (4.17\text{-}6)$$

Thus, since $p \sim r_2^2$, we have proved that a sprinkler acts like a centrifugal compressor. Equivalently, since the sum of the first two terms on the left of Equation 4.17-6 represents p_2^0/ρ, we have

$$p_2^0 = p_i^0 + \tfrac{1}{2}\rho\Omega^2 r_2^2, \qquad (4.17\text{-}7)$$

which states that at any point in the sprinkler arm the total pressure increment varies as the square of the local circumferential speed Ωr_2.

CALCULATION OF THE ROTATIONAL SPEED

A simple procedure allows the computation of the rotational speed Ω. If the mechanical bearing between the stem and the arm can be treated as frictionless, then the resisting torque of the mechanism is zero. Consequently, in ejecting the water flow at the exit station, no net thrust can be developed by the nozzle, and the net torque on the sprinkler arm must be zero. Otherwise, the sprinkler arm would accelerate until the condition of zero torque were attained. Thus, the velocity vector of the flow, *in the frame of* ①, must be radial outward.

Referring to Figure 4.17-1c, the velocity at the exit observed by ① consists of two contributions. The first is $V_2 = V_{2e}e_{se}$, which is measured with respect to the moving sprinkler arm. The second is $\Omega a e_{\vartheta e}$, which is the peripheral rotational velocity of the sprinkler arm at the exit.

In order that the flow be radially outward in the frame of ① we must have

$$u_{\vartheta 1e} \equiv 0 = \Omega a - V_{2e} \sin \alpha_e \qquad (4.17\text{-}8a)$$

or

$$\Omega = (V_{2e}/a) \sin \alpha_e. \qquad (4.17\text{-}8b)$$

When Equation 4.17-8b is combined with Equation 4.17-6 at the exit station ($r_2 = a$), where the pressure is atmospheric (p_a), the following expressions for the jet exit speed (relative to the moving vane) and for the sprinkler rotational speed, are obtained:

$$V_{2e} = \sec \alpha_e \left[2\left(p_i^0 - p_a\right)/\rho\right]^{1/2}, \qquad (4.17\text{-}9a)$$

$$\Omega a = \tan \alpha_e \left[2\left(p_i - p_a\right)/\rho\right]^{1/2}. \qquad (4.17\text{-}9b)$$

These indicate that both the jet exit speed with respect to the sprinkler arm and the rotational speed of the arm approach infinity as $\alpha_e \to \pi/2$, other quantities remaining the same. In practice this cannot be achieved because of the various frictional effects: of the internal flow, in the mechanism bearings, and the air resistance, all of which tend to reduce both V_{2e} and Ω. Furthermore, the attainable rotational speed is also limited by the lowest inlet pressure (highest inlet speed) at which the fluid remains liquid.

Discussion

To observer ① the instantaneous jet shape is a spiral. However, the motion of a fluid particle, after exiting from the sprinkler nozzle, is much simpler. In polar coordinates the radial velocity component is

$$u_{r1} = V_{2e} \cos \alpha_e = \left[2\left(p_i - p_a\right)/\rho\right]^{1/2}. \qquad (4.17\text{-}10)$$

The peripheral velocity is zero by imposition of Equation 4.17-8a. Since there are no horizontal forces acting on the particle (neglecting air resistance), its path, when projected on a horizontal plane, must be a straight, radial line; see Figure 4.17-1c.

However, if the sprinkler-bearing friction, or the wind resistance on the sprinkler arm, were not negligible, as is expected, then the vane would experience a force due to the flow deflection in the bend (see Section 4.12), corresponding to a torque on the vane to overcome that due to the wind or bearing resistance. In this circumstance Ω is reduced below the value predicted in Equation 4.17-8b. Furthermore, although a particle in the jet would still move away in a straight line (as projected), the line would no longer be radial, but would be tilted in the direction of the vane outlet. In the limiting case of "infinite" resistance to rotation of the sprinkler arm, $\Omega \to 0$ and the flow would be ejected from the vane at the fixed angle α_e to a radial line. This can be verified by any reader with such a sprinkler.

In the book, *Surely You're Joking Mr. Feynman* (Feynman et al., 1985), Nobel Laureate, part-time bongo drummer, and investigator of the *Challenger* disaster, posed an interesting alternative question. If a water sprinkler were embedded in an environment of high-pressure water that would be allowed to flow *into* the sprinkler nozzles, assuming that a low pressure were maintained at the conventional inlet, *in which direction would the sprinkler tend to rotate?*

When Feynman decided that only an ad hoc experiment could answer the question, he mounted a sprinkler inside a glass bell jar that he proceeded to pressurize. For the hilarious dénouement of this research effort the reader is referred to the book. For the answer to the question, readers are referred to their own good judgment.

PROBLEMS

4.1. A high-pressure water-pumping facility for a fireboat of the New York City Fire Department is designed for a reservoir pressure of 350 psig with a maximum flow rate of 8800 gal/min. Assuming inviscid flow, determine the discharge speed, the nozzle exit diameter, and the maximum possible jet height if directed vertically upward.

4.2. It is desired to design a venturi meter for use in the flow of water at 20°C within a vertical duct, $D_1 = 4$ in. For a maximum steady-state flow rate of 50 ft^3/min, determine D_2 such that the reading is $h = 18$ in. for a manometer fluid of S.G. = 1.8.

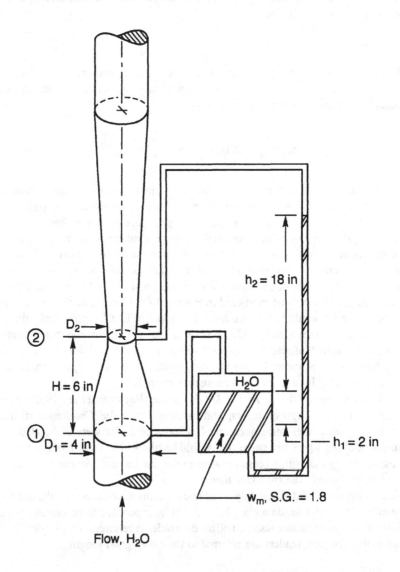

4.3. For the purpose of flow visualization it is desired to inject a dye into a water flow in a pipe fed by a separate reservoir (see figure) of dye. If the local water speed is 3 ft/sec, and the dye is to be emitted at the same speed, determine the required specific gravity of the dye assuming that both flows are inviscid.

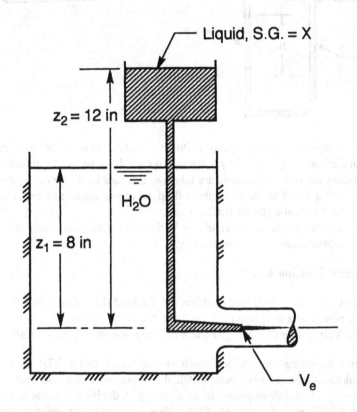

Liquid, S.G. = X

$z_2 = 12$ in

H_2O

$z_1 = 8$ in

V_e

4.4. Consider the ASME venturi meter shown in the figure. The fluid is carbon tetrachloride (CCl_4) whose mass density is $\rho_c = 3.095$ slug/ft³ and whose dynamic viscosity is $\mu_c = 2 \times 10^{-5}$ lb sec/ft², and where, for the "no flow" condition the manometer zeroes out at a level of 13.5 in. below the venturi centerline. The manometer fluid is mercury, S.G. = 13.6.

a. Determine an expression for the flow rate $\dot{Q}$ in terms of the manometer displacement h.

b. Determine the maximum mass rate of flow below which the mercury will not be sucked into the venturi. State all assumptions.

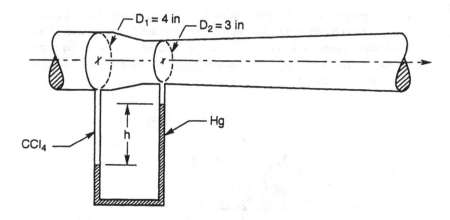

4.5. In an industrial process, using a pitot-static tube, it is necessary to monitor an airflow of $\dot{Q}$ = 20 ft³/sec through a 6-in. ID pipe. At the monitoring station the static conditions are taken as standard atmospheric values.

   a. Using water as the manometer fluid, what reading can be expected at the Pitot-tube station (i.e., at the nose)?

   b. Determine the diameter of the VDI orifice plate that would provide approximately the same reading.

4.6. Verify Equation 4.12-5.

4.7. Carbon tetrachloride (see Problem 4.5 for fluid data) flows through a 3-in. pipe. For a flow rate of $\dot{Q}$ = 0.130 ft³/sec, determine the diameter of the VDI orifice that will provide a mercury manometer reading of 2 ft.

4.8. In reactivating an airflow loop it is discovered that a VDI orifice of unknown area ratio has been installed in a portion of the duct that is not readily accessible to permit its dismantling. A decision is made to install an ASME venturi upstream of the orifice, take pressure measurements on both venturi and orifice for a fixed flow, and then back-calculate to determine the orifice area ratio A_2/A_1.

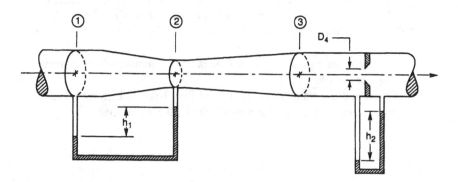

If it is assumed that the air density and temperature stay essentially constant throughout the flow, determine D_2 based on the following data: $D_1 = D_3 = 2$ in., $D_2 = 1$ in., $h = 0.547$ ft Hg, $h = 1.44$ ft Hg, $p_1 = 5p_a$, $T_1 = T_a$, $\mu_1 = 3.745 \times 10^{-3}$ lb sec/ft^2.

4.9. The figure shows a two-dimensional slice of a water jet emitted into the atmosphere by a nozzle from a tank pressurized to 2 atm. The jet (treated as inviscid) impinges on a flat wall and splits in two, as described in Section 4.12. Compute the static pressure at station ② on the wall and the total pressures in the jet at stations ③ and ④.
The whole tank is now moved to the right at 100 ft/sec. Does an observer fixed with respect to the wall observe a steady flow? Compute the total pressures at stations ③ and ④ for this observer.
Note: The creation from a steady flow with a specified, uniform total pressure of a second flow with a pair of jets, one with a higher, the other with a lower total pressure, merely by a change of observer, is an example of a *crypto-steady* flow. It is actually the basis of an invention called the *energy separator.* See Foa (1960).

4.10. A laboratory device involves a "test section" that is a horizontal, circular duct with a linearly decreasing diameter. At ① the flow speed and pressure are μ_1 and p_1, respectively. Show that the diameter variation can be expressed as $D/D_1 = 1 - b\bar{x}$, with $b \equiv (D_1 - D_2)/D_1$ and $\bar{x} \equiv x/L$.

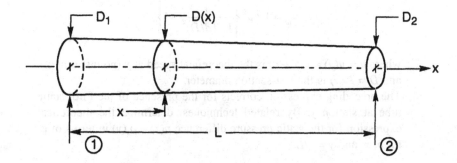

a. For incompressible, inviscid, steady flow, determine expressions for u/u_1, and the pressure coefficient $C_p \equiv (p - p_1)/q_1$, where $q_1 \equiv \frac{1}{2}\rho u_1^2$ is the reference dynamic pressure
b. For the case $D_1 = 2$ in., $D_2 = 1.5$ in., make plots of $\bar{u} = u/u_1$ and C_p vs. $\bar{x}$, for $\bar{x} = 0.0, 0.2, 0.4, 0.6, 0.8, 1.0$. (*Note:* It may actually be more convenient to plot $-C_p$.)
c. Now, referring to the accompanying figure, the flow is to be monitored by the introduction of a Pitot-static tube of diameter d at various stations on the duct centerline such that the total pressure tap is located at station x. Pitot-static tube design requires that the static pressure taps be located at a downstream station c, measured from the nose, to

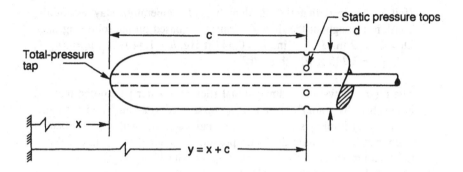

avoid pressure gradient effects associated with the rounded nose. This results in two types of instrument error that must be corrected. In the first type the static pressure detected by the instrument refers to the station $y = x + c$ rather than x. In the second, the presence of the instrument decreases the area available to the flow, causing it to speed up at y, thus generating a static pressure measurement lower than the actual value at y, in the absence of the instrument. Suppose that $p_s \equiv p_s(y)$ is the theoretical pressure at that station in the absence of a Pitot-static tube. Also let $p_{sa} \equiv p_{sa}(y)$ be the actual static pressure sensed after insertion of the Pitot-static tube. By use of the one-dimensional continuity and pressure relations, show that

$$p_s - p_{sa} = \tfrac{1}{2}\rho u^2 \left\{ \frac{1}{1-(d/D)^2} - 1 \right\},$$

where $u = u(y) = u(x + c)$ is the theoretical speed at y (no instrument), and $D = D(y)$ is the test-section diameter.

d. The preceding expression corrects for the presence of the Pitot-static tube at station y. By related techniques, determine the theoretical expression for the static pressure difference $p(x) - p_s(y)$ in terms of ρ, u_1, b, $\bar{x}$, and $\bar{y} = y/L$.

4.11. A New York City fireboat takes water from a reservoir (a river) through an inlet, passes it through a pump, and ejects it from a nozzle at 240 ft/sec, as shown in the figure. The diameter of the nozzle at the exit is $D_2 = 4$ in.

a. Assuming inviscid flow, compute the change of internal energy per unit mass of a particle from the river level to the nozzle exit; compute also the mass flow rate out.

b. By energy considerations, compute the horsepower input to the jet. If the pump has an efficiency of 0.8 and the diesel engine that drives the pump has an efficiency of 0.7, determine the minimum required horsepower rating of the diesel engine.

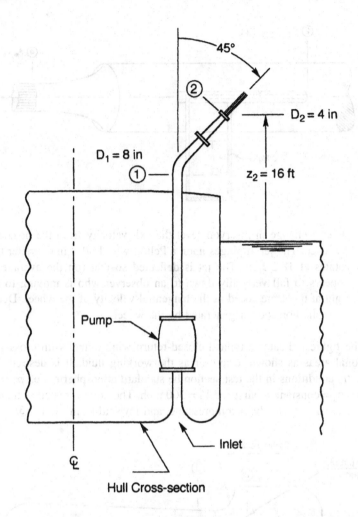

45°

②

$D_2 = 4$ in

$D_1 = 8$ in

①

$z_2 = 16$ ft

Pump

Inlet

₵

Hull Cross-section

c. The inlet pipe, which is 8 in. in diameter rises vertically from the hull bottom to the nozzle control mechanism where the flow is turned through an angle of 45° before it is ejected. Compute the reaction force required to hold the nozzle in place.

4.12.

a. The figure shown is a schematic of a pipeline station of cross section $A = 10$ ft² connected to a water reservoir whose surface level is 75 ft above the nozzle outlet, for which $A_e = 3$ ft². A Pitot-static tube located on the duct centerline just before the outlet station has a manometer reading of $h = 0.45$ ft Hg. Can the flow from reservoir to station ② be treated as essentially inviscid? Explain.

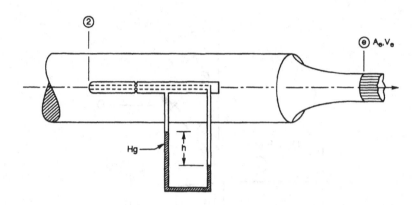

b. After a change in reservoir level the exit velocity from the nozzle is 80 ft/sec. The jet impinges upon a Pelton wheel 30 ft in diameter that rotates at 18.2 rpm. The jet is deflected so that (on the average) it appears to fall vertically down to an observer who is moving to the right at the same speed as the tangential velocity of the wheel. Determine the horsepower generated by the wheel.

4.13. The figure indicates a typical closed-return wind tunnel with cross-sectional areas as shown, using air as the working fluid. It is desired that static conditions in the test section be standard atmospheric, i.e., $p_3 = p_a$, $\rho = \rho_a =$ constant, at airspeed $V_3 = 300$ mph. The flow is treated as though it is adiabatic, steady, incompressible, and inviscid from ① to ③.

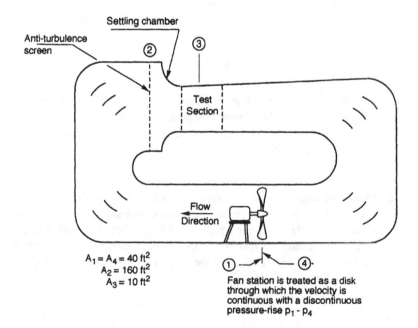

a. Determine conditions in the settling chamber, i.e., $(p_2 - p_3, V_3)$, and the mass rate of flow $\dot{m}$.

b. Due to the higher flow speeds from ③ to ④, the viscous effects are not negligible and, therefore, we cannot use Bernoulli's equation between the two stations. If measurements show that $p_4 - p_3 = 115$ lb/ft^2, determine the useful power output (in horsepower). If the fan efficiency is $\eta_f = 0.62$ and the motor efficiency is $\eta_m = 0.85$ (each value furnished by its manufacturer), determine the required rate of power input to the motor in KW to maintain the flow.

c. If all flow losses are dissipated to heat, thereby increasing the air internal energy, calculate the temperature rise of a particle in one circuit around the tunnel. Use $c_v = R/(\gamma - 1)$.

4.14. You are asked to do a preliminary analysis of launching a toy flight vehicle propelled by a vane driven by a vertical jet of water. The exit jet speed at the nozzle station is V_0 oriented vertically

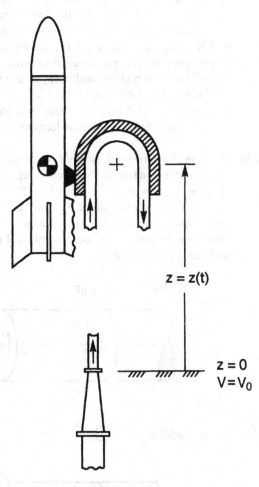

a. Determine the undisturbed speed of the jet $V = V(z)$ as a function of the height z above the launch station.

b. Let the mass flow rate of the jet at the nozzle exit be $\dot{m} = \rho A_0 V_0$. Determine an expression for the cross-sectional area of the jet as a function of z.

c. What is the effective mass flow rate into the vane in the frame of an observer on the vehicle whose vertical velocity is $u = u(z) = u[z(t)]$.

d. According to quasi-steady theory, determine an expression for the thrust force on the rocket.

e. Write down the equation of motion for the rocket–vane assembly (mass M) for vertical motion during the period of acceleration, neglecting aerodynamic drag. Discuss the problem of integrating the resulting differential equation.

4.15. Write the equation of motion (analogous to Equation 4.16-5) for a water-braked rocket sled subject to a solid friction rail force and an aerodynamic drag force.

Hint: The aerodynamic force can be written as $C_D \, (\frac{1}{2}\rho u^2) A_f$, where r is the undisturbed air density, A_f is the projected frontal area of the sled, u is the relative wind velocity, and C_D is a drag coefficient, taken as constant in this problem, but which in practice may depend on the local flow Mach number, the flow Reynolds number, and the shape of the sled body.

a. Determine expressions for the flow speed and the travel distance after initiation of water braking at sled velocity u_i.

b. Then, for $C_D = 1.2$, $A_f = 6$ ft^2, and for a brake inlet diameter of 2 in., compute the distance required to bring the sled to a dead stop, for a solid friction braking coefficient of $\mu = 0.8$, if the sled weights $Mg = 2$ ton and $u_i = 400$ ft/sec.

c. If now the solid friction braking is neglected, compute the speed the sled would have at the same location as obtained in part (b).

4.16. Analyze the horizontal, steady flow of water through two different pipe configurations, each involving an abrupt change of cross section, such that the exit flow at station ② is $V_2 = 20$ ft/sec in each case, with an exit area $A_2 = 1$ ft^2 at atmospheric pressure. If all viscous effects — other than those associated with the area change — are neglected, calculate the following for both cases: the dynamic and total pressure at ②; the total and static pressures at ①.

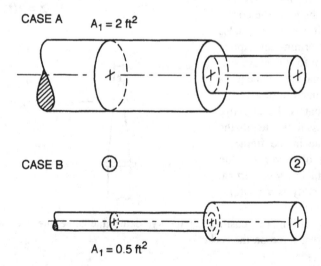

CASE A $A_1 = 2$ ft^2

CASE B   ①   ②

$A_1 = 0.5$ ft^2

4.17. An architect asks for your guidance to help design a fountain device shown in the figure. A vertical, circular, water jet is emitted at station ⓔ that will be used to maintain a 6-in. hollow ball at a height $h = 16$ ft above ⓔ. The mean specific gravity of the ball is S.G. = 0.5. The diameter of the jet just before it reaches the ball must be $D_j = 4$ in. When centered, the jet flows around the ball in a thin axisymmetric layer leaving with an (approximately) horizontal velocity everywhere around the periphery.

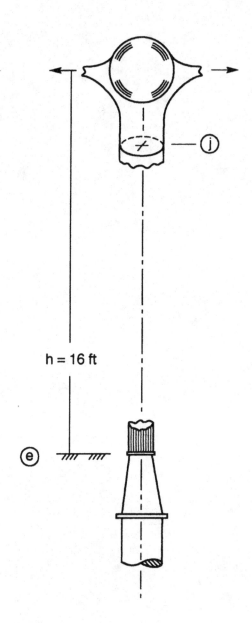

a. If there is some small disturbance that causes the ball to move laterally, the action of the jet tends to return it to the central position; thus the system is stable. Give a qualitative explanation of the stabilizing action of the jet on the ball.

b. Compute the necessary jet velocity required at station ⓔ and the exit diameter of the nozzle required at that station.

$h = 16$ ft

5 Dimensional Analysis, Dynamic Similitude, and Inspectional Analysis

[To utilize dimensional analysis] requires a considerable background of physical experience, and the exercise of a discreet judgement. The untutored savage in the bushes would probably not be able to apply the methods of dimensional analysis to [a] problem and obtain results which would satisfy us.

P. W. Bridgman (1931)

5.1 DIMENSIONS VS. UNITS

It is useful to distinguish between dimensions and units. By *dimension* we mean an intrinsic physical property,* which is independent of the system by which it is measured, i.e., independent of the scheme by which a number is assigned to denote its measure. For example, the edge of a table has the dimension of *length*, and the quantity of matter possessed by a lump of metal has the dimension of *mass*. The notion of dimension is a *primitive concept* and cannot be defined in terms of more fundamental quantities.

The term *unit* refers to the particular, arbitrary procedure by which a number is assigned to denote the magnitude of a physical property. Thus, the table edge can be measured in meters, feet, furlongs, or parsecs, depending on the system of units selected. Usually, the quantity to be measured influences the choice of units to be employed. The essential arbitrariness of units is easily grasped if we recognize that, regardless of what system is employed in making two different measurements of observables having the same dimension, the ratio of their measures remains the same. This is what Bridgman (1931) calls *absolute invariance of relative magnitude*. A discussion of units systems follows in Section 5.2.

BASIC DIMENSIONS

In what is sometimes called the operational approach, a system of dimensioning is built up from the least number of independent dimensions, arbitrarily selected, sufficient to allow the dimensions of all phenomena of interest to be defined. In the science of mechanics it is found that three dimensions, called the *base* or *primary*

---

* This view is not in agreement with that of P. W. Bridgman (1931) who, in his classic treatment of dimensional analysis, states that the concepts of dimensions and units are inextricably linked. Bridgman deals incisively and in more detail with most of the topics of this chapter. With few exceptions I have adopted Bridgman's viewpoint.

dimensions, are the least number possible; the most common trio selected are the *mass*, *length*, and *time*. If thermal changes are to be accounted for in the phenomena under scrutiny, then a fourth, *temperature*, must be added. To complete the set we must add the dimension of *current* to deal with electrical phenomena, and for photometry, or optics, the dimension of *luminous intensity*. We shall restrict considerations to only the first four.

It will be convenient to adopt a special notation for use in dimensional analysis. To denote specification of the dimension of a quantity X we employ a square bracket. Thus, for $[X] = ...$, we read *the dimension of X equals....* For the first four base dimensions, therefore,

$$[\text{mass}] \equiv M, \ [\text{length}] \equiv L, \ [\text{time}] \equiv t, \ [\text{temperature}] \equiv \vartheta.$$

DERIVED DIMENSIONS

The dimensions, called *derived* or *secondary* dimensions, of all other quantities are obtained from physical equations that relate them to quantities expressible in terms of the base dimensions. Thus, the dimensions of some of the more common derived quantities are

$$[\text{area}] \equiv [\text{length} \times \text{length}] = L^2,$$

$$[\text{volume}] \equiv [\text{length} \times \text{length} \times \text{length}] = L^3,$$

$$[\text{velocity}] \equiv [\text{length/time}] = Lt^{-1},$$

$$[\text{acceleration}] \equiv [\text{velocity/time}] = Lt^{-2},$$

$$[\text{density}] \equiv [\text{mass/volume}] = ML^{-3}.$$

According to the present scheme, the dimension of *force** is obtained from Newton's second law, which relates force to mass and acceleration:

$$[\text{force}] \equiv [\text{mass} \times \text{acceleration}] = MLt^{-2}.$$

Some authors suggest that it is equally acceptable in principle to retain force as a base dimension in which case mass would then become a derived dimension, since they are not independent. However, experience has shown that it has not been

---

* In some branches of engineering it is conventional to write Newton's law as $F = (m/g_c)a$, where m is measured in pound-mass units, F in pound-force units, and where g_c is a dimensional constant numerically equal to the uniform gravitational constant at sea level. However, in Newton's second law, a system of units can be employed such that g_c is unity, as well as dimension-free, and the need to carry a dimensional constant is avoided. If this convention were not adopted, then *force, mass, length,* and *time* would be considered as independent dimensions and the law would not be in the form of a *complete physical equation*, a concept that is discussed in a following section. For a further discussion of this point, see Bridgman (1931).

possible to devise a practical laboratory technique employing force as a primary standard; consequently, the M, L, t, ϑ-system is preferable. Several other derived dimensions of particular importance are

$$[\text{pressure}] \equiv [\text{force/area}] = ML^{-1}t^2,$$

$$[\text{work}] \equiv [\text{energy}] \equiv [\text{force} \times \text{distance}] = ML^2t^2.$$

Dimensions of some thermodynamical quantities, in which energies are defined in terms of the preceding mechanical units, are

$$[\text{specific internal energy}] \equiv [\text{energy/mass}] = L^2t^2,$$

$$[\text{specific enthalpy}] \equiv [\text{energy/mass}] = L^2t^2,$$

$$[\text{specific entropy}] \equiv [\text{specific energy/temperature}] = L^2t^2\vartheta^{-1},$$

$$[\text{gas constant}] \equiv [\text{coefficient of specific heat}]$$

$$\equiv [\text{specific energy/temperature}] = L^2t^2\vartheta^{-1}.$$

THE FORM OF DERIVED DIMENSIONS

It is seen that, expressed in terms of the basic dimensions, the derived dimensions appear as products of the basic dimensions raised to a simple power. This result holds in general. For proofs see Bridgman (1931) and Holt (1961).

ANGULAR MEASURE

What are the dimensions of an angle? If the measure of a central angle of a circle is defined as the length of the subtended arc divided by the radius, then it is clear that it has zero dimensions. We call this measure by the name *natural measure* or *radian*. There are alternative measures for angles. In antiquity, the central angle of a circle was divided into 360 parts called *degrees*. We also encounter the *grad*, where a right angle, $\pi/2$ radian, is divided into 100 equal parts called grads. The grad and the degree are also dimensionless. The specification of the numerical value of an angle in degrees or grads is simply an artifice for replacing the value expressed in natural measure by an arbitrary, but more "convenient" scale.

5.2 STANDARDS IN SCIENCE AND TECHNOLOGY

The development of standards of metrology, as it affects the average citizen, is the product of a long history, one which is intertwined with cultural factors and constrained by custom and tradition. In some cases these common uses defy logic. In the U.S., the common system of measurement is an outgrowth of the venerable (and now obsolete) British or Imperial system — sometimes referred to as the *engineering system of units* — which the U.K. abandoned in 1975. Of the civilized nations, the U.S. is joined only by Liberia and Myanmar (formerly Burma) as holdouts from official adoption of the International System of Units (abbreviated SI).

Engineering students, who become thoroughly familiar with SI through their science courses, are sometimes distressed to find that — with the exception of electrical measurements — the metric system is not uniformly employed in engineering practice in the U.S. To compound the confusion, there is not even uniformity among the several branches of engineering, although, for example, the American Society of Mechanical Engineers (ASME) mandates use of the SI in its publications.

Unfortunately, problems involved in any contemplated move to mandate use of the metric system are social and political in nature.* Although Congress passed a Metric Conversion Act in 1975, it failed to provide any regulatory authority over government or business entities, which left the conversion process to the discretion — some think the whim — of these organizations. It seems unlikely that a genuine conversion will be accomplished until well into the third millennium, if ever.

A brief history of the development of legal standards of weights and measures in the U.S. is sketched by Judson (1976). Smith (1958) points out that although the Constitution, ratified in 1787, grants Congress the power to *fix the standards of weights and measures*, in spite of a series of recommendations by committees and commissions, Congress took no effective action until 1836 when a joint resolution authorized fabrication and distribution of a complete set of standard weights and measures to be distributed to each of the several states. Although the metric system had been created over 40 years earlier, its adoption would have been too radical a step in that era as Secretary Adams had foreseen. For a discussion of the extent to which these attitudes still persist, 175 years later, see Browne (1996).

Thus, any decision to ignore the system of units still used in many industries in the U.S. would be negligent. Consequently, a dual approach is employed herein, with applications in the engineering system of units, as well as in SI.

THE INTERNATIONAL STANDARDS

The international standards are set at General Conferences on Weights and Measures under an agreement dating from 1875, of which there are now 46 signatories. The 11th Conference, in 1960, adopted SI (abbreviation for Système International d'Unités) which has, in turn, been adopted by the U.S. via its official representative, the National Institute of Standards and Technology (NIST), formerly the National Bureau of Standards (NBS). NIST has committed itself to use SI in all NIST publications "except where use of these units could obviously impair communication."

In SI, corresponding to the basic dimensions, there are six *base units: kilogram, meter, second, kelvin, ampere,* and *candela.* In addition, to relate the molecular and

---

* "The power of the legislator is limited by the will and action of its subjects. His conflict with them is desperate, when he counteracts their settled habits, their established usages…their prejudices and their wants: all of which is unavoidable in the attempt radically to change, or to originate, a totally new system of weights and measures…" from John Quincy Adams (1821), who was then Secretary of State.

macroscopic scales of matter, a seventh unit the *amount of substance*, designated as the *mole* is defined. Each of the definitions, as listed* by Goldman and Bell (1981) is given below.

Mass. The kilogram (kg) is the mass of a particular cylinder of platinum iridium alloy called the International Prototype Kilogram, which is preserved in a vault at Sévres, France, by the International Bureau of Weights and Measures.

Length. The meter (m) is the length of the path traveled by light in vacuum during a time interval of 1/299 792 458** of a second.***

Time. The second (s) is the duration of 9 292 631 770 periods of the radiation corresponding to the transition between the two hyperfine levels of the fundamental state of the atom of cesium 133.

Temperature.**** The kelvin (K), the unit of thermodynamic temperature, is the fraction 1/273.16 of the thermodynamic temperature of the triple point of water.

Current. The ampere (A) is that constant current which, if maintained in two straight parallel conductors of infinite length, of negligible circular cross section, and placed 1 m apart in vacuum, would produce between these conductors a force equal to 2×10^{-7} newton per meter of length.

Luminous intensity. The candela (cd) is the luminous intensity, in a given direction, of a source that emits monochromatic radiation of frequency 540 $\times 10^{12}$ hertz and that has a radiant intensity in that direction of (1/683) watt per steradian.

Amount of substance. The mole (mol) is the amount of substance of a system that contains as many elementary entities as there are atoms in 0.012 kg of carbon 12. When the mole is used, the elementary entities must be specified and may be atoms, molecules, electrons, other particles, or specified groups of such particles.

The perceptive reader will note apparent anomalies in the SI terminology and definitions. For example, the gram, fundamental to the original (cgs) metric system, is not recognized officially in SI; that is, it is not acceptable SI usage to refer to 0.001 kg as a gram. Furthermore, the definition of current depends on force, which is a derived unit. Even more unexpected is that the definition of the candela depends upon the *steradian*, a supplementary SI-unit, whose use is permitted, but not

---

* In addition, Goldman and Bell give a complete listing of conference actions, as well as valuable discussion of units, a compendium of rules for prefixes and symbols, notes on temporary units, and units outside the SI.
** Note that the denominator of the fraction is the speed of light in vacuum, in units of meters per second.
*** The SI convention of writing numbers in trios, separated by a space, on either side of the decimal, is followed herein. In non-SI units the customary comma is employed, instead of the space.
**** The terminology *degrees Kelvin* (°K) has been dropped in favor of the *kelvin* (K) for the SI. In addition, there is a practical international temperature scale in *degrees Celsius* (°C) defined such that the triple point of water occurs at 0.01°C. The Celsius scale replaces the centigrade scale to which it is numerically identical.

required, to specify solid angles in SI. In practice, none of these creates any difficulty in engineering applications.

It is remarkable that of the preceding units standards, all of which were selected on a somewhat arbitrary basis, only that of mass is defined in terms of a specific and unique quantity of matter. In the event of a catastrophe of the almost inconceivable extent that would be necessary to destroy all standards or devices available to calibrate them, it would be theoretically possible to reconstruct them — except for mass — from new observations of the appropriate physical phenomena. On the other hand, after such a catastrophe it is not clear who would be left to make the observations!

To illustrate one derived unit, the unit force, called the *newton* (N), is obtained from Newton's second law; thus,

$$1 \ \text{N} \equiv 1 \ \text{kg} \times 1 \ \text{m/s}^2 = 1 \ \text{kg} \cdot \text{m/s}^2.$$

In SI the word *kilogram* is reserved for mass. However, in the common usage, and in some branches of engineering in Europe, the kilogram (or kilopond) has also been used for force (1 kg_f = 1 kp = 9.806 65N), so that ambiguity of terminology undoubtedly still exists in some countries even where SI is the legal system for technical communications.

THE MECHANICAL, OR ENGINEERING, SYSTEM OF UNITS*

By law, NIST is charged with fixing standards (but not with their enforcement) in the U.S. Although there is a trend within some industries toward SI (e.g., the automotive industry has adopted SI, no doubt as a result of the global nature of the market), most industries conduct business in some version of the engineering system. For example, as noted in Browne (1996), suppliers to NASA, because of the requirement of supplying one-of-a-kind items, employ mostly the *inch-pound-second* system. In consequence, for homogeneity, engineering calculations in this system require that the gravitational constant be expressed as g = 386 in./sec^2, a relic of the past for most engineers.

Standards in non-SI systems are fixed, as a practical matter, by defining mass, length, time, and temperature through a straightforward conversion from their metrical equivalents. These factors are used internationally when making conversions, and are listed below.

Length	1 foot (ft) $\equiv$ 0.304 8 meter, which is equivalent to
	1 inch (in.) = 2.54 centimeter exactly.
Mass	1 slug** $\equiv$ 14.593 9 kilogram.

---

* There are many references available with practical information on the SI and other systems of units, and on the conversion from one system to another. Three are: Anderton and Bigg (1965), ASME (1981), and ASTM (1980).

** According to this scheme of units the need for the *pound-mass unit* (1 lb_m = 1/32.174 slug), featured in some of the variations of the engineering system employed in the U.S., does not arise.

Temperature 1 degree Rankine (°R) ≡ 5/9 kelvin. The two scales share a common value at absolute zero.

Time   The definition of second (usually abbreviated as sec in engineering usage) remains unchanged.

The derived unit of *force* is called the *pound* (lb or lb_f), such that

$$1 \text{ lb} \equiv 1 \text{ slug ft/sec}^2.$$

Its relation to the newton is obtained by introducing the conversion definitions therein:

$$1 \text{ lb} = 1 \times (1 \text{ slug}) \times (1 \text{ ft}) / (1 \text{ sec}^2),$$

$$= 1 \times (14.593\ 9 \text{ kg}) \times (0.304\ 8 \text{ m}) / (1 \text{ s})^2,$$

$$= 4.448\ 22 \times (1 \text{ kg·m/s}^2),$$

$$= 4.448\ 22 \text{ N}.$$

5.3 COMPLETE PHYSICAL EQUATIONS AND DIMENSIONAL HOMOGENEITY

COMPLETE PHYSICAL EQUATIONS

It is intuitively obvious that for a physical law to be universally valid it must be independent of the scheme by which the observables are measured (i.e., independent of the system of units employed). For example, using the scalar form of Newton's second law, $F = ma$ holds unchanged in a variety of units systems, including the SI, cgs, and British systems. Bridgman (1931) classifies such laws as *complete physical equations*. Any constants appearing in such equations must be dimensionless. Otherwise, the value of the constant would depend upon the system of units employed. In the relation for the area of a triangle $A = \frac{1}{2}$(base × altitude) the constant $\frac{1}{2}$ is dimensionless and the equation is complete. The relation for the speed of a body falling under the action of gravity is $v = gt$, where g is dimensional; the expression fails the test of a complete equation.

On a more instructive level, in the universal law of gravitation $F = Gmm'/r^2$, the constant has the dimensions $[G] = M^{-1}L^3t^{-2}$, rendering the equation not complete. Of course, it is always possible to convert to a different system of units. Thus, if in a second system the reference mass unit is related to that of the first by $M_1 = aM_2$, where a is a (dimensionless) conversion factor and where the length and time units are likewise converted by $L_1 = bL_2$, $t_1 = ct_2$, then $F_1 = abc^{-2}F_2$, $m_1m_1'/r_1^2 = a^2b^{-2} m_2m_2'/r_2^2$. Substituting these expressions into the gravitational law and solving, we get $F_2 = (ab^{-3}c^2G_1)\ m_2m_2'/r_2^2$. Obviously, in the second system the magnitude of G

is $ab^{-3}c^2$ times that in the first. The dependence of the magnitude of the universal gravitational constant on the system of units confirms that, as given, the expression embodying the law is not complete. This does not reduce its utility and for this reason Bridgman calls it an *adequate* equation. It would be possible to choose $G = 1$, thus redefining the force unit, and thereby producing a complete equation. The gain would be illusory, however; Newton's second law would then no longer be expressible as a complete equation since it would require introduction of a dimensional constant. Experience has shown that the former situation is generally preferable.

It does not appear to be possible to define a system of units so that all fundamental laws are expressible as complete equations. When the techniques of dimensional analysis are applied to a problem that is governed by one or more adequate equations, then the appropriate dimensional constants must be included as parameters in the analysis.

DIMENSIONAL HOMOGENEITY

Bridgman proves that if an equation is postulated to be complete it is then *dimensionally homogeneous*; i.e., the dimensions of each of the terms are the same. But every equation that is dimensionally homogeneous is not necessarily complete. For example, in Bernoulli's equation, $p + \frac{1}{2}\rho V^2 + \rho g z =$ constant, each term has the dimensions $ML^{-1}t^{-2}$. However, the value of the gravitational constant g depends upon the units employed. Nevertheless, we expect that all equations governing a physical phenomenon are dimensionally homogeneous.

It might be thought that all relations in which the symbols refer to physical quantities are automatically dimensionally homogeneous. This is false, as Bridgman illustrates with a counterexample in which two equations from particle dynamics, $V = at$, $s = \frac{1}{2}at^2$, are added to produce

$$V + s = at + \frac{1}{2}at^2.$$

The resulting equation, which is clearly dimensionally nonhomogeneous, is valid for any system of units. Because combined relations of this type do not arise in the analysis of physical phenomena, it is safe to discount their importance and to restrict considerations only to dimensionally homogeneous equations.

5.4 A PRIMITIVE EXAMPLE OF DIMENSIONAL ANALYSIS*

All of us who have taken an automobile ride know that if we extend our hands into the external airflow we experience a resisting force. Let us suppose that we are familiar with certain ideas from dimensional analysis — particularly the principle of dimensional homogeneity — but ignorant of the laws of fluid mechanics, and that we wish to obtain some information about the form of the law that governs the aerodynamic force.

---

* Truesdell attributes the first dimensional analysis to Fourier (1822).

Since we know that as the relative speed V increases the force appears to increase in a nonlinear fashion, we can take as a first trial $F \sim V^a$ or $F = kV^a$, in which a is an unknown dimensionless exponent and k is required to be a dimensionless constant. It is obvious that the proposed law is erroneous, but if we pretend ignorance for the purpose of instruction and proceed, then the principle of dimensional homogeneity requires that dimensions on both sides of the equation be the same; i.e., we ask what must be the value of a for this to be true? That is, there is an associated dimensional equation written

$$[F] \stackrel{?}{=} [kV^a],$$

or, equivalently,

$$MLt^{-2} \stackrel{?}{=} L^a t^{-a}.$$

Since there is no value of a that makes the corresponding exponents on either side of the preceding equation equal, there must have been an error in our reasoning. Additional reflection on our experience brings forth the fact that if we move a hand through water, even at very low speeds, a relatively large resisting force is sensed. We conclude that the previously omitted fluid density plays an important role in this phenomenon, and attempt a second trial $[F] = [k\rho^a V^b]$. This produces by the same procedure

$$MLt^{-2} = M^a L^{-3a+b} t^{-b}.$$

This is a little more interesting. Since M, L, and t are independent, and if the preceding equation is to be valid, then the exponents of the base dimensions must be the same on both sides. Thus, we must have, in turn, for the exponents of

$$M: 1 = a; \quad L: 1 = -3a + b; \quad t: -2 = -b.$$

We are then faced with a set of three equations in two unknowns. Since the equations are not linearly dependent, we know that there is no solution and, hence, something must still be wrong with our original assumptions. We next try what actually we knew was true all along, that the force depends also on the area exposed to the flow, i.e., $[F] = [k\rho^a V^b A^c]$. The resulting equations for our third trial are for

$$M: 1 = a; \quad L: 1 = -3a + b + 2c; \quad t: -2 = -b, \qquad (5.4\text{-}1)$$

which have the solution $a = 1$, $b = 2$, $c = 1$; therefore, the desired relation is

$$F = k\rho V^2 A. \qquad (5.4\text{-}2)$$

Of course, the worker in fluid mechanics already knows that the excess pressure developed in decelerating a fluid to rest is proportional to ρV^2 and that the force is produced by this on a surface of area A is proportional to their product. Dimensional analysis cannot reveal the value of the constant k. However, if the resulting relation is correct, then a single experiment would suffice to determine it at least for one set of conditions. So that no erroneous impression is created, it should be noted that in the general situation k, which is called a *force coefficient*, is not in fact a constant, but a variable that depends on the shape of the surface, its orientation to the flow, and on certain viscous and compressibility parameters that are discussed in the following sections. Furthermore, in an actual analysis we would always use all the information available at the start.

5.5   THE ROLE OF DIMENSIONLESS PARAMETERS

Another approach, also based upon the principle of dimensional homogeneity, reveals a new facet of dimensional analysis, specifically, that the relations that disclose information about the behavior of physical systems in the most economical way involve only dimensionless parameters composed of combinations of the dimensional quantities that enter into the analysis. An example helps drive the point home.

EXAMPLE OF FLOW OUT OF A RESERVOIR

Figure 5.5-1 is a schematic of an experimental setup in which a reservoir containing a calibrated fluid volume Q is connected to a capillary tube of fixed length and diameter. The other end of the capillary is connected to a second reservoir that is maintained at a fixed pressure difference $\Delta p = p_1 - p_2$ from the first. It is desired to determine the law governing the time t_d required to transfer a fixed volume Q of a fluid of density ρ between reservoirs. Preliminary indications are that viscous effects (which are presumed to depend on the dynamic viscosity μ) are important, and also that the time t_d is a function of the pressure drop Δp between reservoirs. We can write the assumed relationship in implicit functional form as

$$f_1\left(\Delta p, t_d, Q, \rho, \mu\right) = 0. \tag{5.5-1a}$$

Or, we can "solve" for one of the parameters to obtain an explicit form, e.g.,

$$\Delta p = f_2\left(t_d, Q, \rho, \mu\right). \tag{5.5-1b}$$

The choice of Δp as the explicit function is arbitrary, and the procedure to be demonstrated will also work if any one of the others is chosen.

In any dimensional analysis, one must know the dimensions* of each of the dimensional parameters on which the (unknown) physical relation is assumed to depend. An efficient way to accomplish this is to generate a matrix, as shown, where each element is the exponent of the base dimension for the corresponding dimensional quantity. It appears that in Fourier (1822, p. 160) there is given not only the first example of a dimensions matrix, but also the earliest dimensional analysis.

---

* The dimensions of μ are determined from Newton's law of resistance, i.e., $[\mu] = [\tau]/[\partial u/\partial y] = ML^{-1}t^{-1}$.

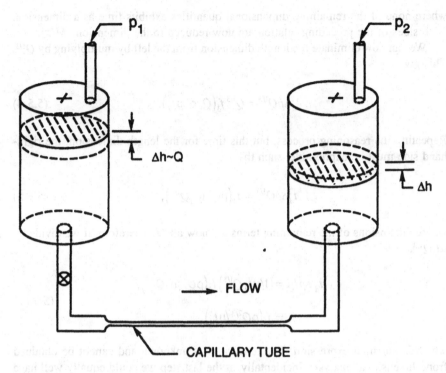

FIGURE 5.5-1 Flow through a capillary tube.

Base Dimension	Quantity				
	Δp	t_d	Q	ρ	μ^*
M	1	0	0	1	1
L	-1	0	3	-3	-1
t	-2	1	0	0	-1

Dimensions of the quantities in Equation 5.5-1 are shown in the preceding table. Now, the principle of dimensional homogeneity requires that the dimensions of both sides of Equation 5.5-1b be the same. If we multiply the entire equation by the square of the discharge time, the dimension of *time* will be eliminated from the left-hand side; thus,

$$t_d^2 \Delta p = t_d^2 f_2(t_d, Q, \rho, \mu). \qquad (5.5-2)$$

But since the left-hand side, which now has the dimensions ML^{-1}, does not involve time, the factor t_d^2 must combine with the dimensional parameters in the function $f_2(t_d, Q, \rho, \mu)$ in just such a way that time is also eliminated from it as a dimension. The only possibility is that Equation 5.5-2 must be capable of being expressed as

$$t_d^2 \Delta p = f_3(Q, \rho, \mu t_d), \qquad (5.5-3)$$

where none of the remaining dimensional quantities exhibits time as a dimension. Both sides of the preceding relation are now reduced to the dimensions ML^{-1}.

We can now eliminate the length dimension from the left by multiplying by $Q^{1/3}$. This gives

$$t_d^2 \Delta p Q^{1/3} = Q^{1/3} f_3(Q, \rho, \mu t_d). \tag{5.5-4}$$

Repeating the reasoning process, but this time for the length dimension, the right-hand side must be expressible, such that

$$t_d^2 \Delta p Q^{1/3} = f_4(\rho Q, \mu t_d Q^{1/3}), \tag{5.5-5}$$

where dimensions of the remaining terms are now all M. Therefore, if we divide by $\mu t_d Q^{1/3}$,

$$t_d \Delta p / \mu = (1/\mu t_d Q^{1/3}) f_4(\rho Q, \mu t_d Q^{1/3}),$$
$$= f_5(\rho Q^{2/3} / \mu t_d), \tag{5.5-6a}$$

which is our final expression. The function f_5 is unknown and cannot be obtained from dimensional analysis. Incidentally, at the last step we could equally well have divided by ρQ to yield the entirely equivalent result:

$$\Delta p t_d^2 / \rho Q^{2/3} = f_6(\mu t_d / \rho Q^{2/3}). \tag{5.5-6b}$$

Which of the two forms is preferable can be decided only on the basis of experiment, or by theoretical analysis in which the basic differential equations governing this phenomenon have been formulated.

The nondimensionalizing procedure leading to Equations 5.5-6, although intended to appeal to intuition, is awkward. In practice, either of the two methods developed in Sections 5.7 and 5.8 is both simpler and more direct.

SIGNIFICANCE OF DIMENSIONLESS PARAMETERS FOR CORRELATION OF EXPERIMENTAL MEASUREMENTS

Equations 5.5-6 illustrate one of the most fruitful consequences of dimensional analysis. If we designate the resulting dimensionless groups as $\pi_1 \equiv t_d \Delta p / \mu$ and $\pi_2 \equiv \rho Q^{2/3} / \mu t_d$, then Equation 5.5-6a becomes

$$\pi_1 = \phi(\pi_2), \tag{5.5-7}$$

where $\phi \equiv f_5$ is unknown. From a mathematical point of view Equation 5.5-7 is clearly simpler than its preceding equivalent dimensional counterpart (Equation 5.5-1), since there remain only two variables instead of the original five. In fact, it states that the particular phenomenon is governed by an ordinary differential equation since there is but one independent variable. If $\phi(\pi_2)$ were known, we would have the complete solution. Although dimensional analysis does not provide the form of the law governing a phenomenon, it does simplify the problem of selecting the significant variables involved.

Furthermore, even if the functional relation is not known, Equation 5.5-7 provides a guide to experimentation on the phenomenon in question. If measurements are made for a variety of conditions, a plot of π_1 vs. π_2 would effectively determine the desired relation. The extraordinary result conveyed by Equation 5.5-7 is that, if a particular set of measurements fixes a pair of π_1 and π_2, then we are assured that any set of dimensional parameters that yields the same π_2 must also produce the same π_1, unless the governing relation is multiple valued. This means that each data point actually holds for an infinite number of possible combinations of the dimensional parameters.

Dimensional analysis thus provides the following simplification:

In any physical problem the simplest description is expressed by a relation (or relations) involving only dimensionless parameters which are always less in number than the number of dimensional parameters involved in the solution.

The example used to lead up to this conclusion has a certain textbook flavor to it. Yet it is actually an example* from real life quoted by the late Theodore von Kármán (1881–1963), scientist, engineer, bon vivant, and raconteur. In his delightful book he relates how, in 1909, the physical chemist E. Bose had measured the discharge times for nine different fluids, all on the same setup and, for each fluid, had plotted the pressure drop vs. the discharge time. Figure 5.5-2 shows plots for four of the liquids. Kármán's contribution was to point out that if the universal relation predicted by Equation 5.5-7 does govern the flow, then the same data ought to plot on a single curve. Figure 5.5-2b shows that such is precisely the case. Knowledge of this fact would have saved Bose a great deal of work, since it would have been sufficient to have made the plot for one fluid and verified it by spot checks with others. For experimentalists who do not have available a complete theory to guide them, this technique is obviously of great utility. For theoreticians the plot of data in their most simple form may provide an important clue either to formulation of the governing basic equations, or to their solution, or both.

5.6 THE BUCKINGHAM Π-THEOREM

In the preceding example it was illustrated, using only the principle of dimensional homogeneity, that it is possible to reduce an assumed relationship between dimensional quantities to one that involves only a lesser number of dimensionless parameters. In attempting to generalize this result, some questions might occur to the critical thinker: For a particular phenomenon, which dimensional parameters are

---

* See von Kármán (1954).

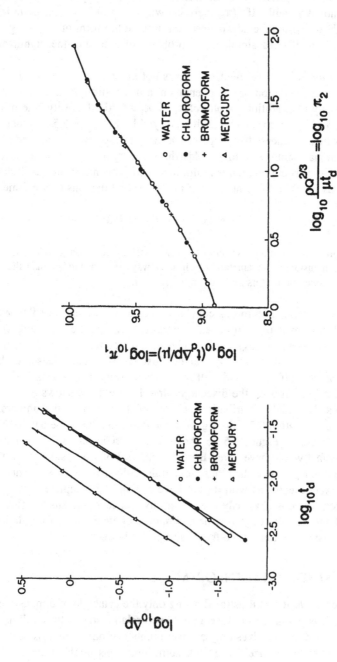

FIGURE 5.5-2   Illustration of how dimensional analysis guides the form in which experimental data are presented. (From von Kármán, T., *Aerodynamics, Selected Topics in the Light of Their Historical Development*, Cornell University Press, Ithaca, NY, 1954, 80–81. With permission.)

important and should be included in the analysis? How many independent dimensionless parameters can be formed of these? Is there a systematic way to determine the dimensionless parameters?

To the first question Bridgman has replied to the effect that dimensional analysis is not a panacea that can be used even by a dilettante to wreak miracles. Rather, in the hands of an experienced person with judgment, it is a powerful tool to make an initial assault on a problem.

The answers to the other questions were first provided in a formal statement by the American physicist E. Buckingham (1914), who propounded what is now called the π-theorem.

Given the existence of a complete physical equation which involves m independent physical (dimensional) quantities Q_i, $i = 1, 2, ..., m$, i.e.,

$$f(Q_1, Q_2, ..., Q_m) = 0,$$

in a system in which there appear n base dimensions, then there exists an associated relation involving k independent dimensionless ratios π_j, $j = 1, 2, ..., k$, such that

$$F(\pi_1, \pi_2, ..., \pi_k) = 0$$

where each π-ratio involves products of the Q_i to some power, and where the number of independent ratios is given by $m - n \le k$.

Proofs of the π-theorem are also given by Bridgman (1931), Langhaar (1951), and Holt (1961). As Langhaar shows in rigorous detail, the actual number of independent π-ratios is $k = m - n'$, denoting the largest number of Q_i from which a π-ratio *cannot* be formed. In most cases $n' = n$, but where they are not equal (i.e., where $n' < n$), this fact can usually be determined from an inspection of the dimensions of the Q_i. The following examples illustrate some of the possibilities.

5.7 APPLICATION OF THE Π-THEOREM

It is assumed that the thrust T developed by a propeller is a function of its diameter D, the average chord* d of a blade, the propeller rotational speed n (in revolutions per second), its forward speed V with respect to the surrounding fluid and the fluid density ρ; i.e., $T = T(D, d, n, V, \rho)$. Dimensions of these quantities are given in the matrix.

	T	D	d	n	V	ρ
M	1	0	0	0	0	1
L	1	1	1	0	1	-3
t	-2	0	0	-1	-1	0

---

* The chord of a propeller is the length of a cross-sectional cut, transverse to the blade axis, measured from the leading to the trailing edge.

We expect that since $m = 6$, $n = 3$, there should be at least $k = 3$ π-ratios. In the simplest situation we might have three quantities whose dimensions are just M, L, t, respectively, and where the dimensions of the other three quantities are various combinations thereof. Because a dimensionless ratio cannot be formed from the first trio, we see that it would take one quantity of the second group combined in some way with the first three to produce a π-ratio.

If we do not have such a trio,* each of whose dimensions is a different one of the base dimensions, it is necessary to select a group of three-dimensional parameters in which all three base dimensions are included, and from which a π-ratio *cannot* be formed. For example, if we choose ρ, V, and D, which are called the working, or repeating, dimensional parameters, then we first attempt to form a π-ratio by multiplying together these parameters, each raised to an arbitrary power, such that the result is dimensionless. Should this be possible, then

$$[\rho^a V^b D^c] = M^a L^{-3a+b+c} t^{-b} = M^0 L^0 t^0,$$

from which, after equating the exponents,

$$a = 0, \quad -3a + b + c = 0, \quad -b = 0.$$

Since this set of equations has no nontrivial solution, we conclude that ρ, V, and D form a suitable set of repeating variables.

The first π-ratio is assumed to be formed from a nonrepeating parameter combined with the repeating variables as before. That is, put

$$\pi_1 = \frac{T}{\rho^{a_1} V^{b_1} D^{c_1}}.$$

Since π_1 must be dimensionless, we obtain

$$M^{1-a_1} L^{1+3a_1-b_1-c_1} t^{-2+b_1} = M^0 L^0 t^0,$$

or $1 - a_1 = 0$, $1 + 3a_1 - b_1 - c_1 = 0$, $-2 + b_1 = 0$, from which $a_1 = 1$, $b_1 = 2$, $c_1 = 2$. Therefore, $\pi_1 = T/\rho V^2 D^2$. The procedure is repeated with the other nonrepeating variables to produce $\pi_2 = d/D$ and $\pi_3 = nD/V$. Our solution takes the form $f(\pi_1, \pi_2, \pi_3) = 0$ or, equivalently, $\pi_1 = G(\pi_2, \pi_3)$.

FORCE COEFFICIENT

It is usually the practice among engineers, when working with a force generated by a steady flow of speed V, to define a (dimensionless) *force coefficient*.*

$$C_F = \frac{F}{\frac{1}{2} \rho V^2 A}, \tag{5.7-1a}$$

---

* In general, we need one working variable for each base dimension appearing in the matrix.

where the force F has been nondimensionalized by dividing by the product of the free-stream dynamic pressure $\frac{1}{2}\rho V^2$ and by an area A appropriate to the geometry of the phenomenon being analyzed. This differs from π_1 only by the factor of $\frac{1}{2}$.

In the case of a propeller, instead of a force coefficient propulsion engineers work with a *thrust coefficient*

$$C_T \equiv T/\rho n^2 D^4, \tag{5.7-1b}$$

where, instead of the free-stream velocity V, the reference velocity selected is nD, twice the propeller tip speed, and where the reference area is chosen as D^2, the propeller disk area multiplied by $4/\pi$. It can be readily verified that $C_T = \pi_1/\pi_3^2$. Since there are only three independent π-ratios, we choose, for the moment, the trio C_T, π_2, π_3.

ADVANCE RATIO

The inverse of π_3 is a dimensionless parameter that has proved useful in propeller theory. It is called the *advance ratio* and is defined by the symbol $J = \pi_3^{-1} \equiv V/nD$. The analysis has now been reduced to the relation

$$C_T = C_T(J, d/D). \tag{5.7-2}$$

Thus, instead of the dimensional relation $T = T(D, d, n, V, \rho)$ involving five independent variables, Equation 5.7-2 governs, which has only two. Clearly, if J and the ratio d/D remain fixed, the thrust coefficient C_T is a constant.

GEOMETRIC SIMILARITY AND MODEL TESTING

It seems intuitively obvious that if we wish to employ the results of testing a model to predict the performance of a full-scale propeller, the tests should involve a model precisely scaled from the prototype. This is achieved by selecting a scale factor λ such that $D_1/D_2 = d_1/d_2 = \lambda$, where the subscripts refer to the prototype and model, respectively. This is equivalent to requiring that the ratio d/D be constant between prototype and model; this requirement holds for *any* pair of length ratios. When achieved, the condition of *geometric similarity* is said to be satisfied.

Assuming that all, and only those, parameters of importance have been included in the analysis, the other requirement, deduced from Equation 5.7-2, for model testing is that results can be compared only if the advance ratio J is the same for model and prototype. This would guarantee that the thrust coefficients be the same in both cases.

We can examine some of the questions facing the test engineer who chooses a scale factor $\lambda > 1$. In order for the advance ratio to be the same in both cases the engineer must have $V_2/n_2D_2 = V_1/n_1D_1$, or $n_2/n_1 = \lambda(V_2/V_1)$ for the ratio of the test rpm values. Since testing in air requires a wind tunnel, it would be economical to test at a lower speed. To keep the example simple, we choose $V_1/V_2 = \lambda$, the same ratio as the scale factor. The rpm ratio then becomes $n_2/n_1 = \lambda^2$. Then, according to

Equation 5.7-2, since the thrust coefficients of model and full-scale must be equal, the generated thrusts are related by $T_2/T_1 = (\rho_2/\rho_1)(n_2/n_1)^2(D_2/D_1)^4 = (1/\lambda^2)(\rho_2/\rho_1)$. The resulting smaller thrusts on the test rig allow for a simplified design of that apparatus.

It is a certainty that the Wright brothers had no knowledge of dimensional analysis in transferring the data from their rudimentary wind tunnel experiments to their full-scale design, but, in 1903, they succeeded in spite of it. Other aviation pioneers were not so lucky.

5.8   ALTERNATIVE METHOD FOR DETERMINING THE Π-RATIOS

In analyzing a particular phenomenon, how can one tell which of the conceivable dimensional quantities are important? Dimensional analysis alone cannot answer this question, but it can provide, as shown in Section 5.5, a guide to presentation of experimental data. A judicious marriage (or, at least liaison) of dimensional analysis and experiment may enable the investigator to rule out certain parameters as unimportant.

Suppose we consider a situation in which a great many dimensional quantities are presumed to enter. The procedure of Section 5.7 allows us to determine a predictable number of π-ratios, but it is rather cumbersome if there are many to be computed, since a set of simultaneous equations has to be solved for each. By an elementary "trick," however, it is possible to establish a procedure with which the ratios can be written down, almost without further analysis.

STATEMENT OF THE PROBLEM

We undertake, by dimensional analysis, the problem, Figure 5.8-1, of determining the drag force D^* on an airfoil of chord c and maximum thickness b, due to the flow of a gas with the following free-stream properties: speed U_1, density ρ_1, pressure p_1, temperature T_1, speed of sound a_1, coefficients of specific heat c_p and c_v, viscosity μ_1, coefficient of heat conduction k_1. In equation form:

$$D = D(U_1, \rho_1, T_1, p_1, a_1, c_p, c_v, \mu_1, k_1, b, c, g), \qquad (5.8\text{-}1)$$

where we have thrown in dependence on the gravitational constant g for good measure.

Of course, the problem is posed with extreme generality. Prior to the analysis we can make two improvements. In the first instance the integrated pressure force is not directly dependent on the free-stream pressure p_1, but on the varying pressure distribution over the airfoil surface. At an arbitrary point the pressure p may be above or below the free-stream value. Thus, we replace p_1 by $\Delta p \equiv p - p_1$. Second, there is a possibility of heat transfer to or from the airfoil by conduction, which

---

* The *drag* is the component of aerodynamic force on the airfoil parallel to the oncoming flow, *lift* being the normal component.

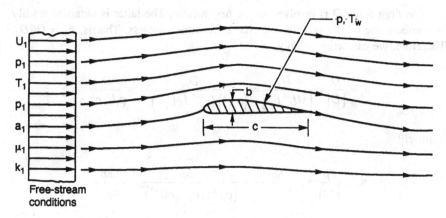

FIGURE 5.8-1   Flow about an airfoil.

conceivably may affect the flow pattern and, hence, the drag. But we know that the rate of heat conduction is proportional to a temperature difference so that it is also reasonable to replace T_1 by $\Delta T \equiv T - T_w$, where T is the fluid temperature adjacent to the airfoil (actually outside the boundary layer), and T_w is the airfoil wall temperature.

Equation 5.8-1 can thus be replaced by

$$D = D(U_1, \rho_1, \Delta T, \Delta p, a_1, c_p, c_v, \mu_1, k_1, b, c, g). \qquad (5.8\text{-}2)$$

There are 13 independent dimensional parameters whose dimensions are given in the matrix:

	D	U_1	ρ_1	ΔT	Δp	a_1	c_p	c_v	μ_1	k_1	b	c	g
M	1	0	1	0	1	0	0	0	1	1	0	0	0
L	1	1	-3	0	-1	1	2	2	-1	1	1	1	1
t	-2	-1	0	0	-2	-1	-2	-2	-1	-3	0	0	-2
ϑ	0	0	0	1	0	0	-1	-1	0	-1	0	0	0

The dimensions of k_1 are computed from Fourier's law of heat conduction $\dot{q} = -k_1 \partial T/\partial y$, where $\dot{q}$ is the heat flux (energy per unit area per unit time). Since there are four primary dimensions, we expect at least nine independent π-ratios.

INTERCHANGING THE ROLES OF THE BASE AND THE REPEATING DIMENSIONS

We select as repeating parameters the quantities U_1, ρ_1, ΔT, and the airfoil chord c. It can be verified that a π-ratio cannot be formed from them. We now interchange the roles of the base and the repeating dimensions. That is, having written $[c] = L$, we write instead $[L] = c$. From the preceding list, by the same procedure, $[t] = [c/U_1] = cU_1^{-1}$. Similarly, $[M] = \rho_1 c^3$, $[\vartheta] = \Delta T$.

The drag force D is involved in the first π-ratio. The latter is obtained readily in consequence of the requirement that it be dimensionless. That is, $\pi_1 = D/[D]$. Therefore, we can write, simply,

$$\pi_1 = \frac{D}{[D]} = \frac{D}{\left[ML^{-1}t^{-2}\right]} = \frac{D}{\left(\rho_1 c^3\right)\left(c\right)^{-1}\left(cU_1^{-1}\right)^{-2}} = \frac{D}{\rho_1 U_1^2 c^2}.$$

Similarly,

$$\pi_2 = \frac{\Delta p}{[\Delta p]} = \frac{\Delta p}{\left[ML^{-1}t^{-2}\right]} = \frac{\Delta p}{\left(\rho_1 c^3\right)\left(c\right)^{-1}\left(cU_1^{-1}\right)^{-2}} = \frac{\Delta p}{\rho_1 U_1^2}.$$

The other ratios are obtained in the same manner:

$$\pi_3 = \frac{a_1}{U_1}, \quad \pi_4 = \frac{c_p \Delta T}{U_1^2}, \quad \pi_5 = \frac{c_v \Delta T}{U_1^2}, \quad \pi_6 = \frac{\mu_1}{\rho_1 U_1 c}, \quad \pi_7 = \frac{k_1 \Delta T}{\rho_1 U_1^3 c}, \quad \pi_8 = \frac{b}{c}, \quad \pi_9 = \frac{gc}{U_1^2}.$$

IDENTIFICATION OF THE SIGNIFICANT Π-RATIOS

There is never any guarantee that the particular π-ratios obtained — which depend very much on the choice of repeating parameters — will be the most useful. However, any product of π-ratios is also dimensionless, so that an infinite number of associated ratios can always be generated. The selection of the most desirable group is a matter partly of choice but more of experience. In the present case the first ratio is obviously a version of the force coefficient defined in Equation 5.7-1. In an actual case, instead of π_1, aerodynamicists would use the drag coefficient

$$C_D = \frac{D}{\frac{1}{2}\rho_1 U_1^2 (\text{planform area of wing})}. \tag{5.8-3}$$

The denominator of Equation 5.8-3 is the product of the free-stream dynamic pressure $\frac{1}{2}\rho_1 U_1^2$ and the planform area, which is the wing area projected upon the horizontal plane.

The ratio π_2, as obtained, is called the *Euler number*. It is the custom in aerodynamics literature to use instead the dynamic pressure in the denominator, resulting in a form called the *pressure coefficient*, i.e.,

$$C_p = 2\pi_2 \equiv \Delta p / \frac{1}{2}\rho_1 U_1^2. \tag{5.8-4}$$

In most cases the magnitude of C_p is the order of unity (i.e., equal, say, to 0.5, or 1.5, or 2.0, but rarely as large as 10.0).

Some of the other ratios are already essentially in a form used by workers in fluid mechanics. Thus, put

$$\pi_3^{-1} = U_1/a_1 \equiv M_1 \qquad = \text{free - stream Mach number};$$

$$\pi_4^{-1} = U_1^2/c_p\,\Delta T \equiv E \qquad = \text{Eckert number};$$

$$\pi_6^{-1} = \rho_1\,U_1 c/\mu_1 \equiv \text{Re}_1 \quad = \text{Reynolds number}; \qquad\qquad (5.8\text{-}5a)$$

$$\pi_8 = b/c \qquad\qquad\qquad = \text{airfoil thickness ratio};$$

$$\pi_9^{-\frac{1}{2}} = U_1/\sqrt{gc} \equiv \text{Fr} \qquad = \text{Froude number}.$$

It turns out that, instead of the remaining ratios, it is more useful in engineering practice to employ ratios formed by combining each with one or more of those already listed. Thus,

$$\pi_4/\pi_5 = c_p/c_v \equiv \gamma \qquad \text{ratio of specific heats};$$

$$\pi_4\pi_6/\pi_7 = \mu_1 c_p/k_1 \equiv \text{Pr}_1 \qquad \text{Prandtl number}.$$

$$(5.8\text{-}5b)$$

The problem has thus been reduced to a relationship between nine π-ratios (there are no other independent ratios), which can be written as

$$C_D = f\big(C_p, M_1, E, \text{Re}_1, b/c, \gamma, \text{Fr}, \text{Pr}_1\big). \qquad\qquad (5.8\text{-}6)$$

Equation 5.8-6 is the end result of applying dimensional analysis to Equation 5.8-2. A principal consequence is that the number of independent variables has been reduced from 12 to 8.

It takes some substantial experience to go further without resorting to analytical techniques. For example, we know that the Mach number is the single most important parameter in characterizing the phenomena due to compressibility effects in gases, whereas the Reynolds number plays a similar role when effects of viscosity enter. As we have seen, the ratio b/c provides us with a criterion for imposing the requirement of geometric similitude when modeling a flow. In heat transfer problems the Eckert and Prandtl numbers appear as important parameters. Finally, the Froude number is significant in flows of liquids with free surfaces, or a flow in which changes of height are of importance. Fortunately, a circumstance in which all of these parameters enter simultaneously is rare. We shall pinpoint the significance of some of the more important parameters in the following chapters.

5.9 EXAMPLE WHERE THE NUMBER OF Π-RATIOS IS GREATER THAN $m - n$

In a small, straight, vertical tube of radius r open at both ends, one end of which is immersed in a liquid of density ρ, the action of capillarity draws the liquid up into a tube to a height h above that of the reservoir surface. It is assumed that the driving force is the surface tension of the fluid characterized by the surface tension coefficient σ.

An attempt to simplify $f(\rho, r, h, \sigma) = 0$ by dimensional analysis fails, since there is no resulting π-ratio involving σ. Hindsight convinces us that the failure is due to the fact that the final position of the capillary surface represents an equilibrium between the surface tension forces and the gravitational force and that we have neglected to include the gravitational constant g. It takes a deeper understanding, however, to recognize that g enters in a special way, that is, as a multiplier of the density, producing the specific weight $w \equiv \rho g$ as an additional dimensional parameter.

We thus revise our assumed functional relation to $f(w, r, h, \sigma) = 0$. Since there are four dimensional parameters and three base dimensions, we might expect there is only one π-ratio. But, obviously, there is a ratio $\pi_1 = h/r$. The dimensions of the two remaining quantities are $[\sigma] = Mt^{-2}$ and $[w] = ML^{-2}t^{-2}$. The tip-off that there will be a difficulty is that the only way in which mass and time appear is in the combination Mt^{-2}, effectively reducing the base dimensions to a total of two. It is simple to form the other ratio by dividing the second of these quantities by the first and multiplying the quotient by the radius squared (this choice of a length parameter is arbitrary; the quantity h^2 would be just as legitimate) to obtain $\pi_2 = wr^2/\sigma$, called the *Weber number*. Thus, we have

$$h/r = G(wr^2/\sigma). \tag{5.9-1}$$

In this example we have obtained by elementary reasoning a relation that is only one step away from the quantitative form of the equation that governs this phenomenon, and that was derived as Equation 2.21-4a. There it was shown that $h/r = 2\sigma/wr^2$; by comparison, we deduce that $G(\pi_2) = 2/\pi_2$.

5.10   KINEMATIC AND DYNAMIC SIMILARITY

There are various kinds of similarity and, when using the word, it is well to note the context and to distinguish carefully between the different types. We have already mentioned geometric similarity, the ratio that must be maintained between any two corresponding lengths when scaling a model.

KINEMATIC SIMILARITY

There is also a *kinematic similarity*, which is usually, although not always, implied* when comparing two flow patterns, for example, Figure 5.10-1.

In flow ①, $V_1 = e_x u_1 + e_y v_1 + e_z w_1$ is the velocity vector at an arbitrary point. If each of the velocity components were taken as parameters in a dimensional analysis of the flow, then we would find that u_1/V_1, v_1/V_1, w_1/V_1 would be dimensionless parameters. That is, the ratio u_1/V_1 which is a direction cosine, must be the same at a corresponding point in flow ② if the two are to be compared. This is the

---

* In applications of the Prandtl–Glauert rule of compressible flow, or the transonic and hypersonic similarity rules discussed in aerodynamics texts, the requirements of geometric and kinematic similarity must be relaxed.

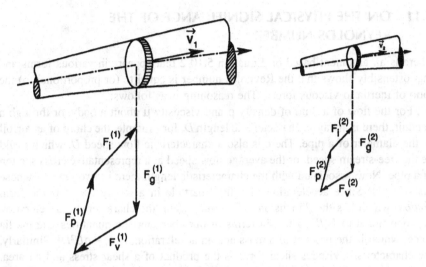

FIGURE 5.10-1 Similarity between two flows.

requirement for kinematic similarity. Stated simply, two flows are kinematically similar if at corresponding points the velocity vectors are parallel and if the ratio of corresponding speeds is everywhere a constant.

DYNAMIC SIMILARITY

Frequently, one encounters the concept of an *inertia force*. This quantity arises when Newton's second law is rewritten in the form called *D'Alembert's principle*, $-ma + \sum F = 0$, where the summation includes viscous, pressure, and body forces. We now define the inertia force as $F_i \equiv -ma$, and thus obtain

$$\Sigma F = 0, \tag{5.10-1}$$

where the summation in Equation 5.10-1 is understood to include, also, the inertia term. But Equation 5.10-1 is of the same form as the equation of equilibrium, of statics. Thus all consequences of the equation of equilibrium also apply as long as the inertia force is included. In particular, this means that the vector force diagram associated with a fluid particle, illustrated in Figure 5.10-1, is a closed polygon. This provides the definition of dynamic similarity:

Two flows are said to be dynamically similar if the force polygons for corresponding fluid particles are geometrically similar.

Denoting the viscous, pressure, and body forces by the subscripts v, p, and b, respectively, then it follows that if flows ① and ② are to be dynamically similar,

$$\frac{F_i^{(1)}}{F_i^{(2)}} = \frac{F_v^{(1)}}{F_v^{(2)}} = \frac{F_p^{(1)}}{F_p^{(2)}} = \frac{F_b^{(1)}}{F_b^{(2)}}. \tag{5.10-2}$$

5.11   ON THE PHYSICAL SIGNIFICANCE OF THE REYNOLDS NUMBER

There is an argument based on Equation 5.10-2 that appears in various forms and that ostensibly shows that the Reynolds number is equal to (or proportional to) the ratio of inertia to viscous forces. The reasoning is as follows:

For the flow of a fluid of density ρ and viscosity μ about a body, or through a conduit, there is always a characteristic length D, for example, the chord of an airfoil or the diameter of a pipe. There is also a characteristic flow speed U, which could be the free-stream speed, or the average flow speed at a representative cross section of a pipe. Now, associated with the characteristic length there is a volume D^3 whose mass is ρD^3 Since acceleration of a fluid particle in a steady flow has the form $u \partial u/\partial x$, which has the dimensions of *speed²/length*, the characteristic acceleration is proportional to U^2/D. Thus, in terms of our characteristic quantities, the inertia force, which is the product of a mass and an acceleration, is $F_i \sim \rho U^2 D^2$. Similarly, the characteristic viscous shear force is the product of a shear stress and an area. That is, $\tau = \mu \partial u/\partial y \sim \mu D^{-1}$, and $F_v \sim \mu U D$. Consequently, we get $F_i/F_v \sim \rho U D/\mu =$ Re, which states that the ratio of inertia to viscous forces is proportional to the Reynolds number.

Now, while most authors note that this is only a proportionality there are those who state flatly that the magnitude of the Reynolds number represents the actual ratio of the two forces at an arbitrary point in the flow. This interpretation has the weight of a most eminent authority behind it, that of L. Prandtl (1875–1953), who is usually considered to be the founder of modern fluid mechanics. In Prandtl (1952, pp. 103–104) he has used essentially the preceding procedure to come to this conclusion.

Yet, there are difficulties. Recalling that pressure forces are proportional to the dynamic pressure $^1/_2 \rho U^2$, we can deduce that the characteristic pressure force has the same form as the inertia force, i.e., $F_p \sim \rho U^2 D^2$. In fact, some writers state that the Reynolds number represents the ratio of pressure to viscous forces. Clearly, one cannot have it both ways.

Apparently, the view of Prandtl just noted was not his ultimate statement on the subject. Kline (1986, p. 91) writes:

> Professor G. F. Wislicenus attributed to L. Prandtl the following opinion: Reynolds number does not always equal the ratio of inertia to viscous stress, hence it is correct to say only that when two systems are geometrically similar and have the same Reynolds number, the ratio of inertia to viscous forces is the same in both flows.

We concur with the latter statement and will return to the subject in Section 5.12 and in Chapter 6.

5.12   INSPECTIONAL ANALYSIS

The technique of *inspectional analysis* dates to the 19th century, but it has been widely exploited only in this century. The name was suggested by Ruark (1935), who also detailed its principal features. In a provocative and somewhat controversial book, Birkhoff (1960) reviewed the method and gave several applications.

The essential aspect of inspectional analysis that distinguishes it from dimensional analysis is that it requires knowledge of the differential equations governing the phenomenon in question. It is not necessary to integrate these equations in order to make them yield certain information that dimensional analysis does not; specifically, it identifies, from the group of possible dimensionless parameters, those upon which the solution of the differential equation actually depends. Furthermore, if initial and boundary conditions are specified (without which, formulation of the problem is incomplete), these can also be made to yield information. Birkhoff calls analysis of only one of the differential equations of a set, and without specification of initial and boundary conditions, *partial* inspectional analysis.

Prior to illustration of the technique, we caution that inspectional analysis, like dimensional analysis, requires judgment tempered by experience to yield the most useful results. To a real extent its conclusions are controlled by the original assumptions, a fact that has occasionally been overlooked.

APPLICATION TO THE DYNAMIC EQUATION

The dynamic equation (Equation 3.12-10) can be simplified slightly. The viscous term involves the factor A/P, which is the ratio of the cross-sectional area to the wetted perimeter of the duct. For a circular cross section $A/P = D/4$, i.e., one fourth the local diameter. For a noncircular duct we also put $4A/P = D_h$, where D_h is called the *hydraulic diameter** and where P is the wetted perimeter.

No important conclusions are lost if, for simplicity, we consider the special case of Equation 3.12-10 for incompressible flow where both density ρ and dynamic viscosity μ are constant. Then, for a gravitational body-force,

$$\frac{DV}{Dt} = -\frac{1}{\rho}\frac{\partial p}{\partial s} - \frac{4\tau_w}{\rho D} - g\frac{\partial z}{\partial s}. \tag{5.12-1}$$

The procedure in inspectional analysis commences by reducing the governing equations — in this case the dynamic equation — to dimensionless form. To accomplish this, we assume, as in the previous section, that the nature of the problem makes evident certain characteristic dimensional quantities. For the characteristic length, we choose the hydraulic diameter at some reference station ①, designated for simplicity as D_1. We also choose a characteristic speed U_1 at the same station. The flow density ρ is constant.

Since pressure forces depend on pressure differences, it is permissible to replace p by $\Delta p \equiv p - p_1$, where p_1 is a (constant) reference pressure. We now define our group of dimensionless variables denoted by a bar: $s \equiv D_1\bar{s}$, $z \equiv D_1\bar{z}$, $D \equiv D_1\bar{D}$, $V \equiv U_1\bar{V}$. The quotient D_1/U_1 provides a characteristic time. Thus, we put $t \equiv (D_1/U_1)\bar{t}$.

---

* In the older technical literature, the hydraulic diameter is sometimes defined as $2A/P$ which, for a circular duct, turns out to be one half the diameter. Perpetuating such a historical incongruity seems to have little purpose.

We next proceed systematically through each term in Equation 5.12-1 to obtain the appropriate transformed relation.

For the left-hand side of Equation 5.12-1, we first expand the substantial derivative prior to nondimensionalizing:

$$\frac{DV}{Dt} = \frac{\partial V}{\partial t} + V\frac{\partial V}{\partial s} = \frac{\partial(U_1\overline{V})}{\partial(D_1/U_1)\overline{t}} + U_1\overline{V}\frac{\partial(U_1\overline{V})}{\partial(D_1\overline{t})} = \frac{U_1^2}{D_1}\left[\frac{\partial\overline{V}}{\partial\overline{t}} + \overline{V}\frac{\partial\overline{V}}{\partial\overline{s}}\right].$$

Thus,

$$\frac{DV}{Dt} = \frac{U_1^2}{D_1}\frac{D\overline{V}}{D\overline{t}}. \tag{5.12-2a}$$

Similarly,

$$\frac{1}{\rho}\frac{\partial p}{\partial t} = \frac{1}{\rho}\frac{\partial(\Delta p)}{\partial(D_1\overline{s})} = \frac{1}{\rho D_1}\frac{\partial(\Delta p)}{\partial\overline{s}}, \quad \frac{4\tau_w}{\rho D} = \frac{4\tau_w}{\rho D_1\overline{D}}, \quad g\frac{\partial z}{\partial s} = g\frac{\partial\overline{z}}{\partial\overline{s}}. \tag{5.12-2b}$$

By substituting Equation 5.12-2 into Equation 5.12-1, and multiplying through by D_1/U_1^2, the dimensionless form of the dynamic equation becomes

$$\frac{D\overline{V}}{D\overline{t}} = -\frac{\partial}{\partial\overline{s}}\left[\frac{\Delta p}{\rho_1 U_1^2}\right] - \frac{4\tau_w}{\rho_1 U_1^2\overline{D}} - \frac{gD_1}{U_1^2}\frac{\partial\overline{z}}{\partial\overline{s}}. \tag{5.12-3}$$

Each term of Equation 5.12-3 is dimensionless. And of the first and third terms on the right, each contains a nondimensional group that is familiar, the first involving the pressure coefficient, the third the Froude number, specifically, $\Delta p/\rho U_1^2 \equiv \frac{1}{2}C_p$, $gD_1/U_1^2 \equiv 1/Fr^2$. The comparable group in the second term is also nondimensional. It involves the *friction factor* of pipe flow, defined by

$$f = \frac{8\tau_w}{\rho U_1^2}. \tag{5.12-4}$$

We will deal at length with the friction factor in Chapter 6. Therefore, the final nondimensional form of the dynamic equation is

$$\frac{D\overline{V}}{D\overline{t}} = -\frac{1}{2}\frac{\partial}{\partial\overline{s}}(C_p) - \frac{f}{2\overline{D}} - \frac{1}{Fr^2}\frac{\partial\overline{z}}{\partial\overline{s}}. \tag{5.12-5}$$

DISCUSSION

We have reduced the differential equation to a form involving only the dimensionless indpendent variables of space and time $(\overline{s}, \overline{t})$, the shape functions $(\overline{z}, \overline{D})$, the two dependent variables $(\overline{V}, C_p)$, and the two dimensionless parameters (f, Fr). Consequently, the solution of Equation 5.12-5 can be represented as an implicit function:

$$f\left(C_p, \overline{V}, \overline{z}, \overline{s}, \overline{D}, \overline{t}; f, \mathrm{Fr}\right) = 0, \tag{5.12-6}$$

where the semicolon separates the parameters from the variables. However, the duct shape and orientation must be representable by independent relations of the form $\overline{D} = \overline{D}(\overline{s}, \overline{t})$, $\overline{z} = \overline{z}(\overline{s}, \overline{t})$. Therefore, $\overline{D}$ and $\overline{z}$ can be eliminated as independent variables, resulting in the further simplification of Equation 5.12-6:

$$f\left(C_p, \overline{V}, \overline{s}, \overline{t}; f, \mathrm{Fr}\right) = 0. \tag{5.12-7a}$$

Equation 5.12-7a can also be rewritten in explicit form, using the pressure coefficient, for example, as a dependent variable:

$$C_p = C_p\left(\overline{V}, \overline{s}, \overline{t}; f, \mathrm{Fr}\right). \tag{5.12-7b}$$

This is as far as one can go* without bringing in additional equations (e.g., the continuity and the energy equations) and boundary and/or initial conditions in which, for example, the flow rate, or the initial velocity at some station, might be specified. Equation 5.12-7b expresses the pressure coefficient at an arbitrary point and time as a function of the independent variables $(\overline{s}, \overline{t})$, the velocity $\overline{V}$, and of the two parameters (f, Fr). Furthermore, we note that the unknown function involves only dimensionless (i.e., independent of scale) variables and is, therefore, valid for the general case. Keeping in mind that this is only a partial inspectional analysis, we see that we have obtained all the information yielded by dimensional analysis and more, since extraneous dimensionless parameters are automatically avoided. Thus, if a complete inspectional analysis is available, the need for a dimensional analysis is obviated.

Since the characteristic dimensional quantities do not appear explicitly, every solution, when obtained, is valid for all combinations of dimensional quantities that produce a given set of the parameters. It also follows that, if dynamic similarity is to be maintained between the two flows, the dimensionless coefficients must be the same in both since, for example, at a given point in a duct, the ratio of inertia to viscous forces is given by either of

$$\frac{F_i}{F_v} = \frac{DV/Dt}{4\tau_w/\rho_0 D} = \frac{D\overline{V}/D\overline{t}}{f/2\overline{D}}. \tag{5.12-8}$$

Now the right-hand term in Equation 5.12-8 exhibits no explicit dependence on the scale of the flow, for example, on the dimensional diameter of the duct.

---

* In Chapter 6 we shall interpret Equation 5.12-7b in light of the solution for steady-state pipe flow.

Thus, since the functions $\overline{V} = \overline{V}(\overline{s}, \overline{t})$ and $\overline{D} = \overline{D}(\overline{s}, \overline{t})$ are independent of the size of the characteristic quantities, the friction factor at a specified pair of $(\overline{s}, \overline{t})$ must also be independent. Therefore, if we are comparing two flows of different scale, and we require them to be geometrically and dynamically similar, the friction factor must be the same at identical values of $(\overline{s}, \overline{t})$ for each. Similarly, the ratios of the other forces — i.e., F_p/F_v and F_g/F_p — must also be independent of the scale. It then follows that C_p and Fr must also be the same in each flow for the same pair of $(\overline{s}, \overline{t})$.

Use of Equation 5.12-8 allows us to make final disposition of the physical interpretation of the Reynolds number as being "equal to" (or, as sometimes stated, "proportional to") the ratio of inertia to viscous forces. We shall see, in the important case of steady, viscous flow through a constant-area pipe, that the particle acceleration $D\overline{V}/D\overline{t} \equiv 0$, and thus the inertia force F_i is also identically zero, as must be the ratio F_i/F_v. However, we shall also show in Chapter 6 that the friction factor for a pipe flow depends importantly on the Reynolds number. Consequently, the interpretation of the Reynolds number as the ratio of two forces is invalid, and perforce meaningless.*

Thus, a rule is hereby proposed that covers all situations: *The Reynolds number can be interpreted as the ratio of inertia to viscous forces except insofar as it is not.*

5.13 DYNAMIC SIMILARITY AND MODELING

It is desired to design a test setup to study the flow at average speed U of a viscous liquid through a uniform duct or channel, dynamically similar to a prototype flow, but which, to keep testing costs down, is of a smaller geometric scale. If the subscripts 1 and 2 denote the prototype and model flows, respectively, then let the scale factor λ be defined by $D_1 \equiv \lambda D_2$, with $\lambda > 1$, where D is the duct diameter. From Equation 5.12-7b inspectional analysis tells us (taking into account the dynamic equation only) for C_p to be the same in both flows, the friction factor and the Froude number must also be the same for both. As we shall see in Chapter 6, since the friction factor is a function of the Reynolds number, the two Reynolds numbers must be equal.

Thus, what is sometimes called Reynolds modeling requires that $\rho_1 U_1 D_1/\mu_1 = \rho_2 U_2 D_2/\mu_2$, or $\lambda \rho_1 U_1/\mu_1 = \rho_2 U_2/\mu_2$. Similarly, Froude modeling requires $U_1^2/D_1 g = U_2^2/D_2 g$, or $U_1^2 = \lambda U_2^2$. From the latter relation, we see that Froude modeling immediately determines the test speed when the prototype speed and model scale are specified. Simultaneous Reynolds and Froude modeling requires that $\lambda^{3/2}\rho_1/\mu_1 = \rho_2/\mu_2$.

---

* It is reasonable, if the author's argument is valid, to inquire why the original statement is so widely encountered, even in the writings of admitted authorities. The answer seems to be, first, that if we select the form "proportional to" rather than "equal to," the statement seems to be "true." Thus, if the ratio of inertia to viscous forces is zero, the constant of proportionality is zero. More important, in actual applications it can be readily verified that the "fact" of supposed F_i/F_v dependence is never explicitly invoked, and nowhere in the analytical result does this fact make its appearance. This conclusion, however, does not negate the requirement that for two flows to be dynamically similar, the Reynolds number must be the same for both.

The last result indicates one of the hard facts of life facing test engineers. If it is specified that the prototype and text fluids be the same liquid at a fixed temperature, it is impossible to design a test that simultaneously satisfies the Reynolds and Froude criteria, since the preceding equation requires that the kinematic viscosity $\mu_2/\rho_2 \equiv v_2 = \lambda^{-3/2}v_1$. Obviously, $v_2 \neq v_1$ unless the scale factor is unity. Even worse, for a prototype flow of water and with $\lambda > 1$, we must have $v_2 < v_1$, a requirement that is virtually unattainable, even by utilization of another liquid in the second (test) flow, since water has one of the lowest kinematic viscosities of all known liquids. In consequence, ship designers find it impossible to conduct tests in water that simultaneously achieve Reynolds modeling (for frictional) drag and Froude modeling (for surface-wave resistance). Instead, tests are conducted for each effect separately, and the combined effects are determined by patching the two sets of experimental data together with plausible semiempirical techniques.

The preceding discussion is, of course, not complete, since of the governing equations (the continuity, the dynamic and energy equations, and the equations of state), we have examined only the dynamic equation for constant-density flow. Inspectional analysis of the continuity equation does not disclose any significant new results. If the question were extended to a problem of high flow speeds for a perfect gas, with heat conduction, the analysis of the energy equation can be expected to show dependence on the Mach number, the ratio of specific heats, and the Eckert number.

In any particular physical situation, to be able to eliminate all nonessential dimensionless parameters, and to retain all those of importance is not a trivial matter, and no general rules can be formulated to cover all situations. For scale-model testing, we have already seen in one instance, the requirements may be mutually incompatible, a difficulty that can be surmounted only with experience and even then not completely. In the following chapters the roles of several of the more important dimensionless parameters will be developed in detail.

5.14 A DIMENSIONAL ANALYSIS AND AN INSPECTIONAL ANALYSIS COMPARED WITH THE COMPLETE SOLUTION

It is instructive to illustrate the significance of the results of dimensional and inspectional analysis in the light of a problem for which the solution is completely known. Consider the familiar, Figure 5.14, case of a linear spring–mass–dashpot system oscillating about its rest position $y = 0$. The applicable dimensions matrix is

	y	m	k	c	F_0	ω	t	g
M	0	1	1	1	1	0	0	0
L	1	0	0	0	1	0	0	1
t	0	0	-2	-1	-2	-1	1	-2

We seek the "steady-state" motion $y(t)$ of the mass m, suspended by a spring (constant k) with a viscous damper (constant c) under a sinusoidal force $F = F_0 \sin \omega t$ indicated in Figure 5.14-1.

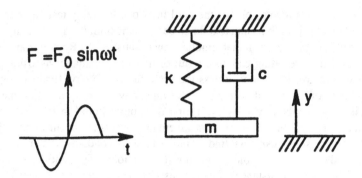

FIGURE 5.14-1   Mass in forced oscillation.

DIMENSIONAL ANALYSIS

If we assume that the solution depends on the gravitational constant g, then, choosing m, F_0, ω as working parameters we have five independent π-ratios: $\pi_1 = m\omega^2 y/F_0$, $\pi_2 = k/m\omega^2$, $\pi_3 = c/m\omega$, $\pi_4 = \omega t$, and $\pi_5 = mg/F_0$. According to the π-theorem these five ratios are linked by a functional relationship.

INSPECTIONAL ANALYSIS

The differential equation governing the motion of the mass is

$$m\ddot{y} + c\dot{y} + ky = F_0 \sin \omega t. \tag{5.14-1}$$

Since the gravitational constant does not appear in Equation 5.14-1, it is apparent that the assumed dependence on it in the dimensional analysis was erroneous and, therefore, the solution cannot involve π_5.

In nondimensionalizing Equation 5.14-1 there is a characteristic time ω^{-1}, but no obvious characteristic length. However, since there are both a characteristic mass m and force F_0, we can generate a characteristic length $F_0/m\omega^2$. We then put $y = (F_0/m\omega^2)\bar{y}$, $t = \omega^{-1}\bar{t}$. Then, similar to the procedure in Equation 5.12-2, $\dot{y} = (F_0/m\omega)\,\dot{\bar{y}}$, $\ddot{y} = (F_0/m)\,\ddot{\bar{y}}$. With these relations and the previously obtained π-ratios, Equation 5.14-1 becomes

$$\ddot{\bar{y}} + \pi_3 \dot{\bar{y}} + \pi_2 \bar{y} = \sin \bar{t}, \tag{5.14-2}$$

where $\bar{y} = \pi_1$ and $\bar{t} = \pi_4$, thus indicating the existence of a functional relationship $f(\pi_1, \pi_2, \pi_3, \pi_4) = 0$.

THE COMPLETE SOLUTION

The general solution of Equation 5.14-1 is the sum of a transient solution (complementary solution, depending upon initial conditions), which vanishes after a time, and a steady-state (particular) solution. The latter is

$$y = \frac{F_0}{\sqrt{(k - m\omega^2)^2 + c^2\omega^2}} \sin(\omega\tau - \psi), \qquad (5.14\text{-}3a)$$

where the phase angle between the force vector and the mass motion is

$$\psi = \tan^{-1} \frac{c\omega}{k - m\omega^2}. \qquad (5.14\text{-}3b)$$

By straightforward rearrangement, Equation 5.14-3 can be made to show its dependence on the π-ratios. Multiplying Equation 5.14-3a by $m\omega^2/F_0$ and rearranging, we obtain

$$\pi_1 = \frac{1}{\sqrt{(\pi_2 - 1)^2 + \pi_3^2}} \sin(\pi_4 - \psi), \qquad (5.14\text{-}4a)$$

where

$$\psi = \tan^{-1} \frac{\pi_3}{\pi_2 - 1}. \qquad (5.14\text{-}4b)$$

Equations 5.14-4 give the explicit relationship among the independent variable π_1, the independent variable π_4, and the two parameters π_2 and π_3, the existence of which is predicted by dimensional analysis. For a given configuration, since π_2 and π_3 are parameters of fixed value for any case, the functional relationship could be written $\pi_1 = f(\pi_4; \pi_2, \pi_3)$, thereby clearly distinguishing the parameters from the variables.

APPLICATION IN TRANSONIC FLOW

The preceding example has the whiff of triviality, since the differential equation is linear and can be readily integrated; hence, one might ask: Why bother with inspectional analysis? An appropriate counterexample involves flow theory far beyond the objectives of this text. Of the various regimes of compressible flow — subsonic, transonic, supersonic, hypersonic — the regime of transonic flow is notoriously more difficult than the others. This regime is encountered when the speed of an aircraft is approximately equal to the critical speed of sound,* which occurs when the aircraft Mach number approaches unity. At a time when aircraft had not yet broken the "sound barrier" von Kármán (1946), using inspectional analysis, deduced the transonic similarity law — which describes the general way that the aerodynamic forces, lift and drag, depend on the flow Mach number. These results were obtained without integrating the governing differential equations, which are nonlinear. This led to the ability to correlate wind tunnel data in a rational way and helped speed up the advent of supersonic flight — a major engineering advance.

---

* When a flowing gas expands, the sound speed decreases. When the flow speed equals the sound speed, *critical conditions* are said to exist.

PROBLEMS

5.1. The velocity of propagation c of surface waves through a liquid is assumed to depend only on the depth of the liquid h, the liquid density ρ, and the acceleration of gravity g. Simplify by dimensional analysis. Compare with the theoretical (linearized) solution of propagating harmonic waves of frequency ω and wavelength L, whose solution is given by Milne-Thomson (1968, p. 447):

$$\omega^2 = \frac{2\pi}{L}\left[g + \frac{\sigma}{\rho}\left(\frac{2\pi}{L}\right)^2\right]\tanh\left(2\pi h/L\right), \quad c^2 = \left[\frac{gL}{2\pi} + \frac{\sigma}{\rho}\left(\frac{2\pi}{L}\right)\right]\tanh\left(2\pi h/L\right),$$

Note: We see that the surface tension coefficient σ enters the problem of wave propagation through a liquid and that the wave frequency and wave length are not independent. A look at some limiting cases is revealing. For $\sigma \to 0$ and for shallow depths, where $2\pi h/L \ll 1$, we can make the replacement $\tanh(2\pi h/L) \approx 2\pi h/L$; thus, $c^2 \approx gh$ and the wave speed is proportional to $\sqrt{h}$. At the other extreme, for deep water waves $2\pi h/L \to \infty$, $\tanh(2\pi h/L) \to 1$, and $c^2 = gL/2\pi$, so that the speed of propagation becomes proportional to $\sqrt{L}$. Hence, in tidal waves, often generated by earthquakes, a single wave of "infinite" wave length (called a *soliton*) may result that travels at enormous speed, sometimes the order of 300 to 400 miles per hour.

5.2. Consider the problem of a sphere of mass density $\bar{\rho}$ and diameter D that is falling at speed U through a very viscous fluid of density ρ and viscosity μ. What other dimensional quantity(s) must enter the problem of determining its terminal rate of fall? Simplify by dimensional analysis using ρ, U, and D as repeating variables. Compare with Stokes' famous solution for the terminal speed in *creep flow*.

$$U = \frac{2}{9}\frac{gD^2(\bar{\rho} - \rho)}{\mu}.$$

Note: This solution was used by Robert Millikan in determining the charge of an electron in his famous oil-drop experiment.

5.3. It is assumed that the thrust F developed by a jet engine depends on the mass rate of flow $\dot{m}$, the cross-sectional area A of the nozzle, the inlet velocity V, the rate of heat addition to the flow $\dot{q}$, given in units of energy per unit mass per unit time, the specific heat at constant pressure of the fluid c_p, and the temperature rise ΔT due to combustion. Simplify by dimensional analysis using $\dot{m}$, A, V, ΔT as repeating variables.

5.4. The figure is a schematic cross section of a sliding bearing in which the slider plate moves over the bearing surface at relative velocity U. It is assumed that the load per unit width $\overline{W}$ that the bearing can support depends on the slider speed U, the average clearance b, the angle of inclination of the slider α, the slider length a, the lubricant viscosity μ, and dp/dx the longitudinal pressure gradient (a constant) in the fluid in contact with the bearing. Simplify by dimensional analysis.

Note: The solution of this problem, which is a part of the hydrodynamic theory of lubrication, can be found in Schlichting (1979, p. 116).

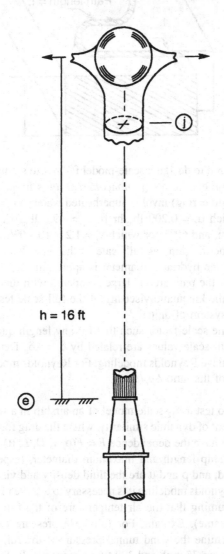

$h = 16 \text{ ft}$

5.5. In a dimensional analysis it is assumed that the overall pressure drop Δp of a flow through a complicated two-dimensional hydraulic path of constant width δ and length L depends on the fluid density ρ, the viscosity μ, and the mean fluid velocity V. Simplify by dimensional analysis using ρ, V, and δ as repeating variables.

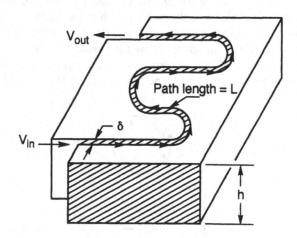

It is next desired to design a scale-model facility to simulate the full-scale flow. Full-scale conditions (in *noncoherent units* frequently encountered in nuclear engineering) involve superheated water at $p_f = 2,000$ psia, $T_f = 600°F$ at which $\mu_f = 0.200$ lb$_m$/hr ft, $\rho_f = 49.6$ lb$_m$/ft^3. The test fluid is water at 1 atm, and 60°F for which $\nu_t = 1.2 \times 10^{-5}$ ft^2/sec. The test speed is chosen to be the same as full-scale so that $V_t = V_f$.

a. Show that the hydraulic diameter is approximately 2δ, as long as the height h of the path area is large compared with that of the width δ.

b. Compute the kinematic viscosity of the full-scale test fluid in the lb$_f$-ft-sec-°R system of units.

c. Let λ be the scale factor such that for any length quantity, say, δ, the test and full-scale values are related by $\delta_f = \lambda \delta_t$. Determine the value of λ to achieve Reynolds modeling. For Reynolds modeling, determine the value of the ratio $\Delta p_t / \Delta p_f$.

5.6. It is desired to test a $1/12$-scale model of an airship in a wind tunnel under the requirement of dynamic similarity, where the drag force F of an airship is assumed to have the dependence $F = F(\rho, V, D, L, \mu)$ and where L and D are the airship length and maximum diameter, respectively, V is the cruising speed, and ρ and μ are the fluid density and viscosity. Show that to achieve Reynolds modeling it is necessary to test the model at a different pressure (assuming that the air temperature of the full-scale and model test are the same). Explain. For full-scale pressure of standard atmospheric, determine the wind tunnel pressure if the full-scale and model test velocities are 75 mph and 300 ft/sec, respectively. Determine also the ratio of the two drag forces. Suppose, instead, the model test were in water at 60°F. What would be the appropriate test speed?

Note: The use of a pressurized wind tunnel to achieve Reynolds modeling was first envisioned by Max Munk, a German, and later American, aerodynamicist, who designed the first pressurized tunnel, which was built at the NACA (now NASA) Langley Laboratory in Virginia.

5.7. A producer of motion pictures engages you as a consultant on the simulation of battle scenes utilizing scale-model ships. Both test and full-scale scenes (subscripts t and f, respectively) take place on water of density ρ. The models are geometrically scaled according to $L_f = \lambda L_t$. The full-scale ships are built of oak, whereas the models are of a low-density plastic. A principal question is how should the model weight W_t (plastic + ballast) be scaled to give the appropriate dynamically similar period τ of a maneuver (say, the period of rolling) due to gravitational buoyancy forces.

a. Would ρ, L, and g form a suitable set of repeating variables? Explain. If not, choose a satisfactory set. Determine the ratio of weights W_t/W_f if the scale factor is $\lambda = 20$.

b. Determine the ratio of the periods of rolling τ_t/τ_f. Note that the period of rolling depends on the mass moment of inertia.

c. If $\tau_t/\tau_f \neq 1$, explain how the film projection speed can be adjusted to make the effective ratio appear to be unity.

d. The model basin in which the filming takes place is a $1/20$-scale model. If dynamic effects of wave action are maintained, how will the speeds (i.e., the ratio V_t/V_f) be related? Then, if the projector speed is modified as suggested in part (c), how will a wave event (e.g., the time for a wave to traverse the length of a ship) appear to the viewer compared with the same full-scale event?

5.8. It is assumed that the peak force exerted on a parachute of cross-sectional area S, due to its sudden inflation, depends on its load mass m_0, the relative airspeed V, the air density ρ and viscosity μ, and the gravitational constant g.

a. Which (if either) of the groups (V, S, g) or (ρ, V, S) would serve as a satisfactory set of repeating variables?

b. Using one of these groups (or one of your own if *not* satisfactory), simplify by dimensional analysis.

c. Experiments show that over the range of interest neither viscosity nor the gravitational constant affects the peak load. Show how you would present the (fictitious) data of drops 1, 2, and 3 on a graph.

d. Using the results of part (c) determine the peak load that would be expected in drop no. 4.

Drop No.	F (lb)	$10^3\rho$ (slug/ft³)	V (ft/sec)	S (ft²)	m_0 (slug)
1	8,000	1	100	100	10
2	8,000	2	50	400	320
3	72,000	1	200	900	810
4	$F_4 = ?$	1	100	100	25

5.9. Frequently, it is the practice to recover small test rockets after launch by allowing them to fall to Earth (usually in a sandy terrain) nose first, the impact being absorbed by a penetration spike as shown in the figure. If properly designed, the spike allows the rocket to decelerate to rest rapidly without catastrophic results.

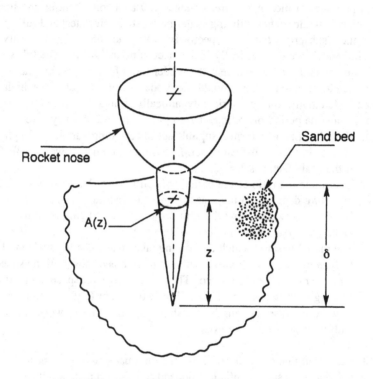

It is assumed that the depth of penetration Δ depends on the rate of change of cross-sectional area $dA/dz \equiv A'$, the total mass m_0 at impact, the impact speed V, the soil density ρ, and the shear modulus of the soil G.

a. Furnish the dimensions of shear modulus.

b. From the quantities Δ, m_0, G, and ρ, determine a suitable set of repeating variables; then simplify by a dimensional analysis.

c. Further reflection indicates that it is probably possible to obtain an immediate simplification by combining mass and impact speed into a single term that is either momentum $P \equiv m_0 V$ or impact kinetic energy $E \equiv m_0 V^2$ (where in a dimensional analysis we can dispense with the factor $1/2$). Using the (fictitious) data provided, determine whether, and if, P or E is the more appropriate parameter on which to correlate the data.

d. Determine the value of the rate of area change required for the specifications of drop no. 6.

Drop No.	m_0 (slug)	V (ft/sec)	Δ (ft)	G (?)	A' (ft)	ρ (slug/ft³)
1	1	2	1	4	1	1
2	4	4	2	4	1	1
3	3	1	1	1	1/3	1
4	8	2	2	1	1/2	1
5	5	1	1	1	1/5	1
6	6	1	2	1	?	1

5.10. In the figure is shown the behavior of an expanding blast wave (i.e., a hemispherical shock wave bounded by the ground) created by an atomic explosion. At $t = 0+$, a quantity of energy E_0, the yield, is released into a small but finite quantity of air, of undisturbed ($t < 0$) pressure and density p_1, ρ_1, respectively. In addition to these quantities, the instantaneous radius $R(t)$ of the hemisphere is assumed to depend on the speed of the shock front c — i.e., the hemispherical boundary — and the ratio of specific heats of air γ. That is, $R = R(t, E_0, \rho_1, p_1, c, \gamma)$.

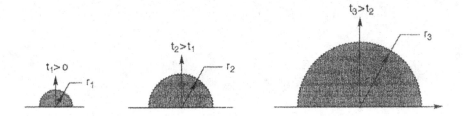

a. Verify that (t, E_0, ρ_1) form a satisfactory set of repeating parameters, and simplify by dimensional analysis.
b. On further reflection it is realized that R, C, and t must be related so that c can be dropped from the analysis. In addition, since the initial pressure of the blast $p = p(0+)$ is enormously greater than p_1, then slight variations in p_1 will have negligible effect on the wave front speed and, therefore, can be dropped out. Last, γ can be taken as a constant for a perfect gas (this is valid only as a first approximation). Applying all these restrictions, use the result of the dimensional analysis for a fixed energy release and initial density to plot the theoretical variation of the shock front radius as a function of time on a log–log plot, and compare with the following (partial) data from the Alamogordo, N.M., explosion of July 16, 1945.

t (ms)	3.5	8.0	10.0	15.0
R (m)	160	240	300	360

Historical Note: This problem was first analyzed by the eminent British fluid dynamicist G. I. Taylor (1950) who, by the use of the techniques of dimensional analysis, and by the method of inspectional analysis, which involves the equations of gas dynamics, was able to back-calculate from the preceding experimental data and estimate the energy release E_0 from the first atomic bomb. This success led to the dropping of similar bombs on the Japanese cities of Hiroshima and Nagasaki on the following August 6 and August 9, respectively. It is now known that the Japanese military dictatorship had decided to sacrifice the entire civilian population rather than to surrender to the Allies. These two bombings, however, caused Emperor Hirohito to demand that the nation surrender, an action that obviated the need for an invasion of Japan proper, and saved countless lives, Japanese as well as American.

6 Flows in Pipes and Conduits

> Let us not forget in this connection that every stream of water, whenever it comes from a higher point and flows into a delivery tank through a short length of pipe, not only comes up to its measure, but yields moreover a surplus; but whenever it comes from a low point, that is under a less head, and is conducted a tolerably long distance, it will actually shrink in measure by the resistance of its own conduit, so that either an aid or a check is needed for the discharge.
>
> Sextus Julius Frontinus (A.D. 40–103), translation by Herschel (1913).

6.1 A HISTORICAL NOTE

The need to create means for the transport of water for human consumption and bathing, for irrigation, and for canals, predates recorded history. There exist relics* of ancient water-distribution systems, some of them dating to 4000 B.C.E., in Egypt, Mesopotamia, and the Indus Valley. Since the ancients had no analytical tools on which to base their designs, their only guide was the technique of trial and error, which was costly in terms of human and material resources. In spite of these deficiencies, the engineers of the time developed remarkably effective systems.

As empirical knowledge accumulated, water-distribution systems in Greece and Rome became more sophisticated. In particular, the Romans, during the first century before, and the two centuries after, the beginning of the common (Judeo-Christian) era, built magnificent aqueducts to supply Rome and several of its colonies. A number of these grand structures still stand** attesting to the knowledge and skill of their designers and builders.

Although the Romans, as Frontinus noted, understood that water required a head to move through a pipe, their understanding of the continuity relation was defective, as Rouse and Ince (1957, p. 31) point out. In consequence, it was not possible to make quantitative measurements of flows, let alone to meter delivery of water to the user. As the quote from Frontinus shows, he recognized that the action of a conduit wall on the flow is to resist the motion. However, the inability to quantify either the

---

* There are two especially noteworthy references that describe the development of *hydraulics* and *hydrology* from the ancient times to the modern era. The first, by Rouse and Ince (1957), treats the history of hydraulics, which is the study of water, or other liquids, in motion. The second, by Biswas (1970), deals with the related, and overlapping, history of hydrology, which is the study of the distribution of water on or near the surface of the Earth, and in the Earth's atmosphere. In addition to the textual material these two volumes provide a compact and valuable source of diagrams, maps, and photographs, as well as a comprehensive list of the primary references that embody the research of the creators of these two subjects.

** The most famous of these is the *Pont du Gard* aqueduct, which supplied the city of Nîmes in southern France.

principle of mass conservation, or the resistance effects, precluded the development of rational design techniques, and restricted the Romans to the use of qualitative, trial-and-error procedures.

The history of hydraulics from the time of ancient Rome to the beginning of the 19th century is fascinating to read, but, excepting the discovery of the continuity equation in the 16th century, it produced little that enabled the designer of water-distribution systems to make rational calculations. This was due, first, to the inability to quantify the effects of flow resistance.

However, there was a second reason almost completely unsuspected either by hydraulics engineers or by the theoreticians who had developed a considerable body of knowledge applicable to flows in which viscosity played a negligible part. This hidden facet of fluid mechanics was the ability of a pipe flow to sustain itself in two completely different internal states, now designated *laminar flow* and *turbulent flow*. The flow losses through a given pipe in these two states must be calculated by quite different relations, neither of which became available until late in the 19th century.

The irrelevance (to hydraulics engineers) of fluid mechanical theory developed up to that point led the practical engineers* to distrust theory and to view theoreticians with disdain, if not antagonism. On the other hand, the *empirical*** and, sometimes, ad hoc relations on which the former relied were ignored by the scientists. This state of mutual contempt has not completely abated even at the end of the 20th century.

6.2 THE EXPERIMENTS OF HAGEN AND POISEUILLE ON FLOW THROUGH CAPILLARY TUBES

At the end of the 18th century there was a significant accumulation of empirical data in hydraulics, mostly based on flows in canals or open conduits, referred to as open-channel flows. We shall not further pursue the question of open-channel flow in this work.

At the beginning of the 19th century Prony in France and Eytelwein in Germany undertook systematic measurements of flows in pipes of the sizes used in large-scale water-distribution systems. The experiments of Prony,*** for example, were conducted in pipes ranging from 1 to 19 in. in diameter, and from 30 to 7500 ft in length. However useful these measurements were to predict the flow losses (i.e., the pressure drops over the pipe length) for comparable pipe segments, they did not result in any improvement in the ability to make general pipe flow calculations.

The measurements that first disclosed genuine scientific information about the nature of the flows in circular pipes were made almost concurrently by G. Hagen (1797–1884), a German engineer, and J. Poiseuille (1799–1869), a French physician, who worked independently of each other. Poiseuille (1846) was interested in the

---

* According to Theodore von Kármán (1954, p. 167) who was one of the great aerodynamicists of the 20th century and a "theoretical engineer" (his expression), when someone told Rankine (himself a pioneer in fluid mechanical theory) that a practical engineer does not need much science, his response was: "Yes, but what you call a practical engineer is one who perpetuates the errors and mistakes of his predecessors."
** Empirical: not founded in science.
*** See Rouse and Ince (1957) for the reference.

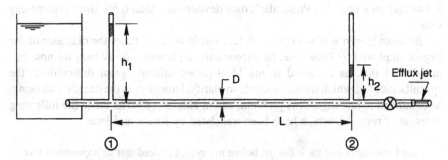

FIGURE 6.2-1 Schematic of the apparatus of Hagen and Poiseuille.

flow of blood, recognizing that the heart was a pump, and that to force blood through capillaries, it must provide sufficiently high upstream pressures. He chose to work with extraordinarily small-diameter glass tubes (diameters ranging from 0.013 to 0.14 mm, and lengths from 1 to 200 mm). In one of Poiseuille's experiments it took over 9 h to achieve a flow output of only 0.36 cc. Hagen* (1839) restricted his experiments to tubes from 2.5 to 6 mm in diameter, and from 47 to 110 cm in length. The experiments of Hagen and Poiseuille both confirmed and supplemented each other, and set a high standard for accurate measurements.

In Figure 6.2-1 is shown only a conceptual view of the apparatus employed by Hagen and Poiseuille. The actual equipment in both cases was quite different from that of the schematic, as well as from each other. A reservoir of the test fluid (distilled water), which can be pressurized as desired, supplies a segment of a (small-bore) tube of diameter D and length L between stations ① and ②, which serves as the test section. The pressure drop $\Delta p \equiv p_1 - p_2$ over L is measured by the heads in the pressure tubes at ① and ②. The flow rate $\dot{Q}$ is computed by measuring the length of time for the efflux jet to pass an accurately determined volume of liquid. Their experimental results, after several years of intense laboratory work, can be summarized in a surprisingly simple relation:

$$\dot{Q} \sim \frac{\Delta p D^4}{L}, \qquad (6.2\text{-}1)$$

Poiseuille gave this relation explicitly. In Hagen's work, it is only implied. If Equation 6.2-1 is converted into an equation, then

$$\dot{Q} = k \frac{\Delta p D^4}{L}. \qquad (6.2\text{-}2)$$

Both investigators found that the factor k is temperature dependent, and Poiseuille concluded that k had the form $A + BT + CT^2$, where T is the temperature in °C.

---

* It is amusing to note that Hagen made his length measurements in *Paris inches* (1 Paris inch = 27 mm), which, in France, had been discarded in favor of the metric system a half century earlier.

Rouse and Ince note that Poiseuille's data deviates less than 0.5% from present-day values.

Because Hagen was working with tubes as large as 42 times the diameter of the largest employed by Poiseuille, he encountered a phenomenon (which we now call turbulence) that, as he noted in his 1839 paper, offered "great difficulties," the "peculiarities" of which he was not able to clarify. However, in the decade following, he continued to investigate the problem, and in Hagen (1854) he made the following noteworthy remarks, which have been translated by Rouse and Ince:

> Since I invariably had the efflux jet before my eyes, I noticed that its appearance was not always the same. At small temperatures it remained immovable, as though it was a solid glass rod. On the other hand, as soon as the water was more strongly heated, very noticeable fluctuations of short period were established, which with further heating were reduced but nevertheless even at the highest temperatures did not wholly disappear...and when I finally made the graphical summary, I found that the strongest fluctuations always took place in that portion of the curve where the velocity decreased with increasing temperature.

> ...Special observations that I made with glass tubes showed both types of movement very clearly. When I let sawdust be carried through the water, I noticed that at low pressure it moved only in the axial direction, whereas at high pressure it was accelerated from one side to the other and often came into whirling motions.

It is clear from Hagen's observations that at low pressure (read: low flow rates in a tube of fixed diameter) the *mechanical* nature of the flow was different from that at high pressures (high flow rates).

6.3   STOKES' SOLUTION FOR HAGEN–POISEUILLE FLOW

In 1845 the British physicist George Stokes derived a set of equations* for three-dimensional flow of a fluid of constant viscosity μ and constant density ρ. One of the several applications he gave in the same paper, Stokes (1845), was to the case of laminar flow in a circular cylindrical pipe inclined to the horizontal, i.e., to what is now known as *Hagen–Poiseuille flow*. This same solution was later given by a number of other investigators including Neumann and Hagenbach.** Peculiarly, the priority of Stokes' solution has frequently been overlooked, even by such diligent researchers as Rouse and Ince.

We will not utilize the Navier–Stokes equations, as they are now known, since we can obtain the same results more simply, merely by using Newton's law of resistance applied to a flow that is axially symmetric about the pipe centerline, Figure 6.3-1. However, for the current purpose we must abandon the strictly one-dimensional approach and take into account the fact that the local flow velocity, although

---

* Rouse and Ince point out that independent derivations of these relations were given by L. Navier in 1824, S. Poisson in 1831, and B. De Saint-Venant in 1843, all of these Frenchmen.
** See Rouse and Ince for the references.

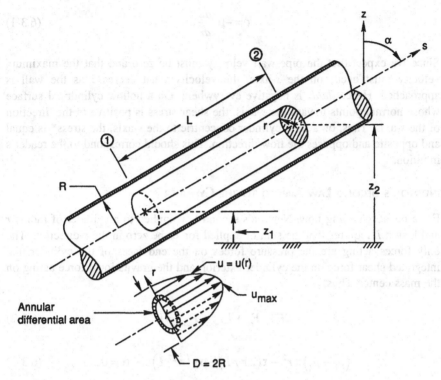

FIGURE 6.3-1   Analysis of Hagen–Poiseuille flow.

everywhere parallel to the centerline, varies in magnitude in the direction normal to the pipe wall.

Consider a cylinder of fluid of radius r concentric to the centerline of a pipe interior of diameter $D = 2R$ and length L between stations ① and ②. We assume that the flow is axisymmetric and laminar. In other words, the only nonzero velocity component is $u_s = u_s(r)$ only, which is everywhere parallel to the s-axis. For simplicity we put $u_s \rightarrow u = u(r)$ in the remainder of this section. The radial and peripheral components are thus identically zero. An illustration of the assumed velocity profile is shown in the figure.

The pipe is assumed to be straight and the angle between the vertical and the pipe axis is denoted α. For the current purpose we take it as evident that the average pressure at any cross-section corresponds to the pressure on the s-axis at that station, which we designate as $p = p(s)$ only. The flow is assumed to be non-time dependent; hence, the acceleration of every particle is zero. The fluid density and viscosity are both constant.

NEWTON'S LAW OF RESISTANCE

For the flow inside a cylindrical pipe we are using polar coordinates, where the origin of the r-coordinate is on the s-axis. Thus, to express the shear stress, originally derived in Equation 1.8-3, we must make the replacement $y \rightarrow -r$. Therefore,

$$\tau = -\mu \frac{du}{dr}. \tag{6.3-1}$$

Since we expect that the pipe-wall velocity must be zero and that the maximum velocity must occur on the s-axis, the velocity must decrease as the wall is approached. Hence du/dr is negative everywhere. On a hollow cylindrical surface whose normal points toward the s-axis, the shear stress is positive in the direction of the s-axis. Thus, on a fluid cylinder concentric to the s-axis, the stress* is equal and opposite and opposes the flow direction. This should correspond to the reader's intuition.

NEWTON'S SECOND LAW APPLIED TO THE CYLINDER

For a nonaccelerating flow, Newton's second law applied to a cylinder of radius r and length L requires that the sum of applied forces be zero in the s-direction. The only forces acting are the pressure forces on the end faces of the cylinder, the integrated shear force on the cylinder exterior, and the gravitational force acting on the mass center. Thus,

$$\Sigma F_s = F_{pr} + F_{visc} + F_{grav} = 0, \tag{6.3-2}$$

$$(p_1 - p_2)\pi r^2 - \tau(2\pi r L) - \rho g(\pi r^2 L)\cos\alpha = 0, \tag{6.3-3}$$

where we have taken into account the fact that the shear stress is positive in the negative s-direction. Noting that $\cos\alpha = (z_2 - z_1)/L$, Equation 6.3-3 can be rearranged to obtain

$$\pi r^2 \left[(p_1 + \rho g z_1) - (p_2 + \rho g z_2)\right] - \rho(2\pi r L) = 0. \tag{6.3-4}$$

We next introduce a special-purpose pressure p^*, which is the sum of the static pressure and the pressure increment due to the head z, i.e.,

$$p^* = p + \rho g z. \tag{6.3-5}$$

Further, since it is pressure differences that drive the flow, we put

$$\Delta p^* \equiv p_1^* - p_2^* = p_1 - p_2 + \rho g(z_2 - z_1). \tag{6.3-6}$$

---

* The shear stress is due to the retarding action of the exterior (surrounding) fluid about a fluid cylinder of radius r, the integration of which results in a force aligned with the s-axis opposing the motion of the cylinder.

Therefore, with Equation 6.3-6, Equation 6.3-4 can be solved for

$$\tau = \tau(r) = \frac{\Delta p^*}{2L} r. \qquad (6.3\text{-}7)$$

The shear stress, as specified by Equation 6.3-7 is positive and is directed in the negative s-direction when applied to the outer boundary of a fluid cylinder concentric with the s-axis. It is also positive, but acts in the $+s$-direction when applied to the inner surface of a hollow cylinder. This seeming confusion is a consequence of the fact that stresses are actually components of a second-order *tensor*, an advanced concept which cannot profitably be explored in an elementary text.

THE VELOCITY PROFILE

We can now combine Equations 6.3-1 and 6.3-7 to obtain the differential equation for the velocity profile:

$$\frac{du}{dr} = -\frac{\Delta p^*}{2\mu L} r. \qquad (6.3\text{-}8)$$

We need one boundary condition to integrate Equation 6.3-8. This is provided by the *no-slip condition of viscous flow*, i.e.,

$$@\, r = R: \quad u(R) = 0, \qquad (6.3\text{-}9)$$

Thus,

$$\int_{u=0}^{u} du = -\frac{\Delta p^*}{2\mu L} \int_{r=R}^{r} r\, dr,$$

which yields

$$u = \frac{\Delta p^*}{4\mu L} \left(R^2 - r^2 \right). \qquad (6.3\text{-}10)$$

It is clear that the profile is parabolic and that the maximum flow speed occurs on the centerline, i.e.,

$$u_{max} = u(0) = \frac{\Delta p^*}{4\mu L} R^2. \qquad (6.3\text{-}11)$$

A convenient nondimensional form results by dividing Equation 6.3-10 by Equation 6.3-11 to produce

$$u/u_{\text{max}} = 1 - (r/R)^2. \tag{6.3-12}$$

CALCULATION OF THE FLOW RATE

Because there is no variation of u in the azimuthal direction, we choose as differential area an annulus of mean radius r and width dr. Thus, $dA = 2\pi r dr$, and

$$\dot{Q} = \int_{r=0}^{R} u(r) \cdot 2\pi r \ dr = 2\pi u_{\text{max}} \int_{0}^{R} \left[1 - (r/R)^2\right] r \ dr,$$

or

$$\dot{Q} = \frac{1}{2}\pi R^2 u_{\text{max}} = \frac{\pi}{8} \frac{\Delta p^*}{\mu L} R^4. \tag{6.3-13}$$

If we eliminate R in favor of the diameter D, then

$$\dot{Q} = \frac{\pi}{128\mu} \frac{\Delta p^* D^4}{L}. \tag{6.3-14}$$

Equation 6.3-14 has exactly the form predicted by Poiseuille (for horizontal flow $\Delta p^* \rightarrow \Delta p = p_1 - p_2$), and, furthermore, it gives an explicit expression for the constant of proportionality, which is inversely proportional to the viscosity.

THE MEAN FLOW SPEED

We can interpret the Hagen–Poiseuille flow in terms of a mean flow speed defined such that

$$V \equiv \dot{Q}/A = \dot{Q}/\pi R^2. \tag{6.3-15a}$$

From Equations 6.3-13 and 6.3-11, we see that

$$V = \frac{u_{\text{max}}}{2} = \frac{\Delta p^*}{8\mu L} R^2, \tag{6.3-15b}$$

a result that is restricted to a circular pipe. Thus, the pressure drop in laminar pipe flow is proportional to the mean flow speed. We shall see that we can apply the one-dimensional flow relations to Stokes' laminar-flow solution where the appropriate one-dimensional flow speed is precisely V.

Note that Equations 6.3-15 provide an experimental means for determining μ. For a given setup R and L are fixed by the apparatus. The only fluid mechanical data needed are the flow rate $\dot{Q}$ and the pressure drop Δp^*.

If Equation 6.3-15b is substituted into Equation 6.3-12, we obtain the alternative expression for the velocity profile:

$$u = 2V\left(1 - r^2/R^2\right). \tag{6.3-16}$$

For future use, we calculate

$$\tau(r) = -\mu \frac{du}{dr} = \frac{4\mu V r}{R^2} \tag{6.3-17a}$$

and

$$\tau_w = \tau(R) = \frac{4\mu V}{R} = \frac{\Delta p^*}{L}\frac{R}{2} = \frac{\Delta p^*}{L}\frac{D}{4}. \tag{6.3-17b}$$

Note in Equation 6.3-17b that the wall shear stress is proportional to $\Delta p^*/L$, which is a pressure drop per unit length or, equivalently, a pressure gradient, in the s-direction. Now $\Delta p^*/L$ must be a constant; otherwise the wall shear stress would be a function of s, i.e., $\tau_w = \tau_w(s)$. But this would violate the original assumption that $u = u(r)$ only, the validity of which is borne out by countless experiments. Thus — other conditions remaining the same — doubling L requires that the static-pressure drop Δp and the change of pressure due to height change $\rho g(z_1 - z_2)$ also be doubled.

THE SKIN-FRICTION COEFFICIENT AND THE FRICTION FACTOR

In aeronautical applications it is customary to nondimensionalize the wall shear stress — which has the dimensions of pressure — by a reference dynamic pressure. For a pipe flow, the appropriate dynamic pressure is $1/2\rho V^2$. The resulting quotient is called the *skin-friction coefficient*, denoted c_f; thus, for Hagen–Poiseuille flow

$$c_f \equiv \tau_w / \tfrac{1}{2}\rho V^2 = 8\mu/\rho VR = 16\mu/\rho VD$$

or

$$c_f = 16/\text{Re}, \tag{6.3-18}$$

where

$$\text{Re} \equiv \rho VD/\mu = VD/\nu, \tag{6.3-19}$$

is the flow Reynolds number, based on the mean flow speed and the pipe diameter. The quantity $\nu \equiv \mu/\rho$ is the kinematic viscosity. To indicate the reference length, the notation $\text{Re} \rightarrow \text{Re}_D$ is sometimes used.

The historical development of hydraulics resulted in the definition of the *friction factor* f, related to the skin-friction coefficient by

$$f = 8\tau_w / \rho V^2 = 4 c_f. \tag{6.3-20}$$

For pipe flow problems, we will utilize the friction factor except where otherwise stated. Thus, for Hagen–Poiseuille flow, we have the famous relation

$$f = 64/\text{Re}. \tag{6.3-21}$$

If we take the logarithmic (to base 10) derivative of Equation 6.3-21, we get

$$\frac{d\left(\log_{10} f\right)}{d\left(\log_{10} \text{Re}\right)} = -1. \tag{6.3-22}$$

Thus, the plot of $\log_{10} f$ vs. $\log_{10}$ Re must be a straight line of negative unit slope.

ON THE INSPECTIONAL ANALYSIS OF SECTION 5.12

We recall that the inspectional analysis of the nonsteady flow through a nonuniform duct of Section 5.12 led to the relation

$$C_p = F\left(\overline{V}, \bar{s}, \bar{t}, f, \text{Fr}\right). \tag{6.3-23}$$

If we restrict the problem to steady flow through a uniform, horizontal duct then $\overline{V}$ = constant, and, hence, $\overline{V}$ and $\bar{t}$ can be eliminated from Equation 6.3-23. Furthermore, since the gravitational term has been absorbed by p^*, the dependency on the Froude number Fr can be omitted; thus reducing the relation to

$$C_p = C_p\left(\bar{s}, f\right). \tag{6.3-24}$$

Consequently, the pressure coefficient in a pipe can depend only on the longitudinal coordinate and the friction factor. For laminar pipe flow, in fact, we have shown that the friction factor depends only on the Reynolds number. We shall see that in turbulent flow an additional parameter enters, namely, the relative roughness of the pipe wall, to be defined in what follows.

6.4   ON THE CORRELATION OF THEORY AND EXPERIMENT

In this era the techniques of dimensional analysis are widely known, and they are routinely applied. However, one has to read the original papers of Poiseuille and

Hagen, particularly the former, to appreciate their problems in presenting the data in a useful, and comprehensible, form.

Recall that the experiments leading to the relation

$$\dot{Q} \sim \frac{\Delta p D^4}{L}$$

required variation of each of the dimensional quantities shown, as well as of the viscosity coefficient, whose only measure was the temperature of the fluid at which the experiment was conducted. In neither of the original papers, Poiseuille (1846) or Hagen (1839), are the data presented in graphical form, although in the later paper, Hagen (1854), he did resort to plots of the head required as a function of the flow rate, using the temperature as a parameter. In Poiseuille's paper he presents his data via an enormous number of tables, which are judiciously used to separate out the effects of Δp, D, L, and temperature, on which $\dot{Q}$ depends.

This gives us an opportunity to grasp the great advantage in the presentation, or correlation, of data in an economical fashion, by using the techniques of dimensional analysis. It was H. Blasius (1913), earlier an assistant to L. Prandtl, who first assembled data on viscous flows from a number of different sources and organized its presentation in a systematic way.

Figure 6.4-1 shows the power of this approach applied to Hagen–Poiseuille flow. The data of Hagen in the laminar regime, correlated in the form suggested by Equations 6.3-21 and 6.3-22, namely, the plot of the friction factor f on a log–log plot, vs. the Reynolds number Re as abscissa. The data points were taken from a similar plot in Prandtl and Tietjens (1934, Vol. II, p. 18), and recalculated according to current usage where the definition of Reynolds number is based on duct diameter (rather than the duct radius).

Equation 6.3-21 is also shown on Figure 6.4-1. Since the agreement is excellent, we conclude that Stokes's theory, which reduces to the relations of Section 6.3 for laminar viscous flow in a circular pipe, is correct. We also conclude that the experiments of Hagen are remarkable for their consistency.

It is worth noting at this point that space limitations prevent adequate discussion of the experimental difficulties faced, and only partially overcome, by both Hagen and Poiseuille. These include the problems of fabricating capillary tubes that are sufficiently circular, uniform in cross section, and straight; the problem of nonuniformities of flow due to the inlet design; and the difficulties of making accurate pressure measurements and maintaining uniform temperature of the fluid during the experiment.

6.5 THE DARCY–WEISBACH EQUATION FOR HEAD LOSS IN PIPE FLOW

Definition of Head Loss in Pipe Flow

If we apply the energy equation (Equation 4.8-3) to the flow in a pipe where $V_1 = V_2 \equiv V$ = constant, then we can solve for the internal energy change between any two stations:

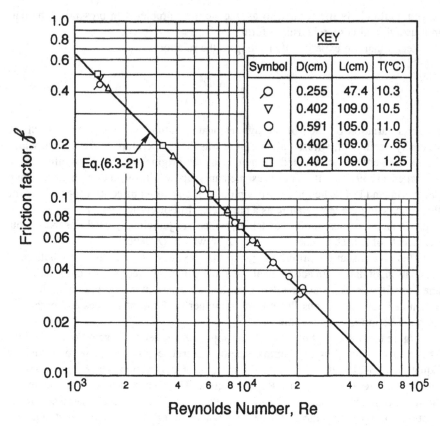

FIGURE 6.4-1 Friction factor as a function of Reynolds number for pipe flow, experimental data from Hagen (1839).

$$\rho \Delta e \equiv \rho \left(e_2 - e_1 \right) = \left(p_1 + \rho g z_1 \right) - \left(p_2 + \rho g z_2 \right) = \Delta p^*. \qquad (6.5\text{-}1)$$

Equation 6.5-1 is valid generally, i.e., for either laminar or turbulent flow. Dividing Equation 6.5-1 by ρg, to convert each term to a length dimension, then we see that a new quantity h_ℓ, called the *head loss*, can be defined such that

$$h_\ell \equiv \frac{\Delta e}{g} = \frac{\Delta p^*}{\rho g}. \qquad (6.5\text{-}2)$$

Thus, the head loss is simply a convenient way of measuring viscous flow losses, which result in an increase in internal energy of a particle and, inevitably, in an increase in entropy* for a constant-density fluid.

---

* See Equation 4.3-5.

THE DARCY–WEISBACH EQUATION

The German engineer Julius Weisbach (1806–1871) brought together in Weisbach (1855) a vast compilation of experimental hydraulics data, including many in pipe flow. The physical scale of the experiments incorporated by Weisbach in his work was so large as to guarantee that the pipe flow data were almost certainly turbulent. There is no evidence to suggest that he was aware of the experiments of Hagen and Poiseuille, or the theory of Stokes. In fact, the energy equation in the form leading to Equation 6.5-1 was still unknown.*

Nevertheless, Weisbach reasoned that the effects of flow resistance resulted in a loss of head. He proposed an empirical relation in which the head loss was assumed to be proportional to the velocity head $V^2/2g$ and to the length of the pipe. In order to keep the factor of proportionality nondimensional, he selected the form

$$h_\ell = f \frac{L}{D} \frac{V^2}{2g},\tag{6.5-3}$$

where, in Weisbach's scheme, f was simply an empirical constant. We will verify, however, that f is, in fact, identical to the friction factor already introduced in Equation 6.3-20. In the turbulent pipe flows described by Weisbach, f had to be determined by experiment, and there was no hint on which flow parameters it depended.

Thus, for the first time, f was identified as the appropriate nondimensional parameter to be used by experimentalists in correlating their data. The question remained open, however, as to what parameter — e.g., flow rate, velocity, etc. — should be used to plot f against. Because Weisbach's contribution was not immediately recognized, it took several decades, at least, before Equation 6.5-3 was widely adopted by hydraulics engineers.

Equation 6.5-3 is called the Darcy–Weisbach equation, although, according to Rouse and Ince, the joining of the name of H. Darcy (1803–1858), an eminent French hydraulics engineer, to Equation 6.5-3 is not entirely justified. Nevertheless, Darcy is justly famous for several original contributions to the hydraulics literature. In the area of pipe flow his experiments clearly showed that losses (in turbulent flow) depended also on the condition of the pipe walls, i.e., on the roughness of the pipe.

For the case of laminar pipe flow, if we eliminate Δp^* in Equation 6.5-2 by means of Equations 6.3-17b and 6.3-19, we obtain

$$h_\ell = \frac{64}{Re} \frac{L}{D} \frac{V^2}{2g}.\tag{6.5-4}$$

---

* See the footnote on page 354 re: Joule and Thomson (1854).

Thus, we confirm that the Darcy–Weisbach equation can be applied to laminar pipe flow, where the multiplying factor is indeed $f = 64/\text{Re}$ and where the Reynolds number is based on the mean flow speed.

6.6 REYNOLDS' EXPERIMENTS ON THE NATURE OF TURBULENCE AND THE DISCOVERY OF A CRITERION FOR THE TRANSITION FROM LAMINAR TO TURBULENT FLOW

Although Hagen had indicated, as early as 1839, that there were two different possible modes of flow, no real insight into the nature of turbulence was given until the British scientist Osborne Reynolds (1842–1912) published his classical investigation of pipe flow, Reynolds (1883).

REYNOLDS' DIMENSIONAL REASONING

Reynolds, who was aware of the experiments of Hagen, Poiseuille, and Darcy and the theory of Stokes, was concerned that the equations of hydrodynamics (the Navier–Stokes equations), which he assumed to be generally valid, seemed to fail in pipe flow when the flow speed exceeded a certain value, i.e., when the flow became turbulent. He started with the assumption that there were "some fundamental principles of fluid motion of which due account has not been taken in the theory." In other words, he assumed that even in the case of turbulent flow, which was characterized by the appearance of large-scale eddies, the Navier–Stokes equations would still be valid if proper account of the turbulence was made.

With a sure intuition he reasoned that the character of the flow should not depend on absolute size (of the pipe, for example) or absolute velocity, but rather on some nondimensional parameter. His approach was, apparently, to perform an inspectional analysis on the Navier–Stokes equations. For reasons unknown, he did not present the details of his analysis in his 1883 paper. Since he is somewhat cryptic in his remarks we offer instead an analysis that applies the procedures of Section 5.12 to the Navier–Stokes equations for steady, two-dimensional, constant-density (in fact, constant-fluid-properties) flow, of which the following is the relation for the x-direction:

$$u\,\frac{\partial u}{\partial x} + v\,\frac{\partial u}{\partial y} = -\frac{1}{\rho}\,\frac{\partial p}{\partial x} + \frac{\mu}{\rho}\left(\frac{\partial^2 u}{\partial x^2} + \frac{\partial^2 u}{\partial y^2}\right). \tag{6.6-1}$$

Equation 6.6-1 is next nondimensionalized, using the techniques of Section 5.12. By selecting the mean flow speed V and the pipe diameter D as characteristic values, the following nondimensional variables are introduced:

$$u' = u/V, \quad v' = v/V, \quad x' = x/D, \quad y' = u/D. \tag{6.6-2}$$

Then, after substituting Equation 6.6-2 in Equation 6.6-1 and dividing through by V^2/D, we obtain

$$u' \frac{\partial u'}{\partial x'} + v' \frac{\partial u'}{\partial y'} = -\frac{\partial p'}{\partial x'} + \frac{\mu}{\rho VD} \left(\frac{\partial^2 u'}{\partial x'^2} + \frac{\partial^2 u'}{\partial y'^2} \right), \qquad (6.6\text{-}3a)$$

$$= -\frac{\partial p'}{\partial x'} + \frac{1}{\text{Re}} \left(\frac{\partial^2 u'}{\partial x'^2} + \frac{\partial^2 u'}{\partial y'^2} \right), \qquad (6.6\text{-}3b)$$

where the dimensionless ratios

$$p' \equiv p / \tfrac{1}{2} \rho V^2 \quad \text{and} \quad \text{Re} \equiv \frac{\rho VD}{\mu} \qquad (6.6\text{-}4)$$

appear* automatically. Reynolds clearly identified the ratio $\rho VD/\mu$ as a nondimensional parameter that governed the flow, and he set as a major goal of his experimental investigations to show whether or not "the birth of eddies depends on some definite value $\rho VD/\mu$." Reynolds further noted that Stokes had hypothesized in 1843 that it was possible that the transition from laminar to turbulent flow was a consequence of an arbitrarily small disturbance, created on the wall, being indefinitely amplified by the action of viscosity in the main flow.

THE EXPERIMENTS OF REYNOLDS ON TRANSITION

The most dramatic experiments of Reynolds had the goal of making turbulence visible to the eye. His setup is shown in Figure 6.6-1, taken from the 1883 paper. The pipe was made of glass and had a trumpet-shape inlet to allow the flow to make a smooth transition from the reservoir to the section of uniform cross section. Just upstream of the inlet was positioned a smaller tube, supplied by a second, smaller reservoir containing water, dyed to make it visible. The experimenter had the ability to control independently the flow rates of the main and the injected fluid. Using Reynolds' chosen words, when the water flow rate was sufficiently low the "streak of colour extended in a beautiful straight line through the tube," Figure 6.6-2a. Reynolds noted that if the water in the reservoir had not quite settled to rest "the streak would shift about the tube, but there was no appearance of sinuosity" (i.e., turbulence**).

"As the flow rate was gradually stepped up,...at some point in the tube..., [Figure 6.6-2b],...always at a considerable distance from the trumpet...the filament of dye...would all at once mix with the surrounding water, and fill the rest of the tube with a mass of coloured water.... On viewing the tube by the light of an electric

---

* Reynolds did not give the ratio $\rho VD/\mu$ a special symbol, and his name was first formally associated with it in a paper by M. Weber (1919), although von Kármán (1954, p. 106) attributes this appellation to the mathematical physicist A. Sommerfeld.

** According to Rouse and Ince, the designation of "turbulence" to describe such flows was first suggested by Lord Kelvin in 1887.

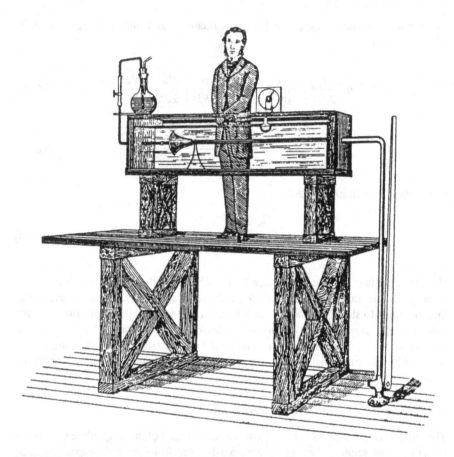

FIGURE 6.6-1 Drawing of Reynolds' original apparatus. (From Reynolds, O., *Philos. Trans. R. Soc.*, 174, 935–982; 1883. With permission.)

spark [thereby "stopping" the motion], the mass of colour resolved itself into a mass of more or less of distinct curls, showing eddies..." as in Figure 6.6-2c.

It is emphasized that the "sinuous motion," or turbulence as it is now called, involves the violent, random motion of discrete masses of fluid (called eddies) superimposed on the main flow. It is a *macroscopic* phenomenon as Hagen had originally observed, confirmed unequivocally by Reynolds, and is unrelated to the molecular agitation, which continues to exist, although which is unobservable on the macroscopic level. In "steady" (i.e., time-averaged) turbulent pipe flow, the flow is locally unsteady, and is three dimensional, although at any point only the velocity component aligned with the pipe axis has a nonzero mean value when averaged over a sufficiently long, finite, time interval.

THE CRITERION FOR TRANSITION

Curiously, in spite of his avowal to explore the role of Re in the transition phenomenon, Reynolds did not undertake this in his 1883 paper. Only in a later

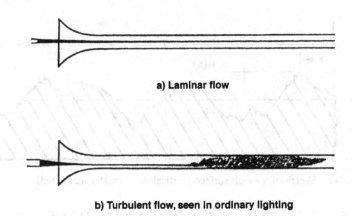

a) Laminar flow

b) Turbulent flow, seen in ordinary lighting

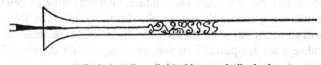

**c) Turbulent flow, lighted by spark disclosing
presence of discrete eddies**

FIGURE 6.6-2 Sketches of laminar and turbulent flow as revealed by the behavior of a lamina of dyed fluid. (From Reynolds, O., *Philos. Trans. R. Soc.*, 174, 1883. With permission.)

paper (Reynolds, 1894) did he conclude that for Re > 2000 the flow would be turbulent.

A SUMMARY OF REYNOLDS' CONCLUSIONS

Reynolds was such a careful experimenter, as well as a perceptive observer, that his principal conclusions, listed below, remain valid today, except for a modest change in the transition Reynolds number based on more-modern experiments.

1. Turbulence consists of violent, random motion of eddies.
2. The sudden appearance of eddies is the result of instabilities in the flow generated at the wall and amplified by the main flow.
3. The flow tends to remain laminar if Re < 1800, and transition will always set in if Re > 2000.
4. The prediction from Stokes's theory for laminar flow, that the pressure drop term is proportional to the first power of the mean flow speed (i.e., that $\Delta p^*/L \sim V$) is verified.
5. For turbulent flow in smooth pipes the pressure-drop term varies as $\Delta p^*/L \sim V^{1.723}$.
6. For turbulent flow in nonsmooth pipes, the pressure-drop term varies as $\Delta p^*/L \sim V^n$, where n ranges from 1.79 to 2.0, depending on the material and the condition of the pipe wall.

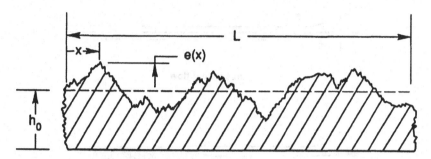

FIGURE 6.7-1   Sketch of a rough surface (vertical scale greatly magnified).

6.7   THE APPLICATION OF DIMENSIONAL ANALYSIS TO PIPE FLOW

We previously noted that Reynolds used dimensional reasoning to determine that transition to turbulence should depend on the dimensionless ratio Re $\equiv \rho VD/\mu = VD/\nu$. This was an enormous scientific advance. By careful experimentation, he was able to formulate a semiempirical criterion involving Re, for the conditions under which a laminar pipe flow can be expected to remain laminar or will be expected to undergo transition to turbulent flow.

However, he apparently did not examine the possibility that losses in a turbulent flow might also depend upon Re. To have achieved this he would have had to have plotted his data for the turbulent regime in dimensionless form. It remained for H. Blasius to recognize this fact and to complete the task some 30 years after Reynolds had published his first paper on pipe flow.

ON PIPE ROUGHNESS

Darcy and other investigators had noted that the mechanical condition of a pipe wall contributed significantly to pipe losses in what later turned out to be the turbulent pipe flow regime. Anyone who has seen the interior of a water-distribution pipe, after a period of service, will have no doubt of this fact. The deposits of solids on a wall, due to minerals coming out of solution over a long period of time, can be enormous and can reduce the effective diameter of a pipe to a fraction of its original value. Even a newly manufactured pipe has some degree of roughness. In order to incorporate pipe roughness within an analytical framework, we need an unambiguous definition of the term.

Consider a cut through a material surface where the cutting plane is normal to the surface exposed to the flow. If the surface were sufficiently magnified, it might appear something like that shown in Figure 6.7-1. A *profilometer* is a device that allows the surface shape to be traced out by a diamond stylus supported by an arm that is moved across the surface to be measured. Transducers sense the motion of the stylus, generating electrical signals that are amplified and recorded on a strip chart, as shown in Figure 6.7-2.

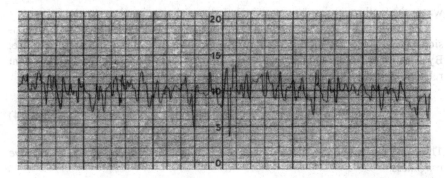

FIGURE 6.7-2 Strip chart produced by a profilometer on a "flat" steel plate. In the sample shown the scale magnification factors are 20 horizontally and 5000 vertically. Each vertical subdivision represents 20×10^{-6} m. (Courtesy of the RPI Tribology Laboratory.)

The local roughness $e(x)$ is measured with respect to a mean line indicated by height h_0, and is positive above the line and negative below. The mean roughness can be computed in two ways. The *root-mean-square* (rms) *roughness* is

$$e_{\mathrm{rms}} \equiv \sqrt{\frac{1}{L} \int_0^L e^2(x) \cdot dx}. \qquad (6.7\text{-}1)$$

The *linear roughness* is defined by measuring the total areas (the areas are both positive), above and below the mean line, adding them, and dividing by L. Thus,

$$e_{\ell \mathrm{in}} \equiv \frac{A_{\mathrm{above}} + A_{\mathrm{below}}}{L}. \qquad (6.7\text{-}2)$$

The two values e_{rms} and $e_{\ell \mathrm{in}}$ may differ from each other by as much as 10%. For standard materials such as cast iron, steel, brass, and cement, the absolute value of the roughness, simply designated as e herein, can be obtained from a handbook or from the manufacturer.

DIMENSIONAL ANALYSIS OF PIPE FLOW

It has previously been noted that it was Blasius (1913) who first put the correlation of pipe flow data on a sound basis by use of the techniques developed in Chapter 5. In the following we preserve the spirit of Blasius' approach, where the details may differ, but not the results.

Since we can adjust to the case of an inclined pipe simply by replacing Δp by Δp^*, we restrict considerations to horizontal flow. Then, we assume that a functional relation exists of the form

$$f\left(\Delta p, \rho, V, D, \mu, \tau_w, L, e\right) = 0, \qquad (6.7\text{-}3)$$

where the wall roughness e is now included as a dimensional parameter.

Equation 6.7-3 involves eight dimensional parameters and three basic dimensions; hence, we expect a minimum of five π-ratios. In this case there are only five. By using the techniques of Chapter 5, Equation 6.7-3 can be reduced to the equivalent relation involving five π-ratios; that is

$$F\left(\Delta p/\rho V^2,\ \tau_w/\rho V^2,\ \rho VD/\mu,\ D/L,\ e/D\right)=0. \qquad (6.7\text{-}4)$$

The first four ratios are already familiar. For the last we introduce the *relative roughness**

$$\varepsilon \equiv e/D. \qquad (6.7\text{-}5)$$

Thus, Equation 6.7-4 can be replaced by

$$F\left(\text{Eu, Re, } f,\ D/L,\ \varepsilon\right)=0. \qquad (6.7\text{-}6)$$

Now, we recall, from the Darcy–Weisbach equation for pipe flow, and from the definition of the friction factor, that

$$f \equiv \frac{\Delta p^{\ast}}{\rho V^2}\,\frac{2D}{L}, \qquad (6.7\text{-}7a)$$

generally, and

$$f = \frac{\Delta p}{\rho V^2}\,\frac{2D}{L} = \text{Eu}\,\frac{2D}{L} \qquad (6.7\text{-}7b)$$

in horizontal flow. Thus, if we retain the friction factor, we may omit the Euler number and the ratio D/L as independent parameters. The resulting explicit form of Equation 6.7-6 then turns out to be

$$f = \left(\text{Re, } \varepsilon\right). \qquad (6.7\text{-}8)$$

Equation 6.7-8 contains the essence of Blasius's paper. Though unprepossessing, for the first time — after more than a century of often fumbling, unguided measurement — it provided a complete rational basis for correlating experimental data of pipe flows, laminar or turbulent.

It happens, for laminar pipe flow, that as long as the Reynolds number is sufficiently low, the wall roughness does not affect the friction factor, which is given

---

* Nikuradse (1933) indicates that the term *relative roughness* for the ratio e/D was first suggested by R. von Mises in 1914.

by the relation $f = 64/Re$. The plot in Figure 6.4-1, which compares the theory of Stokes with the experimental data of Hagen, was first made by Blasius. He could equally well have used the data of Poiseuille.

THE CORRELATION OF EQUATION 6.7-8 FOR SMOOTH-WALL PIPES IN TURBULENT FLOW

As Reynolds had shown, for Re > 2000, the flow was subject to instabilities that led to the development of a fully turbulent flow. He was not able to shed any light on the mechanism of the instability, nor able to quantify the effects of roughness in a general way, except to note that flow losses tended to increase with the wall roughness.

Having verified the agreement between theory and experiment in laminar pipe flow, Blasius turned his attention to the more difficult question of analyzing existing data on the turbulent flow through pipes of varying roughness. Using the theoretically inspired result of Equation 6.7-8, Blasius next sought to uncover how existing experimental data could be correlated within this framework.

Saph and Schoder (1903)* had undertaken an ambitious program of measuring losses in pipe flows involving a large number of pipe sizes, flow rates, water temperatures, and various wall-roughness conditions — with materials ranging from drawn brass tubing to galvanized iron pipe. Because they had no knowledge of dimensional analysis, the best they could do was to plot the head loss vs. the mean flow speed on log–log paper. Although they were aware of Reynolds' 1883 paper, they did not attempt to explore his remarks on the significance of the parameter Re $= \rho VD/\mu$. They were aware of the "critical velocity" phenomenon (i.e., transition to turbulence), but had no idea of a criterion for the same. Furthermore, they made no attempt to offer a quantitative description of the pipe "wall condition," or roughness. A principal finding was actually a verification of the result already obtained by Reynolds that, for turbulent flow in a smooth pipe, the head loss varied as $V^{7/4}$. As Blasius noted, by assuming a law of the form

$$h_\ell/L = AV^{7/4}/D^n, \tag{6.7-9a}$$

they were able to correlate the data and determine A and n, such that

$$h_\ell/L = \left(0.296 \times 10^{-3}\right)V^{7/4}D^{1/4}. \tag{6.7-9b}$$

For nonsmooth pipes their obtained values for A and n were different.

Blasius made a very careful study of the data of Saph and Schoder and, in spite of the great care with which the measurements were made, concluded that there were some important discrepancies. For example, some runs were made with two

---

* These two Americans are among the mere handful listed by Rouse and Ince (1957) as having played a significant role in the history of fluid mechanics (which they refer to as hydraulics) up to the beginning of the 20th century.

abutting lengths of pipe whose diameters were sufficiently different as to negate the results. The elimination of the deficient data, however, did not affect any of their conclusions.

A problem with Equation 6.7-9b is that the constant $A = 0.264 \times 10^{-3}$ is actually a temperature-dependent quantity (since it depends on viscosity). However, since the head loss is related to the friction factor by Equation 6.5-3, i.e.,

$$f = \frac{2gD}{V^2} \frac{h_\ell}{L},$$

                                                                                      (6.7-10)

Blasius converted all data for h_ℓ/L to f-values and plotted them vs. the corresponding Reynolds number. Combining Equations 6.7-10 and 6.7-9b, he obtained

$$f = \frac{2gD}{V^2} \frac{(0.296 \times 10^{-3}) V^{7/4}}{D^{5/4}} = \frac{(0.296 \times 10^{-3})(2g)}{v^{1/4}} \cdot \frac{1}{(VD/v)^{1/4}} = \frac{0.3164}{\mathrm{Re}^{1/4}}. \quad (6.7\text{-}11)$$

In the preceding expression for the friction factor Blasius took advantage of the fact that the exponents of V and D were the same. Thus, by multiplying and dividing by $v^{1/4}$, and grouping the parameters, a Reynolds number dependence can be imposed upon the expression. He was also aware that Saph and Schoder ran their tests at 55°F; at this temperature $v = 1.22 \times 10^{-6}$ m²/s $= 1.313 \times 10^{-5}$ ft²/sec. When evaluated, the constant — which is necessarily nondimensional — was determined to be 0.296 $\times 10^{-3}$ $(2g)$ $v^{1/4} = 0.3164$.

The Blasius replot of the Saph and Schoder data, shown in Figure 6.7-3, is remarkable on several counts. Each data point is identified by a symbol that specifies the pipe diameter in the run. For Re < 2000 the Hagen–Poiseuille theory for laminar flow is once again verified. In the neighborhood of 2200 < Re < 2500, there is a scattering of points that corresponds to a transition to turbulence. For Re > 2500, a vast number of points is seen to lie — with slight scatter — about a straight line of slope –1/4 thereby defining the turbulent regime. It is straightforward to verify that this line is that specified by Equation 6.7-11 on the log–log plot.

We can also verify that the Blasius correlation, Equation 6.7-11, combines with Equation 6.5-2 to produce

$$\Delta p^* \sim V^{7/4}.$$

                                                                                      (6.7-12)

In consequence, as we shall show later in detail, turbulent pipe flow pumping-power varies as $V^{11/4} \approx V^3$ compared with the V^2 of laminar flow.

It is desirable to keep in mind that Equation 6.7-11 — which is the result of a data correlation using the techniques of dimensional analysis — is nevertheless a semiempirical relation. That is, in its formulation — although of great engineering value — no light has been shed thereby on the fundamental mechanism of turbulent flow. As such, the application of Equation 6.7-11 to flows of Reynolds numbers

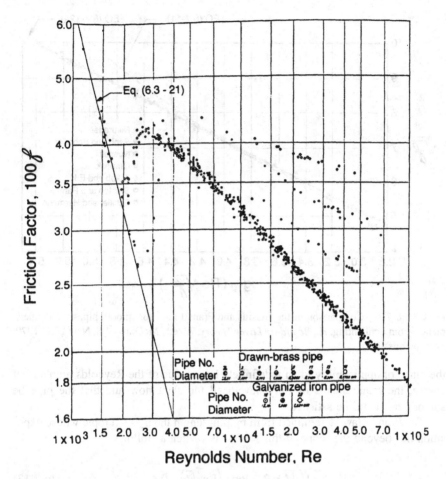

FIGURE 6.7-3 Blasius's (1913) correlation of the Saph and Schoder data for turbulent flow in smooth pipes.

higher than covered by experiment is unjustified and potentially error prone. The upper limit of the Saph and Schoder data was at Re ≈ 10^6, which corresponds to a flow of water at room temperature with a mean flow speed of $V = 15$ ft/sec through a 1-in. diameter pipe.

PRANDTL'S LAW FOR SMOOTH PIPES

Later experiments showed that the correlation of Blasius (i.e., a straight line on a log–log plot with a slope of $-1/4$) deviated from the data for Re > 10^5. As the test Reynolds number was increased, the deviation increased montonically. However, 14 years later Prandtl (1927) established a more accurate but still semiempirical relation, now known as *Prandtl's universal law of friction for smooth pipes*. Not only did the formula correlate the data satisfactorily below Re = 10^5, but it apparently extended

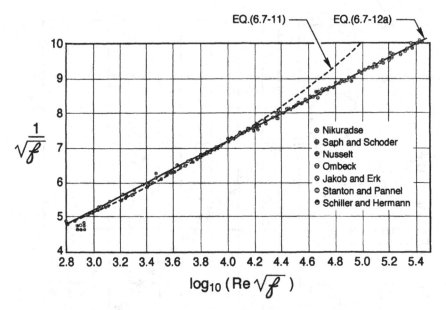

FIGURE 6.7-4 Comparison of the Blasius and Prandtl laws for smooth pipes with experiment. (From Schlichting, H., *Boundary Layer Theory*, 7th ed., McGraw-Hill, New York, 1979. With permission.)

the range of applicability to an indefinitely large value of the Reynolds number. Of course, the requirements of fully developed turbulent flow and that the pipe be smooth, must still be satisfied.

The Prandtl law, determined from mixing-length theory — a topic whose exposition lies beyond the scope of this text — is written as follows:

$$1/\sqrt{f} = 2.0 \; \log_{10}\left(\mathrm{Re}\,\sqrt{f}\right) - 0.8. \tag{6.7-13}$$

A tabulation of f vs. Re is given in Table 6.8-1.* It should be noted that the numerical constants in Equation 6.7-13, as in the Blasius relation, were determined by curve fitting; thus, the formula is semiempirical.

EXPERIMENTAL VERIFICATION OF PRANDTL'S LAW

According to Equation 6.7-13, the graph of $1/\sqrt{f}$ vs. $\log_{10}\left(\mathrm{Re}\,\sqrt{f}\right)$ should be a straight line. Figure 6.7-4 is a plot comparing the theory to data from seven groups of experimenters. The comparison appears to be excellent. However, due to the nature of the logarithmic abscissa of this graph, the true accuracy is obscured.

---

* Although the calculated values in Table 6.8-1 are given to four significant figures, it should be understood that final calculations in which these values are employed should be rounded to only two, or three at most, significant figures.

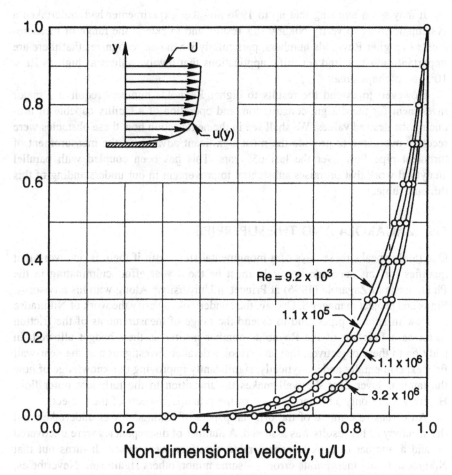

FIGURE 6.7-5 Velocity profiles for turbulent flow in smooth pipes for different Reynolds numbers. (From Nikuradse, J., *Forsch. Arbeit Ing. Wes.*, 356, 1932. With permission.)

For example, using the data at the leftmost point of the graph, where $1/\sqrt{f} = 4.8$, $\log_{10}\left(\text{Re}\sqrt{f}\right) = 2.8$, we can solve for $f = 0.0434$, Re = 79. This is unacceptable on several counts. In the first place, it is known from many experiments that a fully turbulent flow is unsustainable at such a low Reynolds number. More important, the value from the Prandtl law corresponding to $f = 0.0434$ is Re $\approx 3.1 \times 10^3$, raising doubts about the validity of the whole plot.

All the data at the right end of the graph, which corresponds to large values of Re, are the contribution of Nikuradse (1932), who made a landmark study of turbulent flow over the interval $3.2 \times 10^3 \le \text{Re} \le 3.2 \times 10^6$. Figure 6.7-5 shows the velocity profiles for the four values of the Reynolds number measured by Nikuradse, the largest being Re = 3.2×10^6. His technique was to eject the flow from the pipe as a free jet and survey the velocity profile by means of a Pitot tube traversing the jet, the measuring station being very close to the pipe exit.

It may seem amazing that up to 1996 no other experimenter had undertaken a systematic study to verify Nikuradse's results and to extend the range of measurements to higher Reynolds numbers, particularly when one recognizes that there are important industrial and scientific applications that involve values as high as Re = 10^9, and perhaps greater.

However, to extend the results to higher Reynolds numbers requires a major investment for the design, construction, and operation of a facility capable of producing the desired values. We shall see in the next section how these obstacles were recently overcome to provide the most significant advance in the measurement of turbulent pipe flow over the last 65 years. This has been coupled with parallel analytical work that promises substantial improvement in our understanding of this difficult subject.

6.8   ZAGAROLA AND THE SUPERPIPE

One rarely speaks these days of a monumental study. But if there is one work that qualifies for this accolade, it surely must be the 4-year effort culminating in the Ph.D. thesis of Zagarola (1996) at Princeton University.* Along with his advisor A. Smits and project manager S. Orszag, they undertook to verify the work of Nikuradse for flow in smooth pipes, and to extend the range of measurements of the friction factor as a function of the Reynolds number insofar as their budget allowed. In addition to these objectives, they undertook a detailed investigation of the near-wall flow experimentally and analytically, significantly improving our knowledge of how the laminar layer near the wall makes the transition to the turbulent outer flow. Herein we can only summarize some of the principal aspects of the project.

An extensive critique of the original papers of Nikuradse was undertaken and the accuracy of his results was assessed. A number of discrepancies were uncovered — and a number of (now) unanswerable questions were raised. It turns out that Nikuradse's data incorporate errors — some minor, others significant. Nevertheles, the form of the functional dependence of the friction factor on Re, established by Prandtl, Equation 6.7-13, was confirmed.

The experiments were performed in a special facility, a loop, known as the *Princeton Superpipe*. In order to get the highest Reynolds number possible for the available funding (2.5×10^5) it was decided to use air at a maximum allowable pressure (limited by the equipment) of 200 atm at near room temperature. The flow speeds were very low, keeping the Mach number sufficiently small that the flow could be treated as incompressible. Nevertheless, the corresponding maximum specific gravity approximated S.G. ≈ 0.25. The air was circulated by a specially modified turbine pump, driven by a 200 hp variable-speed motor.

The test pipe, which was nominally 5 in. inside diameter and 86 ft long, was made up of a number of segments butted together. The joints were secured by special fittings that ensure that the adjoining segments remain in a fixed relative position. The entire test pipe was honed and polished so that the surface roughness was 6 μ

---

* I would like to acknowledge the courtesy of Professor Ronald Panton of the University of Texas, Austin, in calling my attention to this work.

in. or less. The diameter of any one segment differed from the mean of the others by less than ±0.001 in. The butt joint areas were specially honed and polished *in situ* to eliminate local disturbances. The finished surface of the polished sections (6061 aluminum) was stated to be *almost as smooth as polished glass.*

Measurements (pressures, velocity profiles, and the like) were confined to the downstream 19 ft of the test pipe. This whole segment was confined within a larger pipe (actually a pressure vessel) that was pressurized to ease the problem of sealing instrumentation access holes in the test pipe and protecting the instrumentation. The upstream portion was 67 ft long (160 diameters), sufficient to guarantee that in the measuring section fully developed turbulent flow was attained.

Considerable attention was paid to the development of the state-of-the-art instrumentation and the data acquisition system. For details of all these matters see Zagarola (1996) and the summary paper by Smits and Zagarola (1997).

RESULTS

An important part of the work was devoted to the scientific question of the nature of the "inner flow" adjacent to the wall, which is laminar, and the process by which it blends into the "outer flow," which encompasses the pipe centerline and which is turbulent. We can only gloss over this matter to which some of the greatest names in fluid mechanics have made contributions. For an extensive discussion of the nature of the inner and outer layers the reader is referred to Chapter 6 of White (1974).

We can note, in contrast to the six cases studied by Nikuradse, that Zagarola has provided 26 runs with full documentation* and has extended the available Reynolds number regime by one order of magnitude. His measurements were made by traversing the interior of the test pipe, avoiding the free-jet measurements of Nikuradse, thus enabling him to furnish very accurate and extensive data on the velocity profiles, including the near-wall regime.

Figure 6.8-1 is a combined plot of selected data points for the velocity profiles of Zagarola at $Re = 3.1 \times 10^6$ and of Nikuradse at $Re = 3.2 \times 10^6$, which are the two cases that are most nearly comparable. There is considerable overlap. However, the crucial differences between the two — which lie in the near-wall layer — do not show up on this type of plot, so no inferences are possible. It should be pointed out, in addition, that only a small fraction of Zagarola's data for this run are shown.

Figure 6.8-2, calculated from Zagarola's data, shows a small segment of the velocity profile in the neighborhood of the wall at $Re = 3.1 \times 10^6$. These ten data points all lie within 0.10 in. of the wall, the closest being only 0.035 in. from the surface. These measurements, which produce a very regular curve, represent only a small portion of a very considerable experimental undertaking.

The data were acquired with a Pitot tube mounted on a linear traverse whose position was determinable to within ±0.001 in. and which employed a wall-finding technique accurate to ±0.002 in. Comparable precision was involved in the installation of static-pressure taps and the determination of their streamwise locations.

---

* The entire data set (all Re) of Zagarola (1996) is available on the Web at http://www.princeton.edu/~gasdyn/index.html.

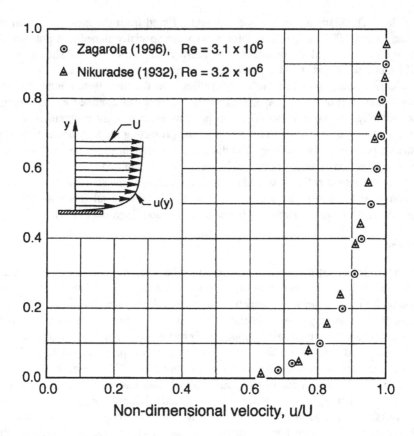

FIGURE 6.8-1 Experimental velocity profile $u/U = f(y)$ for flow through a smooth pipe at Re = 3.5×10^6 obtained by Zagarola (1996) compared to that of Nikuradse (1932) at Re = 3.2×10^6.

The friction factor values were computed from the Darcy–Weisbach relation (Equation 6.7-7) for horizontal flow:

$$\ell = \frac{2\Delta p}{\rho U^2} \frac{D}{L}.$$

(6.8-1)

In the case cited, the run was made with compressed air at a pressure level of $p = 2.830 \times 10^6$ Pa = 27.93 atm, and $\rho = 33.68$ kg/m³ corresponding to S.G. ≈ 0.034. The flow centerline velocity was $U = 15.28$ m/s = 50.1 ft/sec within an actual pipe diameter of $D = 0.129\ 4$ m = 5.094 in.

For engineering purposes, it seems clear that the results for the friction factor are destined to supplant the Prandtl expression as a new standard. We have already pointed out that the form of Equation 6.7-13, as verified by Nikuradse, is retained. However, the constants, again obtained by curve fitting, differ from those of Prandtl. In certain regimes — i.e., at the higher Reynolds numbers — the new correlated

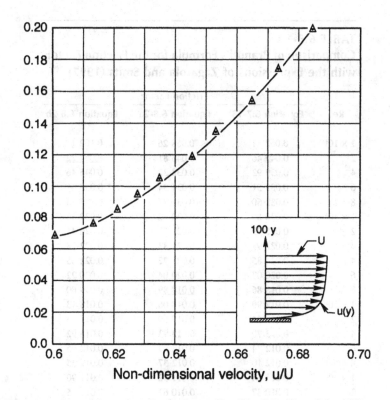

FIGURE 6.8-2   Plot of the near-wall velocity profile $u/U = f(y)$ at Re = 3.5 × 10⁶ in a smooth pipe. Data provided by the courtesy of M. V. Zagarola.

values for the friction factor are as much as 4.5% greater than those of Prandtl's relation. At the highest Reynolds number tested, the Prandtl relation differs from the new experimental value by 5.7%.

Zagarola and Smits (1997) have proposed two new expressions the first is

$$1/\sqrt{f} = 1.889 \; \log_{10}\left(\mathrm{Re}\sqrt{f}\right) - 0.3577, \tag{6.8-2a}$$

Obviously, this is Prandtl's relation with a new set of constants determined by curve fitting with the new data. The second is

$$1/\sqrt{f} = 1.873 \; \log_{10}\left(\mathrm{Re}\sqrt{f}\right) - 0.2631 - 233/\left(\mathrm{Re}\sqrt{f}\right)^{0.9}. \tag{6.8-2b}$$

A tabulation of f vs. Re from these two relations is compared with the values from the Prandtl relation in Table 6.8-1. An exposition of Prandtl's mixing-length theory would take us far beyond the objectives of this text. Interested readers should consult Chapters 19 and 20 of Schlichting (1979), and White (1974) who cite the references.

TABLE 6.8-1
Comparison of Prandtl's Formula for the Friction Factor
with the Expressions of Zagarola and Smits (1997)

	Friction Factor		
Re	Equation 6.7-13	Equation 6.8-2a	Equation 6.8-2b
1×10^3	0.062 61	0.058 26	0.102 1
2	0.049 46	0.046 81	0.064 22
4	0.039 92	0.038 32	0.045 46
6	0.035 51	0.034 34	0.038 61
8	0.032 60	0.031 87	0.034 83
1×10^4	0.030 89	0.030 82	0.032 36
2	0.025 89	0.025 50	0.026 41
4	0.021 97	0.021 83	0.022 18
4.1	0.021 85	0.021 72	0.022 05
6	0.020 07	0.020 04	0.020 22
8	0.018 86	0.018 89	0.019 00
1×10^5	0.017 99	0.018 06	0.018 13
2	0.015 64	0.015 80	0.015 83
4	0.013 70	0.013 93	0.013 92
6	0.012 74	0.012 99	0.012 95
8	0.012 10	0.012 37	0.012 35
1×10^6	0.011 64	0.011 92	0.011 90
2	0.010 37	0.010 67	0.010 65
4	0.009 294	0.009 597	0.009 592
6	0.008 738	0.009 040	0.009 041
8	0.008 372	0.008 677	0.008 678
1×10^7	0.008 104	0.008 408	0.008 411
2	0.007 345	0.007 648	0.007 655
3.5	0.006 806	0.007 105	0.007 115
4	0.006 686	0.006 984	0.006 995
6	0.006 341	0.006 635	0.006 647
8	0.006 111	0.006 401	0.006 414
1×10^8	0.005 941	0.006 229	0.006 243

Note: The heavy lines delineate the range over which the latter formulae lie
within ±1.2% of the experimental data.

The first of these fits the data for f to ±1.2%, or better, over the range $10^5 \leq$ Re $\leq 3.5 \times 10^7$. The second is a modification designed to provide an additional correction at the lower Reynolds numbers. It fits the data to better than ±1.2% over $3.1 \times 10^3 \leq$ Re $\leq 3.5 \times 10^7$.

At this point we note that the Reynolds number range for which experimental data is available has been extended "only" to Re = 3.5×10^7 from Nikuradse's high of Re = 3.2×10^6. Why not go to a higher value? In the first instance this last datum was obtained at the maximum available horsepower of the motor. It is shown in Section 6.11 that pumping horsepower is approximately proportional to V^3. Other conditions being unchanged, this implies the proportionality $P \sim Re^3$. Thus, using the experimental method of Nikuradse to go to $Re = 3.5 \times 10^7$ would have increased the power required by a factor of approximately 1300.

In the Superpipe, to operate at $Re = 10^8$ would necessitate a power increase by a factor of 23, generating the need for a motor-pump assembly capable of producing 4700 hp or so. In addition to the huge costs of such a powerplant and pump, this would call for a major redesign of the system, generating still higher costs. There is also a serious question of whether or not air could be used in such a facility at the desired Reynolds numbers without encountering severe compressibility effects.

6.9 FLOW IN ARTIFICIALLY ROUGHENED PIPES

The correlation of Blasius that demonstrated, for turbulent flow in smooth pipes ($\varepsilon \rightarrow 0$), that the friction factor depends only on the Reynolds number, was a major advance. The requirement, so important to engineers, that the flow losses in smooth pipes (needed, of course, for any pipe) be readily predictable, was realized by Prandtl's discovery of the universal law of friction. However, the problem of predicting the flow losses in nonsmooth pipes remained incomplete since there had been, up to that time, no systematic investigation of the effects of roughness.

This investigation was also undertaken by Nikuradse (1933) who published a remarkable experimental study on the flow losses in pipes with roughness ($\varepsilon > 0$). In order to be able to control the actual roughness in a precise way, Nikuradse used specially prepared pipes to the inside of which was cemented a coat of sand grains. The sand had been carefully sifted to achieve near uniformity of grain size, a condition that was verified by measuring under a microscope the principal diameter of a large number of grains for each batch. Nikuradse's description of the pains required to find a way of holding the grains to the inside wall without the grains rapidly detaching from the wall under flow conditions, on the one hand, and without using a layer of cement (actually a lacquer) so thick that it changed the effective size of the grains, on the other, is indicative of the extraordinarily high quality of the whole experimental investigation.

Nikuradse investigated six values of relative roughness. He chose to plot his data as a function of R/k_s, where R is the pipe radius and k_s the mean size of the grain diameter. Thus, R/k_s is inversely proportional to $\varepsilon = e/D$ selected by Blasius as the relative roughness parameter. The actual values selected by Nikuradse for investigation were $R/k_s = 15, 30.6, 60, 126, 252,$ and 507. The pipe sizes chosen were 2.5, 5.0, and 10.0 cm (about 1, 2, and 4 in., respectively). In a 1-in. pipe $R/k_s = 507$ translates to $k_s \approx 0.001$ in., whereas in a 4-in. pipe $R/k_s = 15$ translates to $k_s \cong 0.133$ in. These two values of k_s indicate the approximate outer limits on the sand grain sizes.

FLOW LOSSES

The principal results of Nikuradse are shown in Figure 6.9-1, which is a replot of his data by Schlichting (1979). As typical, the friction factor (actually $100f$) is plotted on a log–log scale vs. the Reynolds number $Re = VD/\nu$ with R/k_s as parameter. A seventh curve has been added, which is for a manufactured pipe whose natural

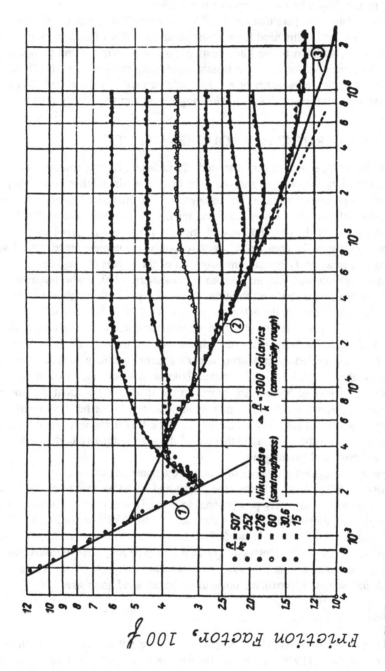

FIGURE 6.9-1 Flow losses in artificially roughened pipes. Replot of Nikuradse's data. (From Schlichting, H., *Boundary Layer Theory*, 7th ed., McGraw-Hill, New York, 1979. With permission.)

roughness is $R/k_s = 1300$ ($\varepsilon = 0.00039$). For the sand-roughened pipes the flow divides itself into three different categories.

The first, for $\mathrm{Re} \leq 2 \times 10^3$, is the Hagen–Poiseuille laminar flow regime, Equation 6.3-21, in which $f = f$ (Re) only. Nikuradse notes, for every flow over a surface, that is adjacent to the wall there exists a laminar sublayer regardless of the outer flow behavior, i.e., whether or not the outer flow is turbulent. The thickness δ of this layer depends in some complicated way on the Reynolds number of the outer flow. Stanton (1911) first pointed this out when his experiments in a turbulent pipe flow showed that near the wall the shear stress obeyed Newton's law of resistance. For a sufficiently low Reynolds number the thickness δ will always be larger than the height of the roughness element k_s. Thus, the effects of roughness are submerged by the boundary layer, and the whole flow tends to remain laminar. In consequence, this regime is usually referred to as *hydraulically smooth*.

In the second regime, designated by Nikuradse as the "region of transition to fully turbulent flow," the boundary-layer thickness δ and the roughness height k_s are of the same order. At the beginning of this regime a few of the roughness elements protrude through the laminar sublayer generating eddies whose creation requires an extraction of energy from the outer flow, thereby initiating an increase in the friction factor. As the Reynolds number continues to increase, the thickness of the laminar sublayer decreases, more roughness elements protrude into the outer region, and, at some point, the interaction between these eddies and the main flow results in a fully developed turbulent flow.

With increasing Reynolds number, the friction factor sharply diverges from the Hagen–Poiseuille type of dependence, over a small range of Reynolds numbers lying between 2160 and 2500. Nikuradse refers to this range as the "critical Reynolds number."* Depending upon the relative roughness, the friction factor, after a sharp rise, tends to follow the Blasius law for the smaller values of relative roughness (large R/k_s), or to continue rising until it reaches, asymptotically, a constant value, for large relative roughness (small R/k_s). However, the curve for each relative roughness value eventually diverges from the Blasius type of dependence and from each other, until each approaches its own asymptotic value of f for sufficiently large Re. In short, in this range, $f = f$ (Re, ε).

The third regime, called fully turbulent flow or the completely rough regime, where each curve approaches a constant f, comprises the region of flow for which the friction factor becomes independent of the Reynolds number and dependent solely on the relative roughness, i.e., $f = f$ (ε) only.

Empirical relations based on Nikuradse's data exist that give the friction factor for the transition and for the fully turbulent regimes, but, since they** are not useful for commercial pipes, they are omitted herein.

DISCUSSION

Nikuradse's experimental study clarified the general type of behavior which can be expected for rough-pipe flow, but the results are not directly applicable to flow in

---

* Based on Nikuradse's experiments, the critical Reynolds number is usually taken to be $\mathrm{Re}_{crit} = 2200$.
** These formulas, with references, can be found in Schlichting (1979) in Chapter 20.

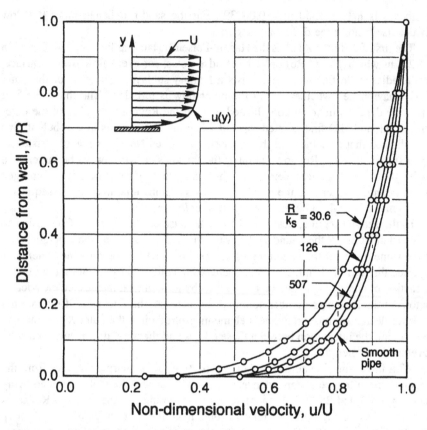

FIGURE 6.9-2 Velocity profiles in artificially roughened pipes at Re $\approx 10^6$ compared with that for a smooth pipe. (After Nikuradse, 1933.)

commercial pipes. Because of the lack of regularity in the way roughness evidences itself in a manufactured pipe, the correlation must be based on measurements from actual commercial pipes. This correlation was made by the American hydraulics engineer, L. F. Moody, and is discussed in Section 6.10.

For a brief discussion of other types of regular roughness patterns, e.g., riveted surfaces, see Schlichting (1979, p. 624).

In addition to the compilation of the f (Re, ε)-plots, Nikuradse made extensive measurements on the velocity distribution, i.e., $u = u(r)$, where u is the time-averaged component of velocity parallel to the pipe centerline, as a function of the radial coordinate. He discovered, at a fixed Reynolds number, that the velocity normalized by U the centerline value, depends on the value of the relative roughness. A plot of u/U vs. the dimensionless distance from the wall y/R, for different values of R/k_s, is shown in Figure 6.9-2, replotted from Nikuradse's data, all for Re $\approx 10^6$. The less the relative roughness, the fuller the velocity profile and, consequently, we see that the flow most nearly approaching the idealization of the one-dimensional approximation is turbulent flow in smooth pipes.

6.10 FLOW LOSSES IN COMMERCIAL PIPES

Although engineers had grappled with the problem of making quantitative prediction of the flow losses in pipes since early in the 19th century, the research required to illuminate the fundamental problems came slowly. The most important contributions prior to World War II were from scientists and engineers in just three countries — France, England, and Germany — although several Americans had made important experimental studies that were highly useful to hydraulics engineers.

One of the later attempts to correlate the experimental data for the flow in commercial pipes is due to Pigott (1933), the same year that Nikuradse published his artificial roughness studies. Although Pigott understood the necessity of correlating data in terms of nondimensional parameters (i.e., the dependence of f on Re), it seems that he was unaware of the work of Blasius (not to mention Nikuradse), the principal result of which is Equation 6.7-8. Although Pigott defined a roughness parameter as the average height of the roughness divided by the pipe radius, his method of correlating the data was not successful in showing a systematic dependence thereon.

Instead, he proposed an empirical relation of the form

$$f = C/(\mathrm{Re})^n, \tag{6.10-1}$$

where the exponent n would depend on the roughness and where both C and n would be calculated from experimental data. For smooth pipes he noted that $n = 0.24$ in which case Equation 6.10-1 reduces (essentially) to the Hagen–Poiseuille law for laminar flow.

It would not be fruitful to explore Pigott's approach to the turbulent flow regime in any detail except to note that no version of Equation 6.10-1 can adequately represent the asymptotic behavior of the friction factor in the completely rough regime, i.e., when $f = f(\varepsilon)$ only, as proved by Nikuradse for artificially roughened pipes. Nevertheless, it appears that the empirical rules proposed by Pigott represented an advance over previous practice.

MOODY'S CORRELATION FOR COMMERCIAL PIPES

Moody (1944) took advantage of the Nikuradse investigation of flow in artificially roughened pipes as well as the nondimensional relation of Blasius governing pipe flow, both of which had become well known to American hydraulics engineers in the ensuing decade. In fact, Moody's contribution was to recognize and plot the data of thousands of experiments in exactly the form deduced by Blasius:

$$f = f\ (\mathrm{Re},\ \varepsilon). \tag{6.10-2}$$

These results, using a log–log plot, appear in Figure 6.10-1.

Note that the dramatic variation of the friction factor in the range of transition to turbulent flow, discovered by Nikuradse for artificially roughened pipes, does not

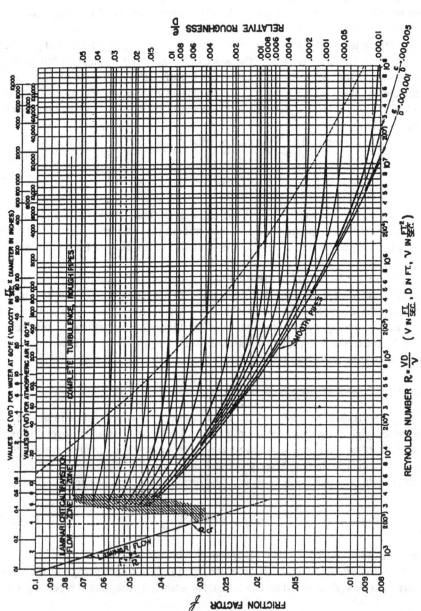

FIGURE 6.10-1 Friction factors for commercial pipes. (From Moody, L.F., *Trans. ASME*, 66, 671–684, 1944. With permission.)

occur in commercial pipes. Because the values of relative roughness (and other data, as well) for the measurements available to Moody were not known with sufficient accuracy, he did not attempt to delineate the transition regime.

It is useful to point out that for rough pipes the extent of the regime with varying friction factor is very small. In the region to the right of the dashed line we have completely rough flow, where the friction factor depends only on ε. Furthermore, there are built-in uncertainties that limit the accuracy of any computations based on this chart. Moody suggests that errors in f for smooth pipes are within $\pm 5\%$. For rough pipes the probable error is within $\pm 10\%$.

Figure 6.10-2, also reproduced from Moody, provides a convenient means for determining the relative roughness for a variety of materials. These plots are based on average values. It should be noted that these values are for new, clean pipes. Because of the action of scale formation, and deposition of materials on a pipe wall from the flow, especially for water, the roughness always increases with time. In the Discussion section of Moody's paper it is pointed out that in one case of water flow in a 3-in. pipe in moderate use it took only 3 years for the roughness to double. On the other hand, in that particular case, the doubling of ε represented only a 20% increase in the friction factor, so that great precision of the roughness is not essential, in any case.

6.11 PUMPING POWER REQUIRED TO MAINTAIN A PIPE FLOW

PUMPING POWER IN A PIPE

Consider a slug of fluid (instantaneously) located between stations ① and ② in a straight pipe that is inclined to the horizontal. The pressure force on the slug (which may be of arbitrary length) is

$$F = (p_1 - p_2)A = \Delta p A. \qquad (6.11\text{-}1)$$

The power absorbed by the slug, whose mean velocity is V, is

$$P \equiv FV = \Delta p A V = \Delta p \dot{Q}. \qquad (6.11\text{-}2)$$

From Equations 6.3-6 and 6.5-2, the static-pressure drop can be related to the head loss and the change of height by

$$\Delta p \equiv p_1 - p_2 = (p_1 + \rho g z_1) - (p_2 + \rho g z_2) + \rho g (z_2 - z_1),$$
$$= \Delta p^* + \rho g (z_2 - z_1),$$
$$= \rho g [h_\ell + (z_2 - z_1)].$$

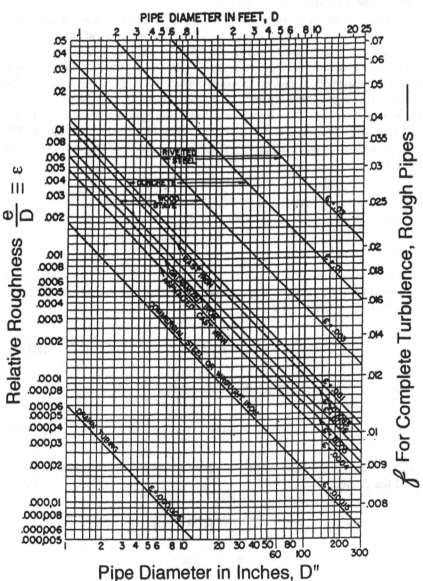

FIGURE 6.10-2 Chart of relative roughness for different pipe materials. (From Moody, L.F., *Trans. ASME*, 66, 671–684, 1944. With permission.)

$$P = \rho g \dot{Q} \left[h_\ell + \left(z_2 - z_1 \right) \right],$$

(6.11-3)

$$= \dot{m} g \left[h_\ell + \left(z_2 - z_1 \right) \right].$$

Note that $\dot{m}g$ is the *weight flow rate* in the pipe. Consequently, power must be supplied to overcome the frictional loss (the h_ℓ-term) and to account for the energy input to the flow when $z_2 - z_1 > 0$. If $z_1 - z_2 < 0$, then work is done on the flow by the gravitational field.

Horsepower Required

In the 18th century, when the industrial revolution began, the process by which machines would relieve horses and, of course people, of the backbreaking toil of producing power, James Watt (1736–1819) found it convenient to introduce a new unit, the *horsepower*,* defined as 1 hp = 550 ft lb/sec. This unit, which still survives, provides a nice touch of irony, since horses are used today for power production only in the most marginally productive societies. Thus,

$$\text{hp} = P/550.$$

(6.11-4)

The expression in Equation 6.11-4 gives the useful horsepower put out by the pump — i.e., absorbed by the flow. Because of inefficiencies (losses in the pump due to the combined effects of friction and leakage past the pump vanes), the power input to the pump must be greater. This is accounted for by the definition of a pump efficiency η_P, where

$$\eta_P \equiv P/(\text{Power Input}).$$

(6.11-5)

The determination of η_P is strictly experimental, and its value is furnished by the pump manufacturer as part of the specifications. There is a second efficiency η_M for the motor (e.g., electric, gasoline, wind, or natural gas powered) that drives the pump, which is furnished by the motor manufacturer. Thus, the power required from whichever source is employed is given by

$$P_{\text{Req}} = \frac{P}{\eta_P \eta_M}.$$

(6.11-6)

---

* The British horsepower is equivalent to 745.700 watts. There is also a metric horsepower, defined such that 1 metric horsepower = 75 $kg_f \cdot$ m/s = 735.499 watts. Neither horsepower unit is officially recognized by the SI.

POWER REQUIREMENTS IN LAMINAR VS. TURBULENT FLOW

For a horizontal, laminar pipe flow, we found in Equation 6.3-15b that $\Delta p \sim V$. For turbulent flow in the Blasius regime, Equation 6.7-2 yields, similarly, $\Delta p \sim V^{7/4}$. Thus,

$$P \sim V^2, \quad \text{in laminar pipe flow}, \qquad (6.11\text{-}7a)$$

$$P \sim V^{11/4}, \quad \text{in turbulent pipe flow}. \qquad (6.11\text{-}7b)$$

If one is condemned to pump a fluid in the turbulent flow regime (as is usually the case in water-distribution systems), it is obviously desirable to use as large a pipe as possible to keep down the velocity, and thus the pumping costs. However, larger pipes require larger capital investments. These factors are only two of those that would enter the process of making a final decision on the pipe diameter for a particular system.

6.12 COMPUTATION OF POWER REQUIRED IN A NONUNIFORM DUCT

INTRODUCTION

The computation of the power required in a nonuniform duct is complicated by the question of whether the flow is laminar or turbulent. In a turbulent flow, one can expect to be forced to invoke semiempirical methods, i.e., a calculation procedure based on appropriate model, or full-scale, measurements. In particular, for *diffusers** — i.e., divergent ducts in which the area increases in the direction of the flow and, thus, the pressure also increases (as long as compressibility effects are negligible) — the difficulty is compounded by the tendency of the boundary layer to separate from the wall, with attendant greater flow losses.

Even in laminar flows, separation may occur in a divergent duct. However, in a slender, convergent duct the pressure decreases in the flow direction, and there is no tendency for the flow to separate. This allows for the flow to be treated analytically (although approximately) by melding together a modification of the laminar pipe flow theory of Section 6.3 with the one-dimensional dynamic equation based on the mean flow speed.

THE DUCT GEOMETRY

Consider the steady, constant-density, nonuniform flow through the truncated, conical duct shown in Figure 6.12-1. Let the duct diameter at station x be $D(x)$; then, for the convergent portion of the duct

$$D(x) = D_1 - \frac{D_1 - D_2}{L} x$$

---

* See, for example, Sovran and Klomp (1967).

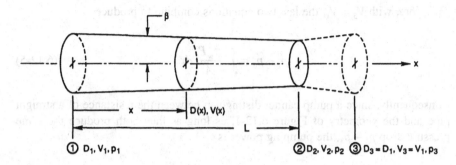

FIGURE 6.12-1 Laminar flow through a slender, convergent duct.

or

$$\frac{D}{D_1} = 1 - \frac{D_1 - D_2}{D_1}\frac{x}{L} = 1 - b\bar{x}, \tag{6.12-1a}$$

where

$$b \equiv (D_1 - D_2)/D_1, \quad \bar{x} \equiv x/L. \tag{6.12-1b}$$

The requirement of a slender duct is met if the cone half-angle is small or, more precisely, if

$$0 < \tan \beta \equiv \frac{D_1 - D_2}{2L} \ll 1. \tag{6.12-2}$$

INTRODUCTION OF THE DYNAMIC EQUATION

From Equation 4.3-4, we obtain the one-dimensional dynamic equation, already in integral form except for the viscous term. For horizontal flow

$$\frac{p_1}{\rho} + \frac{1}{2}V_1^2 - \left(\frac{p_2}{\rho} + \frac{1}{2}V_2^2\right) = \frac{1}{\rho}\int_{x=0}^{L}\frac{\tau_w P}{A}\,dx. \tag{6.12-3}$$

The last term on the right of Equation 6.12-3, which contains the flow losses, can be evaluated by conventional methods if we employ the following artifice. The duct is extended by a hypothetical diffuser (shown in dashed lines in Figure 6.12-1) to ③ where $A_3 = A_1$. In the region of the hypothetical addition, the flow is assumed to be inviscid and there are no flow losses. That is, Bernoulli's equation applies:

$$\frac{p_2}{\rho} + \frac{V_2^2}{2} = \frac{p_3}{\rho} + \frac{V_3^2}{2}. \tag{6.12-4}$$

Therefore, with $V_3 = V_1$, the last two equations combine to produce

$$p_1 - p_3 = \int_0^L \frac{\tau_w P}{A}\, dx.$$

(6.12-5)

Consequently, since a pump cannot distinguish between the resistance of a straight pipe and the geometry of Figure 6.12-1, as long as they both produce the same pressure drop $p_1 - p_3$, the pumping power is

$$P_{1,2} = (p_1 - p_3) A_3 V_3 = (p_1 - p_3)\dot{Q} = \dot{Q} \int_0^L \frac{\tau_w P}{A}\, dx.$$

(6.12-6)

Therefore, the problem is reduced to evaluating the integral in Equation 6.12-6.

INTEGRATION OF THE VISCOUS TERM

We assume that the local velocity profile at station x for the component parallel to the duct axis can be approximated by the parabolic velocity profile of laminar pipe flow, Equation 6.3-16:

$$u(r) = 2V(1 - r^2/R^2).$$

(6.12-7)

However, since the cross-sectional area decreases in the direction of the flow, the local, mean flow speed varies so as to satisfy the one-dimensional continuity equation. Similarly, the (varying) wall shear stress is obtained from Equation 6.3-17b. Thus, with Equation 6.12-1,

$$\tau_w = \frac{4\mu V}{R} = \frac{8\mu V}{D} = \frac{8\mu V_1}{D_1}\cdot\frac{V/V_1}{D/D_1} = \frac{8\mu V_1}{D_1}(1 - b\bar{x})^{-3}.$$

(6.12-8)

Note that Equation 6.12-8 makes use of the continuity equation $V/V_1 = (D/D_1)^{-2}$.
   Therefore, with $P/A = 4/D = (4/D_1)(D/D_1)^{-1}$,

$$\int_0^L \frac{\tau_w P}{A}\, dx = \frac{32\mu V_1 L}{D_1^2}\int_{x=0}^1 (1 - b\bar{x})^{-4}\, d\bar{x} = \frac{32\mu V_1 L}{D_1^2} F(b),$$

(6.12-9)

where we can define a function of the duct parameter b:

$$F(b) \equiv \frac{1}{3b}\left[\frac{1}{(1-b)^3} - 1\right],\quad 0 < b < 1.$$

(6.12-10)

A few values of $F(b)$ are tabulated below:

b	0	0.05	0.10	0.15	0.20	0.25	0.3	0.35	0.4	0.45	0.5
$F(b)$	1	1.109	1.239	1.396	1.589	1.827	2.128	2.516	3.025	3.711	4.667

Therefore, the power required is

$$P_{1,2} = \dot{Q} \frac{32\mu V_1 L}{D_1^2} F(b) = \frac{8\pi}{Re_1} \rho V_1^3 L_1 D_1 F(b). \tag{6.12-11}$$

The station with the highest, local Reynolds number is ②, and that is where we would expect transition to turbulence to occur first. If we use Nikuradse's criterion (although strictly applicable only to pipe flow) for the critical Reynolds number, then the criterion for the avoidance of transition is $Re_2 < 2200$ or, equivalently, $Re_1 < 2200(1 - b)$. This represents only an educated guess, since with the greater complexity of this flow, it seems likely that other factors may enter in with the expectancy of earlier transition than in a pipe. In any case, use of Equation 6.12-11 when $Re_2 > 2200$ would be a dubious practice without experimental verification that the flow remains laminar.

In the following section the approach used in hydraulics calculations is given, which is based on the computation of head losses for the various components of a flow loop. For the case of the slender duct

$$h_{\ell 1,2} = h_{\ell 1,3} = \frac{p_1 - p_3}{\rho g} = \frac{32\mu V_1 L}{\rho g D_1^2} F(b) \tag{6.12-12}$$

or, equivalently,

$$h_{\ell 1,2} = \frac{64}{Re_1} F(b) \cdot \frac{V_1^2}{2g} \frac{L}{D_1}. \tag{6.12-13}$$

Note that for a pipe $b = 0$, $F(b) = 1$; therefore, Equation 6.12-13, which is in the form of the Darcy–Weisbach relation, reduces to the previously obtained expression for the Hagen–Poiseuille flow.

FLOW LOSS COEFFICIENT

In Section 4.8 it was noted that for many hydraulic devices it is convenient to specify their performance by a flow loss coefficient k, defined such that

$$k = \frac{h_\ell}{V^2/2g}. \tag{6.12-14}$$

Obviously, for the slender duct, choosing ① as the reference station,

$$k = \frac{64}{\mathrm{Re}_1} F(b) \cdot \left(L/D_1\right), \qquad (6.12\text{-}15)$$

where $V_1^2/2g$ is the velocity head at the inlet station to the duct.

FLOW LOSSES IN OTHER CONDUIT ELEMENTS

The problem of evaluating flow losses in nonconventional systems is vast and is of great engineering importance. Essentially, the problem is reduced to the collection of data, based on thousands of (usually unrelated) experiments. These experiments are occasionally supported by theory, and they originate from a worldwide distribution of sources. Some examples of configurations for which the resistance data are tabulated are screens of varying mesh, orifices of various cross sections, plates with patterns of holes and slots, unusual pipe fittings, curved ducts, scrolls and shrouds for fans, drag of inserted bodies, and the effects of compressibility.

There are three references that have come to the writer's attention and that appear to be especially noteworthy in this regard. The first two, which are primarily collections of data, are the *Fluid Flow Data Book*, published by the General Electric Company (1981) and the *Handbook of Hydraulic Resistance* by I. E. Idelchik (1986). The third is a comprehensive treatment of the duct flow by A. J. Ward-Smith (1980) entitled *Internal Fluid Flow*.

6.13 HYDRAULIC CALCULATIONS FOR SIMPLE CONDUITS AND FLOW LOOPS

INTRODUCTION

Hydraulics, which is one of the most ancient fields of engineering, is a subject that plays a vital role in the life of every person fortunate enough to live in a civilized country. One need only suffer the loss of an adequate water supply, or the lack of a decent sewage system, to confirm that fact. As previously noted, because hydraulics calculations frequently involve semiempirical or even purely empirical relations, it is sometimes scorned as the "science of variable constants." However, even the scientist who wishes to design a facility in which a fluid circulates must lean on empiricism to design the system.

Thus, a body of specialized techniques has grown up that allows technicians and others with only a superficial knowledge of fluid mechanics to prepare design specifications for fluids (especially for liquids) systems. In this section the standard techniques used in hydraulic computations are introduced.

THE EQUATIONS OF DUCT FLOW

The starting point for all hydraulic calculations is the steady-state energy equation for constant-density flow that we obtain from Equation 4.8-2, i.e.,

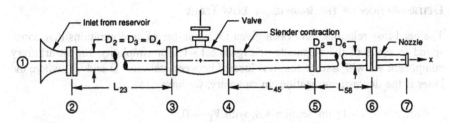

FIGURE 6.13-1 Flow losses in typical ducting components.

$$\rho(e_2 - e_1) = \rho\Delta e = p_1 + \tfrac{1}{2}\rho V_1^2 + \rho g z_1 - \left(p_2 + \tfrac{1}{2}\rho V_2^2 + \rho g z_2\right). \qquad (6.13\text{-}1)$$

Equation 6.13-1 applies to arbitrary changes of cross section or height between stations ① and ②. The head-loss term for the same duct segment is defined as

$$h_{\ell 1,2} = \frac{\Delta e}{g} = \frac{p_1}{\rho g} + \frac{V_1^2}{2g} + z_1 - \left(\frac{p_2}{\rho g} + \frac{V_2^2}{2g} + z_2\right). \qquad (6.13\text{-}2)$$

Note that in a pipe where $V_1 = V_2$, Equation 6.13-2 reduces to the appropriate expression (Equation 6.5-2).

Equation 6.13-2 is now applied to each segment of the duct (see Figure 6.13-1) in turn, and the head losses added. Thus,

$$h_{\ell 1,2} = \frac{p_1}{\rho g} + \frac{V_1^2}{2g} + z_1 - \left(\frac{p_2}{\rho g} + \frac{V_2^2}{2g} + z_2\right),$$

$$h_{\ell 2,3} = \frac{p_2}{\rho g} + \frac{V_2^2}{2g} + z_2 - \left(\frac{p_3}{\rho g} + \frac{V_3^2}{2g} + z_3\right),$$

and

$$h_{\ell n-1,n} = \frac{p_{n-1}}{\rho g} + \frac{V_{n-1}^2}{2g} + z_{n-1} - \left(\frac{p_n}{\rho g} + \frac{V_n^2}{2g} + z_n\right),$$

$$h_{\ell n,n+1} = \frac{p_n}{\rho g} + \frac{V_n^2}{\rho g} + z_n - \left(\frac{p_{n+1}}{\rho g} + \frac{V_{n+1}^2}{2g} + z_{n+1}\right),$$

for which the sum yields

$$h_{\ell 1,n+1} \equiv \Sigma h_\ell = \frac{p_1}{\rho g} + \frac{V_1^2}{2g} + z_1 - \left(\frac{p_{n+1}}{\rho g} + \frac{V_{n+1}^2}{2g} + z_{n+1}\right). \qquad (6.13\text{-}3)$$

Equation 6.13-3 relates the sum of the individual head-loss terms to the inlet and exit flow conditions of the conduit.

Determination of the Individual Loss Terms

The head-loss relations have been given for a number of configurations that correspond to some of the segments of Figure 6.13-1. For simplicity, for an arbitrary component we will assume that it is located between stations ① and ②, where the latter is the downstream station. In summary, we have

Nozzle inlet — From Section 4.8, with $V_1 \to 0$,

$$h_\ell = k V_2^2 / 2g,\tag{6.13-4}$$

where k is an empirical constant determined by experiment and is tabulated in Figure 4.8-1 for representative inlets.

Pipe segment — From Sections 6.1 to 6.5 for laminar flow and Sections 6.6 to 6.10 for turbulent flow, with $V_1 = V_2 \equiv V$,

$$h_\ell = f \, \frac{V^2}{2g} \, \frac{L}{D},\tag{6.13-5}$$

where

$$f = 64/\mathrm{Re}, \quad \text{for laminar flow (i.e., Re} < 2200),\tag{6.13-6a}$$

$$f = 0.3164/\mathrm{Re}^{1/4}, \quad \text{for turbulent flow in smooth pipes,*}$$
$$\mathrm{Re} < 10^5,\tag{6.13-6b}$$

$$1/\sqrt{f} = 2.0 \, \log_{10}\left(\mathrm{Re}\,\sqrt{f}\right) - 0.8, \quad \text{for turbulent flow in smooth pipes,*}$$
$$\mathrm{Re} \text{ arbitrarily large.}\tag{6.13-6c}$$

for turbulent flow in smooth pipes,* Re arbitrarily large. For turbulent flow in commercial pipes $f = f(\mathrm{Re}, \varepsilon)$ where the friction factor is obtained from the Moody diagrams, Figures 6.10-1 and 6.10-2. The same relations apply to all pipe segments as long as proper account is taken of the dependence of both the Reynolds number and the relative roughness on the pipe diameter.

Slender conical contraction — The approximate relations for laminar flow are given in Section 6.12. For nonslender ducts, or for turbulent flow, resort must be made to the standard hydraulics handbooks, or to one of the references mentioned at the end of this Section.

Sudden expansion — This is the Borda–Carnot flow of Section 4.12. The loss is given by Equation 6.13-4, where

---

* If greater accuracy is desired use the formulae of Zagarola and Smits, Section 6.8.

$$k = \left(A_2/A_1 - 1\right)^2. \tag{6.13-7}$$

Sudden contraction — This flow also satisfies Equation 6.13-4. As previously shown

$$k = \left(1/C_c - 1\right)^2, \tag{6.13-8}$$

where the experimentally determined values of the contraction coefficient C_c are tabulated in Table 6.13-1.

TABLE 6.13-1
Representative Values of $(L/D)_{Eq.}$
for Selected Pipe Fittings

Fitting	$(L/D)_{Eq.}$
Globe valve, fully open	340–450
Gate valve, fully open	13
Check valve, fully open	135–150
45° Standard elbow	16
90° Standard elbow	30
90° Long radius elbow	20

Source: Abstracted from *Flow of Fluids*, Technical Paper No. 410, The Crane Co., 1957. With permission.

CONVERSION OF HEAD LOSS TO AN EQUIVALENT LENGTH PIPE

For the purpose of systematizing computations it is a frequent practice to express the loss due to a component with irregular flow geometry (and where the flow is almost certainly turbulent), such as a valve, in terms of an equivalent pipe length. That is, using ② as the reference station, we equate Equation 6.3-4 to the loss term for an equivalent length pipe of the diameter consistent with that of the valve outlet. Thus,

$$h_t = k \frac{V_2^2}{2g} = f \frac{V_2^2}{2g}\left(\frac{L}{D}\right)_{Eq.}, \tag{6.13-9a}$$

where

$$k = f\left(\frac{L}{D}\right)_{Eq.}, \tag{6.13-9b}$$

The value of f in Equation 6.13-9 is taken to be that for the completely rough turbulent flow regime corresponding to the relative roughness of the connecting pipe. This is evidently not a scientific fact, but is nevertheless a conservative value that gives adequate accuracy for design purposes, and is based on representative, measured flow losses.

For an extensive collection of experimentally determined values of k and $(L/D)_{Eq.}$ for various pipe fittings, tables of fluid properties, empirical formulas, samples of worked problems and more, see Technical Paper No. 410, published by the Crane Co., listed in the References as Anonymous (1988). Table 6.13-1 contains a few values of equivalent lengths for pipe fittings abstracted from the 1957 edition. From the same source there is available a software package for hydraulics computations called the *Crane Companion*.* An even more recent, and obviously powerful, library of programs is the *Easy 5 Thermal Hydraulic Library*,** developed by the Boeing Corporation and which interacts with other CAD software.

We can count on the fact that the performance of any flow loop tends to deteriorate with time. Thus, the total flow losses will tend to increase. The wise designer allows for this deterioration, and specifies a pump–motor combination that has sufficient power to handle the increased losses, which, of course, depend on the application.

6.14 STEADY FLOW THROUGH AN ELASTIC TUBE

If the cross-sectional area of a duct varies with changes in the local pressure, the analysis of the flow becomes more complicated than for a fixed duct since the area distribution $A = A [p(x)]$ is initially unknown and must, therefore, be determined as part of the solution. It is useful to keep in mind that the solution obtained will be obtained by perturbing a solution for viscous flow through a straight pipe, which can be either laminar or turbulent, as described in earlier sections of this chapter.

THE PRESSURE–AREA RELATION

Referring to Figure 6.14-1, suppose that an elastic tube of wall thickness $\bar{t}$, radius R is the unstressed condition, and length L is subject to a liquid flow of density ρ, where the inlet pressure is p_b. The tube empties into a plenum at the outlet pressure $p_e = p_a$. The problem is to determine the velocity distribution, the longitudinal distribution of pressure $p = p(x)$, and the deformed area distribution $A(x)$ of the tube.

The local pressure difference $p - p_a$ across the tube wall causes the tube to expand to a deformed local radius $r(x)$, which can be determined by the hoop-stress formula of elementary strength of materials. Let the wall stress at an arbitrary station be $\sigma = \sigma(x)$; thus, by putting a semicircular segment of the tube of length Δx in static equilibrium (Figure 6.14-1), we have

---

* Distributed by AZT, Inc. 4451 Brookfield Corporate Drive, Suite 101, Chantilly, VA 20151, 800-747-7401, FAX 703-631-5282. E-mail: ebz@patriot.net.
** Easy5 Product Support, The Boeing Company, P.O. Box 3703, MS 7L-46, Seattle, WA 98124-2207, 800-426-1443, 425-865-6695, FAX 425-865-2966. E-mail: easy5.sales@boeing.com, http://www.boeing.com/easy5.

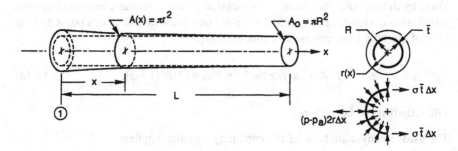

FIGURE 6.14-1 Viscous flow through an elastic tube.

$$(p - p_a)(2r\Delta x) - 2\sigma(\bar{t}\Delta x) = 0 \quad \text{or} \quad \sigma = (r/\bar{t})(p - p_a). \qquad (6.14\text{-}1)$$

From the definition of the strain e (of the circumference)

$$e \equiv \frac{\pi r - \pi R}{\pi R} = \frac{r}{R} - 1. \qquad (6.14\text{-}2)$$

Also, from Hooke's law for a linearly elastic material, $\sigma = Ee$, where E is the modulus of elasticity of the tube wall. Thus, we can combine the numbered equations with Hooke's law and solve for

$$\frac{r}{R} = \frac{1}{1 - (R/E\bar{t})(p - p_a)}. \qquad (6.14\text{-}3)$$

We next define a nondimensional pressure function $\varphi = \varphi(\xi)$, $\xi \equiv x/L$, such that

$$\varphi = (R/E\bar{t})(p - p_a); \qquad (6.14\text{-}4a)$$

consequently, the radius function, from Equation 6.14-3, becomes

$$r/R = (1 - \varphi)^{-1}. \qquad (6.14\text{-}4b)$$

Therefore, at the exit station $\xi_e = 1$, where $p_e = p_a$, $\varphi_e = 0$, $r_e/R = 1$. At the inlet station $\xi_b = 0$, the highest pressure is attained, denoted φ_b. The analysis is restricted to cases such that φ_b is small compared with unity.

It follows, if the unstressed cross-sectional area is $A_e = \pi R^2$, that the area function is

$$A/A_e = (r/R)^2 = (1 - \varphi)^{-2}. \qquad (6.14\text{-}4c)$$

Thus, by determining the pressure distribution in the tube, the corresponding area distribution is yielded. For eventual use it is convenient to note at this point that the ratio of the tube local perimeter to the local area is

$$P/A = 2\pi r/\pi r^2 = 2/(r/R)R = (2/R)(1 - \varphi).\qquad(6.14\text{-}4d)$$

THE CONTINUITY EQUATION

The usual steady-state form of the continuity equation applies:

$$\dot{Q} = uA = u_e A_e = \text{constant}.\qquad(6.14\text{-}5)$$

where $\dot{Q}$ is the volume flow rate. It is desirable, in fact essential, to nondimensionalize our working equations. We choose the exit velocity as the reference value, i.e., $u(L) \equiv u_e$. Thus, the dimensionless velocity is defined as

$$\bar{u}(\xi) = u/u_e,\qquad(6.14\text{-}6a)$$

where $\bar{u}(1) \equiv 1$, in which case the continuity equation can be expressed as

$$\bar{u}(\xi) = (1 - \varphi)^2.\qquad(6.14\text{-}6b)$$

$\bar{u}$ is of order unity, written $\bar{u} = O(1)$, due to the restriction on φ_b.

In what follows we will obtain the solution for the flow through an elastic tube that has the same flow rate and, hence, the same exit velocity u_e, as the flow through a pipe with constant area equal to the tube exit area A_e. This necessarily means that the pressure distribution in the tube deviates from that in the pipe, which is linearly decreasing. We will obtain both the pressure distribution and the velocity distribution in the deformed tube.

THE DYNAMIC EQUATION

The usual form of the dynamic equation for steady, incompressible, viscous, horizontal flow applies:

$$u\frac{du}{dx} = \frac{d}{dx}\left(\frac{u^2}{2}\right) = -\frac{1}{\rho}\frac{dp}{dx} - \frac{\tau_w P}{\rho A}.\qquad(6.14\text{-}7)$$

It is convenient to replace the wall viscous stress by the skin-friction coefficient. Thus, with Equation 6.14-4b,

$$\tau_w = \rho u^2 c_f/2 = \rho(u/u_e)^2 u_e^2 c_f/2 = \rho u_e^2 (1 - \varphi)^4 c_f/2.$$

With this relation and Equation 6.14-4d, the viscous term becomes

$$\frac{\tau_w P}{\rho A} = \frac{u_e^2 (1-\varphi)^5 c_f}{R}.$$

(6.14-8a)

From Equation 6.14-4a, $dp = (E\bar{\imath}/R)\, d\varphi$. The pressure term becomes

$$\frac{1}{\rho}\frac{dp}{dx} = \frac{E\bar{\imath}}{\rho LR}\frac{d\varphi}{d\xi}.$$

(6.14-8b)

By introducing Equation 6.14-6b, the kinetic energy term reduces to

$$\frac{d}{dx}\left(\frac{u^2}{2}\right) = \frac{u_e^2}{2}\frac{d}{d\xi}(\bar{u}^2) = \frac{u_e^2}{2L}\frac{d}{d\xi}(1-\varphi)^4.$$

(6.14-8c)

Substitution of Equations 6.14-8 into Equation 6.14-7 and moving of the first term on the right to the left, yields

$$\frac{E\bar{\imath}}{\rho LR}\frac{d\varphi}{d\xi} + \frac{u_e^2}{2L}\frac{d}{d\xi}(1-\varphi)^4 = -\frac{u_e^2(1-\varphi)^5 c_f}{R}.$$

(6.14-9a)

We now multiply Equation 6.14-9a by the factor $\rho LR/E\bar{\imath}$ to produce

$$\frac{d}{d\xi}\left[\varphi + \frac{u_e^2}{2E\bar{\imath}/\rho R}(1-\varphi)^4\right] = -\frac{\rho L u_e^2 c_f}{E\bar{\imath}}(1-\varphi)^5.$$

(6.14-9b)

Since the left-hand term in Equation 6.14-9b is dimensionless, the principle of dimensional homogeneity requires that each of the other terms be dimensionless also. Thus, in order for the coefficient of the term involving $(1-\psi)^4$ to be dimensionless, the factor $2E\bar{\imath}/\rho R$ must have the dimensions of a velocity squared, a fact that is readily verified. Hence, in a sense, we have discovered a new reference velocity such that $a_{\text{ref}} = \sqrt{2\,E\bar{\imath}/\rho R}$.

This choice would prove to be adequate for the current purpose. However, it turns out that there is in the literature a generally recognized reference value, differing only by a factor of two, which would appear naturally if the corresponding nonsteady flow problem in an elastic tube were undertaken. Thus, we choose, instead,

$$a_0^2 = E\bar{\imath}/2\rho R.$$

(6.14-10)

The quantity a_0,* as specified in Equation 6.14-10, is the speed of propagation of a pressure pulse in a liquid-filled elastic tube. The current application is restricted to situations in which the flow speed u is small compared with a_0.

With Equation 6.14-10, Equation 6.14-9b can be written as

$$\frac{d}{d\xi}\left[\varphi + e^2 (1-\varphi)^4/4\right] = -\varepsilon^2 \frac{L}{2R} c_f (1-\varphi)^5, \qquad (6.14\text{-}11)$$

where the ratio $\varepsilon \equiv u_e/a_0$ is a small parameter that must satisfy the inequality $\varepsilon \ll 1$. We have previously specified out boundary condition that at the tube outlet the pressure is

$$\varphi_e = \varphi(1) = 0. \qquad (6.14\text{-}12)$$

Equations 6.14-11 and 6.14-12 are as far as we can go without specifying the coefficient of skin friction.

APPLICATION TO THE CASE OF LAMINAR FLOW

For flow speeds sufficiently slow that the flow is subcritical (Re < 2200), we expect the flow to be laminar (actually quasi-laminar). Therefore, employing the results of Sections 6.4 and 6.5,

$$c_f = \frac{16}{\text{Re}} = \frac{16v}{u(2r)} = \frac{8v}{(u/u_e)(u_e/a_0)a_0(r/R)R} = \frac{8v}{\varepsilon a_0 R(1-\varphi)}, \qquad (6.14\text{-}13)$$

where v is the fluid kinematic viscosity. By substituting Equation 6.4-13 into Equation 6.4-11, the differential equation governing laminar flow is obtained:

$$\frac{d}{d\xi}\left[\varphi + \varepsilon^2(1-\varphi)^4/4\right] = -\varepsilon\Lambda(1-\varphi)^4, \qquad (6.14\text{-}14)$$

under the boundary condition of Equation 6.14-12, and where a second nondimensional parameter $\Lambda \equiv 4vL/a_0R^2$ is introduced.

SOLUTION OF EQUATION 6.14-14

The equation to be solved is a first-order, nonlinear, nonhomogeneous, ordinary differential equation. It is straightforward to integrate although the resulting form is somewhat unhandy. If we divide through the equation by the factor $(1 - \varphi)^4$, the terms on the left can be rewritten, producing

---

* The quantity a_0 is generally known as the Moens–Korteweg wave speed, for the individuals who rediscovered it in 1878. It was discovered originally by the British physician and scientist Thomas Young (1809).

$$\int_{\varphi=0}^{\varphi} d\left[\tfrac{1}{3}(1-\varphi)^{-3}+\varepsilon^2\ln(1-\varphi)\right]=-\varepsilon\Lambda\int_{\xi=1}^{\varepsilon}d\xi.$$

After substituting in the limits, we have

$$\tfrac{1}{3}\left[(1-\varphi)^{-3}-1\right]+\varepsilon^2\ln(1-\varphi)=\varepsilon\Lambda(1-\xi). \qquad (6.14\text{-}15)$$

There is no apparent way to solve for the explicit form $\varphi = \varphi(\xi)$. Thus, we shall have to settle for an approximate expression, i.e., a series in powers of ε. The term with the logarithm involves the product of ε^2 and $\ln(1-\varphi)$. Since we have previously stated that φ is small compared with unity, we can put $\varepsilon^2\ln(1-\varphi) \approx -\varepsilon^2\varphi$, which is a third-order term, and the product can be neglected.

We next solve for

$$1-\varphi=\left[1+3\varepsilon\Lambda(1-\xi)\right]^{-1/3}.$$

The right side of this expression is then expanded by the binomial expansion, keeping terms only of order ε^2 and larger. This yields

$$\varphi(\xi)=\varepsilon\Lambda(1-\xi)-2\varepsilon^2\Lambda^2(1-\xi)^2+O(\varepsilon^3). \qquad (6.14\text{-}16)$$

This result further justifies the neglect of the term $\varepsilon^2\ln(1-\varphi)$ in Equation 6.14-15.

We can also obtain another partial check on the validity of Equation 6.14-16. If we neglect the second-order term and substitute in the definitions for ε, Λ, and φ, we can then solve for

$$u_e=\frac{(p_b-p_a)R^2}{8\mu L}, \qquad (6.14\text{-}17)$$

which is precisely the mean velocity in Hagen–Poiseuille flow, as given by Equation 6.3-15b. We expect that the second-order term must represent the contribution due to the expansion of the elastic tube, which perturbs the main flow.

Velocity and Area Distributions

The velocity distribution is obtained by substituting Equation 6.14-16 into Equation 6.14-6b, squaring, and discarding terms of order three and higher. This results in

$$\bar{u}(\xi)=1-\varepsilon 2\Lambda(1-\xi)+\varepsilon^2 5\Lambda^2(1-\xi)^2+O(\varepsilon^3). \qquad (6.14\text{-}18)$$

The area distribution is obtained by substituting the pressure function (Equation 6.14-16) in Equation 6.14-4c and, again, expanding in series, disregarding terms of $O(\varepsilon^3)$ and smaller. Thus,

$$A/A_e = 1 + \varepsilon 2\Lambda(1-\xi) - \varepsilon^2\Lambda^2(1-\xi)^2. \qquad (6.14\text{-}19)$$

As a check on the consistency of the expansions, these two relations can be substituted into the (nondimensional) form of the continuity equation to find that $\bar{u}A/A_e = 1 + O(\varepsilon^3)$. As a final thought on perturbation analysis. once one has discarded terms of a given order in any equation, it makes no sense to retain terms of the same or higher order further along in the analysis.

APPLICATION

Consider the example of glycerol at 30°C = 86°F flowing through a horizontal, plastic tube. The dimensions of the tubing are $L = 10$ ft, $R = 0.5$ in., $\bar{t} = 0.010$ in. The modulus of elasticity of the tubing is selected as $E = 1.5 \times 10^5$ lb/in.^{2} = 2.16 $\times 10^7$ lb/ft^2. This value does not correspond to an actual material but is chosen at the lower end of the range for common plastic tubing, in order to produce meaningful (i.e., nontrivial) numerical results. The fluid properties for glycerol at the specified temperature are $\rho = 2.436$ slug/ft^3, $\mu = 7.9 \times 10^{-3}$ lb sec/ft^2, $\nu = 3.423 \times 10^{-3}$ ft^2/sec.

As the first step, we determine the mean flow speed through a pipe of the unstressed cross section, which will be our base flow. Let the pressure drop over the length be 4 atm, i.e., $\Delta p = 4 \times 2117 = 8467$ lb/ft^2. From Equation 6.14-17 we have $u_e = 23.26$ ft/sec; thus Re = 598, and the base flow has been verified to satisfy the laminar criterion.

Then, for the flow through the elastic tube, we compute: $a_0 = 248.1$ ft/sec, $\varepsilon = 0.09374$, $\Lambda = 0.3012$. Thus, $\varepsilon\Lambda = 0.02823$, $2\varepsilon^2\Lambda^2 = 0.00159$. The second of these is the decrement to be applied to the pressure difference for the base pipe flow, which will produce the same flow rate in the elastic tube, i.e., $\Delta p_{dec} = -(E\bar{t}/R) 2\varepsilon^2\Lambda^2 = -477$ lb/ft^2. Thus, the term $2(\varepsilon\Lambda)^2$ represents a reduction of about 5% from the original pressure drop to maintain the flow in a conduit of fixed diameter. Since the mean flow speed in the tube is everywhere less than in the base flow (except at the exit station), the laminar criterion is maintained throughout. At the inlet station, the calculations yield $\bar{u}(0) = 0.9475$, $A(0)/A_e = 1.0553$.

PROBLEMS

6.1. *Couette flow* is the two-dimensional channel flow counterpart of laminar, axisymmetric pipe flow. Consider a segment of a two-dimensional, laminar viscous flow of breadth b. Assuming that $u = u(y)$ only, perform an analysis analogous to that of Section 6.3 and determine the following:

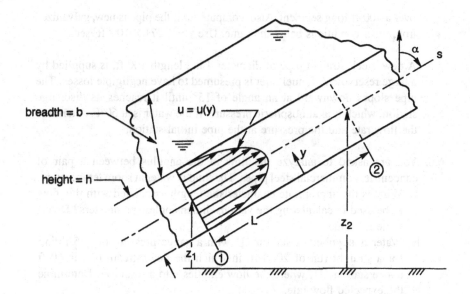

a. $u = u(y)$, u_{max}, and u/u_{max}.

b. $\dot{Q}$, and the mean velocity V as a function of u_{max} and Δp^*.

c. $\tau = \tau(y)$; also $\tau_w(0)$, $\tau_w(h)$ in terms of V and in terms of Δp^*. Why are these two not identical on the opposing walls?

d. The hydraulic diameter D_h, and the Reynolds number based thereon as a function of (ρ, V, h, μ) and as a function of $(\rho, \Delta p^*, h, \mu, L)$.

e. The friction factor $f \equiv 8\tau_o/\rho V^2$ as a function of Re_{D_h}.

Note: The restriction of two-dimensional flow implies that the sidewalls contribute negligible viscous effects to the flow, a condition physically realizable only approximately at best.

6.2. Oil of kinematic viscosity $\nu = 4 \times 10^{-3}$ ft²/sec flows through an inclined tube of 0.5 in. diameter. Find the angle α between the tube and the horizontal plane if the pressure inside the tube is constant for a flow rate of $\dot{Q} = 5$ ft³/hr.

6.3. Making use of Equation 6.8-1 for the velocity profile for turbulent flow in a smooth pipe, show that the relation between the mean flow speed V and the maximum speed U (on the centerline) is given by Equation 6.8-6. Plot V/U vs. $\log_{10}$ Re using the tabulated values in Figure 6.8-2. Then, suppose that it is desired to establish a turbulent flow in a smooth pipe such that $\text{Re} = 2.0 \times 10^6$ for water with $\nu = 1.2 \times 10^{-5}$ ft²/sec. The flow will be monitored by locating a Pitot-static tube on the pipe centerline. What should the centerline velocity be to attain this Reynolds number flow?

6.4. Water at 60°F flows steadily through a smooth, circular horizontal pipe of 6 in. diameter. The discharge rate is 1.5 ft³/sec. Determine the pressure drop over a 400-ft-long segment. Also, compute Δp if the pipe is new, galvanized iron, other conditions being the same. Use $v = 1.271 \times 10^{-5}$ ft²/sec.

6.5. A new, steel, straight pipe of diameter 4 in., length 100 ft, is supplied by a large reservoir. A funnel inlet is presumed to have negligible losses. The pipe slopes downward at an angle of 15° until it reaches its discharge station, which is at atmospheric pressure. If the water is at 60°F, determine the flow rate and the pressure at the pipe initial station.

6.6. You are asked to analyze the flow in the annulus between a pair of concentric, commercial steel pipes. The fluid is water at room temperature.
   a. What is the appropriate characteristic length associated with this flow to be used in calculating the various dimensionless parameters? Determine its value.
   b. Water is supplied at station ① with a gauge pressure of 125 lb/in.² for a straight run of 2000 ft, in which the downstream end is 60 ft above station ①, where the flow empties into a reservoir. Determine the expected flow rate.
   c. If there were a pump located just ahead of ①, with inlet at atmospheric pressure, determine the required pumping horsepower, neglecting other losses.

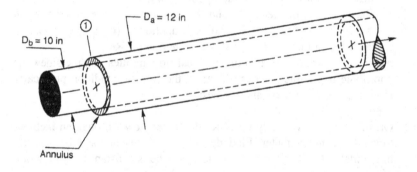

6.7. In a piece of industrial equipment it is required to pump a fluid ($\dot{Q} = 0.1$ ft³/sec) through a horizontal, smooth-wall conical duct of the dimensions shown.
   a. If the liquid is glycerine at 68°F (S.G. = 1.26, $\mu = 3.13 \times 10^{-2}$ lb sec/ft²), determine the pressure drop between stations ① and ②.
   b. Determine the pumping power required in part (a).
   c. If instead of glycerine we use water at the same flow rate, compute the pressure drop at 60°F.

6.8. In an industrial process it is required to pump glycerine ($T = 30°C$, $\rho = 2.436$ slug/ft³, $\mu = 7.9 \times 10^{-3}$ lb sec/ft²) uphill through a pipe of diameter $D = 6$ in., length $L = 2000$ ft at the prescribed flow rate $\dot{Q} = 1000$ gal/hr.

a. Determine the pumping power required for a stainless steel pipe.

b. Determine the minimum thickness of the pipe required to contain the fluid at the pump exit, if the pipe yield stress is $\sigma_y = 280$ MPa.

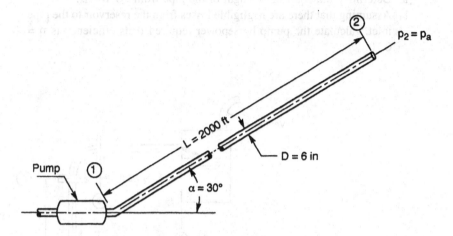

6.9. The purpose of a chimney is to use the ambient pressure decrease with height to induce a vertical flow of a less dense gas within the stack. As shown in the figure, at the bottom of the chimney, standard air enters the inlet. At the station where the diameter becomes constant $D = 4$ ft, heat is added in a constant-pressure process over a negligibly small distance to change the local temperature to $752°R$. If the chimney height is $\Delta z = 200$ ft, determine the flow rate.

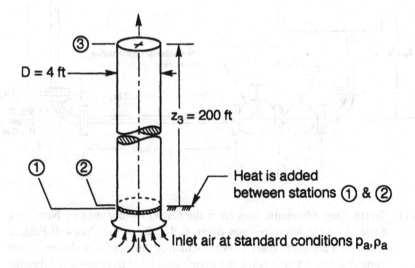

6.10. It is required to select a pump for the purpose of transferring glycerine (see Problem 6.7 for data) from one tank to another at the transfer flow rate of $\dot{Q} = 100$ gal/min.

 a. Determine the equivalent length of the pipe from ① to ⑦.

 b. Assuming that there are negligible losses from the reservoir to the pipe inlet, calculate the pump horsepower required if its efficiency is $\eta = 0.70$.

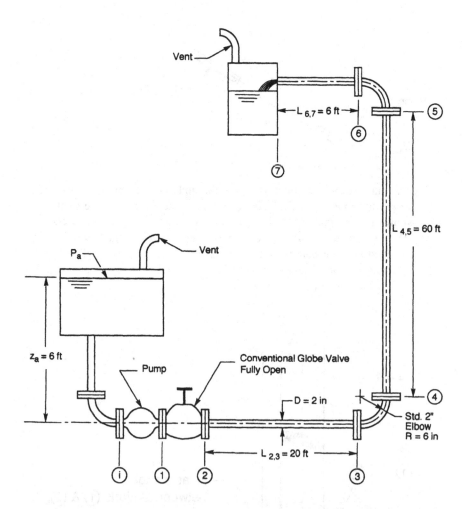

6.11. Storm King Mountain, located in the Catskill Mountains of New York State along the lower Hudson River, is the site of the Cornwall Project, which is a pumped-power storage facility. At times of low electric power consumption, water is taken in through an inlet at river level, fed through electrically driven pumps, and then, by means of a concrete-lined tunnel, directed to a reservoir near the top of Storm King Mountain. When power consumption is high, the flow is reversed and the pumps are converted in to turbines by running them backward. Similarly, the electric motors are converted into generators. Due to various, unavoidable losses the total energy recovered is less than input. This is partly compensated for by the lower cost of energy for pumping during the off-peak periods of consumption.

A schematic is shown in the figure. The maximum average water speed at any station during the inflow is $V_i = 14$ ft/sec. When generating power, the maximum outflow speed is $V_o = 21$ ft/sec. For the purpose of this analysis assume that the concrete surface is finished to $e = 0.1$ ft and that the water is at $T = 50°F$ for which $\eta = 1.271 \times 10^{-5}$ ft²/sec.

a. For the pumping phase at V_i compute $\dot{Q}$, Re, f, $\Delta p^*_{2,3}$, $\Delta p_{2,3}$, and the theoretical power required P_i to pump the flow through the tunnel to the reservoir, neglecting all other losses.

b. Compute the electrical power consumed from the grid if the combined efficiency of the motors and pumps is $\eta_i = 0.82$.

c. Now let the flow be reversed to V_o, and compute the comparable quantities requested in part (a).

d. If the combined efficiency of the turbines and generators is $\eta_o = 0.78$, compute the power returned to the grid.

e. For equal volumes of water exchanged, in and out, determine the energy returned to the grid E_o as a fraction of the energy taken from it.

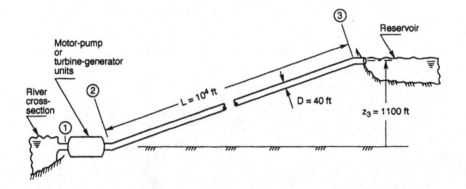

7 Steady Compressible Flow

The essential difference between an incompressible fluid and a compressible fluid is that in the former the propagation of pressure is instantaneous, whereas in the latter the propagation takes place with finite velocity. For example, if we strike the surface of an incompressible fluid, the effect perceived at a great distance is, of course, less than that at a smaller distance, but it reaches even an infinite distance in no time; whereas in a compressible fluid the effect propagates at a finite velocity. The velocity of propagation of a very small pressure change is called the *velocity of sound*.

Theodore von Kármán (1954)

7.1 INTRODUCTORY REMARKS

Up to this point no situations have been considered in which the flow dynamics interact with the fluid medium — that is, only incompressible flows, in which the density is constant, have been treated. The effects of compressibility, however, are profound, but only when the flow speed becomes a significant fraction of the speed at which disturbances are propagated through the fluid, or for internal flows at low speeds through very long ducts where viscous effects and/or heat transfer through the wall have large cumulative effects. In order to grasp these ideas most effectively, it is useful to review some thermodynamic principles, particularly as they apply to gases.

7.2 THERMODYNAMICS OF FLUIDS

First Law

We have previously introduced, Equation 3.15-4, the first law of thermodynamics

$$d\bar{q} \equiv dq + d\psi = de + pdv \qquad (7.20\text{-}1a)$$

written for an observer motionless with respect to the center of mass of a "lump of matter." The term* $d\bar{q}$ represents the energy per unit mass exchanged with its environment, in forms other than work due to pressure forces acting on its boundary. It is further broken into two categories: dq, the energy added in the form of heat transfer, and $d\psi$, the mechanical energy dissipated to heat by the action of viscous forces on its boundary. Each of these differentials depends on the path of the process involved and is, therefore, not an exact differential. On the right side of Equation 7.2-1a each of the variables involved, e, p and $v = 1/\rho$, is a state variable, although the sum $de + pdv$ is also not an exact differential.

---

* The symbol d denotes a nonexact differential, i.e., the differential of a path function.

In fluid dynamics it is customary to work with the specific enthalpy $h \equiv e + pv$ rather than the internal energy e. The first law then becomes

$$d\bar{q} = dh - vdp. \tag{7.2-1b}$$

Specific Heats

It is possible to delineate two different modes by which heat (generated by any process) is added to the material. In a *constant-volume* process we denote

$$c_v \equiv \left(d\bar{q}/dT \right)_v \tag{7.2-2a}$$

and, in a *constant-pressure* process,

$$c_p \equiv \left(d\bar{q}/dT \right)_p. \tag{7.2-2b}$$

Employing Equation 7.2-1 these can be expressed in terms of state variables only, i.e.,

$$c_v = \left(\partial e/\partial T \right)_v, \quad c_p = \left(\partial h/\partial T \right)_p. \tag{7.2-3}$$

Expressing $e = e(p, T)$ and combining the definitions of c_p, c_v into the first law, a classical relation* is obtained:

$$c_p - c_v = \left[\left(\frac{\partial e}{\partial v} \right)_T + p \right] \left(\frac{\partial v}{\partial T} \right)_p. \tag{7.2-4}$$

The ratio of the coefficients of specific heat

$$\gamma \equiv c_p / c_v, \tag{7.2-5}$$

is a combination which frequently appears in thermodynamic relations; in general, it may be a function of any two state variables.

Entropy

Elementary thermodynamics texts show, for a perfect gas, that if the right-hand side of Equation 7.2-1 is divided by the absolute temperature it becomes an exact differential. We postulate that this is true for all substances, i.e., that there is a state function δ, called specific entropy, such that

$$Td\delta = de + pdv = dh - vdp. \tag{7.2-6}$$

---

* See Problem 7.1 for an alternative form of Equation 7.2-4.

EXACTNESS CRITERION

The mathematical conditions* under which differentials multiplied by coefficients combine to form an exact differential is well known. Applying this condition to Equation 7.2-6, we obtain

$$(\partial e/\partial v)_T = T(\partial p/\partial T)_v - p \qquad (7.2\text{-}7a)$$

and

$$(\partial h/\partial p)_T = v - T(\partial v/\partial T)_p. \qquad (7.2\text{-}7b)$$

Equation 7.2-7 gives two forms of what is known as the *exactness criterion* for the entropy function. There are a number of alternative forms, of course, depending on which state variables are regarded as independent. Equation 7.2-7 is employed in determining the form of the internal energy and enthalpy functions of a substance, when the coefficients of specific heat are not constants.

THERMALLY PERFECT GAS

One of the triumphs of 17th-century chemistry was to show experimentally that, to a close approximation, gases obey a law relating their pressure p, temperature T, and volume V, such that pV/T = constant. The value of the constant depends on the volume and the number of moles occupying it. This relation, which is the combined Boyle's and Charles's law, is not in a form particularly useful for application to flow problems since the volume, or the number of the moles, of a gas in a fluid dynamic system (in a wind tunnel, for example) is rarely a known quantity. Even if it were known, it would have little utility since large gradients in temperature or pressure may exist. In that event it would not be possible to assign a single temperature or pressure to the fluid.

Rather than the volume V it is preferable to introduce *mass density* ρ, or its inverse the *specific volume* v. The combined form of Boyle's and Charles's law then leads to the state relation for a *thermally perfect* gas:

$$pv = \left(R^*/\overline{m}k\right)T, \qquad (7.2\text{-}8a)$$

where the *universal gas constant* $R^* = 8.314\,32 \times 10^3\ \text{m}^2 \cdot \text{kg/kmol} \cdot \text{s}^2 \cdot \text{K} = 4.97190 \times 10^4\ \text{ft}^2\ \text{slug/(slug mol sec}^2\ {}^\circ\text{R})$.

In the denominator of Equation 7.2-8a we have the molecular weight $\overline{m}$ (actually a dimensionless number) and the dimensional constant k, of unit magnitude, which specifies the *molar mass* in units consistent with those of the gas constant. That is, k is expressed in kg/kmol or gm/gm mol, slug/slug mol, and so forth. Some molecular weights, based on the carbon isotope $C^{12} \equiv 12.000\,0$ are listed in Table 7.2-1.

---

* For example, $M(x, y)dx + N(x, y)dy$ is exact if $\partial M/\partial y = \partial N/\partial x$.

TABLE 7.2-1
Molecular Weights of Gases

Gas	Mol. Wt. $(\overline{m})$	Gas	Mol. Wt. $(\overline{m})$
Hydrogen (H$_2$)	2.015 9	Oxygen (O$_2$)	31.998 8
Helium (He)	4.002 6	Argon (Ar)	39.948
Methane (CH$_4$)	16.043 0	Carbon dioxide (CO$_2$)	44.010 0
Ammonia (NH$_3$)	17.030 6	Nitrous oxide (N$_2$O)	44.012 8
Neon (Ne)	20.183	Ozone (O$_3$)	44.998 2
Carbon monoxide (CO)	28.010 6	Sulfur dioxide (SO$_2$)	64.062 8
Nitrogen (N$_2$)	28.013 4	Krypton (Kr)	83.80
Air (mean)	28.964 4	Xenon (Xe)	131.20

Source: Dublin, M., et al., *U.S. Standard Atmosphere, 1976*, U.S. Government Printing Office, Washington, D.C., 1976.

It is convenient to introduce the *specific gas constant*

$$R \equiv R^* / \overline{m} k, \tag{7.2-8b}$$

such that the *thermal equation of state* becomes

$$pv = p/\rho = RT. \tag{7.2-8c}$$

For example, the specific gas constant for air, based on its mean molecular weight, is $R_{air} = 287.05$ m^2/s$^2 \cdot$ K $= 1716.6$ ft^2/sec^2 °R. This relation tends to break down either at high temperatures when dissociation and ionization may occur, or at high pressures when intermolecular forces become important.

Substitution of Equation 7.2-8 in Equation 7.2-7 reduces the right-hand side to zero, yielding the useful simplifications that, for a thermally perfect gas, $e = e(T)$, $h = h(T)$, only. With this restriction, Equation 7.2-3 becomes

$$e = \int_0^T c_v(T)dT, \quad h = \int_0^T c_p(T)dT \tag{7.2-9}$$

where, from the preceding result, the coefficients of specific heat are also functions, at most, of the temperature. In Equation 7.2-9 the reference temperature $T = 0$ has been chosen, at which point the internal energy and enthalpy have arbitrarily been set equal to zero. Of course, if flow temperatures become low enough, liquefaction occurs and neither Equation 7.2-8 nor Equation 7.2-9 is valid. For gases commonly found in the atmosphere, liquefaction occurs in a temperature range of 100 to 200 K.

Combining Equations 7.2-4 and 7.2-9, we obtain one of the standard thermodynamic relations for a thermally perfect gas:

$$c_p - c_v = R. \tag{7.2-10}$$

Even for a thermally perfect gas, however, the ratio of specific heats may be a function of temperature.

CALORICALLY PERFECT GAS

The treatment of gas flows under conditions in which it is necessary to account for variations of the coefficients of specific heat is generally beyond the scope of this work. For a treatment of this important area of gas dynamics including such phenomena as chemical reactions, nonequilibrium effects, and radiation, see Vincenti and Kruger (1965). For a comprehensive treatment of gas properties from the viewpoint of a physicist see Tabor (1991). Present considerations are restricted to the case of *calorically perfect gases*, defined such that c_p and c_v are constants. Consequently, their ratio c_p/c_v is also constant. Thus, for a thermally and calorically perfect gas, Equation 7.2-9 yields

$$e = c_v T, \quad h = c_p T. \tag{7.2-11}$$

RELATIONS FOR PERFECT GASES

Unless specified otherwise, we shall mean by a *perfect gas* one which is thermally and calorically perfect. For a perfect gas kinetic theory predicts that

$$c_p = \gamma R / (\gamma - 1), \quad c_v = R / \gamma - 1, \tag{7.2-12a}$$

and

$$\gamma = (n + 2)/n, \tag{7.2-12b}$$

where n is the number of *classical* degrees of freedom of a molecule (excluding vibration of atoms). Typical values of n are given in Table 7.2-2. According to this view, a diatomic molecule is a "dumbbell" and has non-negligible mass moments of inertia about two independent axes. A triatomic molecule still has only two degrees of freedom in rotation if the atoms are in line (linear), but in a nonlinear molecule, such as water, the molecule has three independent mass moments of inertia. By eliminating γ in Equation 7.2-12a by use of Equation 7.2-12b, then

$$c_p = (n + 2) R / 2, \quad c_v = n R / 2. \tag{7.2-13}$$

For a thermally perfect gas, Equation 7.2-6 for the entropy differential can be expressed in the equivalent forms

TABLE 7.2-2
Classical Degrees of Freedom for Various Molecules

Type of Molecule	Degrees of Freedom		n	γ	Examples
	Translation	Rotation			
Monatomic	3	0	3	5/3	He, Ar
Diatomic	3	2	5	7/5	O_2, N_2
Triatomic (linear)	3	2	5	7/5	HCN, CO_2, SO_2
Triatomic (nonlinear)	3	3	6	4/3	H_2O, H_2S

$$d(\delta/R) = \left(c_v/R\right)d(\ln T) + d(\ln v), \tag{7.2-14a}$$

$$= \left(c_p/R\right)d(\ln T) - d(\ln p), \tag{7.2-14b}$$

$$= \left(c_v/R\right)d(\ln p) + \left(c_p/R\right)d(\ln v). \tag{7.2-14c}$$

Equation 7.2-14 can be readily integrated from an initial state, denoted by the subscript 1, to an arbitrary state (no subscript) if the gas is also calorically perfect. By substituting Equation 7.2-12a in Equation 7.2-14c, and integrating, the third form becomes

$$\frac{\Delta\delta}{R} = \frac{1}{\gamma-1} \ln \frac{p}{p_1} + \frac{\gamma}{\gamma-1} \ln \frac{v}{v_1} \tag{7.2-15}$$

or, rearranging,

$$pv^\gamma = p_1 v_1^\gamma \, \exp\left[(\gamma-1)\Delta\delta/R\right]. \tag{7.2-16}$$

Obviously, for an isentropic transformation, the familiar

$$pv^\gamma = \text{constant} \tag{7.2-17}$$

results. It is noted again that all of the preceding formulae refer to a fixed "lump" or "particle" of fluid. No account has yet been taken of any motion of the fluid particle since the equations are written for an observer fixed with respect to the particle center of mass.

LIQUIDS

Although the basic thermodynamic laws, e.g., Equations 7.2-1 through 7.2-7, apply equally to gases and liquids (in fact to solids as well), there is no liquid analog of a perfect gas; hence, compressible liquid flow is less amenable to analytical treatment. Although there are many approximate relations proposed that predict liquid properties over a limited range of state variables, discussion of these is omitted in the following.

7.3 THE EQUATIONS OF STEADY, ONE-DIMENSIONAL, COMPRESSIBLE FLOW

The equations derived in Chapter 3 have no restrictions on the fluid compressibility. In Chapter 4, which deals with incompressible, inviscid flow, there is an enumeration of possible restrictions, many of which are utilized also in the analysis of compressible flows. These will be introduced as required.

CONTINUITY EQUATION

For steady, source-free flow, Equation 3.10-6 reduces to

$$\dot{m} \equiv \rho u A = \text{constant}, \tag{7.3-1}$$

where we employ u, the x-component of velocity, in place of the flow speed V for reasons explained in the next paragraph.

DYNAMIC EQUATION

In the flow of gases under small height changes, the contribution of the gravitational body force is negligibly small, and, consequently, the flow can be treated as though it were horizontal. For steady, horizontal flow, therefore, the dynamic equation (Equation 3.12-10)* becomes

$$u \frac{du}{dx} = -\frac{1}{\rho} \frac{dp}{dx} - \frac{\tau_w P}{\rho A}. \tag{7.3-2}$$

Unlike its incompressible counterpart, only the left-hand term of Equation 7.3-2 is in the form of an exact differential. That is, prior to integrating Equation 7.3-2, a relation in the form $\rho = \rho(p)$ must be available, as well as information on the nature of the viscous wall stress.

ENERGY EQUATION

For steady flows the form of the energy equation (Equation 3.19-7) previously derived, is equally valid for compressible flows:

$$h^0 \equiv h + \tfrac{1}{2} u^2 = h_1^0 + \Delta q + \Delta w. \tag{7.3-3}$$

---

* It is recalled that in this form of the dynamic equation it is assumed that the *normal* viscous stresses are negligible; see Equation 3.12-8.

The term Δq is the energy per unit mass input in the form of heat conduction, radiation, and combustion; i.e., it does not include energy exchanges due to viscous dissipation, whereas Δw, the mechanical work, is energy input in forms other than heat. Specifically, it includes the case of work due to viscous forces exerted on a fluid particle by a wall in relative motion to the observer. It also includes, as noted in Section 3.19, energy input between two flow stations at which the flow is one dimensional and steady, in a process that may actually be two or three dimensional, or even nonsteady, between the two stations. This term is sometimes called *mechanical work*. It also would ordinarily include work done by the gravitational field, an effect that may be neglected for small changes in height.

It is worthwhile to reiterate that dissipation due to viscous forces exerted along a wall fixed with respect to the observer does *not* cause the total enthalpy to vary. It does, however, result in a redistribution between the component terms that make up the total enthalpy $h^0 = e + pv + \frac{1}{2}u^2$.

ADIABATIC FLOW

From the definition of adiabatic flow, $dq \equiv 0$, it is clear that $\Delta q \equiv 0$. However, it is possible, in special circumstances, for $\Delta q \equiv \int dq = 0$, without the requirement that $dq = 0$. For example, if, in a given transformation, heat is first transferred into a fluid particle and then out in just the right proportion, it is possible that the net heat transfer is $\Delta q = 0$. This is actually what happens in the flow through a shock wave. Although some authors refer to this as "adiabatic shock," the terminology is misleading. However, because it is useful to have a term that describes such a situation, we propose to refer to the case where $\Delta q = 0$ as a *globally adiabatic* process, which suggests the nature of the distinction.

A steady flow that is workless and adiabatic is said to be *isoenergetic*.[*] In such a flow, from Equation 7.3-3,

$$h^0 = e + pv + \frac{1}{2}u^2 = h + \frac{1}{2}u^2 = \text{constant}. \qquad (7.3\text{-}4a)$$

Furthermore, since for a thermally and calorically perfect gas $h = c_p T$, $h^0 = c_p T^0$, then Equation 7.3-4a reduces to

$$c_p T^0 = c_p T + \frac{1}{2}u^2 = \text{constant}. \qquad (7.3\text{-}4b)$$

The integrated momentum equation for steady, compressible flow is identical to the incompressible case and is not repeated. Equations 7.3-1 through 7.3-3 plus the momentum equation, and appropriate state relations, form the set of basic relations for one-dimensional, steady, compressible flow.

---

[*] The concept that the total enthalpy — in an adiabatic and workless flow — must be constant was first disclosed in a letter from W. Thomson (later Lord Kelvin) to J. P. Joule in the *Philosophical Magazine*, November, 1850. This disclosure was later formulated in detail, see Joule and Thomson (1854).

CONSTANT-AREA FLOW

The subcase of compressible flows through constant-area ducts is of such special importance that it is useful to summarize the governing equations separately. The continuity equation is reduced to a mass-flux term*

$$\bar{\dot{m}} \equiv \dot{m}/A = \rho u = \text{constant}. \tag{7.3-5}$$

Equation 7.3-5 can be combined with the dynamic equation (Equation 7.3-2) to produce a special form of the momentum equation

$$d\left(p + \bar{\dot{m}}u\right) = -\left(\tau_w P/A\right)dx,$$

or

$$p + \rho u^2 = p_1 + \rho_1 u_1^2 - \int_{x_1}^{x}\left(\tau_w P/A\right)dx. \tag{7.3-6}$$

The energy equation for constant-area flow is unchanged from Equation 7.3-3, or, for isoenergetic flow, from Equation 7.3-4.

7.4 ON THE PROPAGATION OF DISTURBANCES

The concept of an incompressible fluid, though strictly speaking a fiction, is one that is useful and even accurate for many practical purposes, but one that can lead to nonsensical conclusions if applied indiscriminately. In the incompressible flow relations it is implicit that effects of pressure variations at a given point are communicated instantaneously throughout the field. If at every point the flow speed is small compared with the local speed at which disturbances are propagated, then the incompressible assumption leads to results that compare well with experiment, for gases as well as liquids. For flow speeds approaching the disturbance propagation speed, however, an entirely new phenomenon arises. The first task is to define what we mean by propagation speed, and to obtain a relation for it in terms of the usual flow variables: pressure, density, etc.

DISTURBANCE CREATED BY IMPULSIVE MOTION OF A PISTON

Figure 7.4-1 shows a constant-area duct in which, at $t < 0$, the fluid is everywhere at rest, $u_1' \equiv 0$, and at initial pressure p_1 and density ρ_1. Starting at $t = 0$ an impermeable piston is moved impulsively to the left at constant speed $u_2' < 0$ with respect to the duct wall. Since the piston is impermeable, the fluid adjacent to the piston also adjusts to this speed. It is experimentally verifiable that the extent of the

---

* $\bar{\dot{m}}$ is unrelated to the molecular weight $\bar{m}$.

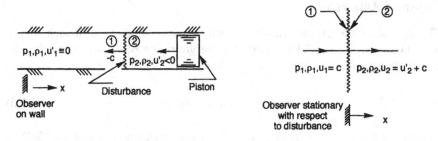

FIGURE 7.4-1   Impulsive motion of a piston in a duct.

region of the fluid in motion grows with time and that the velocity of the disturbance, denoted $-c$, moving into the fluid at rest, region ①, is greater in magnitude than the piston speed. In region ②, after passage of the disturbance, the pressure p_2 and density ρ_2 are changed (actually increased) from their original values.

As seen by an observer on the wall the flow is unsteady. Although it is possible to analyze the disturbance in the unsteady frame, it is simpler to change to an observer to whom the flow appears to be steady, specifically, an observer moving with the disturbance. In the frame of the new observer (unprimed quantities)

$$u_1 = c, \quad u_2 = u_2' + c. \tag{7.4-1}$$

The state variables, which are defined for an observer motionless with respect to the particle mass center, are invariant under the change of observer. We note that *disturbance speed c* is defined for an observer "embedded" in the region of flow into which the disturbance is propagating. Under the change of observer, the fluid moves into the disturbance at a velocity equal and opposite to the disturbance velocity as seen by the embedded observer, i.e., $u_1 = c$.

In this elementary model, the flow is assumed to be uniform on either side of the disturbance. It is easily shown that this is possible only if wall viscous effects are negligible.* Therefore, in the frame of the steady-state observer, the continuity and momentum equations for inviscid flow, together with Equation 7.4-1, become

$$\rho_1 c = \rho_2\left(u_2' + c\right), \tag{7.4-2}$$

$$p_1 + \rho_1 c^2 = p_2 + \rho_2\left(u_2' + c\right)^2. \tag{7.4-3}$$

From a pair of algebraic equations, it is possible to eliminate any one variable; in this case we choose u_2'. By defining $\Delta p \equiv p_2 - p_1$, $\Delta \rho \equiv \rho_2 - \rho_1$, then

---

* For long ducts, viscous effects cannot be ignored. In such cases the entire flow is nonuniform. However, the disturbance speed turns out to be the same as derived in Equation 7.4-4, where subscripts 1 and 2 refer to the conditions immediately upstream and downstream of the disturbance, respectively.

$$\Delta p = c^2 \frac{\rho_1 \Delta p}{\rho_1 + \Delta p},$$

from which

$$c^2 = \frac{\Delta p}{\Delta \rho}\left(1 + \frac{\Delta \rho}{\rho_1}\right), \tag{7.4-4}$$

an equation that is exact.* Thus, it has been shown that the propagation speed of a disturbance can be uniquely determined in terms of the static pressures and densities on either side of the disturbance. This result is quite independent of any equation of state; in fact, Equation 7.4-4 holds also for liquids and elastic solids. Further, since the energy equation has not yet been invoked, it holds for all disturbances separating two flow regions, regardless of the nature of the origin of the disturbance. For example, it also holds for propagating flames in which there is a source of energy input (combustion) located at the disturbance itself.

A sometimes useful relation is obtained if Equation 7.4-2 is solved for

$$\Delta u \equiv u_1 - u_2 = u_1' - u_2' = c\Delta\rho/(\rho_1 + \Delta p). \tag{7.4-5}$$

7.5 THE SPEED OF SOUND

It is remarkable that a concept as venerable as the speed of sound should still be capable of creating a controversy** as late as 1958. The first attempt to calculate the speed of sound is attributed to Sir Isaac Newton (1642–1727). In Book II, Theorem 38, of the *Principia*, first published in 1686, he states, in effect, that the speed of sound (denoted a) is given by $a = (p/\rho)^{1/2}$. Newton compared numerical values from this expression with experiment (computed by measuring the time it takes for an audible noise to traverse a given distance) and concluded that its predictions were low.

Newton made a series of corrections to his original calculation, which seemingly brought experiment and theory into agreement. However, these so-called corrections were completely unscientific and demonstrated, at most, that even Newton was capable of self-delusion.

---

* The possibility of propagating a finite disturbance was first investigated by Stokes in 1848 although the momentum equation was not correctly employed in his analysis. Nevertheless, after the expression in Equation 7.4-4 was derived by Rankine (1870), the possibility of a standing finite wave was denied even by as eminent an authority as Lord Rayleigh (1842–1919). The reason given by Rayleigh was that the preceding result violates the "energy equation" by which was meant the dynamic equation integrated for isentropic flow. Curiously, these views were propagated into the 20th century by Lamb (1932, pp. 477–489) in the various editions of his treatise on hydrodynamics, and long after the existence of standing waves and the validity of Equation 7.4-4 had been verified experimentally.

** See Mirels (1958) for the references. For another, more-protracted controversy, only a decade older, see Truesdell (1951).

The theoretical discrepancy was finally corrected when Laplace* (1816) showed that Newton's expression must be replaced by $a = (\gamma p/\rho)^{1/2}$, where γ is the ratio of specific heats. It turns out that this expression, however, is limited to a thermally perfect gas.

Part of the difficulty is in defining what is meant by the speed of sound. It is apparent from Equation 7.4-4 that, if the strength of a finite disturbance is continuously diminished ($\Delta\rho \to 0$), we should reach some limiting value. To eliminate any ambiguity, it is this limit that we call the *speed of sound*. That is, the speed at which a disturbance — so small that the pressure, density, etc. remain continuous during its passage — moves through a fluid is given by

$$a^2 \equiv \lim_{\Delta\rho\to 0} c^2 = dp/d\rho. \tag{7.5-1}$$

But Equation 7.5-1 is still not an adequate definition since it is not unique, i.e., since, by the postulated form of the equation of state $p = p(\rho, T)$, not enough information is available to obtain from the indicated differentiation a unique result.

It is readily seen, for a thermally perfect gas, that Newton's original expression for the speed of sound is obtained if the additional constraint is imposed that the passage of the infinitesimal disturbance shall not alter the local temperature of the fluid during transit. But this does not correspond to experimental fact. What does occur is that, in transit of an arbitrarily small disturbance, the fluid *specific entropy* does not change. Thus, the *speed of sound* is defined by

$$a^2 \equiv \lim_{\substack{\Delta\rho\to 0 \\ s=\text{const.}}} c^2 \equiv (\partial p/\partial\rho)_s. \tag{7.5-2}$$

The usual explanation offered is that, for a sound wave, the arbitrarily small compressions (or expansions) occur so rapidly there is no time for heat conduction to occur. More precisely, it is required that temperature gradients in the fluid, resulting from transit of the disturbance, be small enough as to render heat conduction negligible. It is also implied that velocity gradients in the direction parallel to the flow due to passage of the wave are sufficiently small that the associated normal viscous stresses create negligible viscous dissipation.

Equation 7.5-2 involves only state variables; thus, the speed of sound is a function of state only. In a general situation it does not depend upon the nature of the particular flow, which may or may not be isentropic. It only requires that in consequence of passage of an infinitesimal disturbance (sound wave) the entropy of a particle be unchanged. An alternative form, derivable from Equation 7.5-2 employing only basic thermodynamic relations, and which is sometimes more convenient, is

$$a^2 = \gamma(\partial p/\partial\rho)_T, \tag{7.5-3}$$

where in general $\gamma = c_p/c_v$ need not be a constant.

For a thermally perfect gas, Equations 7.5-3 and 7.2-8c yield the Laplace result that

$$a^2 = \gamma \mathcal{R} T = \gamma p / \rho. \qquad (7.5\text{-}4)$$

For *air*, with $\gamma = 7/5$, $\mathcal{R} = 1716.6$ ft²/sec² °R) $= 287.05$ m²/(s² · K),

$$a = 49.02 \sqrt{T}, \quad (a \text{ in ft/sec, } T \text{ in } °R),$$
$$= 20.05 \sqrt{T}, \quad (a \text{ in m/s, } T \text{ in } K). \qquad (7.5\text{-}5)$$

The problem of producing experimentally an infinitesimal disturbance, and measuring its speed of propagation, is not trivial. However, by use of a shock tube,* it is possible to produce an isentropic pulse (expansion) of finite spatial extent so that across the pulse there is a finite change of pressure, density, etc., but such that all flow variables (but not necessarily all their derivatives) change continuously throughout the pulse. Glass (1952) has taken schlieren** photographs of a typical pulse, and has verified that the leading wave of this expansion does indeed propagate at the speed predicted by the preceding relations. There are many other experimental results that confirm their validity under the stated restrictions.

7.6 ON THE MAXIMUM SPEED OF A FLUID EXPANDING INTO A VACUUM

One of the experiments that naturally suggests itself, Figure 7.7-1, is to expand a fluid from a tank under pressure p^0 (the reservoir total pressure) into a container at a lower pressure, in the limit a vacuum, and to determine the maximum flow speed possible. For steady, inviscid, horizontal flow of an incompressible fluid, Bernoulli's equation predicts a flow speed, at a point where the pressure is p, of $u = [2(p^0 - p)/\rho]^{1/2}$. The theoretical limit for expansion into a vacuum is $u_{max} = (2 \, p^0/\rho)^{1/2}$. Using this expression with $p^0 = 1$ atm, this results in $u_{max} = 46.67$ ft/sec for water, and $u_{max} = 1335$ ft/sec for air. Of course, for a liquid, vaporization would occur prior to reaching the theoretical limit, at a value in the neighborhood of its vapor pressure, which is 0.0168 atm for water at standard atmospheric temperature (519 °R).

On the other hand, for a gas, it comes as no surprise that use of the incompressible flow relations may lead to erroneous results if serious deviations from the incompressibility restriction occur. If we employ instead the energy equation (Equation 7.3-3) under the restriction of adiabatic, workless (i.e., $\Delta q \equiv \Delta w \equiv 0$) flow,

$$u = \left[2(h^o - h)\right]^{1/2}, \qquad (7.6\text{-}1)$$

---

* See Thompson (1972, p. 423) for an introduction to the theory of the shock tube.
** A special photographic technique used widely in experimental compressible flow studies.

which, for a perfect gas, reduces to

$$u = \left[2c_p\left(T^0 - T\right)\right]^{1/2}. \qquad (7.6\text{-}2)$$

A steady flow in which the total enthalpy (or, for a perfect gas, the total temperature) is constant, is called *isoenergetic*. Obviously, the highest speed is obtained at a point where the static enthalpy, or temperature, approaches zero, i.e.,

$$u_{max} = \left(2h^0\right)^{1/2}, \qquad (7.6\text{-}3a)$$

which, for a perfect gas, becomes

$$u_{max} = \left(2c_p T^0\right)^{1/2} = \left[2\gamma R T^0/(\gamma - 1)\right]^{1/2} = a^0\left[2/(\gamma - 1)\right]^{1/2}, \qquad (7.6\text{-}3b)$$

by introduction of Equations 7.2-11 and 7.2-12a for a perfect gas, and where $a^0 = (\gamma R T^0)^{1/2}$ is the total (or reservoir) speed of sound. As previously noted, as the temperature of a gas drops, condensation eventually sets in. Equation 7.6-3 therefore represents only an upper — and practically speaking, unattainable — limit. For air initially at standard atmospheric temperature, $u_{max} = 2475$ ft/sec, if we ignore condensation.

7.7 DISCREPANCIES BETWEEN THEORY AND EXPERIMENT

Although Equation 7.6-3 actually predicts the correct upper limit on the maximum flow speed, it does not provide information on how to design a device to create a high flow speed, or the requisite pressure distribution required to produce it. The story of the early attempts to investigate these problems is not only fascinating, but in the resolution of the apparent discrepancies encountered the significant distinctions between compressible and incompressible flows are uncovered.

INTEGRATION OF THE DYNAMIC EQUATION

de Saint-Venant and Wantzel (1839) undertook to study the flow, Figure 7.7-1, of a gas contained in a reservoir at conditions p^0, ρ^0, T^0, which emits a jet from a small orifice into a second reservoir, or dump tank, in which the ambient pressure is maintained at a lower value p_r. They assumed the flow to be steady (actually quasi-steady since the reservoir pressure and temperature fall off slowly with time) and that up to, and just beyond, the orifice, viscous effects are negligible. This implies that in the upstream reservoir the flow speed is negligible. Under these assumptions they integrated* the dynamic equation of motion (Equation 7.3-2) to obtain the flow speed at an arbitrary point where the static pressure is p:

---

* The energy equation in its present form was still unknown at that time.

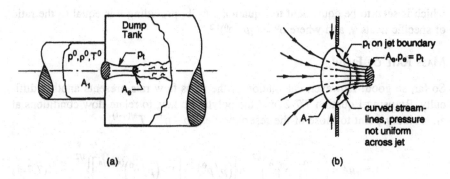

FIGURE 7.7-1   Compressible flow through an orifice.

$$u^2 = -2 \int_{p^0}^{p} dp/\rho. \tag{7.7-1}$$

To evaluate the integral in Equation 7.7-1, an additional relation is required. They pointed out that the previous authors had assumed that during the expansion $p = C\rho$, i.e., Boyle's law (called Mariotte's law in France) holds, from which

$$u^2 = \left(2p^0/\rho^0\right)\ln\left(p^0/p\right).$$

It was noted that Boyle's law implies an isothermal transformation — an unlikely possibility. In addition it had been the practice to assume that the exit pressure at the orifice was equal to the dump tank pressure p_t. Under these assumptions, the exit speed obtained is

$$u_e^2 = \left(2p^0/\rho^0\right)\ln\left(p^0/p_t\right).$$

This last relation, clearly, is not generally valid since in the limit, as the dump tank pressure approaches zero, the exit speed becomes infinite, a possibility ruled out by Equation 7.6-3.

de Saint-Venant and Wantzel suggested that instead of an isothermal law the expansion should follow a polytropic law $p \sim \rho^n$, where n is a constant to be determined. By substituting this in Equation 7.7-1 and integrating,

$$u^2 = \frac{2n}{n-1} \frac{p^0}{\rho^0}\left[1-\left(p/p^0\right)^{(n-1)/n}\right]. \tag{7.7-2}$$

In the limit as $p = p_e = p_t \to 0$, we obtain for the maximum flow speed

$$u_{max}^2 = \frac{2n}{n-1} \frac{p^0}{\rho^0}, \tag{7.7-3}$$

which is seen to be equivalent to Equation 7.6-3b, providing n is equal to the ratio of specific heats γ, and where $a^0 = (\gamma p^0/\rho^0)^{1/2}$.

MASS RATE OF FLOW

So far, so good! However, calculation of the mass flow rate presents another difficulty. By using Equation 7.7-2, and the polytropic law, to relate flow conditions at an arbitrary point to those of the reservoir, i.e., $p \sim \rho^n \sim T^{n/(n-1)}$,

$$\dot{m} = \rho u A = \left(\frac{2n}{n-1}p^0\rho^0\right)^{1/2} A\left\{\left(p/p^0\right)^{2/n}\left[1-\left(p/p^0\right)^{(n-1)/n}\right]\right\}^{1/2}. \qquad (7.7\text{-}4)$$

But, as pointed out by de Saint-Venant and Wantzel, if Equation 7.7-4 is applied to the exit ($p = p_e$, $A = A_e$), the ludicrous result is yielded up that with $p_e = p_t \to 0$, $\dot{m} \to 0$, i.e., the flow rate is zero if the orifice exit is located in a vacuum. On the other hand, since $\dot{m}$ also is zero when the exit and upstream reservoir pressures are equal, Equation 7.7-4 must exhibit an extremum; by the usual process it can be shown that extreme value of the mass flow rate is a maximum and occurs when

$$\left(p_e/p^0\right)\Big|_{\dot{m}_{\max}} \equiv p^*/p^0 = \left[2/(n+1)\right]^{n/(n-1)}. \qquad (7.7\text{-}5)$$

The quantity p^* is usually referred to as the *critical pressure* which, however, is not related to the thermodynamic critical pressure.

THEORY AND EXPERIMENT RECONCILED

The experiments of de Saint-Venant and Wantzel on air showed that this unexpected behavior does occur, to a point. As the dump tank pressure is lowered, they found that the mass flow rate does indeed increase until it reaches a maximum value at an exit pressure corresponding to Equation 7.7-5. Further decreases of the dump tank pressure, however, resulted in no change in the flow rate. From this they concluded that the usual assumption — that the exit pressure and the dump tank pressure were equal — was valid only for dump tank pressures equal to or exceeding that given by Equation 7.7-5.

Furthermore, from the data, the polytropic exponent was calculated to be $n_{air} = 1.42$, which is very close to the perfect diatomic gas value for the ratio of specific heats of $\gamma = 7/5$, and to the actual experimental value of $\gamma_{air} = 1.408$ for moderate temperatures. Thus, except for small discrepancies, no doubt attributable to limitations on the experimental equipment, they showed that to a close approximation the expansion occurs isentropically, i.e., both adiabatically and with negligible viscous dissipation. This refers, of course, only to that portion of the flow up to the exit. Beyond the exit, there are two-dimensional and viscous effects in the jet that cannot be handled by one-dimensional theory.

Thus, de Saint-Venant and Wantzel demonstrated an important fact — that under some conditions compressible flows can be analyzed as though they are isentropic which, for a perfect gas, allows use of Equation 7.2-17. For the orifice flow the appropriate relations are obtained by substituting γ for n in Equations 7.7-2 through 7.7-5. For example, for isentropic flow

$$p^*/p^0 = \left[2/(\gamma+1)\right]^{\gamma/(\gamma-1)}, \tag{7.7-6}$$

which gives the value $p^*/p^0 = 0.528$ for $\gamma = 7/5$.

The Exit Discharge Speed

The story is not yet complete. Making the substitution $n = \gamma$ in Equation 7.7-2 the exit speed at maximum discharge turns out to be

$$\left. u_e \right|_{\dot{m}_{max}} = \left(\frac{2\gamma}{\gamma+1}\frac{p^0}{\rho^0}\right)^{1/2} = a^0\left[2/(\gamma+1)\right]^{1/2}, \tag{7.7-7}$$

which is a good deal less than the value for u_{max} from Equation 7.6-3b. The authors, however, were unable to shed any light on the question why continued lowering of the dump tank pressure had no effect on u_e, or on the discharge rate.

This point was finally clarified by Osborne Reynolds (1842–1912). Reynolds was stimulated by publication of experimental results of Wilde (1885) who was unaware of the work of de Saint-Venant and Wantzel. Wilde's measurement duplicated those of the French investigators, but he was unable to draw the proper conclusions from his data. Reynolds (1885)[*] used Wilde's data to show first that the maximum mass rate of flow was achieved when the tank pressure was equal to, or less than, a certain fraction (about 53%) of the reservoir pressure. He than gave an analysis entirely equivalent to that leading to Equation 7.5-4, except that he assumed isentropic flow at the outset.

Reynolds then went on to identify a previously unknown feature of compressible flow and one which has no incompressible counterpart. Since for a perfect gas $a^2 = \gamma RT$, the isentropic relation can be extended to the speed of sound, i.e.,

$$p \sim \rho^\gamma \sim T^{\gamma/\gamma-1} \sim a^{2\gamma/(\gamma-1)}$$

---

[*] Since Reynolds does not refer to it, he, also, was apparently not familiar with the work of de Saint-Venant and Wantzel. The publication of equivalent scientific results, independently obtained by several writers, usually in different countries, has many examples in history. There was a period when this could have been attributed to poor communications. Today, in spite of the superb advances, the likelihood of duplication continues, but rather because the volume of publication is so enormous that it may be cheaper and faster to do a piece of research from scratch than to search the literature to see if it has already been done. An oft-repeated statement that "knowledge doubles every 10 years" may be nonsense, although the material published may very well grow at that rate. However, all this may be changed by the existence of the internet.

or

$$a_e/a^0 = \left(p_e/p^0\right)^{(\gamma-1)/2\gamma}. \tag{7.7-8}$$

Therefore, if critical conditions obtain at the exit, $p_e = p^*$, and, by combining Equation 7.7-6 with Equation 7.7-8,

$$a_e = a^0\left[2/(\gamma+1)\right]^{1/2} = u_e\big|_{\dot{m}_{max}}, \tag{7.7-9}$$

which means that once maximum discharge has been attained the exit flow speed is equal to the speed of sound at the exit. But small pressure disturbances are propagated at the local speed of sound. And because, for an observer on the wall, the relative speed of a pressure wave moving against the flow is $u - a$, which is zero when the exit pressure reaches the critical value (sonic flow), further decreases of the dump tank pressure have no effect on the flow rate since they cannot be communicated upstream. Thus, achieving sonic conditions effectively "isolates" the upstream flow from downstream (but *not* vice versa, of course). On the other hand, this suggests the possibility that through additional pressure decreases u_{max} can be attained further downstream. A practical means of approaching this goal is discussed in the section on nozzle flow.

FLOW GEOMETRY AT THE EXIT

We have avoided a precise statement of what is meant by the "exit" or, in fact, a description of the flow downstream of the orifice station. We are using one-dimensional relations, but the flow is not truly one dimensional. Like its incompressible counterpart, a compressible jet is highly curved at the orifice and tends to a *vena contracta* downstream where flow conditions across the jet are almost uniform as long as the ambient pressure is in the range $p^* \leq p_t \leq p^0$. The vena contracta is then, effectively, the "exit" as indicated in Figure 7.7-1.

For large values of the Reynolds number it was seen in Chapter 4 that the contraction coefficient of a jet issuing from an orifice approached a fixed value independent of the pressure drop across the orifice. For a compressible flow from a circular, sharp-edged orifice there is no theory, but a semiempirical relation, based on a modification of a two-dimensional theory* of Busemann (1937), gives

$$A_e/A_1 \equiv C_c = \frac{\pi}{\pi + 2\rho_e/\rho^0}. \tag{7.7-10a}$$

where the exit density ρ_e and exit pressure are related by

$$p_e/p^0 = \left(\rho_e/\rho^0\right)^\gamma. \tag{7.7-10b}$$

---

* See Shapiro (1953, Vol. 1, pp. 358–359) for discussion and original reference.

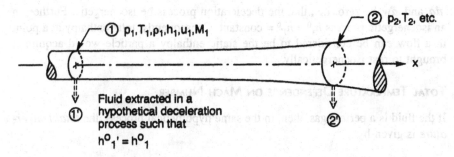

FIGURE 7.8-1 Flow total conditions.

Equations 7.7-10 give satisfactory results for $\rho^* \le \rho_e \le \rho^0$. At the limit $\rho_e \to \rho^0$ the incompressible, high-Reynolds-number result is recovered and, when $\rho_e \to \rho^*$, Equation 7.7-10 yields $C_c \to 0.71$ for $\gamma = 7/5$, which is only a few percent lower than experimental measurement.

For values of the dump tank pressure $p_t < p^*$ the situation is sufficiently complex that the one-dimensional relations fail completely. Alder (1979) has shown that as p_t is reduced the diameter of the vena contracta increases and that, as it moves closer to the orifice station, the local flow departs ever farther from the one-dimensional model. The jet diameter downstream tends to grow and the flow rate actually continues to increase with decreasing p_t, asymptotically approaching a fixed value as $p_t \to 0$. See Brower et al. (1993).

7.8 TOTAL CONDITIONS IN A COMPRESSIBLE FLOW — INTRODUCTION OF MACH NUMBER

TOTAL ENTHALPY

In Equation 3.17-4 total enthalpy, denoted h^0 was introduced as the sum of static enthalpy (the enthalpy of a fluid particle as would be measured by an instrument at rest with respect to the particle mass center) and the kinetic energy per unit mass with respect to an external observer. In Equation 7.3-3 the total enthalpy in a steady flow is related to the total enthalpy at some upstream station and the additional energy acquired by heat transfer and work done on the particle.

Total enthalpy can be made more meaningful, however, if we envision the duct of Figure 7.8-1. At station ① the flow speed is u_1 and the static enthalpy h_1. By definition, the total enthalpy at ① is $h_1^0 = h_1 + \frac{1}{2}u_1^2$. At ① we imagine that a tiny fraction of the flow is extracted and brought to rest in some hypothetical process of deceleration. Equation 7.3-3 states that at the end point of the process, denoted as ①', the fluid will have the total enthalpy

$$h_{1'}^0 = h_1^0 + \Delta q + \Delta w. \tag{7.8-1}$$

We cannot calculate h_1^0 until the values of Δq and Δw corresponding to the process are known. But since the process is hypothetical we can specify that both

dq and dw be zero, i.e., that the deceleration process be isoenergetic. Further, in an isoenergetic process $h_1{}^0 = h_1{}^0 = \text{constant}$. Consequently, total enthalpy at a point in a flow can be considered to be the static enthalpy a particle would acquire if brought to rest isoenergetically.

TOTAL TEMPERATURE DEPENDENCE ON MACH NUMBER

If the fluid is a perfect gas, then, in the same hypothetical process, the *total temperature* is given by

$$c_p T_{1'}^0 = c_p T_1^0 = c_p T_1 + \tfrac{1}{2} u_1^2. \tag{7.8-2}$$

By dividing through by $c_p T_1$, Equation 7.8-2 can be written as

$$T_1^0 / T_1 = 1 + u_1^2 / 2 c_p T_1 = 1 + \tfrac{1}{2}(\gamma - 1)\, u_1^2 / a_1^2, \tag{7.8-3}$$

by use of Equations 7.5-4 and 7.2-12a.

The ratio of the flow speed at any point to the speed of sound at the same point is called the *Mach number*, and is denoted

$$M_1 \equiv u_1 / a_1. \tag{7.8-4}$$

Combining the last two equations, we obtain the highly useful relation*

$$T^0 / T = 1 + \tfrac{1}{2}(\gamma - 1)\, M^2, \tag{7.8-5}$$

where it is permissible to drop the subscripts since Equation 7.8-5 holds for every isoenergetic deceleration of a perfect gas. It is emphasized that Equation 7.8-5 applies to an arbitrary perfect-gas flow regardless of whether or not the flow (in contrast to the hypothetical deceleration process) is isoenergetic.

TOTAL PRESSURE AND TOTAL DENSITY

It is logical next to ask what values of the pressure and density the fluid would acquire at the end of the hypothetical deceleration, and to express them in terms of the original flow Mach number. However, the calculation cannot be made without specifying further restrictions on the process since we must know two state functions (i.e., not merely the temperature) in a perfect gas before it can be said that all state functions are known.

---

* This expression for the ratio of total to static temperature as a function of Mach number is one of the many that are tabulated in a variety of publications, particularly in books devoted to compressible flows. An extensive set of tabulations can be found in Brower (1990). With today's computers every individual can store the requisite formulas and data, to call up the required results routinely.

Since the process is hypothetical, we can specify any restriction compatible with an isoenergetic deceleration. The simplest, as well as the most useful for practical calculations, is to define the *total pressure p^0*, as the pressure a particle would acquire if brought to rest isoenergetically *and* isentropically. Elementary explicit relations are available only for perfect gases. By use of Equation 7.8-5 and the isentropic relations preceding Equation 7.7-8 we can relate the total pressure and total density for perfect gas flow to the Mach number:

$$p^0/p = \left[1 + \tfrac{1}{2}(\gamma - 1)M^2\right]^{\gamma/(\gamma-1)}, \tag{7.8-6}$$

$$\rho^0/p = \left[1 + \tfrac{1}{2}(\gamma - 1)M^2\right]^{1/(\gamma-1)}. \tag{7.8-7}$$

In these last three relations T, p, and ρ are the temperature, pressure, and density, respectively, at the point in the flow where the Mach number is M, prior to initiation of the deceleration process. These relations, under the restrictions cited, apply to an arbitrary point in the flow. Note that the expression for total pressure (Equation 7.8-6) is radically different than that for the case of incompressible flow, Equation 4.5-3.

ENTROPY CHANGES

If ① and ② are two stations in a flow then, for a perfect gas, we can obtain an explicit expression for the entropy change, $\Delta \delta \equiv \delta_2 - \delta_1$ between the two. By combining Equations 7.2-14b and 7.2-12a,

$$\frac{\Delta \delta}{R} = \frac{\gamma}{\gamma - 1}\ln\frac{T_2}{T_1} - \ln\frac{p_2}{p_1}, \tag{7.8-8}$$

where p_1, T_1, p_2, T_2 are the state values for the moving fluid at ① and ②, respectively. But, if we introduce the idea of hypothetical deceleration of the fluid at each of these stations in an isoenergetic and isentropic process, then $\Delta \delta = \delta_2 - \delta_1 = \delta_{2'} - \delta_{1'}$, or

$$\Delta \delta/R = \frac{\gamma}{\gamma - 1}\ln\left[T_2^0/T_1^0\right] - \ln\left[p_2^0/p_1^0\right]. \tag{7.8-9}$$

For the special case in which the *flow* is also isoenergetic

$$\Delta \delta/R = \ln\left[p_1^0/p_2^0\right]. \tag{7.8-10}$$

Thus, in an isoenergetic flow, a category of great practical importance, the total-pressure variation can be used as a measure of the entropy variation of a fluid particle as it travels downstream. But it follows from Equation 3.16-2 that in an adiabatic flow ($dq \equiv 0$), $ds \geq 0$; therefore, we see from Equation 7.8-10 that the total pressure in the isoenergetic flow of a perfect gas must either remain constant or decrease in the flow direction. In the former case the flow is said to be *isentropic*, or *particle isentropic*. Furthermore, since for steady flow $\partial/\partial t \equiv 0$, it follows that $\partial s/\partial x = 0$, and the flow is also *homentropic*, i.e., every particle has the same entropy level in a steady, particle-isentropic flow, assuming that all particles emanated from a region of uniform entropy.

7.9 APPLICATION — TEMPERATURE RISE AT THE NOSE OF A REENTRY VEHICLE

When a space capsule returns from a mission it must pass through Earth's atmosphere. Although there is no discrete height at which the outer boundary of the atmosphere can be precisely fixed (see Chapter 2), the altitude of 250,000 feet, or 75 km, is often chosen as the height below which aerodynamic effects become significant and must be taken into account. At a geometric height of 75 km, Table 2.16-3 gives $T = 208.4$ K, $p/p_a = 2.36 \times 10^{-5}$, $\rho/\rho_a = 3.26 \times 10^{-5}$. Obviously, at this height the air is extremely rarified.

Capsules for manned spacecraft missions acquire a speed the order of 25,000 mph (11.2 km/s) as they near reentry. If the reentry vehicle were to plunge into the atmosphere without prior deceleration, spacecraft temperatures so extreme would result as to guarantee its destruction. To avoid this, a "retro" rocket is fired, decelerating the vehicle, after which the reentry vehicle is then separated from the rocket for the reentry portion of the trajectory.

PROBLEM

Figure 7.9-1 shows a schematic of the flow at the nose of a reentry vehicle that is moving at the speed of 3 km/s as it passes through the 75 km altitude. It is desired to estimate the vehicle nose temperature.

SOLUTION

The complete solution of the flow field for this case is actually very complicated. As shown in the photograph of an aerodynamic test of a reentry vehicle in Figure 7.9-2, the vehicle is preceded by a curved shock that moves with the vehicle. For an earthbound observer, the flow is unsteady. By changing to an observer on the vehicle, however, the flow pattern is essentially fixed and the atmosphere appears to be rushing into it at the steady speed of $u_1 = 3$ km/s, equal to and opposite to the vehicle motion with respect to Earth.

It is advantageous to work in the frame of the vehicle-bound observer since, for that observer, the applicable equations are the simpler steady-flow (actually quasi-steady-flow) relations. In this frame the flow in a thin stream tube on the vehicle centerline is one dimensional as it moves straight through the shock (where it is

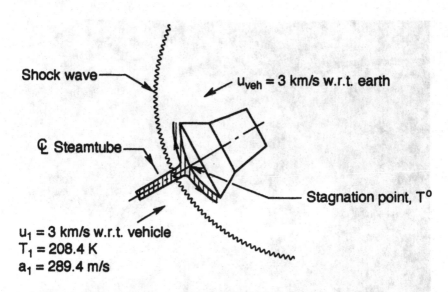

Shock wave

$u_{veh} = 3$ km/s w.r.t. earth

℄ Steamtube

Stagnation point, T^0

$u_1 = 3$ km/s w.r.t. vehicle
$T_1 = 208.4$ K
$a_1 = 289.4$ m/s

FIGURE 7.9-1 Flow pattern for reentry vehicle.

greatly compressed and heated), impinges upon the vehicle nose, and stagnates. It is then swept away along the vehicle surface by the surrounding fluid.

The Mach number of the oncoming flow is $M_1 = 3 \times 10^3/289.4 = 10.4$. According to Equation 7.8-5, the total temperature in the flow is given by

$$T^0/T_1 = 1 + \tfrac{1}{2}(\gamma - 1)M_1^2 = 1 + 0.2(10.4)^2 = 22.6.$$

Thus,

$$T^0 = 22.6 \times 208.4 = 4\,709\ K = 8477°R,$$

where we have treated the atmosphere as a perfect gas with $\gamma = 7/5$. But since there is no energy input to the flow within the stream tube, it is isoenergetic. Therefore, T^0 is also the theoretical stagnation point temperature. Since, in the frame of the observer on the vehicle, the flow there is at rest, it is also the *local* static temperature.

DISCUSSION

There are a number of assumptions in this calculation, some of which lead to significant error when compared with a precise analysis. At temperatures in the neighborhood of the predicted stagnation temperature, the principal constituents of air tend to dissociate, thereby soaking up some of the kinetic energy of the mainstream. One consequence is that the actual stagnation temperature is significantly less than predicted by a perfect gas analysis. Nevertheless, the temperature would still be high enough to require the use of materials on the vehicle surface with extraordinary heat-shielding properties.

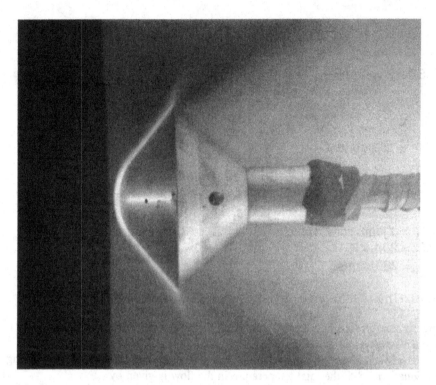

FIGURE 7.9-2 Photograph of a model of reentry vehicle made by H. T. Nagamatsu in a shock tunnel at the General Electric Research & Development Laboratory. The flow Mach number ahead of the vehicle is $M_1 = 7.5$, and the stagnation temperature is 5400 K. At this temperature, the luminescence of the air in the superheated region between the detached shock and the body is sufficient to provide the light source for the photograph. (Courtesy of H. T. Nagamatsu.)

We note that although the flow is isoenergetic, through the shock it is not isentropic, and the stagnation pressure at the nose cannot be calculated until we have developed the equations for the flow through a shock. As we shall see, however, the one-dimensional shock equations lead to a predicted stagnation pressure of $p_2^0 = 140 p_1 \approx 2.7 \times 10^{-3} \, p_a$ for a perfect gas. Although this value is still quite small, the pressure builds up rapidly as the vehicle altitude decreases, resulting in non-negligible aerodynamic forces.

As a last point we ask what is the justification for using continuum flow relations at this altitude? It is recalled from Chapter 1 that if the Knudsen number* is less than unity, then continuum relations apply. If Kn > 10, then one must use free-molecule flow equations. In our case, for an assumed vehicle diameter of $L = 3$ m, then, using the value of $\lambda = 2.035 \times 10^{-3}$ m at 75 km, Kn $\approx 6.8 \times 10^{-4}$, and the continuum relations apply. For the same vehicle at 200 km, Kn ≈ 80, and the relations just employed would be invalid.

---

* The Knudsen number Kn ≡ λ/L is defined as ratio of the mean-free-path length at that altitude to a characteristic body dimension L. See Section 1.7.

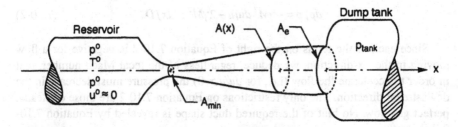

FIGURE 7.10-1 Reservoir supplying nozzle and dump tank.

7.10 NECESSARY CONDITIONS FOR ACCELERATING A FLOW IN AN IDEAL NOZZLE

In Section 7.7 it was shown, for the flow of a gas from a reservoir through an orifice into a dump tank, that when the dump tank pressure was lowered to the critical value (at which point the exit flow speed was equal to the local speed of sound) the flow rate became a maximum and was not affected by further decreases in the dump tank pressure. It was also indicated that one-dimensional flow equations are not adequate to analyze the free-jet behavior downstream of the exit station.

In the late 19th century engineers had demonstrated that controlled supersonic flow (i.e., $u > a$ or $M > 1$) could be achieved by maintaining the flow within a properly shaped duct or nozzle, provided there exists an adequate pressure drop from the reservoir to the supersonic flow station. Figure 7.10-1 illustrates the typical situation. A reservoir of gas is maintained at conditions p^0, T^0, ρ^0. The reservoir cross-sectional area is assumed to be sufficiently large, compared with the (as yet) unspecified duct area $A = A(x)$, that the reservoir flow speed can be neglected. On the downstream end of the duct the flow empties into a dump tank whose pressure is controllable, by pumps, for example. We ask:

1. What duct geometry is required to accelerate the flow continuously from the reservoir to supersonic speeds in the nozzle?
2. What is the corresponding pressure distribution?

PRESSURE DISTRIBUTION

For the general case, including viscous effects, a partial answer can be given. If we multiply through Equation 7.3-2 by $\rho dx/p$, we obtain

$$\frac{dp}{p} = -\frac{\rho u^2}{p}\frac{du}{u} - \frac{\tau_w}{\frac{1}{2}\rho u^2}\frac{\frac{1}{2}\rho u^2}{p}\frac{P}{A}\,dx. \qquad (7.10\text{-}1)$$

For a perfect gas $a^2 = \gamma p/\rho$. Also, we can introduce the *skin-friction coefficient* $c_f \equiv \tau_w/(\frac{1}{2}\rho u^2)$ and relations for a duct of circular cross section, $P/A = 4/D$, where $D = D(x)$ is the duct diameter*; then Equation 7.10-1 becomes

---

* If the actual duct were not circular, then D would be interpreted as the hydraulic diameter $D_h = 4A/P$.

$$dp/p = -\gamma M^2 \, du/u - 2\gamma M^2 \, c_f dx/D. \qquad (7.10\text{-}2)$$

Since each of the terms on the right of Equation 7.10-2 is negative for a flow moving in the +x-direction, we deduce, regardless of the local Mach number, that in order to accelerate the flow (i.e., for $du/u > 0$) the pressure must decrease in the downstream direction. The only restrictions on Equation 7.10-2 are those of steady, perfect gas flow. No hint of the required duct shape is revealed by Equation 7.10-2, however.

THE AREA–MACH NUMBER RELATION

To answer the question on the duct geometry, we must specialize to the case of an adiabatic and inviscid (i.e., isentropic) flow. In this case, Equation 7.3-2 can be written as follows:

$$u\,du = u^2 \, \frac{du}{u} = -\frac{dp}{\rho} = -\frac{dp}{d\rho}\frac{d\rho}{\rho} = -\left(\frac{\partial p}{\partial \rho}\right)_s \frac{d\rho}{\rho} = -a^2 \, \frac{d\rho}{\rho}. \qquad (7.10\text{-}3)$$

By taking the logarithmic differential of the continuity equation (Equation 7.3-1), we obtain

$$d\rho/\rho = -du/u - dA/A. \qquad (7.10\text{-}4)$$

Combining Equation 7.10-3 and 7.10-4, we can solve for

$$\frac{du}{u} = -\frac{dA/A}{1 - u^2/a^2} = -\frac{dA/A}{1 - M^2}. \qquad (7.10\text{-}5)$$

Equation 7.10-5, which was first obtained by Hugoniot (1889), is valid for isentropic flow and does not depend on an equation of state. It is called the differential *area–velocity* relation and leads directly to several far-reaching conclusions. Assuming that we have established the flow at velocity $u > 0$ at a point in a duct of cross section $A = A(x)$, then acceleration of the flow in the downstream direction is possible, according to Equation 7.10-5, only under the following conditions:

a. If the flow is subsonic ($0 < M < 1$) at that point, then $dA < 0$, i.e., the area must decrease in the downstream direction;
b. If the flow is supersonic ($M > 1$), then $dA > 0$, i.e., the area must increase in the downstream direction;
c. If the flow is sonic ($M = 1$), then the only possibility for a finite solution is that locally $dA = 0$, i.e., the area is a minimum* at that station.

---

* If the denominator of the right-hand side of Equation 7.10-5 is zero, then only if the numerator is also zero can the limit of the quotient be finite. Thus, A is an extremum where $M = 1$ and since the area is decreasing upstream of that point A must be a minimum, i.e., a throat.

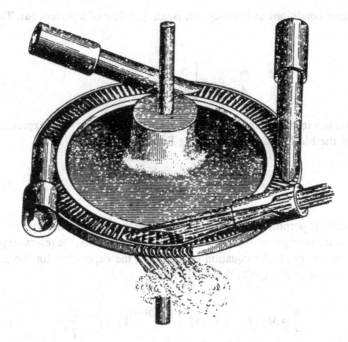

FIGURE 7.10-2   Schematic of de Laval's steam turbine (patented in Sweden in 1888). (From Stodola, A., and Lowenstein, L.C., *Steam and Gas Turbines*, McGraw-Hill, New York, 1945. With permission.)

At a station in a duct where $dA = 0$ and where A is a minimum, the cross section is said to be a *throat*. If the Mach number at a throat is unity, the flow is said to be *choked*. In order to achieve choking, of course, the requisite pressure conditions downstream must exist.

That the cross-sectional area must decrease in order to accelerate a subsonic flow is intuitively what we expect from our knowledge of incompressible flow. However, it is unexpected that in order to go from a subsonic to a supersonic condition it is necessary to shape the duct so as to provide a minimum area after which the area must increase while the pressure continues to decrease.

This fact was recognized at least as early as 1883 by the Swedish engineer de Laval who constructed a stream turbine to power a cream separator. His use of a converging–diverging nozzle to accelerate the flow to supersonic speeds is clearly shown in Figure 7.10-2.

7.11   ISENTROPIC FLOW THROUGH A DE LAVAL NOZZLE

With reference to Figure 7.10-1 we seek to determine relations for the flow in a convergent–divergent nozzle of fixed area distribution $A = A(x)$, such that, for specified reservoir conditions, we can determine completely the distribution of any flow property $u = u(x)$, $p = p(x)$, etc., for an arbitrarily selected dump tank pressure

p_r. We restrict conditions to isoenergetic, isentropic flow of a perfect gas. The mass flux is

$$\frac{\dot{m}}{A} = \rho u = \left[\frac{\rho}{\rho^0}\frac{u}{a}\frac{a}{a^0}\right]\rho^0 a^0.$$ (7.11-1)

Since each ratio in the bracket — under the stated conditions — is expressible as a function of the Mach number, then, using Equation 7.8-5 and 7.8-7,

$$\frac{\dot{m}}{A} = M\left[1 + \tfrac{1}{2}(\gamma - 1)M^2\right]^{-(\gamma+1)/2(\gamma-1)}\rho^0 a^0.$$ (7.11-2a)

For practical purposes it is preferable to eliminate the product $\rho^0 a^0$ in favor of quantities more susceptible to direct measurement, specifically, the reservoir pressure and temperature. From the equation of state and the expression for the speed of sound, we obtain $\rho^0 a^0 = (p^0/\mathcal{R}T^0)(\gamma\mathcal{R}T^0)^{1/2}$; thus,

$$\frac{\dot{m}}{A} = M\left[1 + \tfrac{1}{2}(\gamma - 1)M^2\right]^{-(\gamma+1)/2(\gamma-1)}(\gamma/\mathcal{R})^{1/2}\frac{p^0}{\sqrt{T^0}}.$$ (7.11-2b)

The Mach number corresponds to the station where the area is A.

It is readily shown that when the Mach number is unity $\dot{m}/A$ is a maximum; i.e., we have a throat. To indicate the existence of choking we denote the corresponding area as A^*; correspondingly, the local pressure is the critical value p^*. From Equation 7.11-2b, with $M = 1$,

$$\frac{\dot{m}}{A^*} = (\gamma/\mathcal{R})^{1/2}\left[\frac{2}{\gamma+1}\right]^{(\gamma+1)/2(\gamma-1)}\frac{p^0}{\sqrt{T^0}},$$ (7.11-3a)

or

$$\frac{\dot{m}}{A^*} = C\frac{p^0}{\sqrt{T^0}}, \quad \text{where } C \equiv (\gamma/\mathcal{R})^{1/2}\left[\frac{2}{\gamma+1}\right]^{(\gamma+1)/2(\gamma-1)}$$ (7.11-3b)

Equations 7.11-3 demonstrate that the mass flow rate is proportional to the reservoir pressure and inversely proportional to the square root of the total temperature. For the engineering and SI systems, we have for air, with $\gamma = 7/5$:

AREA–MACH NUMBER RELATIONS

Dividing Equation 7.11-3a by Equation 7.11-2b, we obtain

	Units			
	$\dot{m}/A^*$	p^0	T^0	C
Engineering System	slug/sec ft²	lb/ft²	°R	0.01653
SI	kg/sem²	N/m²	K	0.040 41

$$\frac{A}{A^*} = M^{-1}\left\{\frac{2}{\gamma+1}\left[1+\tfrac{1}{2}(\gamma-1)M^2\right]^{(\gamma+1)/2(\gamma-1)}\right\}, \qquad (7.11\text{-}4)$$

which is a very useful and widely tabulated Mach number function. To use Equation 7.11-4 in an actual calculation, it is not essential that there be a throat in the flow under consideration. In such a case A^* can be thought of as a reference area at some hypothetical station downstream. If there is no throat, if follows from Section 7.10 that the flow must be everywhere subsonic. For purposes of calculation we note that flow conditions at any two stations A_1 and A_2 can be related, since $A_1/A_2 = (A_1/A^*)/(A_2/A^*)$. This point will be clarified in a following example. The isoenergetic and isentropic relation for the area ratio in terms of the pressure distribution at an arbitrary station is obtained by combining Equations 7.8-6 and 7.11-4:

$$\frac{A}{A^*} = \frac{\left[(\gamma-1)/2\right]^{1/2}\left[2/(\gamma+1)\right]^{(\gamma+1)/2(\gamma-1)}}{(p/p^0)^{1/\gamma}\left[1-(p/p^0)^{(\gamma-1)/\gamma}\right]^{1/2}}. \qquad (7.11\text{-}5)$$

The pressure ratio, from Equation 7.11-5 and the area ratio, from Equation 7.8-6, are plotted in Figure 7.11-1 vs. the local Mach number. The pressure function has a minimum at the sonic station; i.e., where $M = 1$, $A = A^*$ and $p \to p^*$, as given by Equation 7.6-6. For every other value of A/A^* the pressure ratio is double valued, one lying on the subsonic branch, the other corresponding to supersonic flow. For a fixed value of A/A^* the pressure ratio for the supersonic case is always less than that for the subsonic case. For both $M \to 0$, $M \to \infty$ the area ratio $A/A^* \to \infty$, neither of which can be achieved in practice.

In a more practical vein, if the pressure distribution is such that the minimum area becomes sonic, then the Mach number at the entrance to the duct from the reservoir will not exceed $M = 0.01$ as long as the local area ratio is equal or greater than 58 times the throat area for $\gamma = 7/5$. This represents a very low flow speed relative to that of the throat, and is a typical example of what we mean when we say that the reservoir flow speed is negligible, i.e., $u^0 \approx 0$.

As indicated, for choked flow, there are two possible Mach numbers for a given area ratio and for each of these there is a corresponding pressure ratio. For example, for $A/A^* = 1.5$, we note that $M_{\text{subsonic}} = 0.43$, $p/p^0 = 0.881$, and $M_{\text{supersonic}} = 1.86$ with $p/p^0 = 0.159$. Note that the subsonic Mach number regime corresponds to a pressure ratio greater than the critical pressure, whereas in the supersonic regime the values are less than critical.

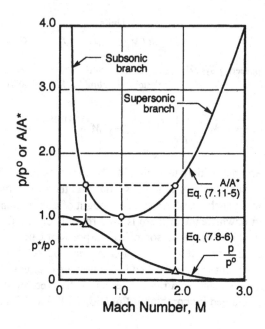

FIGURE 7.11-1 Plot of area ratio and pressure ratio vs. Mach number for isentropic nozzle flow, $\gamma = 7/5$.

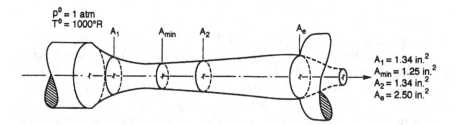

FIGURE 7.11-2 Setup for sample calculations in a convergent–divergent nozzle.

EXAMPLE

Consider a convergent–divergent nozzle as shown in Figure 7.11-2. We assume an isoenergetic, isentropic flow of air, $\gamma = 7/5$, with a reservoir total temperature $T^0 = 1000°R$, reservoir pressure $p^0 = 1$ atm $= 2116$ lb/ft². We shall take advantage of the various flow tables available, e.g., NACA Rep. 1135, listed as a reference under Ames Research Staff.

Case I

Measurement indicates $T_1 = 885°R$. Determine flow conditions at the four stations. Since $T_1/T^0 = 0.952$, the flow tables yield $M_1 = 0.500$. Other conditions at ① follow readily, e.g., $p_1/p^0 = 0.843$ or equivalently, $p_1 = 0.843(2116) = 1784$ lb/ft².

Also, from the tabulation of Equation 7.11-4, $A_1/A^* = 1.34$; thus, $A^* = 1.00$ in.^{2} Since $A_{min} = 1.25$ in.^{2}, A^* is only a hypothetical throat, located at some fictitious downstream station, indicated by the dashed lines. This means that the flow upstream of A^* is everywhere subsonic. Merely by calculating the area ratios $A_{min}/A^* = 1.25$, $A_2/A^* = 1.34$, $A_e/A^* = 2.50$ we find from the tables that for A_{min}, $M = 0.553$, $M_2 = 0.500$, $M_e = 0.240$.

Case II

With the same reservoir conditions, we measure $p_e = 198$ lb/ft^2. Thus, $p_e/p^0 = 0.0935$. Invoking isentropic flow, this means that $M_e = 2.20$ and $A_e/A^* = 2.00$. From this last result, $A^* = 2.50/2.00 = 1.25$ in.^{2} Since $A_{min} = A^*$, the minimum area is thus a throat. Further, since the exit Mach number is supersonic, the flow in the divergent portion of the nozzle must be everywhere supersonic. From the area relations, $A_1/A^* = A_2/A^*$ $= 1.072$; thus, $M_1 = 0.735$ and $M_2 = 1.312$.

Other flow variables are readily calculated. For example, using Equation 7.5-5, $a^0 = 49.02(1000)^{1/2} = 1550$ ft/sec. Thus, $a_2 = (a_2/a^0)a^0 = (T_2/T^0)^{1/2} a^0 = (0.478)^{1/2} a^0$ $= 1072$ ft/sec. Finally, $u_2 = M_2 a_2 = 1.312(1073) = 1406$ ft/sec.

7.12 THE APPEARANCE OF SHOCK WAVES IN A CONVERGENT–DIVERGENT NOZZLE

The relations presented in Section 7.11 were (effectively) known to designers of steam turbines at the end of the 19th century. For a specified convergent–divergent nozzle geometry there are only two possible isentropic solutions in which the minimum area has choked flow. Thus, at the nozzle exit there are only two possible pressures for isentropic, choked flow throughout. In one case the flow in the divergent section decelerates monotonically; correspondingly, the Mach number decreases and the pressure increases, monotonically. In the other case, the flow continues to accelerate monotonically, but the Mach number increases monotonically whereas the pressure decreases.

For a fixed ratio of exit to throat area, these two exit pressures are different, the subsonic-choked case corresponding to the higher of the two. What happens when the exit pressure is maintained (for example, by a large dump tank downstream of the exit) at some value between the two? This question was answered by the investigations of A. Stodola in 1903 who studied the flow in a conical, axisymmetric convergent–divergent nozzle, Figure 7.12-1.

While holding the upstream pressure constant, the exit pressure was varied in steps over a range of values during which an axial pressure survey was made within the nozzle by means of a thin tube inserted along the centerline. Small holes drilled into the tube at one location allowed the pressure to be measured at any longitudinal station by moving the tube relative to the nozzle.

Once choked flow was established at the throat, Stodola found that there were sudden "surges of pressure" occurring over short distances downstream of the throat. Curve E, for example, indicates a 22 psi increase over a distance of 0.12 in. He

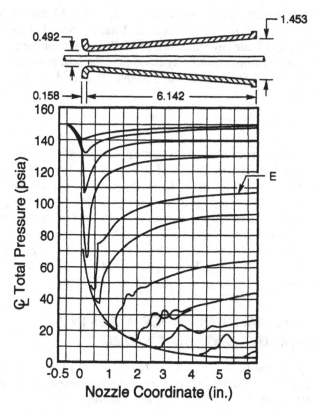

FIGURE 7.12-1 Pressure measurements of flow in a conical, divergent nozzle. (Redrawn from Stodola and Lowenstein, 1945, p. 83.)

attributed these surges to compression shocks, normal to the flow direction, whose possibility had actually been predicted by Rankine (1870) and Hugoniot (1889). As it turned out, the pressure variations in the flow are actually much sharper than revealed by Stodola's measurements.* The boundary layer on the probe (the concept of the boundary layer was proposed by Prandtl only in 1904) tends to smear the pressure rise over a finite length, thus obscuring the fact that the shock is very thin. Nevertheless, Stodola assumed that the shock could be treated on the assumption that it is infinitely thin.

7.13 NORMAL SHOCK WAVES

We have previously introduced the idea of a propagating disturbance in Section 7.4. A standing, normal shock wave is a finite disturbance that propagates into a medium at a velocity equal but opposite to the velocity at which the medium is moving with

---

* For another classical set of measurements of the same phenomenon by a different technique see Ackeret (1927), whose data are also reproduced by Liepmann and Roshko (1957, p. 129).

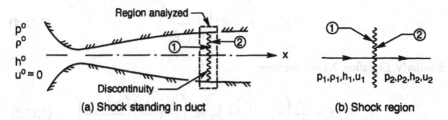

FIGURE 7.13-1   Analysis of a normal shock.

respect to the observer who sees the standing shock. Thus, he sees the flow as steady, a frame of reference that is particularly convenient in which to analyze the flow.

SHOCK RELATIONS FOR A GENERAL SUBSTANCE

For the current purpose we assume that the shock is a finite discontinuity, shown as a wavy line, and that the analysis deals with a constant-area flow (Figure 7.13-1) between stations ① and ②, corresponding to the flow immediately ahead of, and immediately after, the discontinuity. Since the duct length over which the transition occurs is negligible, the flow be treated as though it were a very short length of a constant-area duct. Further, since the duct length is negligible, wall viscous stresses may be neglected. In addition, we suppose that across the shock the flow is isoenergetic, i.e., $\Delta q = \Delta w = 0$. Under the conditions cited, the "conservation" equations are obtained from Equations 7.3-4, 7.3-5, and 7.3-3, respectively.

$$\text{Continuity:} \quad \rho_1 u_1 = \rho_2 u_2 = \dot{m}; \tag{7.13-1}$$

$$\text{Momentum:} \quad p_1 + \rho_1 u_1^2 = p_2 + \rho_2 u_2^2; \tag{7.13-2}$$

$$\text{Energy:} \quad h_1 + \tfrac{1}{2} u_1^2 = h_2 + \tfrac{1}{2} u_2^2. \tag{7.13-3}$$

These involve three state variables and one dynamic variable from which there can be formed four ratios p_{21}, ρ_{21}, h_{21}, u_{21}, where it is convenient to employ the special index notation $p_{21} = p_2/p_1$, $u_{21} = u_2/u_1$, etc. If we regard conditions ahead of the shock (p_1, ρ_1, h_1, u_1) as known, the objective is to determine a relation between these ratios and the known conditions at ①.

Rearranging, Equation 7.13-2 becomes

$$p_2 - p_1 = \rho_1 u_1^2 - \rho_2 u_2^2 = \frac{\rho_1^2 u_1^2}{\rho_1} - \frac{\rho_2^2 u_2^2}{\rho_2}$$

or, with the index notation and Equation 7.13-1,

$$\left(p_{21}-1\right)p_1 = \bar{\dot{m}}^2\left(\frac{1}{\rho_1}-\frac{1}{\rho_2}\right). \tag{7.13-4}$$

Similarly, Equation 7.13-3 becomes

$$\left(h_{21}-1\right)h_1 = \frac{\bar{\dot{m}}^2}{2}\left(\frac{1}{\rho_1^2}-\frac{1}{\rho_2^2}\right) = \frac{1}{2}\,\bar{\dot{m}}^2\left(\frac{1}{\rho_1}+\frac{1}{\rho_2}\right)\left(\frac{1}{\rho_1}-\frac{1}{\rho_2}\right). \tag{7.13-5}$$

We now divide Equation 7.13-5 by Equation 7.13-4 to produce

$$\frac{\left(h_{21}-1\right)h_1}{\left(p_{21}-1\right)p_1} = \frac{1}{2}\left(\frac{1}{\rho_1}+\frac{1}{\rho_2}\right) = \frac{1}{2\rho_1}\frac{p_{21}+1}{p_{21}}$$

or

$$\frac{\left(h_{21}-1\right)\rho_{21}}{\left(p_{21}-1\right)\left(\rho_{21}+1\right)} = \frac{p_1}{2\rho_1 h_1}. \tag{7.13-6}$$

Equation 7.13-6 is remarkable on several counts. In the first place, it involves only ratios of state variables across a finite disturbance and flow conditions ahead of the disturbance; further, it is independent of any dynamic flow variables. As a consequence, it holds for *any* frame of reference. In addition, it can be demonstrated that it also holds for nonsteady flows involving shocks, and for oblique shocks and curved shocks (locally) as well. Furthermore, since it was not necessary to make any statement about the thermal, or caloric, equations of state, it also holds for nonperfect gases, liquids, and solids. Hereafter we refer to it as the *Rankine–Hugoniot equation for general substances*.

PERFECT GASES

Analytical treatment of nonperfect gases may require the use of Mollier charts or numerical treatment using available databases. Unless otherwise specified, we shall deal only with thermally and calorically perfect gases, such that the following relations apply: $p = \rho RT$, $h = c_p T = \gamma RT/(\gamma-1)$, $a^2 = \gamma RT$, $c_p/c_v \equiv \gamma =$ constant. Under these restrictions

$$p_{21} = \rho_{21}T_{21}, \quad h_{21} = T_{21}, \quad p_1/2\rho_1 h_1 = (\gamma-1)/2\gamma. \tag{7.13-7}$$

Introducing Equation 7.13-7 into Equation 7.13-6, we have

$$\frac{\left(T_{21}-1\right)\rho_{21}}{\left(p_{21}-1\right)\left(\rho_{21}+1\right)} = \frac{\left(p_{21}-\rho_{21}\right)}{\left(p_{21}-1\right)\left(\rho_{21}+1\right)} = \frac{\gamma-1}{2\gamma}.$$

This allows us to solve for the density ratio in terms of the pressure ratio. Omitting the algebra, the result is

$$\rho_{21} = \frac{1+\alpha p_{21}}{\alpha + p_{21}}, \quad \alpha \equiv \frac{\gamma+1}{\gamma-1}. \tag{7.13-8}$$

This is a form equivalent to the expressions* derived by Rankine (1870) and Hugoniot (1889), who dealt only with perfect gases. Corresponding relations between any pair of the ratios can be obtained by combining Equation 7.13-8 with Equation 7.13-7 and the continuity equation $\rho_{21} = u_{12}$.

CONDITIONS ACROSS A SHOCK AS A FUNCTION OF M_1

It is particularly convenient to have an expression for, say, the pressure ratio as a function of the upstream Mach number $M_1 = u_1/a_1$. This can be achieved by rewriting Equation 7.13-2 as follows:

$$p_{21} - 1 = (1/p_1)(\rho_1 u_1^2 - \rho_2 u_2^2) = (\rho_1 u_1^2/p_1)(1 - \rho_{21}^{-1}).$$

Then, introducing $a_1^2 = \gamma p_1/\rho_1$,

$$p_{21} - 1 = \gamma M_1^2 (1 - \rho_{21}^{-1}). \tag{7.13-9}$$

The density ratio is next eliminated from Equation 7.13-9 by means of Equation 7.13-8, and after some additional algebra we have**

$$p_{21} = \frac{2\gamma}{\gamma+1} M_1^2 - \frac{\gamma-1}{\gamma+1}. \tag{7.13-10}$$

Corresponding relations for the density, speed, and temperature ratios follow readily, e.g.,

$$\rho_{21} = \frac{(\gamma+1)M_1^2}{(\gamma-1)M_1^2 + 2} = u_{12}. \tag{7.13-11}$$

It is clear that in the procedure leading to Equation 7.13-10 we could have solved equally easily for the pressure ratio in terms of M_2. By interchanging the subscripts in Equation 7.13-10, we have

---

* See Problem 7.9.

** Equation 7.13-10 and many of the following relations have been tabulated in a number of publications as a function of M_1, for various values of γ. In NACA Report 1135 there is a comprehensive tabulation for γ = 7/5, as well as a highly useful collection of formulas and graphs and is referenced under Ames Research Staff (1953).

$$p_{12} = \frac{2\gamma}{\gamma+1} M_2^2 - \frac{\gamma-1}{\gamma+1}. \tag{7.13-12}$$

This allows us to determine the expression for M_2 as a function of M_1. By multiplying Equation 7.13-10 by Equation 7.13-12 and solving,

$$M_2^2 = \frac{(\gamma-1)M_1^2 + 2}{2\gamma M_1^2 - (\gamma-1)}. \tag{7.13-13}$$

We can also get the ratio of total pressures across the shock. Equation 7.8-6 gives the expression relating p_1^0 to p_1, M_1 and γ, for a flow at M_1, p_1, which is brought to rest isentropically and isoenergetically. An equation of the same form relates p_2^0 to p_2, M_2, and γ. Thus, since

$$\frac{p_2^0}{p_1^0} = \frac{p_2^0}{p_2} \frac{p_2}{p_1} \frac{p_1}{p_1^0}, \tag{7.13-14}$$

we can substitute Equation 7.8-6 for the first and third factors, Equation 7.13-10 for p_{21}, and eliminate M_2 by Equation 7.13-13. This yields

$$\frac{p_2^0}{p_1^0} = \left[\frac{(\gamma+1)M_1^2}{(\gamma-1)M_1^2 + 2} \right]^{\gamma/(\gamma-1)} \left[\frac{\gamma+1}{2\gamma M_1^2 - (\gamma-1)} \right]^{1/(\gamma-1)} \tag{7.13-15}$$

It is worthwhile to recall from Equation 7.8-10 that, in the isoenergetic flow of a perfect gas, the entropy change between any two stations is given by

$$\Delta s/R \equiv (s_2 - s_1)/R = \ln\left[p_1^0/p_2^0\right]. \tag{7.13-16}$$

Thus, combining Equations 7.13-15 and 7.13-16, also produces an expression for the entropy change across a shock as a function of M_1.

On Compression Shocks and Expansion Shocks

A shock wave is said to be a *compression* or *expansion* wave depending on whether the pressure *increases* or *decreases* in the direction of flow. In order to demonstrate that only compression shocks are possible, we proceed as follows. By rearranging the terms in Equation 7.13-10, we obtain

$$p_{21} = M_1^2 + \frac{\gamma-1}{\gamma+1}(M_1^2 - 1). \tag{7.13-17}$$

Since $\gamma > 1$ for all perfect gases (in fact, for all gases), the coefficient of $(M_1^2 - 1)$ is a positive number. Therefore, if M_1 is supersonic/sonic/subsonic, p_{21} is greater/equal to/less than unity. That is, for the disturbance to be a compression, the flow Mach number ahead of the shock must be supersonic.

Similarly, from Equation 7.13-13,

$$M_2^2 = 1 - \frac{\gamma+1}{\gamma-1} \frac{M_1^2 - 1}{\frac{2\gamma}{\gamma-1}M_1^2 - 1}, \qquad (7.13\text{-}18)$$

from which we see that if M_1 is supersonic, M_2 is subsonic. By interchanging subscripts again, it is obvious that when M_2 is supersonic, M_1 must be subsonic; and when $M_1 = 1$, $M_2 = 1$ also. Thus, in a compression shock, the Mach number of the flow immediately following the shock is always subsonic. By taking the limit in Equation 7.13-18, we see that

$$\lim_{M_1 \to \infty} M_2 = \sqrt{\frac{\gamma-1}{2\gamma}} \qquad (7.13\text{-}19a)$$

$$= 0.378, \quad \text{for} \quad \gamma = 7/5. \qquad (7.13\text{-}19b)$$

Using similar techniques, it can be shown that both the static temperature and the static density increase in a compression shock. Also of interest is that

$$\lim_{M_1 \to \infty} p_{21} = \frac{\gamma+1}{\gamma-1}, \qquad (7.13\text{-}20a)$$

$$= 6, \quad \text{for} \quad \gamma = 7/5. \qquad (7.13\text{-}20b)$$

Thus, we see that (in a steady flow) the fluid cannot be brought to rest by means of a shock wave and, further, that although the pressure ratio and the temperature ratio across a shock theoretically become infinite at infinite upstream Mach numbers, the density ratio has a finite limit. Thus, the ability to utilize a shock wave as a mechanism to compress gases is sharply limited.

We cannot directly use Equation 7.13-15 to deduce the behavior of the total pressure (or, consequently, the entropy jump), for increasing M_1, since the first factor tends to increase, whereas the second becomes smaller. Of course, in the limit as $M_1 \to \infty$, the total-pressure ratio becomes zero, and the entropy jump becomes infinite. This is a clue to the general behavior of the flow.

We can get around this problem in the following way. Equation 7.2-14c for the entropy jump can be rewritten as

$$\frac{\Delta s}{R} = \ln\left[p_{21}^{1/(\gamma-1)} \rho_{21}^{-\gamma/(\gamma-1)}\right] = \frac{1}{\gamma-1}\left[\ln p_{21} - \gamma \ln \rho_{21}\right]. \qquad (7.13-21)$$

By using a standard series expansion* for the natural logarithm,

$$\ln p_{21} = 2\left[\frac{p_{21}-1}{p_{21}+1} + \frac{1}{3}\left(\frac{p_{21}-1}{p_{21}+1}\right)^3 + \frac{1}{5}\left(\frac{p_{21}-1}{p_{21}+1}\right)^5 + \cdots\right], \qquad (7.13-22)$$

which is convergent for all $p_{21} > 1$. An identical expression exists for the ratio $\ln \rho_{21}$.

Eliminating the pressure ratio by means of the Rankine–Hugoniot relation (Equation 7.13-8), we have

$$\frac{\rho_{21}-1}{\rho_{21}+1} = \frac{1}{\gamma}\frac{p_{21}-1}{p_{21}+1}. \qquad (7.13-23)$$

Combining this result with the expression for $\ln \rho_{21}$, we see that

$$-\gamma \ln \rho_{21} = -2\left[\frac{p_{21}-1}{p_{21}+1} + \frac{1}{3\gamma^2}\left(\frac{p_{21}-1}{p_{21}+1}\right)^3 + \frac{1}{5\gamma^4}\left(\frac{p_{21}-1}{p_{21}+1}\right)^5 + \cdots\right]. \qquad (7.13-24)$$

Finally, by combining Equations 7.13-24 and 7.13-22 the relation for the entropy change across a normal shock becomes

$$\frac{\Delta s}{R} = \frac{2}{\gamma-1}\left[\frac{1}{3}\left(1-1/\gamma^2\right)\left(\frac{p_{21}-1}{p_{21}+1}\right)^3 + \frac{1}{5}\left(1-1/\gamma^4\right)\left(\frac{p_{21}-1}{p_{21}+1}\right)^5 + \cdots\right]. \qquad (7.13-25)$$

The series of Equation 7.13-25 is uniformly convergent. All the coefficients are positive and, since the terms are either all positive or all negative, depending on the magnitude of p_{21}, it follows that

$$\frac{\Delta s}{R} \lessgtr 0, \quad \text{for } p_{21} \lessgtr 1, \quad \text{or, equivalently, } M_1 \lessgtr 1. \qquad (7.13-26)$$

Consequently, we have shown, for a perfect gas, that if the Mach number ahead of the disturbance is subsonic/sonic/supersonic the change of entropy across the disturbance is negative/zero/positive.

However, the analysis involves the adiabatic restriction on the energy equation such that $\Delta q \equiv 0$. But, as indicated by Equation 3.16-3, the second law of thermodynamics stipulates that for an adiabatic process the entropy change may not be negative. Hence, we rule out the possibility of subsonic flow preceding a finite

---

* e.g., Dwight (1961), Formula No. 601.7.

disturbance. This corresponds to eliminating the possibility of an "expansion shock," and in a frame of reference fixed with respect to the shock the upstream flow can only be supersonic. It also follows from Equation 7.13-16 for $\Delta s/R > 0$ that $p_2^0/p_1^0 < 1$, i.e., that in the same reference system there is a corresponding decrease of total pressure across a stationary shock wave. Indeed, the total-pressure loss is the criterion by which we measure losses in the flow through a shock. Finally, we note, in the case of a vanishingly weak shock, defined by $p_{21} \to 1$, that $M_1 \to 1$, $M_2 \to 1$, $\Delta s \to 0$; i.e., all discontinuities vanish.

EXAMPLE

Consider a stationary shock in air, Figure 7.13-1, such that $M_2 = 2$, $p_1 = 2$ atm, $T_1 = 500°R$. Determine conditions in ② behind the shock, using shock tables as available. We will need the speed of sound in ①, $a_1 = 49.02\sqrt{500} = 1096$ ft/sec. Thus, $u_1 = M_1 a_1 = 2(1096) = 2192$ ft/sec. From the tables, $p_{21} = 4.5000$, $T_{21} = 1.688$, $M_2 = 0.5774$. Therefore, $p_2 = 4.5(2) = 9.0$ atm $= 19,040$ lb/ft², $T_2 = 1.688(500) = 844°R$. $a_2 = \sqrt{T_{21}}a_1 = \sqrt{1.688}(500) = 1425$ ft/sec. Also, $u_2 = M_2 a_2 = 0.5774(1424) = 822.2$ ft/sec. The total pressure behind the shock is given by $p_2^0 = (p_2^0/p_1^0)(p_1^0/p_1)p_1 = (p_2^0/p_1^0)/(p_1/p_1^0)^{-1}p_1$, where the last two parentheses are standard, tabulated, isentropic functions of M_1. Thus, we get $p_2^0 = (0.7209)(1/0.1278)(2) = 11.28$ atm. The entropy change across the shock is computed from $\Delta s/R = \ln(T_2^0/T_1^0) = \ln(1/0.7209) = 0.3273$. Note that $p_2^0 = p_2^0 = T_1/(T_1/T^0) = 500/0.5556 = 900°R$, as a consequence of isoenergetic flow.

7.14 ON THE STRUCTURE OF SHOCK WAVES

The relations of Section 7.13, for the flow through a normal shock, were obtained on the premise that the shock wave can be treated as an infinitely thin discontinuity across which the conservation equations (Equations 7.13-10 through 7.13-3) are satisfied. Although this is a completely successful model for predicting the overall shock behavior, the idea of a discontinuity violates the concept of the gas as a continuum.

Up to this point all flows have been analyzed on the assumption that the effect of heat transfer from particle to particle is negligible. However, as Rankine noted in his original analysis of a propagating longitudinal disturbance, the effect of heat conduction in the direction parallel to the duct axis should be examined. Furthermore, for the first time in this text, we must also consider the effect of viscous stresses due to velocity gradients in the flow direction. In treating this problem we closely follow the approach of Becker (1968), Section 4.2.

THE BASIC EQUATIONS FOR FLOW THROUGH A SHOCK TREATED AS A FLOW WITHOUT A DISCONTINUITY

The applicable equations are readily obtained from the equations derived in Chapters 3 and 7. For the case of horizontal, steady, constant-area, mass-conserving flow, the continuity equation is

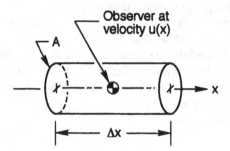

FIGURE 7.14-1 Heat flux and viscous work for a particle as it traverses the shock zone.

$$\dot{m} \equiv \rho u = \text{constant}. \tag{7.14-1}$$

Under these same restrictions the dynamic equation (Equation 3.12-10) reduces to

$$\rho u \frac{du}{dx} = -\frac{dp}{dx} + \frac{d\tau_x}{dx}, \tag{7.14-2}$$

where the longitudinal viscous stress is $\tau_s \rightarrow \tau_x$ and where we neglect the wall viscous stress for the current purpose. Combining these two equations produces the momentum equation in differential form

$$\frac{d}{dx}\left(p + \rho u^2 - \tau_x\right) = 0. \tag{7.14-3}$$

The first law of thermodynamics for an arbitrary (inertial) observer has previously been obtained as Equation 3.15-3. For steady flow this reduces to

$$Tu \frac{ds}{dx} = \dot{q} + \dot{\psi} = u\left(\frac{dh}{dx} - \frac{1}{\rho}\frac{dp}{dx}\right). \tag{7.14-4}$$

Equations 7.14-2 and 7.14-4 are next combined by eliminating the common term dp/dx. This results in expressions for the energy equation and the entropy production relation:

$$Tu \frac{ds}{dx} = \dot{q} + \dot{\psi} = u\left(\frac{dh^0}{dx} - \frac{\mu}{\rho}\frac{d\tau_x}{dx}\right). \tag{7.14-5}$$

EVALUATION OF THE HEAT-CONDUCTION AND DISSIPATION FUNCTIONS IN TERMS OF THE DYNAMIC FLOW VARIABLES

Consider an element of fluid (Figure 7.14-1) for convenience selected as a right, circular cylinder of area A and length Δx, as it traverses the shock region. We shift, temporarily, to an observer traveling with the fluid element mass center at velocity $u = u(x)$ with respect to the inertial observer. We employ Fourier's law of heat conduction, previously introduced in Section 1.9. It is necessary to modify the form of the equation to accommodate the direction of heat conduction, which is along the axis in the current case. In this observer's frame the following calculations would be made:

On the left face:

$$\text{head conduction in} = \left[-k\frac{\partial T}{\partial x} - \frac{\partial}{\partial x}\left(-k\frac{\partial T}{\partial x} \right)\frac{\Delta x}{2} + O(\Delta x)^2 \right] A;$$

On the right face:

$$\text{head conduction out} = \left[-k\frac{\partial T}{\partial x} + \frac{\partial}{\partial x}\left(-k\frac{\partial T}{\partial x} \right)\frac{\Delta x}{2} + O(\Delta x)^2 \right] A.$$

The difference of these two terms divided by the element mass $\rho A\Delta x$ yields the expression for the rate of heat conduction into the particle from its neighbors:

$$\dot{q} = \frac{1}{\rho}\frac{\partial}{\partial x}\left(k\frac{\partial T}{\partial x} \right) + O(\Delta x). \tag{7.14-6}$$

The last term disappears in the limit as the element volume vanishes.

Similarly, we next compute the rate of energy dissipated to heat by the action of the viscous forces on the end faces, which are in motion relative to the center of mass. The right/left face relative velocities, neglecting terms of $O(\Delta x)^2$, are $\pm e_x(\partial u/\partial x)$ $(\Delta x/2)$. The longitudinal viscous stresses (positive in the direction of the outer normal) are $\pm e_x[\tau_x \pm (\partial\tau_x/\partial x)(\Delta x/2)]$. Therefore, the rate of viscous work dissipated to heat (each term of which must be positive) is

$$\left\{ \left[\tau_x + \frac{\partial\tau_x}{\partial x}\frac{\Delta x}{2} \right] e_x \cdot e_x \frac{\partial u}{\partial x}\frac{\Delta x}{2} \right\} A + \left\{ \left[\tau_x - \frac{\partial\tau_x}{\partial x}\frac{\Delta x}{2} \right](-e_x)\cdot(-e_x)\frac{\partial u}{\partial x}\frac{\Delta x}{2} \right\} A$$

$$= \tau_x \frac{\partial u}{\partial x} A\Delta x + O(\Delta x)^3.$$

By dividing the result by the mass and putting $\Delta x \rightarrow 0$, then

$$\dot{\psi} = \frac{1}{\rho}\tau_x\frac{\partial u}{\partial x}, \tag{7.14-7}$$

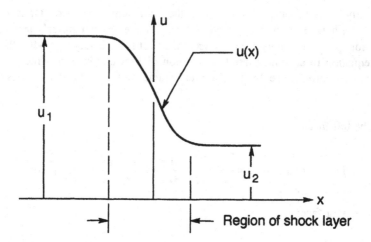

FIGURE 7.14-2   Model for theoretical velocity profile within a shock zone.

exactly. In fact, since the flow is steady, we can make the replacement $\partial/\partial x \to d/dx$ in Equations 7.14-6 and 7.14-7.

In Equation 7.14-5, omitting the left-hand term, we next multiply by ρ and, into the product, substitute Equations 7.14-1, 7.14-6, and 7.14-7 to obtain the following expression:

$$\frac{d}{dx}\left(\dot{\bar{m}}\,h^0 - k\,\frac{dT}{dx} - u\tau_x\right) = 0. \tag{7.14-8}$$

Now the integrated expressions for the velocity, pressure, and temperature functions $u(x)$, $p(x)$, $T(x)$, etc., must be compatible with the Rankine–Hugoniot relations previously obtained. That is, as $x \to \pm\infty$, $u \to u_1$, u_2, $p \to p_1$, p_2, etc., respectively. Furthermore, the gradients du/dx, dT/dx, etc., must vanish in the same limit. Thus, we conceive of a shock essentially as an extremely thin layer (Figure 7.14-2) but whose effects actually extend to infinity on both sides where the flow parameters (u, p, etc.) become asymptotically equal to those in the discontinuous Rankine–Hugoniot model. We can then integrate Equations 7.14-3 and 7.14-8 from $x \to -\infty$, designated by the subscript 1, to an arbitrary point within the shock layer (without subscript). Hence, the integrated momentum equation becomes

$$p + \rho u^2 - \tau_x = p_1 + \rho_1 u_1^2, \tag{7.14-9a}$$

where the upstream viscous stress must vanish. For the other equation we have

$$\dot{\bar{m}}\,h^0 - k\,\frac{dT}{dx} - u\tau_x = \dot{\bar{m}}\,h_1^0, \tag{7.14-9b}$$

in which the temperature gradient far upstream likewise vanishes.

It is noteworthy that this pair of first integrals of the shock differential equations is independent of any state function. Furthermore, it does not involve the introduction of a *constitutive equation* for the viscous stress; that is, the nature of the dependency of the viscous stress on the rate of fluid deformation remains unspecified. On the other hand, the equations are coupled through the appearance of the stress function in both.

INTEGRATION OF EQUATION 7.14-9

As Becker (1968) points out, general solutions of Equation 7.14-9 can be obtained only by numerical methods. To get around this, he instead applies a series of restrictions that allow him to integrate the equations, and then define and calculate a closed-form shock wave "thickness." The solution is therefore applicable only to a restricted class of flows. The restrictions (stated in italics) are applied serially.

1. *The normal viscous stress is specified by the linear stress-rate-of-strain relation* (i.e., by the constitutive relation* that leads to the Navier–Stokes equations for dilute gases):

$$\tau_x = \frac{4}{3}\,\mu\,\frac{du}{dx}. \tag{7.14-10}$$

After dividing Equation 7.14-9b by $\bar{m} = \rho u$, this viscous stress relation** is introduced into both equations to yield

$$p + \rho u^2 - \frac{4}{3}\,\mu\,\frac{du}{dx} = p_1 + \rho_1 u_1^2, \tag{7.14-11a}$$

$$h^0 - \frac{k}{\bar{m}}\,\frac{dT}{dx} - \frac{4}{3}\,\frac{\mu}{\rho}\,\frac{du}{dx} = h_1^0. \tag{7.14-11b}$$

2. *Every particle is in (instantaneous) thermal equilibrium.*** Consequently, any two independent thermodynamic variables, say ρ and T, suffice to specify the thermodynamic state of the gas. Also, with Equation 7.14-1, we can eliminate ρ in favor of u. Thus, if we can solve for u and T as functions of x, we can solve for all of the thermodynamic variables, including the entropy.

3. *Calorically perfect gas relations apply.* Therefore, $h^0 = h + u^2/2 = c_pT + u^2/2$, and Equation 7.14-11b becomes

---

* Strictly speaking, the form of the viscosity law in Equation 7.14-10 is limited to monatomic gases. For more-complex molecules there should be an additive second term called the *bulk viscosity*, which is generally small compared with μ. For the current purpose it can be neglected.

** Note that with Equation 7.14-10 the steady-state form of Equation 7.14-7 becomes $\dot{\psi} = (4\mu/3\rho)$ $(du/dx)^2$. This meets the requirement that the dissipation function be a positive, definite form.

*** This is a straightforward instance of the concept of thermal equilibrium as introduced in Section 3.14.

$$h^0 - h_1^0 = \frac{4}{3} \frac{\mu}{\rho u} \frac{u du}{dx} + \frac{k}{\rho u} \frac{d(c_p T)}{c_p dx},$$

$$= \frac{\mu}{\dot{m}} \left[\frac{4}{3} \frac{d}{dx} \left(\frac{u^2}{2} \right) + \frac{k}{\mu c_p} \frac{dh}{dx} \right].$$

We recognize the coefficient of dh/dx in the preceding relation as the inverse of the Prandtl number, defined as $\text{Pr} \equiv \mu c_p / k$. Therefore, the energy equation becomes

$$h^0 - h_1^0 = \frac{\mu}{\dot{m}} \left[\frac{4}{3} \frac{d}{dx} \left(\frac{u^2}{2} \right) + \frac{1}{\text{Pr}} \frac{dh}{dx} \right]. \qquad (7.14\text{-}12)$$

4. *The Prandtl number is equal to 3/4.* This value was selected because Becker realized that it would provide a convenient analytical simplification for the following, where we have introduced the definition of total enthalpy and arbitrarily inserted h_1^0 into the parenthesis. Since h_1^0 is a constant, the equation is not altered thereby. Using the last equality and cross-multiplying, Equation 7.14-12 reduces to

$$\frac{d(h^0 - h_1^0)}{h^0 - h_1^0} = \frac{3\dot{m}}{4} \frac{dx}{\mu}. \qquad (7.14\text{-}13)$$

The right-hand side of (Equation 7.14-13) is obviously finite for all x. However, as $x \to \pm\infty$, we must have $h^0 \to h_1^0 = h_2^0$. The only possibility for a solution is that the left side must be in the indeterminate form 0/0; thus,

$$h^0 = h_1^0 = \text{constant}, \qquad (7.14\text{-}14)$$

and the flow is isoenergetic throughout.*
5. *The gas is also thermally perfect.* In such a case,

$$h = c_p T = (c_p / R)(p/\rho) = \frac{\gamma}{\gamma - 1} \frac{p}{\rho}.$$

Equivalently,

$$p = \frac{\gamma - 1}{\gamma} \rho h = \frac{\gamma - 1}{\gamma} \rho \left(h^0 - \frac{1}{2} u^2 \right) = \frac{\gamma - 1}{\gamma} \frac{\dot{m}}{u} \left(h^0 - \frac{1}{2} u^2 \right). \qquad (7.14\text{-}15)$$

---

* In consequence, the heat transfer and viscous terms are equal and opposite; i.e., $\dot{q} \equiv -\dot{\psi}$ and the flow is globally adiabatic.

With Equation 7.14-15 we eliminate p in Equation 7.14-11a and solve for

$$\frac{4}{3} \mu \frac{du}{dx} = \frac{\gamma-1}{\gamma} \dot{m}h^0 + \frac{1}{u}\left[\frac{\gamma+1}{2\gamma} \dot{m}u^2 - \left(p_1 + \rho_1 u_1^2\right)u\right].$$

This devolves to

$$\frac{1}{\dot{m}} \frac{2\gamma}{\gamma+1} \frac{4\mu}{3} \frac{du}{dx} = \frac{1}{u}\left[u^2 - \frac{2\gamma}{\gamma+1} \frac{\left(p_1 + \rho_1 u_1^2\right)}{\dot{m}} u + \frac{2(\gamma-1)}{\gamma+1} h^0\right],$$

$$= \frac{1}{u}\left(u - u_1\right)\left(u - u_2\right).$$

(7.14-16)

where u_1, u_2 are roots of the quadratic in the brackets and where we proceed to verify that they are the asymptotic values of the velocity components on either side of the shock that satisfy the normal shock relations.

The constant term of the quadratic is equal to the product of the roots; thus, $u_1 u_2 = 2(\gamma - 1)h^0/(\gamma + 1)$. Now, a point where the flow speed is equal to the local speed of sound is designated $u^* = a^*$. From the energy equation, under the stated restrictions $h^0 = [(\gamma + 1)/2(\gamma - 1)]a^{*2}$. When substituted into the previous relation, we obtain $u_1 u_2 = a^{*2}$, which is Prandtl's relation.* Similarly, it can be shown that the coefficient of the first-order term is $u_1 + u_2$.

6. *The viscosity varies according to the power law** $\mu \sim T^\omega$. From the definition of total enthalpy, we can write $h^0 = c_p T + u^2/2 = c_p T^0 = u_m^2/2$, where $u_m = \sqrt{2c_p T^0}$. Therefore, $T/T^0 = [1 - (u/u_m)^2]^\omega$, according to the power law, which means that this relation is correlated with experiment at the total temperature T^0. Therefore, Equation 7.14-16 becomes

$$\frac{1}{\dot{m}} \frac{8\gamma\mu_0}{3(\gamma+1)} \frac{du}{dx} = \frac{\left(u - u_1\right)\left(u - u_2\right)}{u\left[1 - \left(u/u_m\right)^2\right]^\omega}.$$

Now, put

$$\lambda \equiv \frac{8\gamma\mu_0}{3(\gamma+1)\dot{m}},$$

(7.14-17)

where λ has the dimension of length. Our differential equation can now be written

---

* See problem 7.9.
** This expression is an approximation to Sutherland's viscosity law, which gives satisfactory results over a moderate temperature range near T^0. Typical values of the exponent ω range from 0.5 to 1.0. See Schlichting (1979, p. 329) for further details.

$$\frac{dx}{\lambda} = \frac{u\left[1-\left(u/u_m\right)^2\right]^{\omega}}{\left(u-u_1\right)\left(u-u_2\right)} \, du. \qquad (7.14\text{-}18)$$

Becker presents the solution for several special cases of Equation 7.14-18, of which we shall consider only one.

7. *Viscosity is directly proportional to the temperature, i.e.,* $\omega = 1$. This version of the power law is reasonably accurate for temperatures near T^0, but for lower temperatures tends to underestimate the actual value — the greater the deviation from T^0, the greater the inaccuracy. For the integration it is convenient to make a variable substitution in Equation 7.14-18, such that $v \equiv u/u_m$. We then obtain

$$\frac{dx}{\lambda} = \frac{v\left(1-v^2\right)}{\left(v-v_1\right)\left(v-v_2\right)} \, dv, \qquad (7.14\text{-}19)$$

for which the integral is

$$\frac{x}{\lambda} = \frac{1}{v_1-v_2}\left[v_1\left(1-v_1^2\right)\ln\left(1-v/v_1\right) - v_2\left(1-v_2^2\right)\ln\left(v/v_2-1\right)\right] \qquad (7.14\text{-}20)$$

$$-\left[v^2/2 + \left(v_1+v_2\right)v\right].$$

The constant of integration is omitted since it has only the effect of moving the origin of the x/λ-axis left or right at an arbitrary amount.

A NUMERICAL EXAMPLE

Consider the case of a normal shock in air such that far upstream of the shock: p_1 = 1 atm = 0.101 3 MPa, T_1 = 288.15 K, μ_1 = $1.789\ 9 \times 10^{-5}$ Pa·s, a_1 = 340.294 m/s. Choose M_1 = 2.0, $\gamma = 7/5$; then, from tabulated values, T_1/T^0 = 0.1728; therefore, T^0 = 1252 K at which μ_0 = $4.999\ 9 \times 10^{-5}$ Pa·s. Using the power law for viscosity with $\omega = 1$, we calculate μ_1 = 1.153×10^{-5} Pa·s, which is 35.6% less than the actual value. This means that there will be some unknown error in our results for the shock profile, no doubt less than this percentage. From the tabulated values and the specified data it is routine to obtain the following results: u_m = 102 1 m/s, u_1 = 680.6 m/s, u_2 = 255.2 m/s, v_1 = 0.666 7, v_2 = 0.250 0.

Figure 7.14-3 is a plot of v vs. x/λ in the subject case. Since the curve approaches the values of v_1, v_2 asymptotically, the "thickness" of the shock is truly infinite. However, it is clear that most of the velocity change occurs in a limited region. Therefore, it is customary to define a nondimensional thickness $\Delta(x/\lambda)$, such that $\Delta(v_1 - v_2) / \Delta(x/\lambda) = (du/dx)_{max}$, illustrated on the plot by a straight-line tangent to the velocity function v at its point of maximum slope.

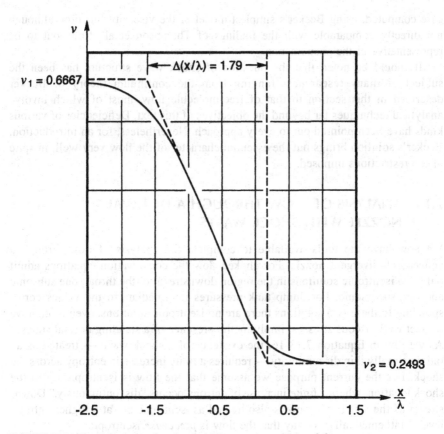

FIGURE 7.14-3 Calculation for shock thickness, case of $M_1 = 2.0$.

In the current case, taken directly from the plot, $\Delta(x/\lambda) = 1.79$. Thus, calculating in turn, the mass flux is $\dot{m} = 833.7$ kg/m²s, and $\lambda = 9.329 \times 10^{-8}$ m. Hence, the shock thickness is $\Delta x = 1.670 \times 10^{-7}$ m, which is about 2.5 times the upstream mean-free-path length (temporarily designated λ') and which, from Table 2.16-3, is $\lambda' = 6.663 \times 10^{-8}$ m. Typically, shock thicknesses are specified in multiples of a mean-free-path length.

It is characteristic that as the upstream density decreases, the shock thickness increases. Also, as the upstream Mach number increases, the thickness tends to decrease. In the limit, as the upstream Mach number tends to unity, the shock thickness tends to infinity.

NOTE ON EXPERIMENTAL STUDIES

To avoid plotting the infinity in the neighborhood of $M_1 \rightarrow 0$ it is customary to plot the inverse of the shock thickness as a function of the upstream Mach number. Thompson, et al. (1983) undertook a comprehensive comparison of theory (using numerical integration of the exact equations) with the experimental measurement of the shock thickness in argon, over a range of Mach numbers. Our result for the one case computed, using Becker's simplest model of the viscosity function, although not directly comparable with the findings of Thompson et al., turns out to be representative of the phenomenon.

It should be noted that the question of shock wave structure has been the subject of many researchers ranging from the continuum theory of Becker described in this section to that of free-molecule flow, most of which involve analytical techniques far beyond the objectives of this text. Deficiencies of various kinds have been pointed out to every approach. Nevertheless, for an introduction, Becker's solution brings out the essential character of the flow very well, in spite of the restrictions imposed.

7.15 ANALYSIS OF FLOW THROUGH A DE LAVAL NOZZLE WITH SHOCK WAVES

We now have the tools available to complete the analysis of flow through a convergent–divergent nozzle. For choked flow the conservation equations admit only two isentropic solutions in the region downstream of the throat, one subsonic and one supersonic. For dump tank pressures intermediate to the values corresponding to these two solutions there are no isentropic solutions, even though we neglect wall viscous stresses, implying the presence of a standing normal shock.* As we saw in Equation 7.13-16, the existence of a shock wave — treated as an infinitesimally thin discontinuity — requires a finite increase in entropy across the shock. For the current purpose we assume that the flow is isentropic up to the shock location, where a finite increase of entropy occurs "discontinuously." Downstream of the shock, the flow is also treated as isentropic but at a higher entropy level. Mathematically, we say that the flow is *piecewise* isentropic.

RELATION FOR SECOND-THROAT AREA

The experiments of Stodola indicate that, for certain values of the downstream pressure, shocks appear in the diverging part of the nozzle. As shown in Section 7.13, the flow upstream of a shock must be supersonic. We use the model of Figure 7.15-1, in which the analysis is restricted to the isoenergetic flow of a perfect gas. The exit pressure p_e — fixed by the dump tank — is such that a normal shock wave is located between the throat A_1^* and the exit A_e. For isoenergetic flow, the total temperature T^0, and the total speed of sound a^0, are constant throughout. However, as shown in Section 7.13, immediately ahead of the shock the flow is supersonic, i.e., $M_1 > 1$, and immediately behind the shock $M_2 < 1$. Furthermore, in passing through the shock the total pressure decreases such that $p_2^0 < p_1^0$. Since the flow is treated as piecewise isentropic, the entire flow downstream of ② is at total pressure p_2^0.

It is simple to show that downstream of the shock the Mach number decreases in the divergent portion of the nozzle. Since

---

* Experiments on short, well-designed nozzles show that wall viscous effects alter the details but not the main features of such flows.

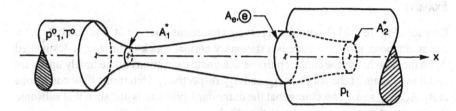

FIGURE 7.15-1 Convergent–divergent nozzle with (fictitious) second throat.

$$\frac{a^0}{a} = \left[1 + \frac{1}{2}(\gamma - 1)M^2\right]^{\frac{1}{2}}, \tag{7.15-1}$$

we can take its logarithmic differential to produce

$$-\frac{da}{a} = \frac{\frac{1}{2}(\gamma - 1)M^2}{1 + \frac{1}{2}(\gamma - 1)M^2} \frac{dM}{M}. \tag{7.15-2}$$

From the Mach number definition

$$dM/M = du/u - da/a. \tag{7.15-3}$$

If Equations 7.15-2 and 7.15-3, and the area–Mach number relation (Equation 7.10-5), are combined by eliminating the terms on the right of Equation 7.15-3, we obtain

$$\frac{dM}{M} = -\frac{1 + \frac{1}{2}(\gamma - 1)M^2}{1 - M^2} \frac{dA}{A}. \tag{7.15-4}$$

Consequently, since the area is increasing, starting from ② the downstream side of the shock, where $M_2 < 1$, the Mach number must continuously decrease.

Now suppose that downstream of ⓔ the duct is modified by the addition of a hypothetical convergent segment, indicated in the figure by dashed lines. At what station (i.e., at what value of the area) would the flow again become sonic? Denote this area by $A_2{}^*$. Then, by applying Equation 7.11-3 on either side of the shock, and noting that $\dot{m}$ = constant throughout, we have

$$A_2^*/A_1^* = p_1^0/p_2^0. \tag{7.15-5}$$

Since the total pressure across a shock always decreases, the area of the second throat must always be greater than the first. In fact, it is not necessary that $A_2{}^*$ be an actual station in the duct. Its introduction provides a convenient reference area that enables one to use the standard tables in calculating the flow in the region behind the shock.

EXAMPLE

Consider Figure 7.11-2 with the reservoir conditions with air at $T = 1000°R$, $p_1^0 = 1$ atm. Suppose, now, that there is a stationary normal shock at the station designated A_2 on the sketch. For the current purpose designate the stations immediately upstream and downstream of the shock as A_{2u} and A_{2d}, respectively. Determine flow conditions at A_1, A_{2u}, A_{2d}, and A_e, assuming that the dump-tank pressure is just such that subsonic flow is maintained downstream of the shock.

Then with $A_{min} = A_1^* = 1.25$ in.^{2}, $A_{2u} = A_{2d} = 1.34$ in.^{2}, $A_e = 2.5$ in.^{2} and since the flow must be subsonic at A_1, $A_1/A_1^* = 1.34/1.25 = 1.072$; thus $M_1 = 0.733$ and $p_1/p_1^0 = 0.696$. Since the flow at A_{2u} must be supersonic, $M_{2u} = 1.313$, $p_{2u}/p_1^0 = 0.354$ and $p_{2d}^0/p_{2u}^0 = A_1^*/A_2^* = 0.977$, where A_2^* is fictitious. From shock calculations $M_{2d} = 0.779$, $p_{2d}/p_{2u} = 1.845$; therefore $p_{2d} = 1.845 \times 0.354 = 0.653$ atm. Also, $A_e/A_2^* = (A_e/A_1^*)(A_1^*/A_2^*) = 1.954$, for which the corresponding (subsonic) Mach number is $M_e = 0.314$. Hence, $p_e = (p_e/p_{2d}^0)(p_{2d}^0/p_{2u}^0)p_{2u}^0 = 0.933 \times 0.977(1) = 0.912$ atm.

EFFECT OF VARYING THE DUMP TANK EXIT PRESSURE

Consider a series of dump tank settings identified by the letters a, b, c,.... The representative axial pressure and Mach number distributions are illustrated in Figure 7.15-2. The flow first becomes choked for the dump tank pressure corresponding to curve c, and is isentropic throughout. For the next setting d, there is no isentropic solution and a shock wave appears; its position depends on the exit pressure, which in turn is governed by the dump tank pressure. As long as the shock wave stands between the throat and the exit, these two pressures must be the same.

As the dump tank pressure is further lowered, curves e and f, the shock wave location moves further downstream. Eventually, the shock stands directly at the exit, where the pressure immediately downstream of the shock, point g, is equal to that of the tank. In all of these solutions the flow enters the tank as a subsonic jet. Figure 7.15-3, reproduced from Howarth (1953), is a presentation of research performed at the National Physical Laboratory in Great Britain. It resulted in a series of schlieren photographs, which can be compared with the preceding illustrations. In cases a through g, there is a rapid mixing on the jet boundary and, possibly, the appearance of secondary shocks, neither of which situation can be handled by one-dimensional theory.

The curve corresponding to i designates the case of purely supersonic flow in the nozzle, without the involvement of shocks, and where the tank pressure and exit pressure are equal. We see that the supersonic jet retains its definition longer than in the subsonic case, but, even so, mixing on the jet boundary rapidly alters the nature of the flow.

For dump tank pressures between the values corresponding to curves g and i the flow adjusts to the dump tank pressure by means of a complex system of oblique shock waves, examples of which can be seen in Figure 7.15-3. The treatment of these mechanisms, which may involve *triple shock intersections* and *slip lines* — lines bounding two regions of parallel flow with different entropy histories — are beyond the scope of this text.

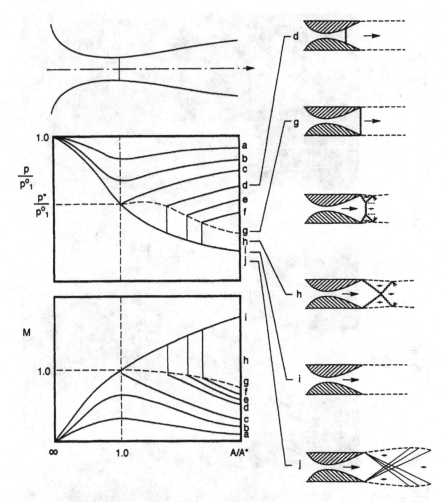

FIGURE 7.15-2 Pressure and Mach number distributions for de Laval nozzle flow at various values of the ratio p_e/p_1^0.

Finally, when the dump tank pressure is less than that of case i, the flow must expand to adjust to the tank pressure. The expansion occurs through a pair of symmetrical fan-shaped regions — called *Prandtl–Meyer expansion fans* — whose study is also a part of two-dimensional, compressible flow theory; typical expansion fans are visible in the lowest photograph of Figure 7.15-3. Such a flow is said to be *underexpanded*.

NOTE ON THE FORMATION AND STABILITY OF NORMAL SHOCK WAVES

It is not possible to explain the process of formation of shock waves by steady-state theory. The process is inherently nonsteady, and its analysis requires the use of the method of characteristics of one-dimensional, nonsteady flow. In a remarkable arti-

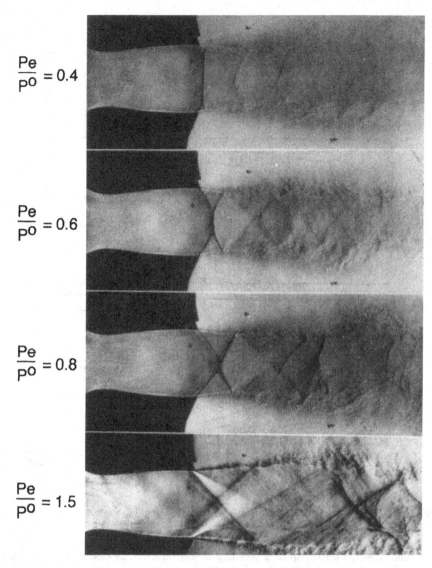

$\dfrac{Pe}{p0} = 0.4$

$\dfrac{Pe}{p0} = 0.6$

$\dfrac{Pe}{p0} = 0.8$

$\dfrac{Pe}{p0} = 1.5$

FIGURE 7.15-3 Schlieren photographs of flow from a two-dimensional supersonic nozzle into a dump tank for different values of the ratio of the exit pressure (p_e) to the upstream reservoir pressure (p_t). (From Howarth, L., Ed., *Modern Developments in Fluid Dynamics*, Vol. I, Clarendon Press, Oxford, U.K., 1953. With permission.)

cle, Kantrowitz (1958) gives a complete description of the process for one configuration.

Kantrowitz also examines the question of shock wave location due either to disturbances swept downstream from ahead of the shock or to disturbances propagated upstream from the subsonic flow behind the shock. In either case, he demonstrates that a shock displaced from its equilibrium position in a divergent duct by an arbitrarily small disturbance is stable; that is, it tends to return to its original location as the disturbance dies down.

On the other hand, he also shows that if it were possible to establish a normal shock in the convergent section of a nozzle — which situation would not violate the conservation equations — that it would be unstable under receipt of a small disturbance. Whichever way the shock were displaced, it would tend to move off to infinity in that direction. Practically speaking, however, a standing normal shock has never been observed in a convergent duct.

ON THE MOMENTUM EQUATION FOR STEADY, COMPRESSIBLE FLOW

For steady, horizontal, mass-conserving flow the momentum equation (Equation 3.20-3) reduces to

$$F_{1,2} = (pA + \dot{m}u)_2 - (pA + \dot{m}u)_1, \tag{7.15-6}$$

where $F_{1,2}$ is the force due to pressure and viscous effects of the boundaries on the fluid contained between stations ① and ②. The quantity $pA + \dot{m}u$ is usually called the *stream force*. Obviously, the force $F_{1,2}$ is given by the change in the stream force between the two stations.

For perfect gases the stream force can be written as a function of the local Mach number, i.e.,

$$pA + \dot{m}u = \dot{m}\left(\frac{pA}{\dot{m}} + u\right) = \dot{m}\left(\frac{p}{\rho u} + u\right). \tag{7.15-7}$$

Therefore, with $u = aM$, $p/\rho = a^2/\gamma$,

$$\frac{p}{\rho u} + u = \frac{a^2}{\gamma}\frac{1}{aM} + aM = \frac{a^0}{\gamma}\frac{a}{a^0}\frac{1+\gamma M^2}{M}. \tag{7.15-8}$$

Thus, since $a^0/a = \left[1 + \frac{1}{2}(\gamma-1)M^2\right]^{1/2}$, the stream force reduces to

$$pA + \dot{m}u = \dot{m}a^0 F(M;\gamma), \tag{7.15-9a}$$

where

$$F(M;\gamma) \equiv \frac{1+\gamma M^2}{\gamma M\left[1 + \frac{1}{2}(\gamma-1)M^2\right]^{1/2}} \tag{7.15-9b}$$

is a tabulated function of Mach number.

Consequently, for isoenergetic flow for which a^0 = constant, Equation 7.15-6 reduces to

$$F_{1,2} = \dot{m}a^0 \left[F_2(M;\gamma) - F_1(M;\gamma) \right]. \tag{7.15-10}$$

Note that Equation 7.15-10 includes viscous effects as long as the flow is isoenergetic. For flows with heat addition between ① and ② the applicable expression is simply

$$F_{1,2} = \dot{m} \left[a_2^0 F_2(M;\gamma) - a_1^0 F_1(M;\gamma) \right], \tag{7.15-11}$$

where the energy equation (Equation 3.19-7) must be invoked to compute a_2^0.

APPLICATION OF THE MOMENTUM EQUATION TO A CONVERGENT–DIVERGENT NOZZLE

Consider the application of Equation 7.15-10 to the portion of the duct shown in Figure 7.11-2 from the station designated A_1 to the station A_e. The reservoir conditions are the same as in the Example immediately following Equation 7.15-5 and uses the same nomenclature. There are three cases: (1) Isentropic, subsonic, choked flow; (2) Isentropic flow which is supersonic and shock-free in the divergent section; (3) Supersonic flow in the divergent section up to a standing normal shock at the station designated A_e.

We need first the Mach numbers at stations A_1 and A_e, which are calculated by now familiar methods and whose values are listed in the following table. In addition we need the mass flow rate and the reservoir speed of sound which are given by Equations 7.11-3b and 7.5-5, respectively. Thus,

$$\dot{m} = A^* C p^0 / \sqrt{T^0} = (1.25/144)(0.01653)(2116)/\sqrt{1000} = 9.601 \times 10^{-3} \text{ slug/sec}.$$

$$a^0 = 49.02 \sqrt{T^0} = 49.02\sqrt{1000} = 1550 \text{ ft/sec},$$

and,

$$\dot{m}a^0 = 9.601 \times 10^{-3} (1550) = 14.88 \text{ lb}.$$

Consequently, from Equation 7.15-10, the force of the duct on the fluid is given by

$$F_{1,e} = 14.88 \left(F_e - F_1 \right).$$

The values for the three cases are shown in the last column. Note that the force exerted by the fluid on the duct is $- F_{1,e}$.

Case	M_1	M_e	F_e	$F_{1,e}$ (lb)
Isentropic choked subsonic flow	0.733	0.309	2.596	14.9
Isentropic supersonic flow in divergent section	0.733	2.20	1.799	2.65
Inviscid flow with shock at station ②	0.733	0.314	2.564	14.0

At first glance, it might appear that if one attempts to use a convergent–divergent nozzle to obtain thrust (which would require $-F_{1,e}$ to be negative) the most desirable of the three is the case of isentropic, choked, subsonic flow. Such an interpretation is not correct, however. In every thruster there is an energy input to the flow (including the water rocket discussed in Section 4.12) and the force on the entire duct configuration from inlet to exit must be analyzed, and not merely a portion as selected in the current example. In other words, neither the directions nor the relative magnitude of $F_{1,e}$ in the current examples are indicative of the thrust obtained in a jet engine, or in a rocket.

7.16 FANNO PROCESSES

We have not yet dealt with the case of wall friction in the flow of a compressible fluid. Except for special cases, such flows can be analyzed only by one-dimensional theory and, then, only approximately. One important case amenable to analysis is a *Fanno process*, which treats the steady flow of a perfect gas through a constant-area duct under the restrictions that it be workless and adiabatic and, therefore, isoenergetic. The adiabatic restriction implies that the pipe is insulated, such that heat conduction through the pipe wall is negligible.

WORKING EQUATIONS

The working relations are readily written down. The continuity equation is given by Equation 7.3-5:

$$\dot{m} = \rho u = \text{constant}. \tag{7.16-1}$$

The most convenient form of the dynamic equation is obtained by rearranging Equation 7.10-2, i.e.,

$$du/u = -\left(1/\gamma M^2\right)dp/p - 2c_f\, dx/D. \tag{7.16-2}$$

For a perfect gas, the thermal equation of state (Equation 7.2-8b) can be combined with the expression for sound speed (Equation 7.5-4) to give

$$p = \rho RT = \rho a^2/\gamma. \tag{7.16-3}$$

For isoenergetic flow of a perfect gas, Equation 7.8-2 becomes, after introducing $c_p T = a^2/(\gamma - 1)$,

$$a^2/(\gamma - 1) + u^2/2 = c_p T^0 = \text{constant.} \qquad (7.16\text{-}4)$$

MACH NUMBER RELATIONS

As might be expected, it is desirable to introduce the Mach number as the independent variable. Taking the logarithmic differentials of Equations 7.16-1, 7.16-3, and $M = u/a$, we have

$$d\rho/\rho + du/u = 0, \qquad (7.16\text{-}5a)$$

$$dp/p - d\rho/\rho - 2da/a = 0, \qquad (7.16\text{-}5b)$$

$$dM/M - du/u + da/a = 0. \qquad (7.16\text{-}5c)$$

Also, differentiating Equation 7.16-4, we obtain

$$\frac{2}{\gamma - 1} \frac{da}{a} + M^2 \frac{du}{u} = 0. \qquad (7.16\text{-}5d)$$

Since Equations 7.16-5 involve four independent relations and five ratios* (e.g., du/u), we can eliminate any trio desired. Retaining M as the independent variable,

$$\frac{du}{u} = \frac{1}{1 + \frac{1}{2}(\gamma - 1)M^2} \frac{dM}{M}, \qquad (7.16\text{-}6a)$$

$$\frac{dp}{p} = -\frac{1 + (\gamma - 1)M^2}{1 + \frac{1}{2}(\gamma - 1)M^2} \frac{dM}{M}. \qquad (7.16\text{-}6b)$$

From Equation 7.16-6 it is seen that in an accelerating/decelerating flow the Mach number is increasing/decreasing in the direction of the flow. Similarly, the pressure is decreasing/increasing. However, these relations do not disclose the conditions under which it is possible to accelerate or decelerate a flow in the presence of wall friction.

---

* The coefficients of dM/M in Equation 7.16-6 are functions of M and γ only. They are referred to as *influence coefficients*, a designation introduced by Shapiro (1953, Vol. I, pp. 226–232). Shapiro treats a more general category of flows and finds it convenient to employ a wider range of dependent variables. For Fanno processes and several other flows, he tabulates comprehensive sets of influence coefficients.

THE INFLUENCE OF FRICTION

We note that in Equation 7.16-2 the *skin-friction coefficient*** is defined as $c_f = 2\tau_w/\rho u^2$. Since both terms in the denominator are variables — in contrast to incompressible flow — this opens the possibility of a nonconstant c_f, a possibility which would impose a heavy analytical burden. It turns out, for the current purpose, that use of the mean value of c_f, over the length of the duct involved, usually provides adequate engineering accuracy.

We next eliminate dp/p in Equation 7.16-2 by introducing Equation 7.16-6b. Thus,

$$\frac{du}{u} = \frac{2\gamma M^2}{1 - M^2} c_f \frac{dx}{D}. \tag{7.16-7a}$$

By similar procedures, we obtain

$$\frac{dM}{M} = \frac{2\gamma M^2\left[1 + \tfrac{1}{2}(\gamma - 1)M^2\right]}{1 - M^2} c_f \frac{dx}{D}, \tag{7.16-7b}$$

$$\frac{dp}{p} = -\frac{2\gamma M^2\left[1 + \tfrac{1}{2}(\gamma - 1)M^2\right]}{1 - M^2} c_f \frac{dx}{D}, \tag{7.16-7c}$$

$$\frac{dp^0}{p^0} = -2\gamma M^2 c_f \frac{dx}{D}. \tag{7.16-7d}$$

Equation 7.16-7a has a limited similarity to the area–velocity relation for a de Laval nozzle. For c_f and dx positive, we conclude that $M \gtrless 1$ implies $du \gtrless 0$; i.e., at a point in a Fanno process where the flow is subsonic/supersonic it is accelerating/decelerating. Combined with the conclusion following Equations 7.16-6 it follows, surprisingly, that in a Fanno process the flow always tends toward the sonic condition, assuming, of course, that appropriate requirements on the downstream pressure are maintained. Particularly, it is unexpected in a supersonic Fanno process that the pressure should increase in the flow direction.

We summarize the trends indicated by the preceding relations in the following table.

	If		$M < 1$	$M > 1$
Then	u		Increases	Decreases
	M		Increases	Decreases
	p		Decreases	Increases
	p^0		Decreases	Decreases
	s/R		Increases	Increases

---

* The skin-friction coefficient c_f is related to the friction factor of hydraulics by $f = 4c_f$.

THE INTEGRATED EQUATIONS OF FANNO FLOW

We now seek to determine the Mach number distribution and pressure distribution in a Fanno flow. By combining Equations 7.16-6a and 7.16-7a and rearranging,

$$c_f \frac{dx}{D} = \frac{1-M^2}{4\gamma M^4 \left[1 + \frac{1}{2}(\gamma - 1)M^2\right]} dM^2,$$

(7.16-8)

a form in which the right-hand side is readily integrated by use of partial fractions.

Choosing, arbitrarily, $x = 0$ as the point with Mach number M, we integrate to the reference station x^*, where the Mach number is unity, i.e.,

$$\int_{x=0}^{x^*} c_f \frac{dx}{D} = \frac{1}{4\gamma} \int_{M^2}^{1} \frac{1-M^2}{M^4 \left[1 + \frac{1}{2}(\gamma - 1)M^2\right]} dM^2,$$

$$= \frac{1}{4\gamma} \left[\frac{1-M^2}{M^2} + \frac{\gamma + 1}{2} \ln \frac{(\gamma + 1)M^2}{2\left[1 + \frac{1}{2}(\gamma - 1)M^2\right]}\right] \equiv G(M;\gamma)$$

(7.16-9)

The left-hand side of Equation 7.16-9 is not integrable without specific information about c_f. In incompressible flow, c_f depends on the Reynolds number and the dimensionless roughness of the pipe wall surface. For a compressible flow we might expect it to depend also on M and γ. Experience has shown — particularly for short pipes — that an acceptable engineering approximation is to replace c_f by its mean value $\bar{c}_f$, as determined by experiment. Other conditions being equal, this guarantees that theory will reproduce the experimental data, at least within the range of parameters (Reynolds number and Mach number) over which the data were taken. Therefore, with

$$\bar{c}_f \equiv \frac{1}{x^*} \int_{x=0}^{x^*} c_f dx,$$

(7.16-10)

Equation 7.16-9 becomes

$$\bar{c}_f x^*/D = G(M;\gamma),$$

(7.16-11)

where x^* is the length of run required to go from a Mach number M to the sonic flow condition, and where $G(M;\gamma)$ is a universal function of Mach number, with γ as parameter, and which can be tabulated once and for all.

For the pressure distribution we have, after integrating Equation 7.16-6b,

$$\frac{p}{p^*} = M^{-1} \left[\frac{(\gamma + 1)/2}{1 + \frac{1}{2}(\gamma - 1)M^2}\right]^{1/2},$$

(7.16-12)

where p and p^* refer to the pressures at $x = 0$ and x^*, respectively. The use of p^* to denote the pressure at the downstream station merely indicates that the local Mach number is unity. The value of p^* at that station is unrelated to the critical pressure value in a de Laval nozzle, which employs the same symbol.

Note that in the limit of infinite upstream Mach number

$$\lim_{M \to \infty} G(M;\gamma) = \frac{1}{4\gamma}\left[\frac{\gamma+1}{2}\ln\frac{\gamma+1}{\gamma-1} - 1\right] \qquad (7.16\text{-}13a)$$

$$= 0.205\,38, \quad \text{for } \gamma = 7/5, \qquad (7.16\text{-}13b)$$

and

$$\lim_{M \to \infty} p/p^* = 0. \qquad (7.16\text{-}13c)$$

The static temperature in a Fanno flow can be obtained from the isoenergetic temperature relation (Equation 7.8-5). With that, and the pressure relation Equation 7.16-12, the flow density can be obtained form the perfect-gas equation of state.

ON THE SKIN-FRICTION COEFFICIENT

There is no theoretical procedure by which c_f can be calculated from first principles in turbulent flow. For subsonic, fully developed, turbulent flow through a smooth pipe, measurements* show that Prandtl's semiempirical, *universal law of friction*,

$$\frac{1}{\sqrt{c_f}} = 4.0\,\log_{10}\left(\text{Re }\sqrt{c_f}\right) - 0.396, \qquad (7.16\text{-}14)$$

derived originally for incompressible flow, provides adequate accuracy. The Reynolds number $\text{Re} = \rho u D/\mu$ is based on u, the local mean flow speed. For most gases Sutherland's law (see Chapter 1) provides an adequate relation for computing the dynamic viscosity, as long as the gas temperature falls within a specified range, and if extremely high pressures are not involved.

EXAMPLE 1

At a specified station ① (see Figure 7.16-1) in a smooth pipe the flow is air at $M_1 = 0.1$, $p_1 = 10$ atm, $T_1 = 500°R$. Determine the length of pipe required to reach $M_2 = 0.2$, and the flow conditions at ②. It is assumed that tables of the various Fanno functions, such as $G(M;\gamma)$, p/p^* are available.

---

* See Shapiro (1954, Vol. II, pp. 1130–1136), who also gives the original references. See Section 6.7 for a discussion of the more recent measurements of the friction factor by Zagarola. The friction factor f is related to the skin-friction coefficient by $f = 4c_f$.

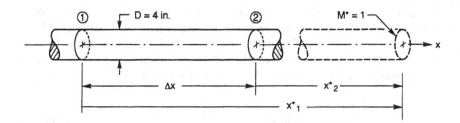

FIGURE 7.16-1   Setup for Fanno flow analysis, subsonic case.

Since $T_1 = 500°R$, $a_1 = 49.02 \sqrt{500} = 1096$ ft/sec. Therefore, $u_1 = 109.6$ ft/sec. From the isoenergetic relation (Equation 7.8-5) $T^0 = 500/0.998 = 501.0°R$, which is constant throughout the flow. Thus, $\rho_1 = p_1/RT_1 = 10(2116)/(1716)(500) = 0.02466$ slug/ft^3. From the isoenergetic, isentropic relation (Equation 7.8-6), the *local* total pressure is $p_1^0 = p_1/(p_1/p_1^0) = 10.0/0.9930 = 10.07$ atm.

For air, Sutherland's viscosity law, Equation 1.11-1, can be put into the form $10^8\mu = 2.270T^{3/2}/(T + 198.6)$, where T is in °R and μ is in lb sec/ft^2. Thus, $\mu_1 = 3.633 \times 10^{-7}$ lb sec/ft^2. Hence, $\mathrm{Re}_1 = \rho_1 u_1 D/\mu_1 = 2.48 \times 10^6$. By use of Equation 7.16-14 $c_{f1} = 0.00251$. We shall use this as a preliminary estimate of $\bar{c}_f$.

In order to determine the pipe length required, we make a hypothetical extension of the pipe to station x_1^*, where the local Mach number is unity. Then, by applying Equation 7.16-11 twice,

$$\Delta x = x_1^* - x_2^* = \left[G(M_1;\gamma) - G(M_2;\gamma) \right] D/\bar{c}_f$$

$$= (G_1 - G_2) D/\bar{c}_f.$$

This yields $\Delta x = (16.73 - 3.633)(1/3)/0.00251 = 1739$ ft. Also, $p_2 = (p_2/p^*)(p^*/p_1)p_1 = (5.456)(10.94)^{-1} 10 = 4.987$ atm. Since $T^0 = 500°R$ throughout, $T_2 = 0.9921(500) = 496.1°R$. As at station ①, therefore, we can calculate $\rho_2 = 0.01237$ slug/ft^3, $u_2 = 218.6$ ft/sec, $p_2^0 = 5.128$ atm. The drastic loss in total pressure is the price we must pay for moving gas against the wall friction forces.

It is important in any of these calculations to justify the estimated value of $\bar{c}_f$. The simplest approach is to note that, since $\rho u D$ is constant, $\mathrm{Re}_2/\mathrm{Re}_1 = \mu_2/\mu_1$. But, since viscosity in a dilute gas depends only on temperature, and since temperature is essentially constant over Δx, Re is also constant. Consequently, there is no significant variation of c_f between ① and ②.

DISCUSSION

In the preceding example the length of pipe required to reach $M^* = 1$ from $M_1 = 0.1$ is $x_1^* = G_1 D/c_f = 16.73(1/3)/0.00251 = 2220$ ft. Invariably, the question surfaces: What happens if the pipe is extended beyond the length at which $M = 1$? It turns out that this question has no answer; in fact, it is the wrong question. Rather we must first answer the question of how to establish the flow at station ①.

If we assume that the flow originates in a reservoir of stagnation temperature T^0, stagnation pressure p_r^0, and negligible velocity, we can then visualize the upstream end of the pipe connected to the reservoir by a short, well-designed, convergent nozzle that has negligible losses. Then, we can ask what length of pipe connected to the nozzle is required to establish a given *subsonic* Mach number at the pipe entrance — assuming that the appropriate downstream pressure is imposed — such that the downstream end of the pipe is just sonic? If we require the upstream station also to be sonic, the answer is zero pipe length. This is intuitively obvious from our studies of a de Laval nozzle. At the pipe inlet the total temperature and total pressure would remain at T^0 and p_r^0, respectively, but the static pressure would be the critical pressure for isentropic flow through a de Laval nozzle, given by Equation 7.7-6.

Then, to achieve a lower (subsonic, of course) entry Mach number, the length of pipe required is given by Equation 7.16-11. Correspondingly, the pressure p^* at the downstream station must be decreased according to Equation 7.16-12 to maintain sonic conditions. This could be achieved by means of a variable-pressure dump tank. As the entry Mach number decreases, and approaches zero, the length of pipe required approaches infinity. Simultaneously, the dump tank pressure required approaches zero.

Therefore, the flow between any two subsonic Mach numbers $M_1 < M_2 < 1$ can be thought of as a segment of an entry flow at specified M_1, with a dump tank pressure at station ② adjusted appropriately. In any case, it should be realized that the one-dimensional relations utilized in this analysis are only approximate, and that real gas effects, and the neglected effects such as heat transfer, will modify the actual flow attained, particularly when Mach numbers near unity are involved.

EXAMPLE 2

It is desired to establish a supersonic Fanno flow of nitrogen in a smooth, straight pipe such that $M_1 = 4.0$, $p_1 = 1$ atm, $T_1 = 200$ K, $D_1 = 0.2$ m. Determine the inlet geometry required to establish such a flow, and compute the corresponding reservoir conditions assuming that the pipe is completely insulated. Determine, also, the longest possible run of the pipe to maintain supersonic flow, and the exit flow conditions.

To establish an initial supersonic flow we employ a short convergent–divergent nozzle, in which viscous effects are assumed to be negligible. Thus, from reservoir to ①, shown in Figure 7.16-2, the flow is assumed to be isentropic as well as isoenergetic. For $\gamma = 7/5$, $M_1 = 4.0$, $A_1/A^* = 10.72$. The throat diameter, therefore, is $D^* = D_1/(10.72)^{1/2} = 0.061\ 09$ m.

The reservoir conditions are determined by the isoenergetic and isentropic relations. $T^0 = T_1/(T_1/T^0) = 200/0.2381 = 840.0$ K. $p_1^0 = p_1/(p_1/p_1^0) = 1/(0.006\ 586) = 151.8$ atm.

For nitrogen $\mathbf{R} = 8.314 \times 10^3/28.01 = 296.8$ m²/s² · K. Therefore, $\rho_1 = 1.013 \times 10^5/296.8(200) = 1.707$ kg/m³, $a_1 = [1.4(296.8)(200)]^{1/2} = 288.3$ m/s. Hence, $u_1 = 1\ 153$ m/s. For N_2 the Sutherland law constants in SI are $S = 106.7$ K, $T_r = 273.1$ K, $\mu_r = 16.63 \times 10^{-6}$ Pa·s. These result in the expression $10^6\mu = 1.400T^{3/2}/(T + 106.7)$,

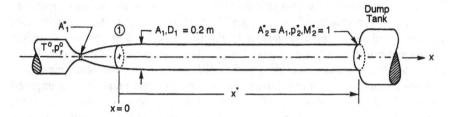

FIGURE 7.16-2 Setup for Fanno flow analysis, supersonic case.

where T is in K and μ is in Pa·s. Thus, $\mu_1 = 12.91 \times 10^{-6}$ Pa·s, $\text{Re}_1 = 1.707(1\ 153)(0.2)/12.91 \times 10^{-6} = 3.05 \times 10^7$, for which $c_f = 0.001\ 74$. Again using this as an estimate for $\bar{c}_f$, Equation 7.16-11 yields $x^* = (0.158\ 3)(0.2)/(0.001\ 74) = 18.2$ m.

At the exit, where the flow is sonic, $p^* = p_1/(0.133\ 6) = 7.485$ atm $= 7.583 \times 10^5$ Pa, $T^* = [2/(\gamma+1)]T^0 = 700$ K, $\mu^* = 32.14 \times 10^{-6}$ Pa·s, and $\text{Re}_2 = \text{Re}_1(32.14/12.91) = 7.59 \times 10^7$, for which $c_f = 0.001\ 538$. By using as mean the average of the two c_f values, $\bar{c}_f = 0.001\ 639$, for which $x^* = 19.3$ m, a change of only 6%.

DISCUSSION

Because of the high flow losses (reflected by the decrease in total pressure), it is generally undesirable to utilize supersonic flow for the transport of a gas. Even when necessary, there are limitations. Rearrangement of Equation 7.16-11 produces

$$x^*/D = G(M;\gamma)/c_f,$$

where x^*/D is the number of pipe diameters to reach sonic conditions from an initial flow Mach number M. For $\gamma = 7/5$, and for a representative skin-friction coefficient $\bar{c}_f = 0.002$, then the upper limit on possible pipe lengths is given in the following table for supersonic flow.

M	1.5	2.0	3.0	5.0	10	∞
x^*/D	17.0	38.1	65.25	86.75	98.35	102.7

7.17 ISOTHERMAL FLOW

For the steady, workless, isothermal (defined as T = constant) flow of a perfect gas through a constant-area, horizontal duct, the analysis follows the pattern of Fanno flow. Leaving details to the reader, the following relations are obtained:

$$\frac{du}{u} = \frac{dM}{M} = -\frac{dp}{p} = -\frac{d\rho}{\rho} = \frac{2\gamma M^2}{1-\gamma M^2} c_f \frac{dx}{D}. \qquad (7.17\text{-}1)$$

Analogous to Equation 7.16-7a, Equation 7.17-1 defines an asymptotic Mach number

$$M_f \equiv 1/\sqrt{\gamma}$$

$$= 0.845, \quad \text{for } \gamma = 7/5,$$

(7.17-2a)

to which every isothermal flow must tend. From Equation 7.17-1 we see that for $0 < M < M_f$ the flow accelerates, the Mach number increases, and the pressure and density decrease in the flow direction. By combining the state, and continuity equations, we find that

$$u/u_f = M/M_f = p_f/p,$$

(7.17-2b)

where the subscript f denotes the value at the station corresponding to M_f. On the other hand, for $M > M_f$ the flow decelerates, and the pressure and density increase.

From the definition of T^0, we have

$$\frac{dT^0}{T^0} = \frac{(\gamma-1) M^2}{1+\frac{1}{2}(\gamma-1)M^2} \frac{dM}{M},$$

(7.17-3)

so that T^0 increases/decreases for $M \lessgtr M_f$. Under the restrictions cited, the energy equation (Equation 3.18-2) becomes

$$\frac{Dh^0}{Dt} = c_p u \frac{dT^0}{dx} = \dot{q},$$

(7.17-4)

where $\dot{q}$ is the rate of heat addition per unit mass due to conduction from the pipe wall. If Equations 7.17-3, 7.17-4, and 7.17-1 are combined, the following expression for $\dot{q}$ is obtained:

$$\dot{q} = \frac{2\gamma a^3 M^5}{1-\gamma M^2} \frac{c_f}{D}.$$

(7.17-5)

Thus, heat is transferred to/from the gas for $M \lessgtr M_f$. However, the heat transfer rate required by Equation 7.17-5 to maintain isothermal flow grows so large as M increases, even for Mach numbers as small as $M = 0.1$, that its realization is highly improbable, except under extraordinary conditions. Thus, in practical cases, isothermal flow is restricted to very small Mach numbers.

Integration of Equation 7.17-1 leads to the definition of x_f, the pipe length required to reach M_f from a prescribed Mach number M. By defining

$$H(M;\gamma) = \tfrac{1}{4}\left[\left(1 - \gamma M^2\right)/\gamma M^2 + \ln \gamma M^2\right], \tag{7.17-6}$$

then

$$\bar{c}_f x_f / D = H(M;\gamma). \tag{7.17-7}$$

As usual, $H(M; \gamma)$ is a standard Mach number function that has been tabulated.

Discussion

In the transmission of natural gas, which is essentially pure methane (CH_4) since the denser hydrocarbons are separated out early in the distribution network,* the pipe is usually buried about 4 ft below the surface. At this depth the ambient temperature of the ground is essentially constant in a range between 40 and 60°F, depending on the region and the season. Consequently, the gas temperature, due to heat conduction between the ground and the pipe, quickly adjusts (approximately) to the ambient value, and the flow can be analyzed as isothermal, to a high degree of accuracy.

To reduce flow losses (which have to be made up by pumping), it is desirable to keep the mean flow speed as small as possible for a specified mass flow rate. This means that high pumping pressures are required. Typical values range from 1000 psia (68 atm) to 4000 psia (272 atm) for piping near the well head. At these high pressures, gas imperfections, due to the close packing of the molecules, become significant, and must be considered if high accuracy in the analysis is required.

Assuming a flow temperature of 40°F = 500°R, and neglecting gas imperfections, the speed of sound of CH_4 (for which $\gamma = 4/3$, $\overline{m} = 16.04$, $R = 4.972 \times 10^4/16.04 = 3100$ ft²/sec² °R) is $a = [(4/3)(3100)(500)]^{1/2} = 1438$ ft/sec. The upper range of acceptable speeds for gas transmission is 10 to 20 ft/sec. The corresponding Mach number range is 0.0070 to 0.0139. By no means, however, can such a low-Mach-number flow be considered to be incompressible** as is demonstrated in the following problem.

Example

Methane (treated as a perfect gas, with $\gamma = 4/3$, $\overline{m} = 16.043$) leaves pumping station ① at $p_1 = 1000$ psia $= 6.895 \times 10^6$ Pa, $u_1 = 4$ m/s, $T_1 = 278.0$ K. Determine the distance to pumping station ②, and the pressure p_2, if the flow speed at ② is $u_2 = 6$ m/s. The pipe diameter is $D = 0.4$ m and the pipe walls are smooth. At ② pumps

---

* For example, a chromatograph analysis, reported by Diehl (1983) private communication, taken in the Niagara Mohawk Power Corp. natural gas system, at its Lawrenceville, PA, station on May 5, 1982, consisted of the following components with the percentage in parentheses: methane (96.6231), ethane (1.8852), propane (0.2272), isobutane (0.0577), n-butane (0.0537), isopentane (0.0267), n-pentane (0.0166), hexane (0.0159), oxygen (0.0068), nitrogen (0.4129), carbon dioxide (0.6742).

** This condition may be contrasted with an aircraft flying at 200 mph in air where $a = 1117$ ft/sec, i.e., at $M = 0.26$. The errors involved by treating such a flow as incompressible are truly negligible.

restore the original pressure in a polytropic compression, for which pv^n = constant, $n = 1.2$. Determine the temperature of the gas as it leaves the pumps. The gas is then passed through heat exchangers, at constant pressure, to restore the original temperature. Determine the required rate of heat removal.

For the conditions specified $R = 518.25$ m²/s² · K, $a_1 = 438.3$ m/s, $M_1 = 0.009$ 126, $M_2 = 0.013$ 69. In solving this problem it is useful to note, first, that at these low Mach numbers there is no significant difference between the local total pressure and the local static pressure. For example, at the station where $M_2 = 0.013$ 69, $p_2^0/p_2 \approx$ 1.0001. Like the static pressure, the total pressure decreases in the flow direction.

For methane treated as a dilute, perfect gas, the appropriate form of Sutherland's law* is $10^7\mu = 9.473T^{3/2}/(T + 137.7)$, with T in K and μ in Pa·s. Thus, $\mu_1 = 105.63 \times 10^{-7}$ Pa·s. The density is $\rho_1 = 6.895 \times 10^6/(519.6)(278.0) = 47.85$ kg/m³. Therefore, $Re_1 = 47.85(4)(0.4)/105.63 \times 10^7 = 7.25 \times 10^6$, for which $c_{f1} = 0.002$ 122. Since the flow is isothermal, Re is constant throughout; hence, $\bar{c}_f = c_{f1}$.

From Equation 7.17-7, written for M_1 and M_2, then $\Delta x = (2\ 248.8 - 998.1)0.4/0.002\ 122 = 235.8$ km $= 146.5$ mi. At ②, by Equation 7.17-2b, $p_2 = (M_1/M_2)p_1 = (2/3)p_1 = 4.597 \times 10^6$ Pa.

The mass flow rate through the pipe is constant at $\dot{m} = \rho_1 u_1 A = 47.85(4)(\pi)(0.4^2)/4 = 24.05$ kg/s. The work done,** per unit mass, to compress gas at state p_2, $T_2 = T_1$, to pressure $p_3 = p_1$, in a polytropic process, is

$$-\int_{v_1}^{v_2} p\,dv = RT_1\left(1 - p_{21}^{(n-1)/n}\right)/(n-1)$$

$$= 518.25(278.0)\left[1 - (2/3)^{1/6}\right]/0.2$$

$$= 4.707 \times 10^4 \ m^2/s^2.$$

This value multiplied by the mass flow rate gives the required power; i.e., $P = 4.707 \times 10^4 (24.05) = 1.132\ 1 \times 10^6$ J/s, which is slightly less than 1600 hp.

The temperature after compression is given by

$$T_3 = T_2\ p_{32}^{(n-1)/n} = T_1\ p_{12}^{(n-1)/n} = 278.0(1.5)^{1/6} = 297.4\ K,$$

which is also the total temperature at ③ to a close approximation. Thus, the rate of heat removal is

$$\dot{m}c_p\left(T_3 - T_1\right) = \dot{m}\left(T_3 - T_1\right)\gamma R/(\gamma - 1) = 24.05(19.4)(4)(518.25)$$

$$= 9.672 \times 10^6\ J/s.$$

---

* According to the data in Stephan and Lucas (1979, p. 161), the effect of treating methane as a perfect gas at this pressure level yields a viscosity coefficient which is about 20% low. The corresponding value for c_f is only about 3% low.

** See any standard text in engineering thermodynamics, e.g., Van Wylen and Sonntag (1976, p. 224).

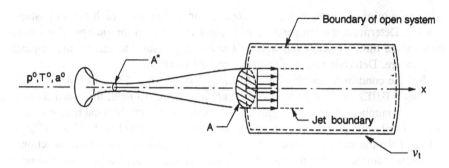

FIGURE 7.18-1 Setup for supersonic inflow problem.

That the required rate of heat removal is greater than the energy input during the compression process is confirmation that heat is transferred to the flow from the ambient heat source (i.e., the ground) during its passage from ① to ②.

7.18 THE INFLOW PROBLEM

Referring to Figure 7.18-1, consider a setup consisting of a convergent–divergent nozzle supplying a tank of fixed volume V_t. The inlet to the nozzle has available an infinite supply of gas (e.g., the atmosphere) at a known temperature and pressure, denoted p^0, T^0, respectively. The corresponding speed of sound is a^0. We suppose that the tank has been evacuated to some low, initial pressure p_i, such that when a valve (not shown) is opened, and that after the starting transients die down (experimentally, an extremely short period), steady choked flow of constant-mass flow rate $\dot{m}$ is established in the nozzle. This flow rate will last for some definite interval, during which a shock system is established in the jet within the tank, which is initially underexpanded. As the tank pressure builds up the jet shock system continuously changes until the point is reached when a normal shock stands at the nozzle exit. This process corresponds to the flow configurations shown in Figure 7.15-3, progressing from the bottom to the top figure. Then, within a relatively short time, the shock moves upstream to the throat and the whole flow becomes subsonic. For practical purposes the *supersonic running time* is defined as the interval between starting and the time at which the normal shock stands at the nozzle exit.

APPLICATION OF ENERGY BALANCE FOR AN OPEN SYSTEM

Let $m = m(t)$ be the instantaneous mass in the tank and m_i the initial mass. For simplicity, we assume that after evacuating the tank there is sufficient time to establish thermal equilibrium with the surroundings. In other words, the initial tank temperature is also T^0. During the inflow, however, there is no heat conduction through the tank walls so that the process is adiabatic.

To determine the behavior of the gas within the tank we need appeal only to the principle of conservation of energy, in this case applied to an *open system* where the only exchange with the surroundings is the influx of mass and energy received from the nozzle. Since it is evident that the incoming gas must compress the initial volume, it follows that the temperature within the tank must rise. Clearly, the mean gas velocity in the tank is zero, although the incoming gas may generate eddies and swirls within.

A word statement of the energy balance follows:

$$\begin{Bmatrix} \text{Rate of change of} \\ \text{internal energy of} \\ \text{instantaneous mass} \\ \text{within tank} \end{Bmatrix} = \begin{Bmatrix} \text{Rate of influx of} \\ \text{energy received} \\ \text{from nozzle} \end{Bmatrix}. \qquad (7.18\text{-}1a)$$

In equation form this becomes, for a perfect gas,

$$d\left(mc_vT\right)/dt = \dot{m}c_pT^0. \qquad (7.18\text{-}1b)$$

An objection can be raised to this last equation on the basis that the mixing process would not occur rapidly enough to ensure instantaneous thermal equilibrium within the tank. This is undoubtedly true, and thus the temperature $T = T(t)$ represents the uniform temperature that the fluid within the tank at any given time *would* acquire if the flow were to be terminated and the mixing process were allowed to continue until attainment of equilibrium, without exchanging energy with the environment by heat conduction.

Equation 7.18-1b can be rewritten in the form of an exact differential, and integrated from an initial tank condition m_i, $T_i = T^0$; i.e.,

$$\int_{m_i,T^0}^{m,T} d\left[m\left(\gamma T^0 - T\right)\right] = 0, \qquad (7.18\text{-}2)$$

which becomes

$$m\left(\gamma T^0 - T\right) = m_i\,T^0\left(\gamma - 1\right).$$

By rearranging,

$$T/T^0 = \gamma - \left(\gamma - 1\right)m_i/m, \qquad (7.18\text{-}3)$$

where T and m are both functions of time.

A remarkable result emerges from Equation 7.18-3. If the tank is initially completely evacuated, $m_i = 0$, and the tank equilibrium temperature* becomes $T =$

---

* In the 1960s a mode of high-speed ground transportation was proposed involving a vehicle traveling in an evacuated tube. Were the vehicle — within which was maintained its own artificial atmosphere — to suffer damage and lose its airtightness, it was proposed to flood the tunnel with outside air through quick-opening valves. Although this might prevent suffocation, the inside temperature would rise to a maximum of γT^0. If the outside air were 40°F, this raises the possibility of attaining inside temperatures as high as $1.4 \times 500 = 700°R$, or 240°F, a less-than-happy alternative.

γT^0 for all time. To one familiar with the famous Joule experiment, in which gas compressed in one chamber at a fixed temperature is allowed to flow into an adjacent evacuated chamber, resulting in a final equilibrium state with both chambers at the same (lower) pressure but at the *original* temperature, the prediction in Equation 7.18-3 of a higher temperature in the inflow problem may seem surprising. The difference is that in the Joule experiment, which involves a closed system, energy conservation requires that the internal energy per unit mass of the initial and final equilibrium states be the same. For the inflow problem, which involves an open system, the inrushing air does work on the gas already within the tank, resulting in a higher final temperature.

DETERMINATION OF TANK PRESSURE

Since during choking, the mass rate of flow is a constant, the instantaneous tank mass is given by

$$m = m_i + \dot{m}t. \tag{7.18-4}$$

For a perfect gas

$$m_i = \rho_i V_t = \left(p_i / R \, T^0\right) V_t, \quad m = \left(p / RT\right) V_t;$$

thus,

$$m_i / m = p_i T / p T^0, \tag{7.18-5}$$

which can be combined with Equation 7.18-3 to produce

$$p / p_i = (\gamma - 1) / \left[\gamma T^0 / T - 1\right]. \tag{7.18-6}$$

From Equation 7.11-3 the mass flow rate during choking is

$$\dot{m} = A^* p^0 \sqrt{\frac{\gamma}{RT^0}} \left(\frac{2}{\gamma + 1}\right)^{(\gamma+1)/2(\gamma-1)} = C_1 A^* p^0 / a^0, \tag{7.18-7}$$

where, for convenience, we put

$$C_1 \equiv \gamma \left(\frac{2}{\gamma + 1}\right)^{(\gamma+1)/2(\gamma-1)} \tag{7.18-8}$$

When Equations 7.18-3 through 7.18-8 are combined, the following expression results for the tank pressure, as long as the flow is choked:

$$p/p^0 = p_i/p^0 + \left(C_1 A^* a^0 /V_t\right)t. \qquad (7.18\text{-}9)$$

APPLICATION — RUNNING TIME FOR A VACUUM-EXHAUST, SUPERSONIC WIND TUNNEL

Figure 7.18-1 illustrates a useful scheme for an intermittent, supersonic wind tunnel design that requires only several low-power vacuum pumps to operate, compared with the same test section in a closed-circuit, continuously operating tunnel that would require an impressive bank of compressors.

The running time is fixed (approximately) at the instant when the tank pressure given by Equation 7.18-9 increases to the value of the total pressure behind a normal shock standing at the nozzle exit. This value, of course, depends on the nozzle area ratio A/A^*, which corresponds to the Mach number in the nozzle ahead of the normal shock. The appropriate expressions for the area ratio in terms of the Mach number and pressure are given by Equations 7.11-4 and 7.11-5, respectively. The total pressure p_2^0 behind a normal shock is given by Equation 7.13-15; this must be equated to p in Equation 7.19-9. Correspondingly, p^0 in Equation 7.18-9 is equal to p_1^0 for the shock flow.

EXAMPLE

The RPI vacuum-exhaust tunnel has an $M = 3$ test section and an exhaust tank volume of $V_t = 2278$ ft^3. Its throat area is $A^* = 1 \times 4$ in.^{2} If the tank is pumped down to 29.0 in. Hg of vacuum, compute the running time assuming standard atmospheric conditions at inlet ($p^0 = 29.92$ in. Hg, and $T^0 = 519°$R).

We have the following:

$$a^0 = 49.02\sqrt{T^0} = 117 \text{ ft/sec};$$

$$p_i/p^0 = (29.92 - 29.0)/29.92 = 0.0307;$$

$$C_1 = 1.4/(1.2)^3 = 0.8102, \quad \text{for } \gamma = 7/5;$$

$$p/p^0 = p_2^0/p_1^0 = 0.3283, \quad \text{from tables of Equation } 7.13\text{-}15.$$

Thus,

$$t = 2278(0.3283 - 0.0307)/(4/144)(0.8102)(1117)$$

$$= 26.7 \text{ sec, maximum.}$$

In practice, viscous flow losses tend to decrease the attainable running time, perhaps 10% or more. On the other hand, if a properly designed second throat and diffuser were introduced between the test section and the tank, significant increases in the running time could be attained. The penalty one pays for the intermittent design are the requirements of a very large vacuum tank and a relatively large pumping time to evacuate it for each run.

PROBLEMS

7.1.

    a. In Equation 7.5-2 the expression for the speed of sound $a^2 = (\partial p/\partial \rho)_s$ was derived. Starting from first principles (e.g., first law of thermodynamics, entropy equation, definitions of specific heats, etc.), show that an equivalent — but sometimes more convenient — expression is

$$a^2 \equiv (\partial p/\partial \rho)_s = \gamma(\partial p/\partial \rho)_T,$$

    where

$$\gamma \equiv c_p/c_v = 1 + (T/c_v)\,(\partial v/\partial T)_p\,(\partial p/\partial T)_v,$$

    and where no restrictions are involved on the form of the equation of state or on the coefficients of specific heat. (In other words, the restrictions of a thermally or calorically perfect gas are not invoked.)

    b. In the analysis of a flow that is neither thermally nor calorically perfect, a certain combination of thermodynamic terms resulted. Show that the combination is nothing more than the sound speed squared, i.e.,

$$a^2 \equiv (\partial p / \partial \rho)_s = pv\{1 - [\partial(pv) / \partial h]_s\}^{-1}.$$

7.2. Various equations have been proposed for the thermal equation of state of a gas that reproduce more accurately the properties in the neighborhood of the critical point. Several are

$$p = \frac{RT}{v - b_1}\, e^{-a_1/Tv} \text{ (Dieterici)}, \quad p = \frac{RT}{v - b_2} - \frac{a_2}{Tv^2} \text{ (Berthelot)},$$

$$p = \frac{RT}{v - b_3} - \frac{a_3}{v^2} \text{ (van der Waals)}.$$

    a. Determine the constants a_n, b_n for the three cases in terms of the critical point values v_c and T_c.

    b. By combining the relations in Problem 7.1a show that an expression for the critical point speed of sound a_c can be obtained in the form $a_c^2/RT_c = \text{(const.)}R/c_v$ and determine the numerical value of the ratio in each case.

    Note: In evaluating certain of the derivatives, indeterminate forms may arise; if so, they must be treated accordingly.

7.3. In an attempt to identify an unknown gas (assumed to be thermally and calorically perfect) certain measurements were made and the following data obtained:

1. At p_1 = 2116 lb/ft^2, T_1 = 510°R, the specific volume is v_1 = 358.6 ft^3/slug.

2. In an isentropic transformation from state ① we have p_2 = 4232 lb/ft^2, v_2 = 213.2 ft^3/slug. Determine the following: ratio of specific heats, specific gas constant, coefficient of specific heat at constant volume, molecular weight, speed of sound at T_1, stating the units of each.

7.4. A gas obeys (over a limited range of pressures and temperatures) the thermal equation of state $p(v - b) = RT$ with R = 1.243 × 10^3 ft^2/sec^2 °R, c_v = 1.864 × 10^3 ft^2/sec^2 °R, where all quantities are expressed in engineering units.

a. Measurements show when p = 2000 lb/ft^2, T = 200°R that v = 126 ft^3/slug. Determine b and give its units.

b. Does the relation $c_p - c_v = R$ apply? If so, prove it. If not, derive the applicable relation.

c. Does $\gamma = c_p/c_v$ apply? If not, determine the appropriate relation. In either case, determine the value of γ.

d. Determine the appropriate expression for the speed of sound, and compute its value when p = 2000 lb/ft^2, T = 500°R.

7.5. In the flow of a perfect gas (γ = 7/5) through a duct, instruments monitor the pressure and density at a fixed point as a function of time. These functions, for the current purpose, can be approximated as $\rho = \rho_0 e^{-at}$, $p = p_0(1 + bt)$, where a and b are constants with dimension 1/*time*. For the tabulated data at the fixed point, calculate the speed of sound when t = 1 sec.

t (sec)	ρ (slug/ft^3)	p (lb/ft^2)
0	0.002 ≡ ρ_0	2000 ≡ p_0
2	0.001	1000

7.6. You are asked to design a de Laval nozzle to produce a supersonic jet (free of shocks) that has the highest possible exit speed u_e. Specifications call for an upstream reservoir that is restricted to 150 atm at standard atmospheric temperature. Ambient pressure of the jet is to be 1 atm. For a preliminary analysis it is decided to consider four different gases (treated as perfect), H_2, He, N_2, and Kr.

a. At the exit station which gas produces the highest total temperature?

b. Which gas produces the highest jet speed? Compute its value.

c. For the gas in part (b), determine the required throat area if A_e = 10 in.^{2}

d. Would H_2 or N_2 involve the higher mass flow rate? Explain.

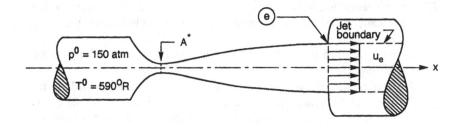

7.7. A rocket-launch booster design is proposed that involves evacuating the launch tube to a low pressure with the tube sealed at the top with a frangible (easily broken) diaphragm. At the tube bottom a rocket weighing 32,000 lb sits within a sabot (a low-friction, sliding, hollow piston that separates from the rocket after emerging from the tube). At launch, air is allowed to flow freely into the nozzle at the bottom. If the air pressure atop the vehicle is treated as negligible, the pressure on the bottom causes it to accelerate to moderately high speeds within a tube of moderate length. The advantage of such a system, of course, is that a relatively low powered pump could evacuate such a tube by working over a long period — a day perhaps — whereas the launching process would be able to extract a large amount of energy from the atmosphere in a short time interval thereby reducing the energy required to be produced by the rocket motor itself.

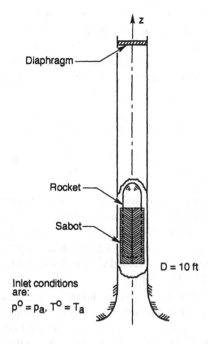

a. Write down the equation of motion of the rocket–sabot combination.

b. Assume that the inflow (truly a nonsteady flow) can be treated as quasi-steady, isentropic, and isoenergetic. Then, determine an expression for the pressure on the lower face of the rocket–sabot.

c. Determine also the rocket acceleration in "g's."

7.8. A rocket is being propelled horizontally to the left such that its acceleration is constant and equal to $-B$ ft/sec (B being a positive constant) relative to an inertial observer. Show that in the frame of an observer who moves with the rocket and for whom the flow is one dimensional, steady, isentropic, and shock-free, without mass creation, that choking cannot occur at the minimum area. Also determine the relative position of the choking station with respect to A_{min}.

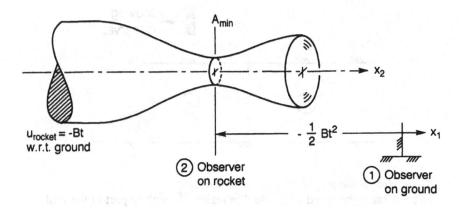

Hint: Use Equation 3.3-4 to obtain the appropriate form of the dynamic equation in the frame of ②.

7.9.

a. Show that alternative forms of the Rankine–Hugoniot equation (Equation 7.13-6) are

$$h_2 - h_1 = \tfrac{1}{2}(v_2 + v_1)(p_2 - p_1),$$

$$e_2 - e_1 = \tfrac{1}{2}(p_1 + p_2)(v_2 - v_1).$$

b. For a thermally and calorically perfect gas flow through a normal shock prove *Prandtl's relation*:

$$u_1 u_2 = a^{*2}.$$

7.10. Using Equation 7.4-5 and the relations developed in Section 7.13 for a perfect gas, show that the disturbance speed c of a shock wave is related to the speed of sound ahead of the shock a_1 and the velocity jump $\Delta u \equiv u_1 - u_2$ by

$$c/\Delta u = \frac{\gamma+1}{4}\left\{1+\sqrt{1+\left[4a_1/(\gamma+1)\Delta u\right]^2}\right\}.$$

7.11. An observer is moving to the right with respect to the wall of a constant-area duct at 300 ft/sec. The observer observes a normal shock that is moving to the left with respect to the observer at the speed of 2400 ft/sec. In regions ① and ②, the temperatures of the gas (air) are 504°R and 851°R, respectively.

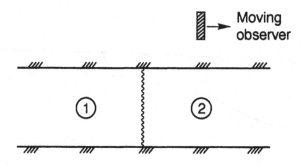

a. What is the propagation speed of the shock?
b. What is the speed of the fluid in region ① with respect to the wall?
c. What is the total pressure of region ① which would be computed by the observer if $p_1 = 1000$ lb/ft²?
d. What is the static pressure in region ②?

7.12. At a certain instant of time a uniform duct containing air is subjected to the passage of a pair of shock waves that divides the flow into three distinct regions as shown.
a. Is the flow steady or unsteady for an observer fixed on the wall? Explain.
b. For the shock dividing regions ① and ②, compute: the shock speed with respect to the wall, and indicate the direction as well and furnish an explanation; the shock propagation speed; also the flow speed in ② relative to the wall.
c. For the shock dividing regions ② and ③, compute the related quantities as in part (b) as well as the total pressure in ③ for an observer moving with the second shock.

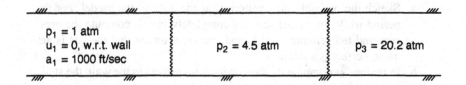

7.13. In a uniform duct at $t < 0$ the piston and the fluid (air at standard atmospheric conditions) are at rest. At $t = 0+$ the piston is moved impulsively to the left at 100 ft/sec with respect to the wall, thereby creating a shock wave. At $t = 0.1$ sec the piston speed is increased to 200 ft/sec, thereby creating a second shock. Neglecting wall friction, how far will the second shock travel before it overtakes the first, or will it tend to lag behind? If so, prove it.

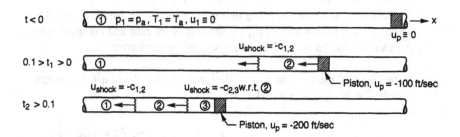

7.14. A gun tunnel is a laboratory device used to attain high-pressure and high-enthalpy stagnation conditions on a model at moderate to high Mach numbers. It consists of a shock tube where the flow is accelerated to moderate supersonic speeds in one direction by propagation of a normal shock, and by simultaneously firing (from the gun) a small model in the direction opposed to the shock motion, to achieve high relative flow speeds. Referring to the figure, in region ① the gas is at rest within the tube. After firing, the model velocity is $u_m = -5000$ ft/sec with respect to the wall.

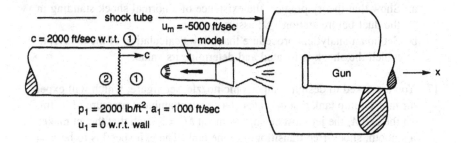

a. Sketch the probable flow pattern near the nose of the model. For the period while the model still lies completely in ①, compute the pressure and temperature on the model nose assuming that the gas is air, using perfect gas relations.

b. In region ② compute the flow speed with respect to the wall, the static temperature, and pressure.

c. After the model has passed completely through the shock between ① and ②, calculate the new values for the pressure and temperature on the model nose. Neglect any deceleration of the model.

7.15. The Mach number in the flow ahead of a stationary normal shock can be written as $M_1 \equiv u_1/a_1 = (a_1 + \Delta u)/a_1 = 1 + \Delta u/a_1$, where $\Delta u \equiv u_1 - a_1$ and where a_1 is the speed of sound ahead of the shock. If the shock is very weak, then $\Delta u/a_1 \ll 1$ and we can neglect $(\Delta u/a_1)^2$ with respect to unity. Hence, to first-order, $M_1^2 = (1 + \Delta u/a_1)^2 \approx 1 + 2\Delta u/a_1$.

Determine the corresponding first-order expressions for T^0/T_1, p_{21}, M_2. Then, if $M_1 = 1.05$, compute p_{21} for the first-order case and compare with the exact result for $\gamma = 7/5$ and compute the percent error.

7.16. In the flow of nitrogen through a de Laval nozzle, data on the pressure distribution is shown in the sketch.

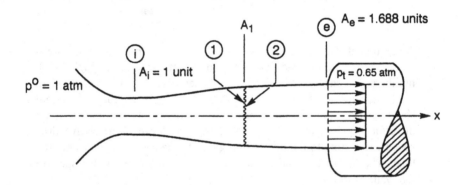

a. Show that theory predicts the existence of a normal shock standing in the duct before station ⓔ; assume $p_e = p_t$.

b. Set up an analytical procedure that allows calculation of the area A_1 at which the shock is located and determine its value.

7.17. You are asked to design a supersonic nozzle, see figure, which will expel air into a dump tank that operates at a constant ambient pressure of 2 atm. At the exit A_e the jet must be supersonic at $M_e = 2$, such that the jet makes a smooth, shock-free transition into the tank. The exit speed is to be $u_e = 2000$ ft/sec.

  a. Determine the upstream reservoir pressure. If the exit area is 14.24
     in.^{2}, compute the mass rate of flow.
  b. Now the dump tank pressure is changed, but not the reservoir condi-
     tions, such that the following temperature readings are obtained: for
     $A_3 = 10.57$ in.^{2}, $T_3 = 495°R$; for $A_4 = 12.66$ in.^{2}, $T_4 = 710°R$. Determine
     whether or not the mass flow has changed from (a) and by how much.
     Determine the pressure at station ④, if not precisely at least give an
     upper and a lower estimate, and explain. Determine the pressure and
     temperature at ① where $A_1 = 10.57$ in.^{2}

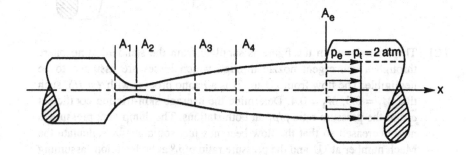

7.18. A de Laval nozzle is to be designed for steady isoenergetic flow of air so
      that the test section Mach number (maximum Mach number) is $M_{test} =$
      2.5. If the upstream minimum area is $A_1 = 1$ ft^2, determine the test section
      area A_2 according to one-dimensional theory.
   a. The test section is followed by a second converging–diverging section
      (called a *diffuser*) so that shocks formed in the test section can be
      "swallowed," permitting a finite length in which supersonic flow can
      be maintained in the test section. What is the smallest possible theo-
      retical value of A_3 that will permit the shock to be swallowed?
   b. After swallowing, determine the flow speed in the test section.

7.19. Show that Hugoniot's differential area–velocity relation, Equation 7.10-
      5, for the conditions at a throat $A*$, for steady, isentropic flow of a perfect
      gas, can be extended* to obtain expressions for the critical conditions for
      the gradients of the throat Mach number and the throat pressure, i.e.,

$$(dM/dx)* = -[(\gamma + 1)/2\gamma p*](dp/dx)*,$$

$$(dp/dx)* = \pm[(\gamma p*)/(\gamma + 1)^{1/2}][(d^2A/dx^2)*/A*]^{1/2}.$$

7.20. The setup consists of a short de Laval nozzle to which is attached a length
      of constant-area duct. Determine the shortest length of duct such that the
      flow would be choked at both stations ① and ③, assuming that from

---

* See Weinbaum and Garvine (1969).

② to ③ a Fanno process occurs. If needed, use $\gamma = 7/5$, $\bar{c}_f = 0.002$. Then calculate the ratio $p_3/p_1{}^0$. If p_{tank} were decreased below this value of p_3, what would happen to the flow within the straight portion of the duct? Explain.

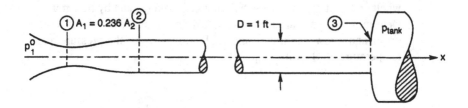

7.21. The duct shown in the figure is supplied from the standard atmosphere through a convergent nozzle through which losses are assumed to be negligible. The flow from ① to ② is a Fanno process with $\gamma = 7/5$, such that $M_1 = 0.3$, $M_2 = 0.4$. Determine the average skin-friction coefficient $\bar{c}_f$ and the pressure ratio $p/p_1{}^0$ at both stations. The dump tank pressure is now decreased so that the flow becomes just sonic at ②. Calculate the Mach number at ① and the pressure ratio $p/p_1{}^0$ at both stations assuming that $\bar{c}_f$ remains the same.

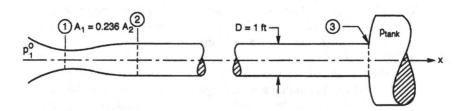

7.22.

    a. Referring to Section 7.17, verify the relations given in Equations 7.17-5 through 7.17-7.

    b. Consider a horizontal, isothermal flow of air such that at station ① $T_1 = 500°R$, $p_1 = 450$ psia, $u_1 = 11$ ft/sec, $D = 1$ ft, $\bar{c}_f = 0.002$. Determine M_1 and the length of pipe required for the Mach number to increase by 10%. Designate this station as ②, and compute the local Mach number and pressure ratio p_{21}.

    c. Now restrict considerations to the flow from ① to ②. Let a new coordinate x be defined such that $x_1 = 0$, $x_2 = L$. Then let the change (increase) of Mach number from 1 to $x > 0$ be denoted as ΔM. Apply Equation 7.16-7 twice (at x_1 and at x) to obtain the first-order expansion of the Mach number–distance relation in terms of the small parameter $\Delta M/M$. For example,

$$\left(1+\Delta M/M_1\right)^{-1}=1-\Delta M/M_1+O\left[\left(\Delta M/M_2\right)^2\right].$$

This should lead to

$$M/M_1 \equiv 1+\Delta M/M_1 = 1+2\gamma M_1^2 \bar{c}_f(x/D),$$

$$= 1+2\gamma M_1^2 \bar{c}_f(L/D)(x/L).$$

d. Use this result to derive the first-order expression for p/p_1. Compare the numerical values at ② for M_2 and p_{21} from the approximate expressions with the exact values.

Note: This approximation thus yields a linear variation of the Mach number with the duct longitudinal coordinate, a considerable simplification without significant loss of accuracy. It is straightforward to verify numerically that, for the coefficient multiplying x/L, we have $2\gamma M_1^2 \bar{c}_f(L/D) \ll 1$, which justifies the use of a first-order approximation as long as the Mach number remains small.

7.23. It is necessary to supply a de Laval nozzle from a compressor located 1200 ft from the nozzle entrance by means of a pipe of diameter $D = 4$ in. The pipe is heavily insulated so that heat transfer can be neglected. It is further required that at the nozzle entrance, station ②, the Mach number and static pressure, respectively, shall be $M_2 = 0.2$ and $p_2 = 1$ atm, with $T_2 = 519°R$.

a. Determine the ensuing mass flow rate.

b. If the mean skin-friction coefficient is $\bar{c}_f = 0.0025$, determine the Mach number and static pressure at the pipe inlet station ①.

c. Determine the total pressure (in atm) at stations ② and ① and in the reservoir (a pressurized tank) just upstream of ①, assuming that from the reservoir to pipe inlet the flow is isentropic.

d. Determine the Reynolds number at stations ① and ②, and the corresponding skin-friction coefficients for a smooth pipe. Is $\bar{c}_f = 0.0025$ a reasonable approximation? If not, determine the appropriate value.

8 Nonsteady Flow

W. B. Brower: It appears that the great majority of flows are steady.
J. V. Foa: On the contrary! Most real flows are unsteady.

8.1 INTRODUCTION

In Section 4.19 we undertook to find an approximate solution of a nonsteady flow by utilizing a method in which terms involving partial time derivatives (i.e., those involving the operator $\partial/\partial t$) are assumed to be negligibly small compared with other terms in the applicable equations. The solution involving such an approximation is said to be *quasi-steady*.

In fact, in a truly nonsteady flow, these terms can generally not be neglected, and it happens that some quasi-steady solutions, for practical purposes, may be severely deficient, even meaningless.

In the (one-dimensional) flow of a constant-density fluid through a duct of fixed geometry, the governing equations can usually be combined to produce a single, time-dependent, ordinary, differential equation. The equation may be first- or second-order, and it may be linear or nonlinear. In such a flow, if one knows the velocity at any one point and time, the entire velocity field is immediately calculable from the continuity equation at every other point at the same fixed time. This is usually not the case in most nonsteady flows. In the first few following sections we give several examples of the first type, proceeding from a simple case to those more complex.

We then tackle a more complicated class of flows by relaxing the restriction on either the duct geometry or the fluid medium. This results in a system of coupled, partial differential equations. We then apply standard mathematical techniques to these equations to show that the governing equations fall into one of three categories — *elliptical, parabolic,* or *hyperbolic* — a distinction that is vital for the analyst to comprehend and which should be ascertained early in the analysis. These equations may be linear or nonlinear. Hyperbolic differential equations are of special importance in fluid mechanics, giving rise to the existence of *characteristic lines*, which play a major role in analyzing the nonsteady flow of a gas, for example.

8.2 ANALYSIS OF STARTING FLOW IN A PIPE SUPPLIED BY AN INFINITE RESERVOIR

DESCRIPTION

Consider the problem (Figure 8.2-1) of a straight, horizontal pipe (cross-sectional area A) supplied by an "infinite" reservoir of liquid, of density ρ, which is maintained at a constant pressure level p^0. Initially (i.e., with the flow control valve shut), the

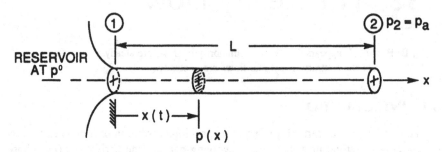

FIGURE 8.2-1 Starting flow in a pipe.

flow velocity is everywhere zero and the pressure p^0 persists throughout. The reservoir cross-sectional area is sufficiently large compared with that of the pipe that we may put the upstream velocity equal to zero for all time without significant error.

At $t = 0+$ the valve is "instantaneously" opened and the pressure at the exit immediately adjusts to the ambient value p_a. Let the coordinate of the left end of the slug of fluid, initially located between stations ① and ②, be denoted as $x = x(t)$. Since the pressure at the left end, denoted as $p = p(x)$, is higher than ambient, the fluid lying between the two stations is subject to a force $(p - p_a)A$, causing the slug to accelerate.

Although viscous effects should eventually be accounted for, we can neglect them in the first approximation in order to determine a lower bound for the time required for the flow to approach a *steady-state* condition. The problem is to determine the time for the pressure at the slug left end to adjust to the ambient value.

ANALYSIS

As the slug accelerates, the pressure at the left end must decrease. In this first approximation, we make the assumption of quasi-steady flow from the reservoir to the station in the pipe at the slug left end. Denoting the slug velocity — which must be uniform over the entire slug length — as $u = u(t)$, Bernoulli's equation, written between the two stations, produces

$$p^0 = p + \tfrac{1}{2}\left(\rho u^2\right). \tag{8.2-1}$$

The dynamic equation for any particle in the slug is

$$\frac{Du}{Dt} \equiv \frac{\partial u}{\partial t} + u\,\frac{\partial u}{\partial x} = \frac{\partial u}{\partial t} + \frac{\partial}{\partial x}\left(\frac{u^2}{2}\right) = -\frac{1}{\rho}\,\frac{\partial p}{\partial x}. \tag{8.2-2a}$$

This can be rewritten as

$$\frac{\partial u}{\partial t} + \frac{\partial}{\partial x}\left(\frac{p}{\rho} + \frac{u^2}{2}\right) = 0. \tag{8.2-2b}$$

We now multiply the entire equation by ∂x and integrate over the slug length for a fixed, but arbitrary, time. That is,

$$\int_0^f \frac{\partial u}{\partial t}\, \partial x + \int_0^L \frac{\partial}{\partial x}\left(\frac{p}{\rho}+\frac{u^2}{2}\right)\partial x = 0. \qquad (8.2\text{-}3)$$

Since $u = u(t)$ only, the integrand of the first integral can be taken right through the integral sign, and we can put $\partial u/\partial t \rightarrow du/dt$; therefore,

$$\int_0^L \frac{\partial u}{\partial t}\, \partial x = \frac{du}{dt}\int_0^L \partial x = L\,\frac{du}{dt}. \qquad (8.2\text{-}4)$$

The second integral is in the precise form of a definite integral of an exact partial differential. However, we must supply the limits on the variables $p(x)$ and $u(x)$, corresponding to the upper and lower limits on $x(t)$. Thus,

$$\int_0^L \frac{\partial}{\partial x}\left(\frac{p}{\rho}+\frac{u^2}{2}\right)\partial x = \left[\frac{p}{\rho}+\frac{u^2}{2}\right]_{p,u(x)}^{p_a,u(x)} = \frac{p_a-p}{\rho}. \qquad (8.2\text{-}5)$$

The flow speed term vanishes because all points within the slug have the same speed at any given time.

The relation formed from the last equality is next combined with Equation 8.2-3. Then, p_a is eliminated from the result by Equation 8.2-1 to produce the following ordinary differential equation:

$$L\,\frac{du}{dt} = \frac{p^0 - p_a}{\rho} - \frac{u^2}{2} = \frac{\Delta p}{\rho} - \frac{u^2}{2}, \qquad (8.2\text{-}6)$$

where $\Delta p \equiv p^0 - p_a$. It is straightforward to integrate Equation 8.2-6 by separating the variables; thus,

$$\int_{t=0}^t dt = 2L \int_{u=0}^u \frac{du}{2\Delta p/\rho - u^2}.$$

This yields

$$t = \sqrt{\frac{2\rho L^2}{\Delta p}}\ \tanh^{-1}\sqrt{\frac{\rho}{2\Delta p}}\ u,$$

or, inverting,

$$u = \sqrt{\frac{2\Delta p}{\rho}}\ \tanh\sqrt{\frac{\Delta p}{2\rho L^2}}\ t.$$

The maximum exit speed is $u_m \equiv \sqrt{2\Delta p/\rho}$ which is the steady-state inviscid value. The last relation can therefore be written in the form

$$u/u_m = \tanh\left(u_m t/L\right). \tag{8.2-7}$$

Thus, we see that the maximum speed is attained only asymptotically, after an infinite time. Further, with $dx/dt = u$, the distance traveled by the slug after the valve is opened is given by

$$
\begin{aligned}
x &= \int_0^x dx = \int_0^t u\,dt, \\
&= u_m \int_0^t \tanh\left(u_m t/L\right) dt, \\
&= L \int_0^{u_m t/L} \tanh\left(u_m t/L\right) d\left(u_m t/L\right), \\
&= L \,\text{lncosh}\left(u_m t/L\right).
\end{aligned}
\tag{8.2-8}
$$

These results were first obtained by Poisson (1833).

SAMPLE COMPUTATION

A numerical example helps put matters in perspective. Suppose the fluid is water at 20°C, where $\rho = 998.3$ kg/m^3 and $\Delta p = 2$ atm $= 2.026\ 5 \times 10^5$ Pa. Let the constant-area pipe length be $L = 1.5$ m; then, $u_m = 20.15$ m/s. We next calculate the time t_f for the exit velocity to reach $u/u_m = 0.99$. By inverting Equation 8.2-7, $t_f = (L/u_m)$ $\tanh^{-1}(0.99) = 0.197\ 0$ s. Obviously, the terminal flow is essentially established in a very short time interval. The distance traveled by the slug in the same time interval is $x_f = 1.5$ lncosh $(2.646) = 2.937$ m.

8.3 NONSTEADY LIQUID FLOW THROUGH AN ORIFICE IN A RESERVOIR

INTRODUCTION

The problem of liquid ejected from a nozzle, or an orifice of a reservoir, under the action of gravity, is one of the oldest in fluid mechanics. In Section 4.6 we dealt with the quasi-steady treatment of nozzle flow, making use of Torricelli's equation to relate the pressure, velocity, and height at an arbitrary point in the emitted jet.

This method bypasses the phenomenon of the *starting transient*, which occurs prior to the establishment of the "main flow." When the main flow is established, Torricelli's law can be applied to many configurations with only minor error. Note, however, for applications involving Torricelli's law, as well as for the theory of this section, viscous effects are treated as negligible.

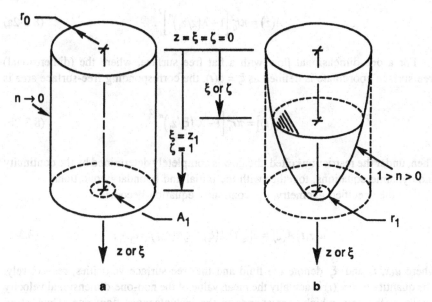

FIGURE 8.3-1 Liquid flow from a reservoir through an orifice.

We now undertake to analyze the flow under gravity from a cylindrical reservoir that has a circular orifice in the bottom. This problem was considered by some of the earliest and most renowned investigators, including Newton (1686). They recognized that existing theory was not able to predict the flow rate in the emitted vertical jet due to the lack of one dimensionality in the neighborhood of the orifice.

Some investigators attempted to estimate, apparently by visual observation, the contraction coefficient in a jet emitted from an orifice in the vertical wall of a reservoir, before gravitational effects generate significant jet curvature. Almost needless to say, the results varied over a wide range of values, with little success. For example, Newton estimated a value of $C_c = 12/17 = 0.706$, instead of the value of 0.591 obtained relatively recently by Bloch (1968), which we accept as correct. See Brougham and Routh (1855) for a critique of Newton's work.

THE ANALYTICAL MODEL

The setup is shown in Figure 8.3-1. Considerations are restricted to the flow of a constant-density, inviscid liquid. Although the actual geometry is shown on the left, it will be convenient to start with the more general configuration on the right, where the positive z-axis is directed down. The cylinder is of height z_1, and radii at the top and bottom are r_0, r_1 respectively. We denote the radius function as

$$r/r_0 = 1 - \lambda(z/z_1)^n, \tag{8.3-1}$$

where $\lambda \equiv 1 - r_1/r_0$, $1 > \lambda \geq 0$. The cross-sectional area at an arbitrary station is, perforce,

$$A(z) = \pi r_0^2 \left[1 - \lambda (z/z_1)^n \right]^2.$$

(8.3-2a)

For a one-dimensional flow with a flat free surface, where the (dimensional) free-surface coordinate is defined as $\xi = \xi(t)$, the corresponding free-surface area is

$$A[\xi(t)] = \pi r_0^2 \left[1 - \lambda (\xi/z_1)^n \right]^2.$$

(8.3-2b)

Then, under the restrictions cited, the flow is completely determined by the continuity and dynamic equations, together with the initial and boundary conditions.

For the specified geometry, the continuity equation becomes

$$u(z,t) A(z) = u(\xi,t) A(\xi) = \dot{\xi} A(\xi,t) = u_1 A_1,$$

(8.3-3)

where $u(x, t)$ and $\dot{\xi}$ denote the fluid and the free-surface velocities, respectively. The quantity $u_1 = u_1(t)$ is actually the mean value of the non-one-dimensional velocity profile at the exit, which corresponds to the instantaneous flow rate. Thus, at an arbitrary station within the portion of the reservoir occupied by fluid and, at any instant,

$$u(z,t) = \dot{\xi} A(\xi)/A(z) = \dot{\xi} \left[1 - \lambda (\xi/z_1)^n \right]^2 \Big/ \left[1 - \lambda (z/z_1)^n \right]^2.$$

(8.3-4)

In a one-dimensional flow a "particle" is conceived to be an infinitesimally thin disk transverse to the duct axis. With only pressure and gravitational terms to drive the flow, the dynamic equation (Euler's equation) is

$$Du/Dt = \partial u/\partial t + (\partial/\partial z) (u^2/2) = -(1/\rho) \, \partial p/\partial z + g,$$

(8.3-5)

where ρ is the fluid mass density. Equation 8.3-5 can be rearranged, and integrated at a fixed time, from the free surface to the exit. As boundary condition, we impose the requirement that at the free surface, and at the orifice station, the pressure be constant at p_a. Thus, multiplying by ∂z, and integrating, we have

$$\int_{z=\xi}^{z_1} \frac{\partial u}{\partial t} \, \partial z + \int_{p=p_a, u=\xi, z=\xi}^{p_a, u_1, z_1} \frac{\partial}{\partial z} \left(\frac{p}{\rho} + \frac{1}{2} u^2 - gz \right) \partial z = 0.$$

or,

$$\int_{z=\xi}^{z_1} \frac{\partial u}{\partial t} \, \partial z + \frac{1}{2} \left(u_1^2 - \dot{\xi}^2 \right) + g(\xi - z_1) = 0.$$

(8.3-6)

Since the integrand in the second integral is in the form of an exact partial differential, the result of Equation 8.3-6 follows directly.

It is convenient to nondimensionalize Equation 8.3-6 before proceeding with the remaining integration. We introduce nondimensional variables defined as follows: $x \equiv z/z_1$, $\zeta = \zeta(\tau) \equiv \xi/z_1$, $\tau \equiv t(g/z_1)^{1/2}$, $w(x, \tau) \equiv u(z, t)/(gz_1)^{1/2}$. Then, dividing out a common factor, Equation 8.3-6 becomes

$$\int_{x=\zeta}^{1} \frac{\partial w}{\partial \tau} \, \partial x + \frac{1}{2} \left(w_1^2 - \dot{\zeta}^2 \right) + \zeta - 1 = 0. \tag{8.3-7}$$

Similarly, Equation 8.3-4 becomes

$$w(x, \tau) = \dot{\zeta}(1 - \lambda\zeta^n)^2 / (1 - \lambda x^n)^2 = A_1 w_1 / A(x). \tag{8.3-8}$$

From Equation 8.3-8 we compute the partial derivative

$$\frac{\partial w}{\partial \tau} = \ddot{\zeta}(1 - \lambda\zeta^n)^2 / (1 - \lambda x^n)^2 - 2\lambda n\zeta^{n-1}\dot{\zeta}^2 (1 - \lambda\zeta^n)^2 / (1 - \lambda x^n)^2. \tag{8.3-9}$$

THE GOVERNING DIFFERENTIAL EQUATION

We have now gone as far as we can without specializing to the particular geometry to be analyzed. For the case of a constant-area circular cylinder with a circular orifice at the bottom station, which is shown in Figure 8.3-1a, we put $n \to 0$. It is precisely at this point that an important error is introduced, one that is not always recognized. For an exit with a sharp edge the flow detaches from the orifice edge, initiating the jet; the local direction of the jet at separation is horizontally inward. In the jet interior at this station the streamlines are curved, and the one-dimensional approximation breaks down. We shall have more to say about this in the ensuing discussion.

Ignoring this for a moment, the one-dimensional model yields for the area distribution:

$$\lim_{n \to 0} A(\zeta) = A_0 = \pi r_0^2, \quad \text{for } 0 \le \zeta \le x < 1; \quad \lim_{n \to 0} A = A_1 = \pi r_1^2, \quad \text{for } \zeta = x = 1.$$

Also, for $0 \le x \le \zeta < 1$,

$$\lim_{n \to 0} w(x, \tau) = w_1 A_1 / A_0 = \dot{\zeta}, \tag{8.3-10a}$$

and

$$\lim_{n \to 0} \frac{\partial w}{\partial \tau} = \ddot{\zeta}. \tag{8.3-10b}$$

Therefore,

$$\int_{x=\zeta}^{1} \frac{\partial w}{\partial \tau}\, \partial x = (1-\zeta)\,\ddot{\zeta}. \qquad (8.3\text{-}10c)$$

Combining the last five relations produces the following differential equation:

$$(1-\zeta)\ddot{\zeta} + \tfrac{1}{2}(\alpha^2-1)\dot{\zeta}^2 + \zeta - 1 = 0, \qquad (8.3\text{-}11)$$

where we define the area ratio $\alpha \equiv A_0/A_1 \geq 1$. The initial conditions on the free-surface coordinate and velocity are

$$\zeta(0) = \dot{\zeta}(0) = 0. \qquad (8.3\text{-}12)$$

Equation 8.3-11 is nonlinear, although not horribly so. The book by Polyanin and Zaitsev (1995) provides an extensive listing of solutions of ordinary differential equations. However, we can integrate Equation 8.3-11 in terms of gamma functions by proceeding as follows. Put $y \equiv 1 - \zeta$, $s = \alpha^2 - 1$; then

$$y\ddot{y} - \tfrac{1}{2}s\dot{y}^2 + y = 0. \qquad (8.3\text{-}13)$$

with initial conditions

$$y(0) = 1; \quad \dot{y}(0) = 0. \qquad (8.3\text{-}14)$$

If Equation 8.3-14 is multiplied by the integrating factor $y^{-(1+s)}$, an expression for the free-surface velocity is obtained as a first integral; i.e.,

$$\dot{y}^2 = 2(y - y^s)/(s-1), \quad s \neq 1. \qquad (8.3\text{-}15)$$

In Equation 8.3-15 we must take the negative root in order to give the proper sign to $\dot{y}$. Thus,

$$-\dot{\zeta} = \dot{y} = -\left[2(y-y^s)/(s-1)\right]^{1/2}, \quad s \neq 1. \qquad (8.3\text{-}16)$$

To integrate Equation 8.3-16, put $y = v^2$; then

$$\dot{v} = dv/d\tau = -\left[(1 - v^{2s-2})/2(s-1)\right]^{1/2}, \quad s \neq 1, \qquad (8.3\text{-}17)$$

with

$$v(1) = 0. \tag{8.3-18}$$

Defining a nondimensional discharge time $\tau_d \equiv t_d \, (g/z_1)^{1*2}$, we can solve for $d\tau$ and integrate over the container length. Thus,

$$\tau_d = 2\sqrt{s-1} \int_0^1 \frac{dv}{\left(1 - v^{2s-2}\right)^{1/2}}, \quad s \neq 1, \tag{8.3-19}$$

which is a canonical form in terms of gamma functions, e.g., Dwight (1961) formula 855.34. See Abramovitz and Stegun (1964) for tabulated values.* Hence,

$$\tau_d = \sqrt{\frac{\pi}{2(s-1)}} \, \frac{\Gamma\left[1/2(s-1)\right]}{\Gamma\left[s/2(s-1)\right]}, \quad s > 1. \tag{8.3-20}$$

Appropriate expressions can be similarly derived for $s = 1$ and $0 < s < 1$. In the limit as $s \to 0$, $\alpha \to 1$, $A_0 = A_1$, and the entire body of fluid is in free fall, producing the familiar result

$$\tau_d = \sqrt{2}. \tag{8.3-21}$$

THE QUASI-STEADY SOLUTION

If Torricelli's law for the exit velocity is used, we have

$$u_1 = \sqrt{2g(z_1 - z)}, \quad \text{or} \quad w_1 = \sqrt{2(1 - \zeta)} = \sqrt{2y}, \tag{8.3-22}$$

and

$$\dot{y} = -\alpha^{-1} \sqrt{2y}. \tag{8.3-23}$$

Equation 8.3-23 produces

$$\tau_d = \sqrt{2} \, \alpha = \sqrt{2} \, \left(A_0/A_1\right), \tag{8.3-24}$$

for the quasi-steady discharge time.

DISCUSSION

The quasi-steady expression for the discharge time, Equation 8.3-24, is compared with exact result, Equation 8.3-20, in Figure 8.3-2. For $\alpha \geq 4$ the two coincide for

---

* For an even more remarkable source of integral tables, see also Prudnikov et al. (1986).

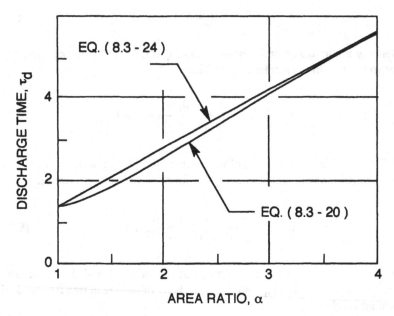

FIGURE 8.3-2 Comparison of discharge times for exact and quasi-steady theories.

practical purposes. Unexpectedly, the two theories produce the same discharge times for the case of the free fall where $\alpha = 1$.

This can be explained by comparing the discharge velocities in the two cases, plotted in Figure 8.3-3a, as a function of the free-surface coordinate y. The quasi-steady relation is given in Equation 8.3-22. From the continuity expression,

$$w_1 = -\alpha^{-1}\dot{y}. \tag{8.3-25}$$

Thus, with Equation 8.3-16, the exact expression is

$$w_1 = \left[\left(y - y^{\alpha^2-1}\right)\left(2\alpha^2\right)\Big/\left(\alpha^2 - 2\right)\right]^{1/2}, \quad 1 \le \alpha \ne \sqrt{2}. \tag{8.3-26}$$

The plot discloses an interesting facet of the flow. For $\alpha > 5$, the starting effects are confined to a small region near $\zeta = 0$ (i.e., near $y = 1$), after which the quasi-steady behavior is (nearly) attained. Note, also, in the quasi-steady solution, that the initial velocity is theoretically nonzero, a physical impossibility.

For $\alpha > 5$, the starting flow has the character of a (mathematical) *boundary-layer* problem. In fact, this type of starting behavior is encountered in a variety of inviscid, tank-draining geometries. We will examine a more complicated case in the next section.

At the other extreme, where $\alpha \to 1$ corresponds to free fall, we observe that the quasi-steady solution gives the wrong discharge velocity at every instant except one, and that it varies in the wrong sense at all times. The fact that it gives the correct

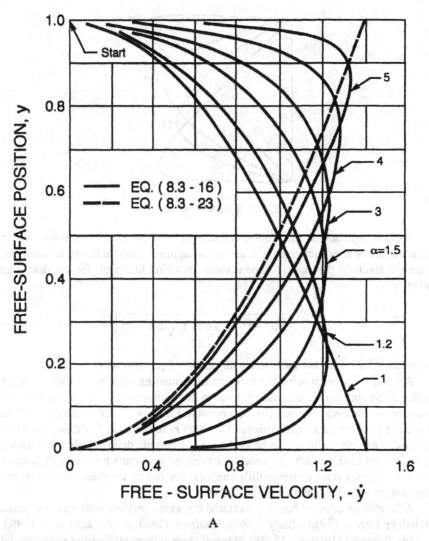

FREE-SURFACE POSITION, y (vertical axis)

EQ. (8.3 - 16)
EQ. (8.3 - 23)

FREE - SURFACE VELOCITY, -ẏ (horizontal axis)

A

FIGURE 8.3-3A (a) Comparison of free-surface velocities for exact and quasi-steady theories. (b) Plot of free-surface velocity made by D. Bernoulli (1738).

discharge time is fortuitous. The explanation is that, for $\alpha = 1$, the Torricelli and the exact velocity expressions are reflections of each other about $y = 0.5$.

NOTE ON PREVIOUS THEORETICAL WORK

In Figure 8.3-3b there is reproduced a graph due to D. Bernoulli (1738) that is almost indistinguishable in character from our own plot, a quarter of a millennium later. It appears that Bernoulli was the first to undertake the analysis of a nonsteady flow. I have previously noted in Section 8.2 that in 1833 Poisson gave the solution for the starting transient in a pipe supplied by an infinite reservoir.

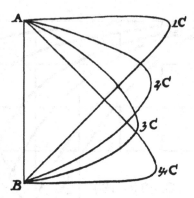

FIGURE 8.3-3B *(Continued.)*

Poisson also tackled the problem of the starting flow from a reservoir with an orifice at the bottom such that the reservoir is supplied continuously to maintain the reservoir level. In the current dimensionless notation his result for the discharge velocity is

$$w_1 = \sqrt{2(\alpha^2 - 1)/\alpha^2} \ \tanh\left[\tau / \sqrt{2(\alpha^2 - 1)}\right]. \tag{8.3-27}$$

Equation 8.3-27 does not incorporate any effect of jet contraction.

Poisson went on to solve the problem of a draining circular reservoir with an orifice in the bottom (i.e., the case explored in this section). He obtained a differential equation for the discharge velocity equivalent to Equation 8.3-26, and evaluated the discharge time for the explicit values of the area ratio $\alpha = 1, 2, 3$. He indicated that the discharge time could always be evaluated in terms of "definite Eulerian integrals of the second kind," which, presumably, are related to gamma functions. Peculiarly, Poisson does not refer to Bernoulli's analysis, nor did he consider whether or not his solution was substantiated by experiment.

A handful of other authors has explored this same problem with varied success, including Kozeny (1954), Stary (1962), Kaufman (1963, p. 56), Bird et al. (1963, p. 239), Burgreen (1960, pp. 13–28). None of these authors offered any experimental verification.

COMPARISON WITH EXPERIMENT

Not having located any reported measurements, in 1964 John Way and I (Brower and Way, 1964) undertook a series of experiments in a tank 25.75 in. high and 13.88 in. inside diameter, with tap water as fluid. We quickly observed that the free surface does not tend to remain flat. Instead the center region descends more quickly than the fluid near the walls, an effect that is not related to the fluid viscosity, but to the inherent three-dimensionality of the flow. To avoid this, and to maintain the desired one-dimensionality, except near the orifice station, we floated a thin, circular, balsa disk on the free surface.

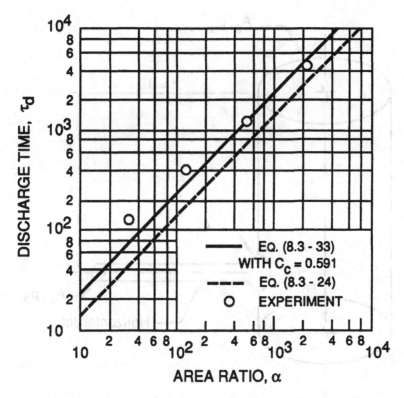

FIGURE 8.3-4 Comparison of discharge times from quasi-steady theory with experiment.

Several orifice diameters were tested resulting in area ratios such that $\alpha \geq 100$ in all test cases. In consequence there is no significant difference between the calculated values for the discharge time for the exact and the Torricelli expressions. The quasi-steady theory is compared with our measurements in Figure 8.3-4, indicating a severe deficiency in the theory.

It can be readily verified that it is not possible to correct the theory by assuming that the actual discharge rate is the theoretical value predicted by Equation 8.3-20, say, multiplied by Bloch's contraction coefficient $C_c = 0.591$. This would produce another line lying only slightly above the dashed line in the figure, and parallel to it.

APPROXIMATE THEORY TO CORRECT FOR ORIFICE EFFECTS

The magnitude of the deviation from experiment of both the exact and Torricelli theories would seem to raise questions about the value of the theoretical approach. The actual difficulty is the failure to account for the behavior of flow from the orifice in any realistic way. It is proposed to remedy this defect at once.

It should be obvious, for a flow that is essentially one dimensional, that the behavior of the free surface should not be particularly sensitive to the location of the orifice, as long as it is maintained at station z_1. Thus, it is proposed to replace

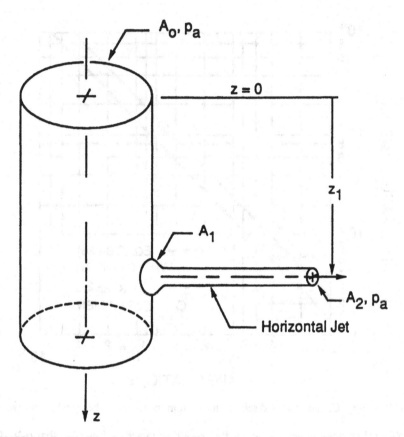

FIGURE 8.3-5 Modification of flow model to account for jet contraction.

the actual flow model by the fictitious one shown in Figure 8.3-5, where the orifice is installed in the side of the reservoir at station ①, with its axis horizontal.

We assume that the flow within the reservoir supplies the orifice, the fictitious flow leaving in a horizontal jet. The fact that gravitational effects in the jet are neglected should not affect the analysis of the portion of the flow within the reservoir. The system will be treated as a quasi-steady flow allowing us to apply the Bernoulli equation to relate conditions at the free surface to conditions at station ②, where the fictional jet becomes one dimensional.

The Bernoulli equation is

$$p_a + \tfrac{1}{2}\rho\dot{\xi}^2 - \rho g\xi = p_a + \tfrac{1}{2}\rho u_2^2 - \rho g z_1 \qquad (8.3\text{-}28a)$$

or

$$u_2^2 = \dot{\xi}^2 = 2g(z_1 - \xi). \qquad (8.3\text{-}28b)$$

From the continuity equation

$$\dot{Q} = u_2 A_2 = u_2 (A_2/A_1) A_1 = C_c u_2 A_1 = \dot{\xi} A_0, \qquad (8.3\text{-}29)$$

where $C_c \equiv A_2/A_1 = 0.591$ is the contraction coefficient for an axisymmetric jet. Therefore,

$$u_2 = \dot{\xi} \frac{A_0/A_1}{C_c} = \dot{\xi} \alpha/C_c. \qquad (8.3\text{-}30)$$

This can be substituted into Equation 8.3-29 and solved for

$$\dot{\xi}^2 = \left(\frac{d\xi}{dt}\right)^2 = \frac{2g(z_1 - \xi)}{(\alpha/C_c)^2 - 1}. \qquad (8.3\text{-}31)$$

Taking the positive root and separating variables yields the definite integrals:

$$\int_{\xi=0}^{z_1} \frac{d\xi}{\sqrt{z_1 - \xi}} = \sqrt{\frac{2g}{(\alpha/C_c)^2 - 1}} \int_{t=0}^{t_d} dt. \qquad (8.3\text{-}32)$$

Substituting in the limits and nondimensionalizing, we have the discharge time

$$\tau_d \equiv t_d \sqrt{g/z_1} = \sqrt{2\left[(\alpha/C_c)^2 - 1\right]}. \qquad (8.3\text{-}33)$$

The discharge time computed from Equation 8.3-33 is plotted in Figure 8.3-4. It appears that, for values of α not too small, it compares favorably with experiment. For the smaller range of α the starting transient becomes important and the accuracy of the quasi-steady theory deteriorates. It is probable that the fictitious model of the horizontal jet could be patched together with the exact theory to handle that case, if necessary.

APPROXIMATE THEORY FOR A RESERVOIR WITH A SHORT NOZZLE AT THE OUTLET

If the sharp-edged orifice is replaced by a short nozzle of length z_2, then, referring to Figure 8.3-6, and for a properly shaped nozzle contour, we can patch together a simple intuitive theory. We assume that the dimensional discharge time, for the free surface to fall to the entrance station of the nozzle, is the time, based on the Torricelli formula, required to empty a reservoir of height z_1, less the time for a reservoir of height z_2. That is,

$$t_d = \sqrt{2}\,\alpha \left[(z_1/g)^{1/2} - (z_2/g)^{1/2}\right],$$

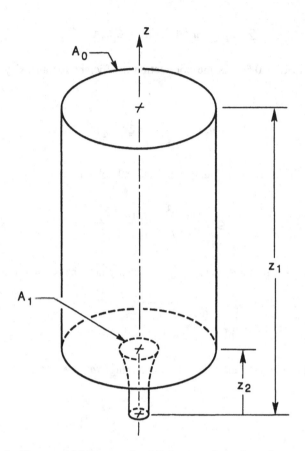

FIGURE 8.3-6   Flow model for reservoir with short nozzle at the outlet.

or

$$\tau_d \equiv t_d \sqrt{g/z_1} = \sqrt{2}\,\alpha\left(1 - \beta^{1/2}\right), \quad \beta \equiv z_2/z_1. \tag{8.3-34}$$

The comparison with experiment, Figure 8.3-4, from Brower and Way (1964), shows good agreement. The nozzle contour must be smoothly contoured so as to avoid flow separation from the nozzle wall. Also, it is useful to note, even for very short nozzles, that the factor $1 - \beta^{1/2}$ may be substantially less than unity. In our setup, with $\beta = 0.0497$, $1 - \beta^{1/2} = 0.777$, for example.

8.4   THE DRAINING OF A CONICAL RESERVOIR

THE GOVERNING DIFFERENTIAL EQUATION

If we put $n = 1$ in Equations 8.3-7 through 8.3-9, the case of a conical reservoir is obtained, Figure 8.4-1. To ensure that the *vena contracta* effect in the jet be negli-

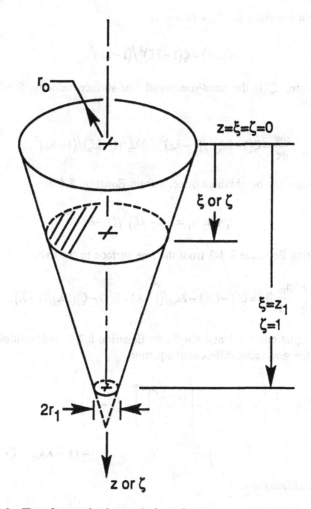

FIGURE 8.4-1 Flow from a slender, conical nozzle.

gible, we impose the additional condition that the cone be slender, i.e., that the cone vertex semi angle be small. For our purpose 15° is a small angle. Since we are actually working with a truncated cone, an additional condition is imposed that the exit radius of the reservoir be small compared with the radius of the cone at the top. This guarantees that the exit radius is also small compared with the truncated height. The latter condition is specified as follows:

$$\varepsilon \equiv r_1/r_0 \ll 1. \tag{8.4-1}$$

The parameter λ, defined after Equation 8.3-1, is related to ε by

$$\varepsilon = 1 - \lambda. \tag{8.4-2}$$

Then, from Equation 8.3-8, with $n = 1$,

$$w(z,\tau) = \dot{\zeta}(1-\lambda\zeta)^2/(1-\lambda z)^2, \tag{8.4-3}$$

where, as before, $\dot{\zeta}$ is the nondimensional free-surface velocity. Similarly, from Equation 8.3-9,

$$\frac{\partial w}{\partial \tau} = \ddot{\zeta}(1-\lambda\zeta)^2/(1-\lambda z)^2 - 2\lambda\dot{\zeta}^2(1-\lambda\zeta)/(1-\lambda z)^2. \tag{8.4-4}$$

The exit velocity can be obtained directly from Equation 8.4-3:

$$w(1,\tau) \equiv w_1 = \dot{\zeta}(1-\lambda\zeta)^2/(1-\lambda)^2. \tag{8.4-5}$$

Also, integrating Equation 8.4-3 from the free surface to the exit,

$$\int_\zeta^1 \frac{\partial w}{\partial \tau}\, \partial z = \ddot{\zeta}(1-\zeta)(1-\lambda\zeta)/(1-\lambda) - \dot{\zeta}^2(1-\zeta)(2\lambda)/(1-\lambda). \tag{8.4-6}$$

Substituting Equations 8.4-5 and 8.4-6 into Equation 8.3-7, and multiplying by $(1-\lambda)^4$, yields the governing differential equation:

$$(1-\lambda\zeta)(1-\zeta)\ddot{\zeta} - \frac{1}{2}\left\{(4\lambda)(1-\zeta) - \left[\frac{(1-\lambda\zeta)^4}{(1-\lambda)^3}\right] + (1-\lambda)\right\}\dot{\zeta}^2 \tag{8.4-7}$$

$$- (1-\lambda)(1-\zeta) = 0.$$

The initial conditions are

$$\zeta(0) = \dot{\zeta}(0) = 0. \tag{8.4-8}$$

We are faced with the problem of determining the solution of a highly nonlinear equation. Other than resorting to numerical integration, we have little recourse but to seek an approximate solution requiring the expansion of the original function ζ as a series involving a small parameter. The parameter is provided by Equation 8.4-1. By eliminating $\lambda = 1 - \varepsilon$ from Equation 8.4-7 and simplifying, the equation becomes

$$\varepsilon^3(1-\zeta)(1-\zeta+\varepsilon\zeta)\ddot{\zeta} + \tfrac{1}{2}\left[(1-\zeta+\varepsilon\zeta)^4 - 4\varepsilon^3(1-\varepsilon)(1-\zeta) - \varepsilon^4\right]\dot{\zeta}^2 \tag{8.4-9}$$

$$- \varepsilon^4(1-\zeta) = 0.$$

The initial conditions are unchanged.

The procedure for the expansion, although routine, is tedious because of the huge volume of algebra involved. However, with symbolic mathematical programs now a commonplace, the work is feasible, even for those who "hate" algebra. The first step is to assume that the dependent variable $\zeta = \zeta(\tau; \varepsilon)$ can be expanded in a series in terms of the small parameter ε, as follows:

$$\zeta(\tau) = \zeta_0(\tau) + \varepsilon\zeta_1(\tau) + \varepsilon^2\zeta_2(\tau) + \cdots. \qquad (8.4\text{-}10)$$

Equation 8.4-10 is next substituted into Equation 8.4-9, and then all the terms arranged in an ascending series in ε. Since the right-hand side of the equation must be zero, the only way that the equation can be satisfied is for the coefficient of each term to be zero. This yields a series of ordinary differential equations of order not higher than 2.

Note that the first term on the right side of Equation 8.4-10, for consistency, is ostensibly multiplied by the factor $\varepsilon^0 = 1$. For this reason the function $\zeta_0(\tau)$ is said to be the *zeroth-order* term in the series. In many cases, the zeroth-order equation can readily be obtained merely by putting $\varepsilon \to 0$. However, by inspection, it is evident that both the first and third terms in Equation 8.4-9 would vanish, reducing our problem to a nullity. This occurrence is actually a warning that the variable τ is probably not of order unity, expressed as $\tau \ne O(1)$. [Physically, in the nondimensional scheme employed, it is evident that the dependent variable $\zeta = O(1)$.] Therefore, we need some kind of estimate for the upper limit on τ to be able to proceed.

THE QUASI-STEADY SOLUTION FOR THE DISCHARGE TIME

On the basis of the solution of the cylindrical reservoir, Section 8.3, we expect, for small values of ε, that the duration of the (so-called) starting transient should be a small fraction of the discharge time. Thus, it is reasonable to expect that the discharge time calculated from the Torricelli relation, although the starting transient is completely neglected, should give a value of the same order as the (still unknown) solution of the exact equation (Equation 8.4-9).

According to quasi-steady theory, the exit velocity in dimensional terms is related to the free-surface velocity by

$$u_1 = \sqrt{2g(z_1 - \xi)} = \frac{\dot{\xi}\left[1 - \lambda(\xi/z_1)\right]^2}{(1-\lambda)^2}. \qquad (8.4\text{-}11)$$

The corresponding nondimensional relation is

$$w_1 = \sqrt{2}(1-\zeta)^{1/2} = \dot{\zeta}(1-\lambda\zeta)^2(1-\lambda)^{-2}. \qquad (8.4\text{-}12)$$

Thus, by integrating to obtain a first estimate of the discharge time,

$$\int_0^{\tau_d} d\tau = \frac{\sqrt{2}}{\varepsilon^2} \int_{\zeta=0}^1 \frac{(1-\lambda\zeta)^2}{(1-\zeta)^{1/2}} \, d\zeta$$

or

$$\tau_d = \frac{\sqrt{2}}{\varepsilon^2} \left[\frac{(1-\varepsilon)^2}{5} + \frac{2}{3}\,\varepsilon(1-\varepsilon) + \varepsilon^2 \right] = \frac{\sqrt{2}}{5\varepsilon^2} \left[1 + \frac{4}{3}\,\varepsilon + \frac{8}{3}\,\varepsilon^2 \right]. \qquad (8.4\text{-}13)$$

We can see that the largest term in the expression is $\sqrt{2}/5\varepsilon^2$. In the lingo of the perturbation analyst one says that

$$\tau_d = 0\left(1/\varepsilon^2\right). \qquad (8.4\text{-}14)$$

It makes no difference in the order of the term to replace the multiplier $\sqrt{2}/5 = 0.2828...$by unity. The order is determined in powers of ε, a quantity that may be large or small, depending on the sign of the exponent.

It may bother engineers to note the following, but, in a mathematical sense, the theory becomes exact in the limit as $\varepsilon \to 0$ and then only if we compute an arbitrarily large number of terms in the series. In a practical sense, experience has shown, for a vast array of different applications, to provide satisfactory engineering accuracy, solutions require only a few orders of ε (sometimes only one, perhaps two, and rarely three orders or more). Otherwise, one may have to settle for a numerical solution of the exact equation. The latter, while very valuable, does not provide the ease of studying the effect of changing a parameter as readily and at as low a cost as does a solution in closed form.

THE OUTER SOLUTION OF EQUATION 8.4-9

On the basis of the Torricelli analysis, we now propose to make a further change of variables, still nondimensional, such that both the new independent variable and the new dependent variable are at most order unity. We know that ζ is already of order unity (since the vertical coordinate was nondimensionalized on the height of the cone z_1). Thus, there can be no mathematical gain in introducing a new independent variable, but we do so anyway to emphasize that a transformation has occurred. Thus, we put

$$Z \equiv \zeta, \quad T \equiv \frac{\tau}{1/\varepsilon^2} = \varepsilon^2\tau. \qquad (8.14\text{-}15)$$

This operation is generally referred to as "stretching" the independent variable, although in this case it is obvious that we have actually shrunk the time variable. Consequently, the free-surface derivatives become

$$\dot{\zeta} = \frac{d\zeta}{d\tau} = \frac{dZ}{d\left(\dfrac{T}{1/\varepsilon^2}\right)} = \varepsilon^2 \frac{dZ}{dT} = \varepsilon^2 \dot{Z}, \tag{8.4-16}$$

and

$$\ddot{\zeta} = \frac{d}{d\tau}\,\dot{\zeta} = \varepsilon^2 \frac{d}{dT}\,\dot{\zeta} = \varepsilon^4 \frac{d^2 Z}{dT^2} = \varepsilon^4\,\ddot{Z}.$$

When these two equations are substituted into Equation 8.4-9, a common factor of ε^4 can be divided out to produce

$$\varepsilon^3 (1 - Z)(1 - Z + \varepsilon Z)\,\ddot{Z} + \tfrac{1}{2}\,[(1 - Z + \varepsilon Z)^4$$

$$- 4\varepsilon^3 (1 - \varepsilon)(1 - Z) - \varepsilon^4]\,\dot{Z}^2 - (1 - Z) = 0. \tag{8.4-17}$$

We now see in the limit as $\varepsilon \to 0$ that once again the first term vanishes, but the other two do not. This is a hopeful sign. This produces a first-order, ordinary differential equation:

$$\tfrac{1}{2}\left(1 - Z_o\right)^3 \dot{Z}_o^2 - 1 = 0. \tag{8.4-18}$$

The subscript denotes the fact that we are now seeking the *outer* solution of the differential equation, a concept that we try to clarify below. To avoid confusion there ought to be also a separate notation to identify the order (exponent of ε) in Equation 8.4-10 of the solution. However, since we will not go beyond the zeroth-order problem in the cone flow application, notation for the order has been omitted.

The identical procedure must next be applied to Equation 8.4-8 to derive the transformed initial conditions, which are

$$Z_o(0) = 0; \quad \dot{Z}_o(0) = 0. \tag{8.4-19}$$

Evidently, we cannot expect Equation 8.4-18 to satisfy both of these initial conditions. Lacking any reason to prefer one of the conditions above the other, we choose instead to integrate Equation 8.4-18 with a constant of integration to be determined later. This results in

$$\sqrt{2}\int_0^T dT = \int_{z_o=0}^C \left(1 - Z_o\right)^{3/2} dZ_o,$$

or

$$Z_o = 1 - \left(C - 5T/\sqrt{2}\right)^{2/5}. \qquad (8.4\text{-}20)$$

If the value of the constant were $C = 1$, Equation 8.4-20 would, in fact, satisfy the first initial condition, but not the second. This brings us to an aspect of analysis not previously encountered herein. That is, the simplification offered by putting $\varepsilon \to 0$ has created a novel situation: since the order of the governing differential equation has been reduced, it is no longer possible to impose both of the initial conditions. Such case is categorized as a *singular perturbation problem*, one in which the solution of the equation of reduced order is not uniformly valid (i.e., not valid for all values of the time variable). An effective test of this last phrase is whether or not the solution violates one or more of the initial (or boundary) conditions. Alternatively, the solution is singular if some physical variable becomes infinite in the domain of interest. In the current case the singularity involves a nonzero free-surface velocity when $z_0 = 0$, or equivalently, when $\zeta = 0$. Equations that are not singular are said to be *regular*.

On the other hand, Equation 8.4-20 appears to describe the flow behavior quite well in the region not too close to the origin. This, then, explains the designation of Equation 8.4-20 as the outer solution, in this case meaning away from the origin. We next seek to find a way to analyze the flow in the neighborhood very close to the origin where resides the mathematical boundary layer previously noted.

The problem is to find a way to modify our fundamental equation (Equation 8.4-9) by stretching the variables, such that, while simplifying the equation by putting $\varepsilon \to 0$, the term with the second derivative is retained. Thus, we seek still another transformation, where we want each of the transformed coordinates to be of order unity over the boundary layer, which we can now identify as the distance moved by the free surface during the time period over which the starting transient occurs. For a mathmetician's view of perturbation methods see Hinch (1991).

The Inner Solution

We have no hint at this point of how to estimate the thickness of this layer. So we will introduce another technique, called *rescaling* by Lin and Segel (1974, p. 292), which will help us do the job. We define new variables as follows:

$$\zeta = \sigma_1(\varepsilon)Z_i, \quad \tau \equiv \sigma_2(\varepsilon)T. \qquad (8.4\text{-}21)$$

Here Z_i is the new space variable that results from a stretching of ζ by the factor $\sigma_1(\varepsilon)$. This factor depends on our small parameter ε in a way to be determined, and similarly for σ_2. The subscript i denotes the *inner* dependent variable.

We repeat the process that led to Equation 8.4-17, except that this time it is complicated by the presence of two stretching factors. Since the term with the highest-order derivative must be retained, we divide through by the factor multiplying the first term to produce the following equation:

$$(1-\sigma_1)Z_i\left[1-(1-\varepsilon)\sigma_1 Z_i\right]\ddot{Z}_i + \frac{1}{2}\left\{\left[1-(1-\varepsilon)\sigma_1 Z_i\right]^4 - 4\varepsilon^3(1-\varepsilon)(1-\sigma_1 Z_i)-\varepsilon^4\right\}\frac{\sigma_1}{\varepsilon^3}\dot{Z}_i^2$$

$$-\varepsilon\frac{\sigma_2^2}{\sigma_1}(1-\sigma_1 Z_i)=0. \qquad (8.4\text{-}22)$$

Since we have introduced two unknown parameters σ_1 and σ_2, we can thus impose two arbitrary conditions to determine them. As first trial we require that their values be such that none of the three terms in the equation shall vanish as $\varepsilon \to 0$. We see that this will be the case if the multiplying factors σ_1/ε^3 and $\varepsilon\sigma_2^2/\sigma_1$ are both equal to unity, i.e., if $\sigma_1 = \varepsilon^3$, $\sigma_2 = \varepsilon$. With this selection we now let the remaining $\varepsilon \to 0$, and Equation 8.4-22 reduces to

$$\ddot{Z}_i + \tfrac{1}{2}\dot{Z}_i^2 - 1 = 0. \qquad (8.4\text{-}23a)$$

The new initial conditions become

$$Z_i(0) = \dot{Z}_i(0) = 0. \qquad (8.4\text{-}23b)$$

The solution of Equations 8.4-23 is given by

$$Z_i = 2 \ln \cosh\left(T/\sqrt{2}\right), \qquad (8.4\text{-}24a)$$

$$= \sqrt{2}T + 2 \ln\left(\frac{1+e^{-\sqrt{2}T}}{2}\right). \qquad (8.4\text{-}24b)$$

We shall select the form in Equation 8.4-24b as the zeroth-order inner solution

If the requirement that both multiplying factors be unity had failed to yield a useful result, then we would have had to investigate other possibilities following the procedures outlined by Lin and Segel. Whatever the choice, the term involving the second-order derivative must be retained, since, otherwise, there would be no hope of imposing the initial conditions.

MATCHING THE INNER SOLUTION TO THE OUTER SOLUTION

We now have an outer solution (Equation 8.4-20), with an undetermined constant, and an inner solution (Equation 8.4-24b). The problem of matching them is somewhat akin to taking an aerial photograph of some topography from 20,000 ft, say, and a second shot of a special portion of the area bounding the first, from 1000 ft, with a different scale of magnification. We would like to combine the second with the first, but scale factors obviously intervene, particularly if both photos include portions that overlap, thus creating the necessity of avoiding duplication. In the laboratory the photographer must change the enlargement scale of one, or both, in order to piece them together to obtain a composite picture. We can extend the analogy only slightly further by pointing out that, as in photography, determining the composite solution of singular perturbation equations is an art, one that has been brought to a high state by the applied mathematicians.

We shall use the technique expounded by Van Dyke (1975, p. 70). This book was one of the earliest on perturbation methods and one of the best. Another, that of Lin and Segel (1974), written from the viewpoint of an applied mathematician, employs a related but different matching technique. In the 20 years following, books on perturbation theory have proliferated.

The matching procedure can be considered as following a recipe with frequent switching between inner and outer variables, particularly if higher-order solutions (in ε) are sought. First, we note the following relations between the (nondimensional) physical variables and the inner and outer variables (also nondimensional):

$$\zeta = Z_o = \varepsilon^3 Z_i, \quad \tau = T/\varepsilon^2 = T/\varepsilon \tag{8.4-25}$$

Since we are proceeding no farther than the zeroth-order matching, the situation is greatly simplified and the reader is cautioned that much of the art of perturbation analysis remains unexposed. In the following the zeroth-order solution is the "first" term. The first-order solution would involve an additional term, multiplied by ε to some power, which would make it a two-term expression. We proceed by furnishing the relations specified:

1-term outer solution:

$$Z_o = 1 - \left(C - 5T/\sqrt{2}\right)^{2/5};$$

Rewritten in inner variables:

$$\varepsilon^3 Z_i = 1 - \left(C - 5\varepsilon^3 T/\sqrt{2}\right)^{2/5};$$

$$= 1 - C^{2/5} \left\{ 1 - \varepsilon^3 \frac{5T}{\sqrt{2}C} \right\}^{2/5};$$

Expanded (in a series) for small ε:

$$\varepsilon^3 Z_i = 1 - C^{2/5} \left\{ 1 - \varepsilon^3 \frac{2}{5} \frac{5T}{\sqrt{2}C} + O(\varepsilon^6) \right\};$$

1-term inner of 1-term outer:

$$\varepsilon^3 Z_i = 1 - C^{2/5} + \varepsilon^3 \frac{\sqrt{2}T}{C^{3/5}};$$

Rewritten in outer variables:

$$Z_o = 1 - C^{2/5} + \frac{\sqrt{2}T}{C^{3/5}}. \tag{A}$$

1-term inner solution:
$$Z_i = \sqrt{2}T + 2 \ln \left\{ \frac{1 + e^{-\sqrt{2}T}}{2} \right\};$$

Rewritten in outer variables:
$$Z_o/\varepsilon^3 = \sqrt{2}T/\varepsilon^3 + 2 \ln \tfrac{1}{2} \left\{ 1 + e^{-\sqrt{2}T/\varepsilon^3} \right\};$$

Expanded for small ε:
$$Z_o = \sqrt{2}T + 2\varepsilon^3 \ln \tfrac{1}{2} \left\{ 1 + e^{-\sqrt{2}T/\varepsilon^3} \right\};$$

1-term outer of 1-term inner:
$$Z_o = \sqrt{2}T. \tag{B}$$

After completing the matching process we must have eliminated everything from the outer solution Equation A, and the inner solution Equation B, except the portion that is common to both, in which case the solutions are said to be matched. The only way for this requirement to be satisfied is for the constant term in Equation A to be $1 - C^{2/5} = 0$, or $C = 1$. Thus, the zeroth-order outer solution is

$$Z_o = 1 - \left(1 - 5T/\sqrt{2} \right)^{2/5}. \tag{8.4-26}$$

THE COMPOSITE SOLUTION

The composite solution, designated by the subscript c, is the sum of the inner and outer solutions, expressed in a common set of variables, less the portion of either that is common to both. In the current problem the common part was obtained during the matching process as $z_o = \sqrt{2}T = \varepsilon^2 \sqrt{2}\tau$. Returning to the original, nondimensional physical variables, we have the following expressions for the position and velocity of the free surface:

Inner:
$$\zeta_i = \varepsilon^2 \left\{ \sqrt{2}\tau + \varepsilon 2 \ln \tfrac{1}{2} \left[1 + e^{-\sqrt{2}\tau/\varepsilon} \right] \right\}, \tag{8.4-27a}$$

$$\dot{\zeta}_i = \varepsilon^2 \sqrt{2} - \frac{\varepsilon^2 2\sqrt{2}}{1 + e^{\sqrt{2}\tau/\varepsilon}}; \tag{8.4-27b}$$

Outer:
$$\zeta_o = 1 - \left[1 - \varepsilon^2 5\tau/\sqrt{2} \right]^{2/5}, \tag{8.4-28a}$$

$$\dot{\zeta}_o = \frac{\varepsilon^2 \sqrt{2}}{\left(1 - \varepsilon^2 5\tau/\sqrt{2} \right)^{3/5}}; \tag{8.4-28b}$$

Composite:
$$\zeta_c = \zeta_i + \zeta_o - \varepsilon^2 \sqrt{2}\tau,$$

or,

$$\zeta_c = 1 - \left[1 - \varepsilon^2 5\tau/\sqrt{2}\right]^{2/5} + \varepsilon^3 2 \ln \tfrac{1}{2}\left[1 + e^{-\sqrt{2}\tau/\varepsilon}\right]; \qquad (8.4\text{-}29a)$$

with

$$\dot{\zeta}_c = \varepsilon^2 \sqrt{2}\left(1 - \varepsilon^2 5\tau/\sqrt{2}\right)^{-3/5} - \varepsilon^2 2\sqrt{2}\Big/\left[1 + e^{\sqrt{2}\tau/\varepsilon}\right]. \qquad (8.4\text{-}29b)$$

For a comparison to follow we also give the free-surface position, and velocity formulas for the quasi-steady solution, designated by the subscript q. The velocity has previously been given in Equation 8.4-12. Its integral furnishes the position. Thus,

$$\varepsilon^2 \sqrt{2}\tau = \tfrac{2}{3}\left(1 + \tfrac{4}{3}\varepsilon + \tfrac{8}{3}\varepsilon^2\right) + \left(1 - \zeta_q\right)^{1/2}\left\{-\tfrac{2}{3}\left(1 + \tfrac{4}{3}\varepsilon\tfrac{8}{3}\varepsilon^2\right)\right.$$

$$\left. + \tfrac{4}{3}\left(1 - \tfrac{1}{3}\varepsilon - \tfrac{2}{3}\varepsilon^2\right)\zeta_q - \tfrac{2}{3}(1 - \varepsilon)^2\zeta_q^2\right\}; \qquad (8.4\text{-}30a)$$

$$\dot{\zeta}_q = \varepsilon^2 \sqrt{2}\left(1 - \zeta_q\right)^{1/2}\left[1 - (1 - \varepsilon)\zeta_q\right]^{-2}. \qquad (8.4\text{-}30b)$$

DISCUSSION

This example provides a good opportunity to assess the relative accuracy, as well as the deficiencies, of the several approximate solutions obtained, and to compare them with both the exact numerical solution* (designated by the subscript e), and experiment, for the representative case $\varepsilon = 0.1$. Figure 8.4-2 presents plots of the free-surface position, as a function of time, for the outer, composite, quasi-steady, and exact solutions. It turns out that the plot of the outer solution differs so little from the composite solution that it cannot be illustrated on this graph. Note that both of these solutions terminate at the point where $\tau \to \tau_m \equiv \sqrt{2}/5\varepsilon^2 = 28.28...$, $\zeta_m \to 0.9986...$, for $\varepsilon = 0.1$. For $\tau > \tau_m$ both functions turn imaginary.

The exact and quasi-steady solutions both terminate as $\zeta \to 1$. For the latter, the discharge time (τ_{dq}) has previously been given in Equation 8.4-13. The numerical values are $\tau_{de} = 32.56...$ and $\tau_{dq} = 32.81...$, respectively, which are remarkably close to each other, and both are significantly greater than the value of τ_m for the outer and composite cases.

---

* The results presented in this section for the exact numerical integration of Equation 8.4-9 were obtained by Mr. Yang-Zhou Cheng using the RPI IBM 3090-200S computer. The computation employed the MTS differential-equation solver.

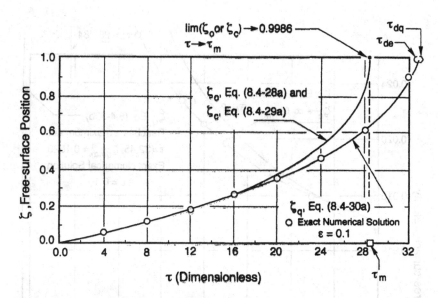

FIGURE 8.4-2 Plot of free-surface position vs. time for outer solution (ζ_o), composite solution (ζ_c), and quasi-steady solution with exact numerical solution, $\varepsilon = 0.1$.

Plots of the free-surface velocity distributions in Figure 8.4-3 provide new insight into the nature of a mathematical boundary layer. In the first place, note that the quasi-steady solution has a finite free-surface velocity at $\tau = 0$ — a violation of the initial condition. On the other hand, the composite and exact solutions (and, in fact, the inner solution that dominates the outer solution in this region) start from zero, but rise extremely rapidly to a value near to that of the initial quasi-steady value. This rapid rise is more dramatically exhibited in Figure 8.4-4, and is the reason for the earlier reference to a "starting transient." In the (dimensionless) time increment of $\Delta\tau \approx 0.3$, which is about 1% of the discharge time, the velocity rises from zero to become equal to the local quasi-steady value. This is not unlike the change in velocity in a viscous boundary layer from the no-slip value on the wall, adjusting to (almost) the free-stream value outside the boundary layer. Indeed, the analysis of the viscous boundary layer, conceived by Prandtl, and carried out by Blasius in 1907, was the first successful attempt to obtain a meaningful approximate solution for a boundary layer, viscous *or* mathematical, in their case, both.

Within the still short time increment of $\Delta\tau \approx 2$, the curve for the exact velocity parts company with that of the composite expression to join with that of the quasi-steady relation for most, but not all, of the remaining interval. The curve for the exact solution is monotonic throughout, increasing rapidly to its final value of $\dot{\zeta}_e = 0.3087$ at τ_{de}. This surge in the free-surface velocity is the type of behavior that one expects from most emptying reservoirs in the final stage.

As shown in Figure 8.4-5, the quasi-steady solution diverges from the exact only near $\tau = 30$, increasing monotonically to $\dot{\zeta}_e = 0.1529$ at $\tau = 32.45$, after which it precipitately falls to zero, a flourish that is physically impossible.

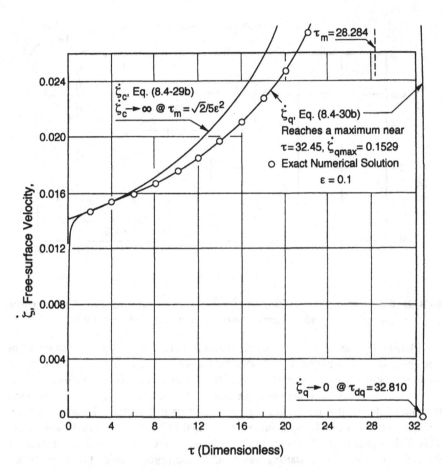

FIGURE 8.4-3 Plot of free-surface velocity vs. time for composite, quasi-steady, and exact solutions, $\varepsilon = 0.1$.

The inner and outer solutions never intersect. Yet, in the matching process, from which arose the zeroth-order composite expansion, it is remarkable that the inner solution blends seamlessly into the outer solution to cover (most of) the regime with a single function that satisfies the initial conditions. However, by comparison with the exact solution, as shown in Figures 8.4-3 and 8.4-4, it is evident that the composite solution is still defective. Not only does the composite velocity become infinite as $\tau \to \tau_m$, a physical impossibility, but τ_m provides only a crude approximation to the discharge time, understating the exact value by 13%, for the case of $\varepsilon = 0.1$.

The infinity of the outer velocity is a tip-off that there is another boundary layer, this time at the other end of the interval, i.e., near τ_d. Such a situation in singular perturbation problems, although unusual, can be anticipated in situations where the phenomenon — in this case the exit flow — terminates abruptly. A second analysis analogous to that leading to the discovery of the inner solution should allow the

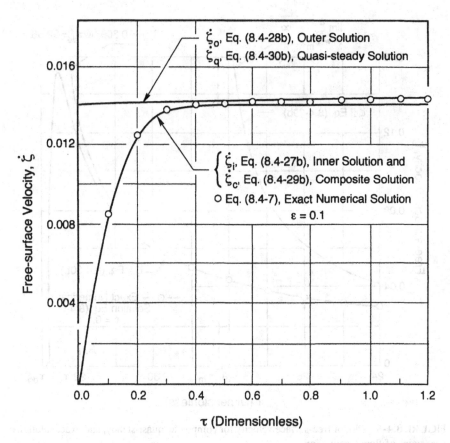

FIGURE 8.4-4 Plot of free-surface velocity vs. time near $\tau = 0$ for all solutions, $\varepsilon = 0.1$.

deficiency to be removed. Attempts to complete this analysis have so far not been successful.

COMPARISON WITH EXPERIMENT

A series of measurements of conical reservoir flow were reported in Brower and Way (1964). The nominal height of the cone was 30 in. The exit end had removable segments, which allowed the choice of any of three different outlet diameters. By varying the outlet, and by partially filling the reservoir, measurements over a wide range of the parameter ε were attained.

The data, using tap water as fluid, are compared with exact theory and quasi-steady theory in Figure 8.4-6. A plotting error in the original report has been corrected, and the original data have been replotted with $\varepsilon^2 \tau_d 5 / \sqrt{2}$ as the ordinate vs. ε. The runs with the higher values of the discharge time τ_d are on the left. The scatter in the data is attributed (in hindsight) to inability to measure the initial free-surface height with sufficient precision, an effect that is magnified in the plot for small values of ε.

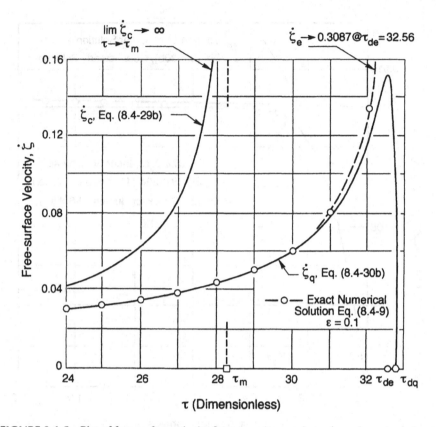

FIGURE 8.4-5 Plot of free-surface velocity for composite, quasi-steady, and exact solutions near point of flow termination.

Although the experimental data lie almost entirely above both of the theoretical curves, it is unlikely that this is due to viscous effects not accounted for in the theory, because visual observations failed to detect any noticeable deviation from a flat free-surface in any of the runs.

As to the theoretical curves, the faired line drawn through the points obtained by numerical integration does not pass through the point for $\varepsilon = 0.06$ — a calculation that was double-checked. The only explanation offered is to note that computations for $\varepsilon < 0.04$ all resulted in overflow, and we conjecture that the anomaly represents an early warning signal by the computer.

Most important, it is remarkable that quasi-steady theory, except in the vicinity of the starting and terminal transients, provides reasonable accuracy for values of the free-surface velocity and, consequently, the discharge time, as long as $\varepsilon < 0.2$.

8.5   ON THE NOTION OF CHARACTERISTICS

We next turn to an aspect of unsteady flow not encountered in situations involving a constant-density fluid moving in a duct of fixed geometry. In the previous sections

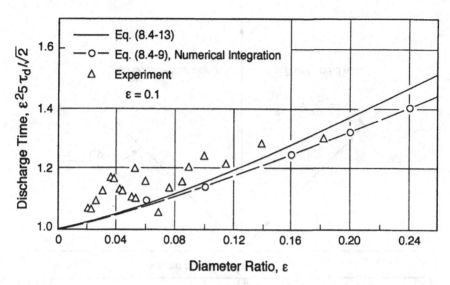

FIGURE 8.4-6 Comparison of discharge times for exact and quasi-steady solutions with experiment.

we demonstrated that if the fluid velocity is known at any point at a given time, the velocity at every other point at the same time can be immediately calculated from the continuity equation. Furthermore, the analysis of such flows devolves to the question of solving an ordinary differential equation.

However, if the restriction of either a constant-density fluid or a fixed duct-shape is removed, a situation arises that is accompanied by new analytical complexities, one that demands a more sophisticated view of the physics of the flow.

Suppose that we have a horizontal tube of fixed cross-sectional area, Figure 8.5-1a, filled with a perfect gas of uniform temperature which, initially, is at rest. At the center of the tube there is a gas particle (i.e., a disk of fluid), which, by a mechanism we need not specify, causes a short-duration sound pulse to be emitted at $t = 0+$. This could be thought to be generated by a person within the tube who unleashes a shout. We are all familiar with the way a tube conducts sound pulses. In fact, two pulses would necessarily be created, one moving to the right, the other to the left, both moving at the local sound speed a.

The motion of the pulses through the gas can be represented on what is known as an x,t-diagram. Since, for the current purpose, the speed of sound is constant, the plot of the propagating pulses is a pair of inclined lines with slopes $d^{\pm}t/dx = (d^{\pm}x/dt)^{-1} = \pm 1/a$, to the right and left, respectively. These represent the simplest case of *characteristics*, which are the lines along which "information" is transmitted acoustically from one part of a fluid to another.

Now suppose that the fluid is moving to the right at constant velocity $u < a$ with respect to the wall of the tube. The differential equation for the particle pathline is $Dx/Dt = u$, and the slope of the pathline is $Dt/Dx = 1/u$. The pathline for the particle, which is located at $x = 0$ when $t = 0$, is shown as a solid line. In the equation for a pathline the differentials Dx and Dt are treated no differently from any other differ-

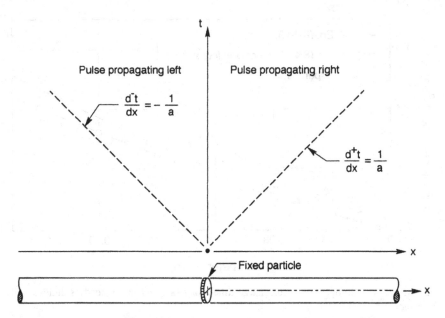

FIGURE 8.5-1a Sketch of pathlines for pulse generated on a particle within a quiescent tube of gas at a fixed temperature.

entials; the capital letter simply emphasizes that the differential is taken along the pathline.

The Differential Equation for a Characteristic Line

The velocities of the disturbances in the moving fluid are now $u \pm a$ with respect to the wall. Because we must be concerned with calculating the paths of these disturbances, i.e., the characteristic lines, we introduce a new, special symbol to specify the differential equations of the propagating disturbances. That is, we put

$$\frac{\delta^{\pm}x}{\delta t} = u \pm a. \tag{8.5-1}$$

Analogous to the differential Dx along a pathline, the symbols $\delta^{\pm}x$ signify that the differentials are to be taken along the characteristic lines. Their slopes are given by the inverse of the preceding expression; thus $\delta^{\pm}t/\delta x = 1/(u \pm a)$. A sketch of the pathline for the original particle, and the characteristics emanating therefrom, is given in Figure 8.5-1b. The left-hand characteristic is said to be propagating upstream. In all cases shown, the values of u and a are supposed to remain constant. In general, characteristics are likely to be curved due to spatial and temporal variation of both the flow speed and the speed of propagation of the disturbances.

In the third case, Figure 8.5-1c, $u > a$; therefore, $u - a$ is positive, corresponding to a flow that is supersonic with respect to the wall. The slopes of the pathline and

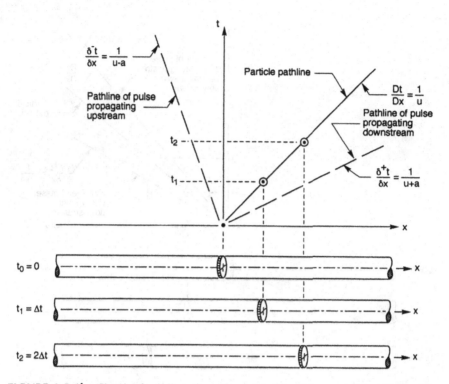

FIGURE 8.5-1b Sketch of pathlines for pulse generated on a gas particle within a tube, moving at uniform velocity $u < a$.

characteristics are all positive. In such a case we say that the left-hand disturbance is propagating upstream, but is being swept downstream by the flow.

DIFFERENTIATING AN ARBITRARY FUNCTION ALONG A CHARACTERISTIC

In Section 3.5 we discussed the significance of the substantial derivative that was determined from the expression for the time derivative of a function taken in an arbitrary direction. We can repeat the process for a derivative along a characteristic. Thus, for $f = f(x, t)$, we have

$$\left.\frac{d}{dt}\right|_{char} = \frac{\partial f}{\partial t} + \left.\frac{dx}{dt}\right|_{char} \frac{\partial}{\partial x}$$

or

$$\frac{\delta^{\pm}}{\delta t} = \frac{\partial}{\partial t} + \frac{\delta^{\pm}x}{\delta t}\frac{\partial}{\partial t},$$

$$= \frac{\partial}{\partial t} + (u \pm a)\frac{\partial}{\partial t}.$$

(8.5-2)

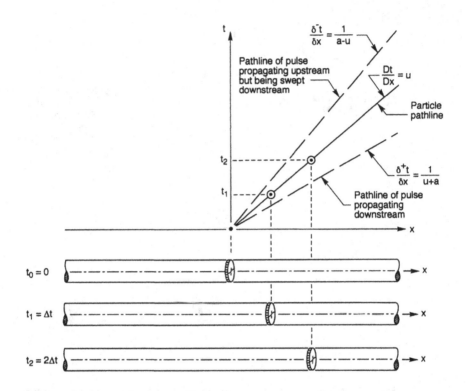

FIGURE 8.5-1c   Sketch of pathlines for pulse generated on a gas particle moving within a tube at velocity $u > a$.

If these simple, physical ideas are firmly grasped, the analytical techniques introduced in the next section will be much easier to digest.

8.6   THEORY OF HYPERBOLIC EQUATIONS FOR FUNCTIONS OF TWO INDEPENDENT VARIABLES

INTRODUCTION

The subject of partial differential equations is a vast field. Although we only scratch the surface herein, it is possible to present a technique that has proved immensely valuable to analysts and that is fundamental for attacking a wide variety of problems in fluid mechanics. This section leans heavily on the presentation of Courant and Friedrichs (1948, pp. 40–45). We shall put the techniques developed to immediate use in the following sections.

BASIC EQUATIONS

Consider a pair of first-order differential equations with dependent variables (u, v) and independent variables (x, y); i.e, $u = u(x, y)$, $v = v(x, y)$. Then we denote a pair of differential equations by

$$
\left.
\begin{aligned}
L_1 &= A_1 u_x + B_1 u_y + C_1 v_x + D_1 v_y + E_1 = 0 \\
L_2 &= A_2 u_x + B_2 u_y + C_2 v_x + D_2 v_y + E_2 = 0
\end{aligned}
\right\},
\qquad (8.6\text{-}1)
$$

where $A_1, A_2, B_1, \ldots E_2$ are specified functions of (x, y, u, v). It is assumed that there is no point at which $A_1/A_2 = B_1/B_2 = C_1/C_2 = D_1/D_2$. Furthermore, all functions are assumed to possess as many (noninfinite) derivatives as needed. Since the dependent variables appear in both equations, they are said to be *coupled*.

The following definitions will help classify the nature of the equations:

1. If $A_1, A_2, B_1, \ldots$ are all constants, or functions of (x, y) only, the system is said to be *linear*. Otherwise, they are *nonlinear*, which makes for greater analytical difficulty.
2. If $A_1, A_2, B_1, \ldots$ are functions only of (u, v), the equations are said to be *quasi-linear*.
3. If $E_1 = E_2 \equiv 0$, the equations are said to be *homogeneous*; otherwise, they are *nonhomogeneous*.
4. If the system of equations is both quasi-linear and homogeneous, then it is said to be *reducible*, that is, reducible to a set of linear equations by interchanging the roles of the dependent and independent variables. This may lead to a substantial simplification. For example, if (x, y) are the usual space variables and (u, v) the corresponding velocity components, the process of reduction throws the problem into the *hodograph* plane, a technique that has been applied with good effect in transonic flow. In accomplishing this simplification, however, the difficulty in applying the boundary conditions, which are specified in the physical plane, may increase.

We will not take the question of reducibility further in this text. Therefore, the following is devoted to the question of the general theory of quasi-linear, partial differential equations.

THE NOTION OF A CHARACTERISTIC DIRECTION

A brief attempt to explain the physical significance of characteristics is given in the previous section. From a mathematical viewpoint, quoting Courant and Hilbert (1962, p. 408), "a direction is characteristic at a point P if there exists a linear combination of the differential equations for which all of the unknowns [read 'dependent variables'] are differentiated at P only in this direction." Thus, characteristics are intimately associated with directional derivatives, which are discussed in any good book on advanced calculus.

With regard to Figure 8.6-1, a directional derivative of the function $f = f(x, y)$ in the direction s, which makes an angle α with the x-axis at an arbitrary point, is given by

$$
df/ds = \cos \alpha \, f_x + \sin \alpha \, f_y,
\qquad (8.6\text{-}2)
$$

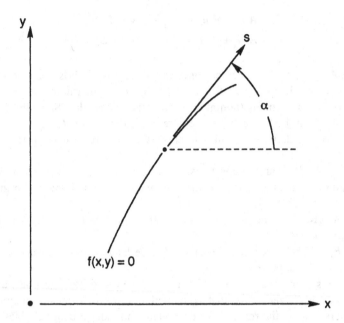

FIGURE 8.6-1   Definition of directional derivative along a curve $f(x,y) = 0$.

where the subscripts are shorthand notation for partial derivatives. But any combination $af_x + bf_y$ is a directional derivative, since we can always define an angle $\alpha = \tan^{-1}(b/a)$ by factoring out $(a^2 + b^2)^{1/2} \equiv \surd$; thus,

$$af_x + bf_y = \surd \left\{ \frac{a}{\surd} f_x + \frac{b}{\surd} f_y \right\} = \surd \left[\cos \alpha \, f_x + \sin \alpha \, f_y \right] = \surd \, (df/ds). \quad (8.6\text{-}3)$$

Consider next some curve specified by a parametric variable σ, so that $x = x(\sigma)$, $y = y(\sigma)$, then, $dy/dx = y_\sigma/x_\sigma = b/a$, where a and b are the results of the partials. Consequently, $af_x + bf_y$ represents a directional derivative along the curve. If, in L_1, L_2 we consider that $u = u(x, y) = u[x(\sigma), y(\sigma)]$, $v = v(x, y) = v[x(\sigma), y(\sigma)]$, then we seek a linear combination of the two equations such that the dependent variables combine in characteristic form. That is, we seek the pair of constants λ_1, λ_2 such that, in forming $L \equiv \lambda_1 L_1 + \lambda_2 L_2$, the u and the v terms, when combined, have exactly the characteristic form just described. By forming the prescribed combination,

$$L = \left(\lambda_1 A_1 + \lambda_2 A_2 \right) u_x + \left(\lambda_1 B_1 + \lambda_2 B_2 \right) u_y$$

$$+ \left(\lambda_1 C_1 + \lambda_2 C_2 \right) v_x + \left(\lambda_1 D_1 + \lambda_2 D_2 \right) v_y + \lambda_1 E_1 + \lambda_2 E_2 = 0. \quad (8.6\text{-}4)$$

We now require that the multiplying factors play the roles of a or b, respectively. Thus, we must have $(\lambda_1 A_1 + \lambda_2 A_2) = a = x_\sigma$, $(\lambda_1 B_1 + \lambda_2 B_2) = b = y_\sigma$, and similarly for the terms in v. The necessary conditions are

$$\frac{\lambda_1 B_1 + \lambda_2 B_2}{\lambda_1 A_1 + \lambda_2 A_2} = \frac{\lambda_1 D_1 + \lambda_2 D_2}{\lambda_1 C_1 + \lambda_2 C_2} = \frac{y_\sigma}{x_\sigma}. \tag{8.6-5}$$

Equation 8.6-5 provides two independent relations. We next solve for the two factors $\lambda_1 B_1 + \lambda_2 B_2$, $\lambda_1 D_1 + \lambda_2 D_2$, and substitute them in Equation 8.6-4. After multiplying through by x_σ, we have

$$\left(\lambda_1 A_1 + \lambda_2 A_2\right)\left[u_x x_\sigma + u_y y_\sigma\right] + \left(\lambda_1 C_1 + \lambda_2 C_2\right)\left[v_x x_\sigma + v_y y_\sigma\right]$$

$$+ \left(\lambda_1 E_1 + \lambda_2 E_2\right) x_\sigma = 0. \tag{8.6-6}$$

By the chain rule, $u_x x_\sigma + u_y y_\sigma = u_\sigma$, and $v_x x_\sigma + v_y y_\sigma = v_\sigma$. Thus, Equation 8.6-6 becomes

$$\left(\lambda_1 A_1 + \lambda_2 A_2\right) u_\sigma + \left(\lambda_1 C_1 + \lambda_2 C_2\right) v_\sigma + \left(\lambda_1 E_1 + \lambda_2 E_2\right) x_\sigma = 0. \tag{8.6-7}$$

Similarly, if we eliminate the other two parentheses, we get

$$\left(\lambda_1 B_1 + \lambda_2 B_2\right) u_\sigma + \left(\lambda_1 D_1 + \lambda_2 D_2\right) v_\sigma + \left(\lambda_1 E_1 + \lambda_2 E_2\right) y_\sigma = 0. \tag{8.6-8}$$

Equations 8.6-5, 8.6-7, and 8.6-8 form a set of four linear, algebraic equations in λ_1, λ_2; consequently, they must be *linearly dependent*. They are

$$\lambda_1\left(A_1 y_\sigma - B_1 x_\sigma\right) + \lambda_2\left(A_2 y_\sigma - B_2 x_\sigma\right) = 0,$$

$$\lambda_1\left(C_1 y_\sigma - D_1 x_\sigma\right) + \lambda_2\left(C_2 y_\sigma - D_2 x_\sigma\right) = 0,$$

$$\tag{8.6-9}$$

$$\lambda_1\left(A_1 u_\sigma + C_1 v_\sigma + E_1 x_\sigma\right) + \lambda_2\left(A_2 u_\sigma + C_2 v_\sigma + E_2 x_\sigma\right) = 0,$$

$$\lambda_1\left(B_1 u_\sigma + D_1 v_\sigma + E_1 y_\sigma\right) + \lambda_2\left(B_2 u_\sigma + D_2 v_\sigma + E_2 y_\sigma\right) = 0.$$

Since these equations are homogeneous and linearly dependent in λ_1, λ_2, for a solution to exist the determinant of the coefficients of any pair must be zero. From the first two:

$$\begin{vmatrix} A_1 y_\sigma - B_1 x_\sigma & A_2 y_\sigma - B_2 x_\sigma \\ C_1 y_\sigma - D_1 x_\sigma & C_2 y_\sigma - D_2 x_\sigma \end{vmatrix} = 0. \tag{8.6-10}$$

This yields the equivalent equations:

$$\bar{a} y_\sigma^2 - 2\bar{b} x_\sigma y_\sigma + \bar{c} x_\sigma^2 = 0,$$

or

$$\bar{a}(dy/dx)^2 - 2\bar{b}(dy/dx) + \bar{c} = 0, \tag{8.6-11}$$

where, introducing the special notation $[XY] \equiv X_1Y_2 - X_2Y_1$,

$$\bar{a} \equiv A_1C_2 - A_2C_1 \equiv [AC],$$

$$2\bar{b} \equiv \qquad\qquad \equiv [AD] + [BC], \tag{8.6-12}$$

$$\bar{c} \equiv \qquad\qquad \equiv [BD].$$

The solution of Equation 8.6-11 is

$$\frac{dy}{dx} = \frac{\bar{b} \pm \sqrt{\bar{b}^2 - \bar{a}\bar{c}}}{\bar{a}} \equiv \zeta^{\pm}. \tag{8.6-13}$$

This analysis requires that $\bar{a} \neq 0$, and that $\bar{b}$ and $\bar{c}$ not be simultaneously zero.

The sign of the discriminant $\Delta \equiv \bar{b}^2 - \bar{a}\bar{b}$ allows us to make a crucial classification of the differential equations. Thus, if

$$> 0, \quad \text{equations are hyperbolic};$$

$$\Delta = 0, \quad \text{equations are parabolic}; \tag{8.6-14}$$

$$< 0, \quad \text{equations are elliptical}.$$

For the current purpose we are interested only in the case where the equations are hyperbolic. If the discriminant is real and nonzero, then a pair of characteristic lines exists, and the equations are said to be *hyperbolic*. A discussion of the parabolic and elliptical cases does not fall within the current objectives.

For emphasis we note, for the hyperbolic case, that the expressions for $\zeta^{\pm}$ in Equation 8.6-13 represent a pair of directional derivatives (necessarily unequal) at an arbitrary point P in the field, where $d^{\pm}y/dx = y_\sigma/x_\sigma = \zeta^{\pm}$. Further, Equation 8.6-13 is in the form of a pair of ordinary differential equations the integration of which provides the pair of characteristic curves passing through P. The two families of the curves thus generated are often denoted by $C^{\pm}$. Note that if $\zeta^{\pm}$ depends upon (u, v) as well as (x, y), the equations are nonlinear and are said to be *coupled*. In most cases the characteristic curves form an intersecting, curvilinear net.

THE INITIAL DATA LINE

The solution of these differential equations must proceed from an initial data line along which the dependent variables $u(x, y)$, $v(x, y)$, and, hence, $\zeta^{\pm}$ are specified.*

---

* The last result follows from the fact that the coefficients A_1, A_2, B_1...are known functions of (x, y, u, v). Consequently, the same must be true for $\bar{a}$, $\bar{b}$, $\bar{c}$,...and, thus, for $\zeta^{\pm}$.

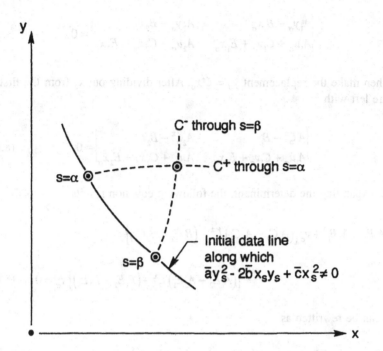

FIGURE 8.6-2 Plot illustrating intersection of C^+-characteristic from $s = \alpha$ with C^--characteristics from $s = \beta$, both points lying on an initial data line.

In principle any line that is not a characteristic will do, e.g., as indicated in Figure 8.6-2. In physical problems the most appropriate choice of the initial data line will usually be evident. If the independent variables on this line can be expressed in terms of a parameter s, i.e., $x = x(s)$, $y = y(s)$ — where the use of s is unrelated to its previous use in directional derivatives — then the requirement to be satisfied on the line is $\bar{a}y_s^2 - 2\bar{b}x_s y_s + \bar{c}x_s^2 \neq 0$.

Suppose that we choose a pair of points on the curve designated by $s = \alpha$, $s = \beta$. Then, if the points are sufficiently close together — i.e., if $|\alpha - \beta|$ is sufficiently small — the appropriate characteristics, C^+ through α and C^- through β, will intersect as shown.

DETERMINATION OF THE CHARACTERISTIC PARAMETERS

The determination that a given system of equations is hyperbolic and the differential equations that fix the shape of the characteristics are both provided by Equation 8.6-13. We next determine the equations that must be satisfied along the two families of characteristics.

We bypass the determination of the parameters λ_1, λ_2 — since they are linearly dependent, only their ratio can be determined in any case — in favor of determining the equations governing the dependent variables along the two families of characteristics. We use the first and third equations of Equation 8.6-9 and put the determinant of their coefficients equal to zero, as before. Thus,

$$\begin{vmatrix} A_1 y_\sigma - B_1 x_\sigma & A_2 y_\sigma - B_2 x_\sigma \\ A_1 u_\sigma + C_1 v_\sigma + E_1 x_\sigma & A_2 u_\sigma + C_2 x_\sigma + E_2 x_\sigma \end{vmatrix} = 0.$$

We then make the replacement $y_\sigma = \zeta^{\pm} x_\sigma$. After dividing out x_σ from the first row, we are left with

$$\begin{vmatrix} A_1 \zeta^{\pm} - B_1 & A_2 \zeta^{\pm} - B_2 \\ A_1 u_\sigma + C_1 v_\sigma + E_1 x_\sigma & A_2 u_\sigma + C_2 x_\sigma + E_2 x_\sigma \end{vmatrix} = 0. \qquad (8.6\text{-}15)$$

Then, expanding the determinant, the following equation results:

$$u_\sigma \left(A_1 B_2 - A_2 B_1 \right) + v_\sigma \left[\left(A_1 C_2 - A_2 C_1 \right) \zeta^{\pm} - \left(B_1 C_2 - B_2 C_1 \right) \right]$$

$$+ \left[\left(A_1 E_2 - A_2 E_1 \right) \zeta^{\pm} - \left(B_1 E_2 - B_2 E_1 \right) \right] x_\sigma = 0. \qquad (8.6\text{-}16)$$

This can be rewritten as

$$T u_\sigma + \left(\bar{a} \zeta^{\pm} - S \right) v_\sigma + \left(K \zeta^{\pm} - H \right) x_\sigma = 0, \qquad (8.6\text{-}17a)$$

where

$$\begin{aligned} T &\equiv [AB], & K &\equiv [AE], \\ S &\equiv [BC], & H &\equiv [BE]. \end{aligned} \qquad (8.6\text{-}17b)$$

An alternative form to Equation 8.6-17a that omits the parameter σ and that will be useful for future purposes can be obtained by dividing through by x_σ. Then, since $u_\sigma/x_\sigma = du/dx$, $v_\sigma/x_\sigma = dv/dx$, the equations for the dependent variables take the form:

$$T \frac{du}{dx} + \left(\bar{a} \zeta^{\pm} - S \right) \frac{dv}{dx} + \left(K \zeta^{\pm} - H \right) = 0. \qquad (8.6\text{-}17c)$$

Equation 8.6-17c forms a pair of coupled, ordinary differential equations that are applicable along the respective characteristic lines $C^{\pm}$, whose shape is determined by the integration of Equation 8.6-13. A practical application of this theory is provided in Section 8.7.

Equations 8.6-13 and 8.6-17 have thus become our working relations for a system of quasi-linear, partial differential equations in two independent variables. There has been some agony in getting there, but if the reader treats the current results as a recipe, to be followed with care, the reader may verify that the effort has been worthwhile. We will put these relations to work in the following sections.

The following quotation from Courant and Friedrichs (1948) may be helpful:

> The ... particular case ... mentioned before should be noted: when the differential [Equation 8.6-1] are linear, then $\zeta^{\pm}$ are known functions of x, y; and [Equations 8.6-13] are not coupled with [Equation 8.6-17], and thus [Equation 8.6-13] determines two families of characteristic curves $C^{\pm}$ independent of the solution [of Equations 8.6-17].

Otherwise, the equations are coupled and are nonlinear. If so, the equations may have to be integrated numerically, or by some approximate technique such as perturbation analysis.

NOTE

The engineer-analyst — most probably of the number-crunching variety — will ignore the lessons of this section at his or her own peril. The engineer may end up like the very real graduate student who — with a limited analytical background — was analyzing the plastic deformation of a rectangular plate. He applied standard finite-difference techniques to the differential equations. After many fruitless sessions on the mainframe, he finally appealed to a mathematician for guidance.

While listening to the agonies his confessor elicited the fact our protagonist had applied boundary conditions consistent with the equations of elasticity on the four walls of the plate. He was quickly informed that therein lay the source of his dilemma. The equations of plasticity are hyperbolic, whereas the equations of elasticity are elliptic, requiring entirely different boundary and/or initial conditions. By imposing conditions on all four walls, the problem had become overprescribed, and the computation forever doomed to nonconvergence.

> Folly and Innocence are so alike,
> The diff'rence, though essential, fails to strike.

William Cowper as quoted by Bradford Stevenson (1948)

8.7 THE GENERAL EQUATIONS FOR ONE-DIMENSIONAL, NONSTEADY GAS FLOW IN A CONSTANT-AREA DUCT

THE GOVERNING EQUATIONS

We will now utilize the methods expounded in Section 8.6 to derive the equations of nonsteady duct flow for a general gas, although all applications herein will be restricted to the perfect gas case. The objective is to obtain a set of two equations in terms of appropriate pairs of dependent and independent variables. For a horizontal duct, it is clear that the independent variables should be the pair (x, t). For one of the dependent pair the local velocity u is an obvious choice. It turns out that the most versatile choice for the second coordinate, which characterizes the state of the gas, is the static enthalpy h.

We start with the continuity equation for constant-area, mass-conserving flow (Equation 3.10-18):

$$\frac{D}{Dt}\left(\ln \rho\right)+\frac{\partial u}{\partial x}=0.$$

Eliminating the density in favor of the specific volume $v = 1/\rho$, this becomes

$$-\frac{D}{Dt}\left(\ln v\right)+\frac{\partial u}{\partial x}=0. \tag{8.7-1}$$

We next call on the first law of thermodynamics expressed in terms of the entropy function, Equation 3.15-4, which can be written in several equivalent forms:

$$Td\boldsymbol{s} = de + pdv = dh - d(pv) + pdv. \tag{8.7-2}$$

If we solve the second of these for pdv, divide by pv, the following form is obtained:

$$d(\ln v) = \frac{T}{pv}\,d\boldsymbol{s} - \frac{dh}{pv} + d(\ln pv). \tag{8.7-3}$$

Equation 8.7-3 is written for an observer fixed with respect to the mass center of the particle. We can shift to an inertial observer by dividing by dt and making the replacement of $d/dt \rightarrow D/Dt$. If we also multiply through by negative unity, the result is

$$-\frac{D}{Dt}(\ln v)=-\frac{T}{pv}\frac{D\boldsymbol{s}}{Dt}+\frac{1}{pv}\frac{Dh}{Dt}=\frac{Dg}{Dt},$$

where, for convenience, we put $g \equiv \ln pv$. When substituted into Equation 8.7-1, we have

$$\frac{\partial u}{\partial x}+\frac{1}{pv}\frac{Dh}{Dt}-\frac{T}{pv}\frac{D\boldsymbol{s}}{Dt}-\frac{Dg}{Dt}=0. \tag{8.7-4}$$

Thermodynamic equilibrium is assumed to prevail throughout. Thus, we can express g as a function of any two independent thermodynamic variables, i.e., $g = g(h, \boldsymbol{s})$. We can then write

$$dg = \frac{\partial g}{\partial h}\bigg|_{\boldsymbol{s}} dh + \frac{\partial g}{\partial \boldsymbol{s}}\bigg|_{h} d\boldsymbol{s},$$

where the subscripts denote the thermodynamic variable being held constant. Therefore, using the observer-change artifice again, this becomes

$$\frac{Dg}{Dt} = \frac{\partial g}{\partial h}\bigg|_{\mathscr{S}} \frac{Dh}{Dt} + \frac{\partial g}{\partial \mathscr{S}}\bigg|_{h} \frac{D\mathscr{S}}{Dt}. \qquad (8.7\text{-}5)$$

Substituting Equation 8.7-5 into Equation 8.7-4 produces

$$\frac{\partial u}{\partial x} + \left[\frac{1}{pv} - \frac{\partial g}{\partial h}\bigg|_{\mathscr{S}}\right]\frac{Dh}{Dt} - \left[\frac{T}{pv} + \frac{\partial g}{\partial \mathscr{S}}\bigg|_{h}\right]\frac{D\mathscr{S}}{Dt} = 0. \qquad (8.7\text{-}6)$$

It is left as an exercise* to show that

$$\left[\frac{1}{pv} - \frac{\partial g}{\partial h}\bigg|_{\mathscr{S}}\right]^{-1} = \left(\frac{\partial p}{\partial \rho}\right)_{\mathscr{S}} = a^2, \qquad (8.7\text{-}7a)$$

$$\left[\frac{T}{pv} + \frac{\partial g}{\partial \mathscr{S}}\bigg|_{h}\right] = \frac{\partial(\ln v)}{\partial \mathscr{S}}\bigg|_{h}. \qquad (8.7\text{-}7b)$$

The final form of the continuity equation is obtained by substituting Equations 8.7-7 in Equation 8.7-6 and expanding the factor Dh/Dt; it is displayed as Equation 8.7-8b.

We now manipulate the dynamic equation to put it into appropriate form. Starting from Equation 3.12-10 we expand the derivative, replace $1/\rho = v$, and then add and subtract the quantity $p\partial v/\partial x$ and combine terms, to yield

$$\frac{\partial u}{\partial t} + u\frac{\partial u}{\partial x} + \frac{\partial}{\partial x}(pv) - p\frac{\partial v}{\partial x} + \frac{\tau_w P}{\rho A} = 0.$$

We next utilize the relation provided by the first two terms of Equation 8.7-2; by solving for $pdv = Td\mathscr{S} - de$ and using a procedure analogous to the observer-change artifice, it is converted into a partial derivative, specifically,

$$-p\frac{\partial v}{\partial x} = -T\frac{\partial \mathscr{S}}{\partial x} + \frac{\partial e}{\partial x}.$$

This is substituted into the preceding relation to produce

$$\frac{\partial u}{\partial t} + u\frac{\partial u}{\partial x} + \frac{\partial}{\partial x}(pv + e) - T\frac{\partial \mathscr{S}}{\partial x} + \frac{\tau_w P}{\rho A} = 0.$$

After putting $h \equiv e + pv$, the final form of the dynamic equation is displayed as Equation 8.7-8a:

---

* See Problem 7.1b.

$$\frac{\partial u}{\partial t} + u\frac{\partial u}{\partial x} + \qquad \frac{\partial h}{\partial x} - T\frac{\partial \omega}{\partial x} + \frac{\tau_w P}{\rho A} = 0. \qquad (8.7\text{-}8a)$$

$$\frac{\partial u}{\partial x} + a^{-2}\frac{\partial h}{\partial t} + a^{-2}u\frac{\partial h}{\partial x} - \frac{\partial(\ln v)}{\partial \omega}\bigg|_h \frac{D\omega}{Dt} = 0. \qquad (8.7\text{-}8b)$$

DERIVATION OF EQUATIONS GOVERNING THE SHAPE OF THE CHARACTERISTIC LINES

Equations 8.7-8 are now in the form required to apply the analytical techniques of Section 8.6. The coefficients corresponding to Equation 8.6-1 are tabulated:

$$A_1 = 1, \qquad\qquad\qquad A_2 = 0,$$

$$B_1 = u, \qquad\qquad\qquad B_2 = 1,$$

$$C_1 = 0, \qquad\qquad\qquad C_2 = a^{-2}, \qquad\qquad (8.7\text{-}9)$$

$$D_1 = 1, \qquad\qquad\qquad D_2 = a^{-2}u,$$

$$E_1 = -T\frac{\partial \omega}{\partial x} + \frac{\tau_w P}{\rho A} \qquad E_2 = -\frac{\partial(\ln v)}{\partial \omega}\bigg|_h \frac{D\omega}{Dt}.$$

By using these, the coefficients of the quadratic of Equation 8.6-11 can be evaluated:

$$\bar{a} = [AC] = a^{-2},$$

$$2\bar{b} = [AD] + [BC] = 2u/a^2, \qquad\qquad (8.7\text{-}10)$$

$$\bar{c} = [BD] = u^2/a^2 - 1.$$

It is vital — to avoid confusion in the notation — to note that the independent coordinates of Section 8.6 have been changed for the current application; i.e., $(x, y) \rightarrow (t, x)$. Similarly, the dependent coordinates are given by $(u, v) \rightarrow (u, h)$. Therefore, the differential equations for the characteristic lines, from Equation 8.6-13, are

$$\zeta^{\pm} \equiv \frac{\delta^{\pm}x}{\delta t} = \frac{-\bar{b} \pm \sqrt{\bar{b}^2 - \bar{a}\bar{c}}}{\bar{a}} = u \pm a. \qquad (8.7\text{-}11)$$

Equation 8.7-11 determines two families of intersecting lines, denoted $C^{\pm}$, respectively. The equation depends on the flow speed u, and the local speed of sound a, both of which are initially unknown. These are the same equations as obtained in Section 8.5 by physical reasoning.

EQUATIONS GOVERNING THE DEPENDENT VARIABLES (u, h)

We next compute the quantities listed in Equation 8.6-17b:

$$T = [AB] = 1,$$

$$S = [BC] = u/a^2,$$

$$K = [AE] = -\frac{\partial(\ln v)}{\partial \mathbf{s}}\bigg|_h \frac{D\mathbf{s}}{Dt}, \tag{8.7-12}$$

$$H = [BE] = -u\frac{\partial(\ln v)}{\partial \mathbf{s}}\bigg|_h \frac{D\mathbf{s}}{Dt} + T\frac{\partial \mathbf{s}}{\partial x} - \frac{\tau_w P}{\rho A}.$$

Then, for the terms in Equation 8.6-16a

$$\bar{a}\zeta^{\pm} - S = \pm 1/a,$$

$$K\zeta^{\pm} - H = \mp a\frac{\partial(\ln v)}{\partial \mathbf{s}}\bigg|_h \frac{D\mathbf{s}}{Dt} + T\frac{\partial \mathbf{s}}{\partial x} = \frac{\tau_w P}{\rho A}. \tag{8.7-13}$$

Finally, therefore, the equations governing the dependent variables u, h, which apply along the lines $C^{\pm}$, respectively, are

$$\frac{\delta^{\pm}u}{\delta t} \pm a^{-1}\frac{\delta^{\pm}h}{\delta t} = \mp a\frac{\partial(\ln v)}{\partial \mathbf{s}}\bigg|_h \frac{D\mathbf{s}}{Dt} - T\frac{\partial \mathbf{s}}{\partial x} + \frac{\tau_w P}{\rho A}. \tag{8.7-14}$$

Since the speed of sound, a quantity related to the specific enthalpy, appears in both Equations 8.7-11 and 8.7-14, the equations are said to be coupled. Furthermore, the nonhomogeneous terms involve the entropy and the specific volume functions that are related to the specific enthalpy, adding to the complexity of the problem and increasing the extent of nonlinearity.

We note that this set of nonlinear equations is not complete. Needed is an additional state relation (or relations) that enable us to calculate the speed of sound as a function of two independent thermodynamic variables, say, $a = a(T, \mathbf{s})$. In fact, without imposing additional restrictions, the problem is hopelessly nonlinear, and the only possibility for a solution is a numerical integration of the differential equations, one that relies upon a substantial database of real gas properties.

THE EQUATIONS FOR NONSTEADY, HOMENTROPIC, PERFECT GAS FLOW

We impose the requirement that the gas be thermally and calorically perfect. It therefore follows that $h = \gamma RT/(\gamma - 1) = a^2/(\gamma - 1)$ and

$$a^{-1} \frac{\delta^{\pm} h}{\delta t} = \frac{\delta^{\pm}}{\delta t} \left(\frac{2}{\gamma - 1} a \right). \qquad (8.7\text{-}15)$$

Furthermore, since $h = h(T)$, $e = e(T)$ only, and utilizing Equation 8.7-2 once more, then $\partial (\ln v)/\partial \omega_h = 1/R$. Therefore, invoking all these additional restrictions, Equation 8.7-14 reduces to

$$\frac{\delta^{\pm}}{\delta t} \left[\frac{2}{\gamma - 1} a \pm u \right] = \frac{a}{R} \frac{D\omega}{Dt} \pm T \frac{\partial \omega}{\partial x} \mp \frac{\tau_w P}{\rho A},$$

$$= \frac{a}{R} \frac{D\omega}{Dt} \pm \frac{a^2}{\gamma R} \frac{\partial \omega}{\partial x} \mp \frac{\tau_w P}{\rho A}, \qquad (8.7\text{-}16)$$

where T has been eliminated in favor of the speed of sound $a^2 = \gamma RT$.

The entropy function shows up twice in the nonhomogeneous terms. If we impose the further restriction that the flow be adiabatic and inviscid, we see from Equation 3.15-10 that the flow must be particle isentropic, i.e., that $D\omega/Dt \equiv 0$. Further, we consider only flows that have originated from a state of uniform entropy which, with the particle-isentropic restriction, requires that $\partial \omega/\partial x \equiv 0$. Thus, the flow is also homentropic. In such case, the nonhomogeneous terms in Equation 8.7-16 vanish and Equations 8.7-16 further simplify to

$$\frac{\delta^{\pm}}{\delta t} \left[\frac{2}{\gamma - 1} a \pm u \right] = 0. \qquad (8.7\text{-}17)$$

It is customary to denote $P, Q \equiv 2a/(\gamma - 1) \pm u$. The quantities P, Q are referred to as the *Riemann parameters*, thereby honoring the mathematician who discovered them; see Riemann (1858). The significance of Equation 8.7-17 is that the Riemann parameters P, Q are constant on each and every characteristic of the families $C^{\pm}$, respectively, determined by the integration of Equation 8.7-11. On the other hand, for both families, the values of P, Q may vary from one characteristic to another. In fact, in most problems, both P and Q vary in any direction transverse to its own family of characteristics.

As a final step we can utilize a simplification due to Kantrowitz (1958) that eases the algebra slightly in the following applications. For temperatures not so high that the vibrational modes of molecules are excited, the ratio of specific heats can be written as $\gamma = (n + 2)/n$, where n is the number of classical degrees of freedom of the molecule. For example, $n = 3$ for monatomic gases, $n = 5$ for diatomic gases, etc. By solving, $n = 2/(\gamma - 1)$. It follows therefore that the Riemann parameters in Equation 8.7-17 can be written as:

$$\binom{P}{Q} = na \pm u. \qquad (8.7\text{-}18)$$

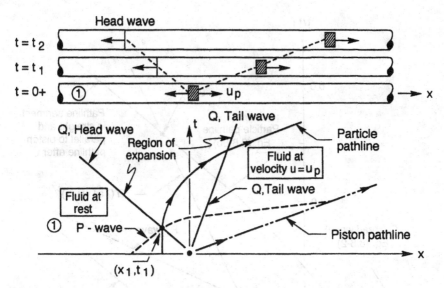

FIGURE 8.8-1 Schematic of wave diagram for impulsive motion of a piston.

8.8 IMPULSIVE MOTION OF A PISTON IN A DUCT

PROBLEM STATEMENT

At $t < 0$ a piston, Figure 8.8-1, is located in a duct at the origin $x = 0$. To the left of the piston there is a semi-infinite reservoir of thermally and calorically perfect gas at rest, at conditions T_1, p_1 or, equivalently, a_1, p_1. At $t = 0+$ the piston is caused to move impulsively to the right at constant velocity $u_P > 0$. The equation of the piston pathline is evidently

$$x_p = u_p \, t \qquad (8.8\text{-}1)$$

and is shown in Figure 8.8-1 as a solid line interrupted by dashes. We shall not concern ourselves at this point with the behavior of fluid to the right of the piston.

This motion generates an expansion that propagates left through the reservoir generating motion in the gas, which is accelerated from its initial motionless state, moving right, in the direction of the piston. We seek to analyze the motion of the gas and determine the distribution of velocity and the state functions in the gas as a function of time and space.

PRELIMINARY ANALYSIS

There must be a Q-characteristic emanating from the origin that signals to the stagnant gas the initiation of the expansion. The equation of this line is

$$x = -a_1 t. \qquad (8.8\text{-}2)$$

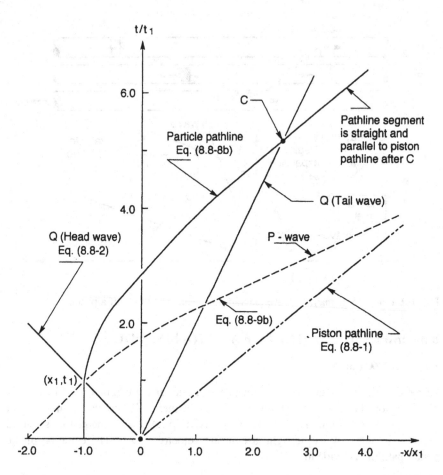

FIGURE 8.8-2 Example of wave diagram for an expansion due to a piston impulsively accelerated in a uniform duct.

This line is referred to as a *head wave*. All the other Q-characteristics emanating from the negative x-axis must be parallel to this characteristic, effectively indicating, in the region of unaccelerated gas, that no signal has yet been received tending to change the velocity from its initial state of $u_1 = 0$.

If the impulsive velocity of the piston does not exceed the theoretical maximum value (to be discussed in the following) to which the gas can be accelerated, then we expect all particles of gas in the reservoir eventually to accelerate until they move at the speed of the piston u_p. As the piston continues along its track, it must be accompanied by a continuously increasing body of gas, at a uniform state, also moving with the piston at the terminal velocity u_p. Thus, there must be another Q-characteristic, the tail wave, a straight line emanating from the origin, which denotes the time (and position) at which an arbitrary particle reaches its terminal speed. The other Q-characteristics emanating (but not shown) from the piston pathline must be parallel to the latter Q-wave.

The head and tail waves define the expansion regime in which we seek the analytical solution. This presents a novel situation for this text. It is a problem with neither a characteristic (i.e., representative) length nor, correspondingly, a characteristic time. A new idea must be invoked.

A clue is the fact that at least two Q-characteristics emanate from the origin. This indicates that the origin is a singular point.* Obviously, all the Q-characteristics lying between the head and tail waves must pass through the origin — generating what is referred to as a *centered expansion*. In fact, it seems almost certain that all these characteristics are also straight. This means that instead of seeking a solution in the form $f(x, t)$ independently, we must seek a solution of the form $f(x/t)$, since x/t = constant is the equation of a straight line through the origin.

It turns out that problem is one of a class said to be *self-similar*. In such a case the solution must be invariant on any radial line from the origin — an idea that is tied to the lack of a characteristic length or time. A way of visualizing such a solution is to imagine that a photograph is made of the completed wave diagram. If an enlargement is laid down on the original, the two correspond point by point, precisely. This unsteady problem is the analog of the *Prandtl–Meyer* expansion in steady, two-dimensional, supersonic flow, which is discussed in any text in two-dimensional, compressible flow.

SOLUTION

The equation of a Q-characteristic is specified by Equation 8.7-11. If we further require the wave to satisfy the self-similar requirement, then the only possibility is that

$$\frac{\delta^- x}{\delta t} = u - a = \frac{x}{t}, \tag{8.8-3}$$

which is the equation of a radial line. Now, we know, from Equation 8.7-17, on a P-characteristic, that $P = na + u$ = constant. Since every P-characteristic (in this problem) must originate on the negative x-axis, along which $u_1 \equiv 0$, then, at an arbitrary point thereon,

$$P = na + u = na_1.$$

But, since there is no variation in the radial direction, this includes the entire field of points in the expansion fan. Thus, within the expansion fan, at any point

$$na = na_1 - u. \tag{8.8-4}$$

Therefore, by combining Equations 8.8-3 and 8.8-4,

---

* The intersection of two characteristics of the same family always indicates the existence of a singular point.

$$\frac{x}{t} = u - \left(a_1 - \frac{u}{n}\right) = \frac{n+1}{n}u - a_1, \qquad (8.8\text{-}5)$$

and we can solve for

$$u = \frac{n}{n+1}\left(a_1 + x/t\right), \qquad (8.8\text{-}6a)$$

$$a = \frac{1}{n+1}\left(na_1 - x/t\right). \qquad (8.8\text{-}6b)$$

Equations 8.8-6 are valid at any point on, or interior to, the fan-shaped region bounded by the head and tail waves.

Now, referring to Figure 8.8-2, consider any point on the head wave designated by (x_1, t_1) where, $x_1 = -a_1 t_1$. The equation for the segment of the P-characteristic originating on the x-axis that passes through that point is

$$x = x_1 + a_1\left(t - t_1\right), \qquad (8.8\text{-}7)$$

which is a straight line. For the portion of the characteristic lying within the expansion, the differential equation* is

$$\frac{\delta^+ x}{\delta t} = u + a = \frac{1}{n+1}\left[2na_1 + (n-1)\left(x/t\right)\right]. \qquad (8.8\text{-}8a)$$

This is a linear, nonhomogeneous, ordinary differential equation. Integrating from (x_1, t_1) to an arbitrary point on the characteristic, the result is

$$\frac{x}{x_1} = (n+1)\left(\frac{t}{t_1}\right)^{(n-1)/(n+1)} - n\left(\frac{t}{t_1}\right) \qquad (8.8\text{-}8b)$$

within the expansion fan.

Similarly, the differential equation for the particle pathline is

$$\frac{Dx}{Dt} = u = \frac{n}{n+1}\left(a_1 + x\right), \qquad (8.8\text{-}9a)$$

whose integral passing through (x_1, t_1) is

---

* Equations 8.8-7 and 8.8-8a allow the illustration of one of the fine points of the method of characteristics. From the first it turns out that $-d(x/x_1)/d(t/t_1) = 1$ at (x_1, t_1), where x_1 is a negative number. From the second we have $-d(x/x_1)/d(t/t_1) = (n-1)/(n+1)$ at the same point. Thus, while the function x/x_1 is continuous at the characteristic line, its derivative is not. Similar discontinuities may occur at the tail wave.

$$\frac{x}{x_1} = (n+1)\left(\frac{t}{t_1}\right)^{n/(n+1)} - n\left(\frac{t}{t_1}\right). \tag{8.8-9b}$$

With these relations we can turn to an application of the theory.

APPLICATION

Consider the setup of Figure 8.8-2. The tube reservoir is filled with air ($n = 5$) at $p_1 = 5$ atm = 5.065×10^5 Pa, $T_1 = 300$ K. Therefore, referring to Table 2.15-2, the reservoir speed of sound is $a_1 = a_a\sqrt{T_1/T_a} = 340.3\sqrt{300/288.15} = 347.2$ m/s. We choose, arbitrarily, the piston velocity to be $u_p = 450$ m/s. The flow conditions on the tail wave are designated by the subscript 2 — which also corresponds to the conditions of the fluid between the tail wave and the piston pathline. From Equation 8.8-4 we can then write $a_2 = a_1 - u_2/n$. But $u_2 = u_p$. Therefore, $a_2 = 347.2 - 450/7 = 282.9$ m/s and $a_{21} = 0.814\,8$. On the tail characteristic the local Mach number is $M_2 = 450/282.9 = 1.59$. Also, from the equations leading to Equation 7.7-8, the isentropic relation yields $p_{21} = a_{21}^{(\gamma-1)/2\gamma} = a_{21}^{n+2} = 0.158\,3$ and $p_2 = p_1 p_{21} = 0.801\,8 \times 10^5$ Pa.

DISCUSSION

The idea of an impulsive acceleration of a piston is a fiction, but one that could conceivably be achieved approximately, within a short distance, using a sufficiently strong electromagnetic accelerator. There is a limit, however, to the speed to which the fluid can expand. Applying Equation 8.8-4 again, this time on the assumption that the piston is moving just fast enough that there would be a vacuum on the back face of the piston, where the pressure, temperature, and speed of sound all approach zero, then the terminal velocity must be $u \equiv u_e = na_1 = 2\,431.0$ m/s and the Mach number $M_e \to \infty$.

The subscript e corresponds to the depiction *escape speed*, used by some authors, e.g., Courant and Friedrichs (1948, p. 101). This is apparently an analogy to the speed required for a particle to escape Earth's gravitational field. If the piston speed exceeds u_e the fluid does not have enough energy to overtake it. Before reaching that speed the fluid would condense into a liquid, and even if condensation were delayed by some quasi-static effect, the mean-free-path length would grow so large that continuum flow theory would fail, and free-molecule theory would be required to complete the analysis.

On first glance it might seem possible to utilize this solution to generate a high-velocity, supersonic jet from a long tube reservoir with a partition at the tube exit bounding a low-pressure plenum, by suddenly removing the partition. Here plausibility fails, however. With the partition removed, the flow accelerates until sonic conditions are just attained. This condition then precludes any additional expansion waves from being propagated upstream so that the flow never becomes supersonic in the tube interior.

We emphasize that the ability to obtain a closed form analytical solution in this case depended upon the fact that the Q-characteristics were straight lines, all emanating from a fixed point. In most practical cases, neither set of characteristics is composed of straight lines. In such a situation, even for a homentropic flow, a numerical solution is the only hope. See Rudinger (1955) for a large number of worked problems by numerical methods. Foa (1960) has a complete chapter devoted to the subject, with many worked problems, and a number of useful charts for gases of different ratios of specific heats.

The shock tube, which produces extremely short-duration, high-stagnation-temperature supersonic flows, utilizes a centered expansion to initiate the process of creating the ultimate high-temperature pulse. See Thompson (1972, pp. 423–430) for a short review of the principles of a shock tube.

8.9 PROPAGATION OF AN ISENTROPIC FINITE-AMPLITUDE COMPRESSION PULSE

INTRODUCTION

In Section 8.8 we dealt with a centered expansion in which the only point involving an intersection of two characteristics of the same family was at the vertex of the fan. This point can be ignored as being a mere mathematical curiosity.

In the current section we will deal with a more instructive singularity, one which has important fluid mechanical significance, namely, the inception of the creation of a shock wave. This analysis is due to Kantrowitz (1958). We first review the applicable relations from the previous two sections.

- Differential equation for pathline:

$$Du/Dt = u; \tag{8.9-1a}$$

- Differential equations for $C^{\pm}$-characteristics:

$$\delta^{\pm} x / \delta t = u \pm a; \tag{8.9-1b}$$

- Equations for Riemann parameters:

$$(P, Q) \equiv na \pm u = \text{constants, on } C^{\pm}. \tag{8.9-2}$$

From Equation 8.9-2 we can solve for

$$a = \tfrac{1}{2n} (P + Q), \quad u = \tfrac{1}{2} (P - Q). \tag{8.9-3}$$

Substitution of Equation 8.9-3 into Equation 8.9-1b produces

$$\frac{\delta^{\pm} x}{\delta t} = \tfrac{1}{2n} (P + Q) \pm \tfrac{1}{2} (P - Q) = \tfrac{1}{2} P \left(\tfrac{1}{n} \pm 1 \right) + \tfrac{1}{2} Q \left(1 \mp \tfrac{1}{n} \right). \tag{8.9-4}$$

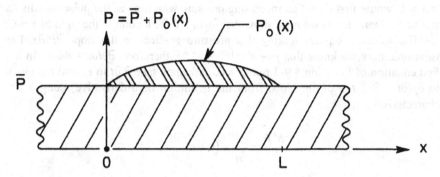

FIGURE 8.9-1 Distribution of function $P_0(x, 0)$ for a finite-amplitude P-wave.

THE PULSE DEFINITION

In Section 8.8, the initial data line was the negative x-axis, but because we had to integrate along only one P-characteristic, which was representative of the entire flow, the real significance of an initial data line was obscured. In this section we will deal with a pulse of finite amplitude, propagating in a uniform tube, which is chosen so that only one of the two families of Riemann parameters is actively involved in the analysis. The initial data line will be the x-axis at $t = 0$.

Consider a pulse such that

$$Q(x, 0) \equiv 0, \tag{8.9-5a}$$

$$P(x, 0) = \left.\begin{array}{ll} \overline{P}, & x < 0, \\ \overline{P} + P_0(x), & 0 \le x \le L, \\ \overline{P}, & x > L. \end{array}\right\} \tag{8.9-5b}$$

The pulse is specified by the function $P_0(x)$, which varies over the interval $0 \le x \le L$, superimposed on the otherwise uniform distribution $P = $ constant. Therefore, from Equation 8.9-3, and over the same intervals as in Equation 8.9-5b,

$$u(x, 0) = \begin{array}{l} \frac{1}{2}\overline{P}, \\ \frac{1}{2}(\overline{P} + P_0), \\ \frac{1}{2}\overline{P}, \end{array} \quad a(x, 0) = \begin{array}{l} \frac{1}{2n}\overline{P}, \\ \frac{1}{2n}(\overline{P} + P_0), \\ \frac{1}{2n}\overline{P}. \end{array} \tag{8.9-6}$$

An example typical of the pulse under consideration is shown in Figure 8.9-1.* The pulse contour could be half the cycle of a sine curve, although any convex curve would be satisfactory for our purpose. Since $Q = 0$, $\delta Q/\delta t \equiv 0$. Therefore, we are dealing with a pure P-pulse that propagates to the right. We first show that the right-hand portion of the pulse is a compression. That is, a sensor, fixed in space, located

---

* How one would generate such a pulse in red life is a more complex question.

at $x > L$, would first detect an increasing pressure with time as the pulse transits its station. In literal terms we must show that $\partial p/\partial t > 0$ for a point to the right of $x = C$.

The solution requires linking this pressure gradient to the slope $\partial P/\partial x$. For isentropic flow we know that $p \sim a^{(\gamma - 1)/2\gamma} \sim a^{n + 2}$; therefore, $\partial p/\partial t \sim \partial a/\partial t$. In the first equation of Equation 8.9-3 put $Q = 0$, and again take $\partial/\partial t$ to extend the chain to $\partial p/\partial t \sim \partial a/\partial t \sim \partial P/\partial t$. Now, from the definition of a derivative along a C^+-characteristic,

$$\frac{\delta^+ P}{\delta t} = \frac{\partial P}{\partial t} + (u + a) \frac{\partial P}{\partial x};$$

since $P =$ constant on a C^+-characteristic, $\delta^+ P/\delta t = 0$, then

$$\frac{\partial P}{\partial t} = -(u + a) \frac{\partial P}{\partial x}.$$

The sum $u + a$ is now eliminated from the last relation by Equation 8.9-6 to produce

$$\frac{\partial P}{\partial t} = -\tfrac{1}{2} P(1/n + 1) \frac{\partial P}{\partial x}. \qquad (8.9\text{-}7)$$

On the right-hand, or "front" side of the pulse envelope, the slope of the curve $\partial P/\partial x$ is negative; therefore, $\partial P/\partial t \sim \partial p/\partial t > 0$, and the proposition is proved — the front of the curve is, indeed, a compression. Of course, since $P = P_0(x) + \overline{P}$, we could make the replacement $\partial P/\partial x = \partial P_0/\partial x$, leaving the conclusion unaltered. It should not come as a surprise that the back of the pulse turns out to be an expansion.

THE WAVE DIAGRAM FOR A PULSE

From Equation 8.9-4, with $Q \equiv 0$, in the regime affected by the pulse, the differential equation of a C^+-characteristic is

$$\frac{\delta^+ x}{\delta t} = \frac{n+1}{2n} P = \frac{n+1}{2n} \left[\overline{P} + P_0(x) \right]. \qquad (8.9\text{-}8)$$

Since the value of P on a C^+-characteristic is constant and equal to its initial value, Equation 8.9-8 defines a straight line. However, since the value of $P_0(x)$ increases from the head wave to the middle of the pulse, the slope of the C^+-characteristics,

$$\frac{\delta^+ t}{\delta x} = \frac{2n}{n+1} \frac{1}{\overline{P} + P_0(x)},$$

tends to decrease in the same interval — the larger the value of $P_0(x)$, the smaller the slope. Figure 8.9-2 indicates the behavior for a hypothetical flow. In the expansion regime the characteristics diverge. Conversely, in the compression regime the char-

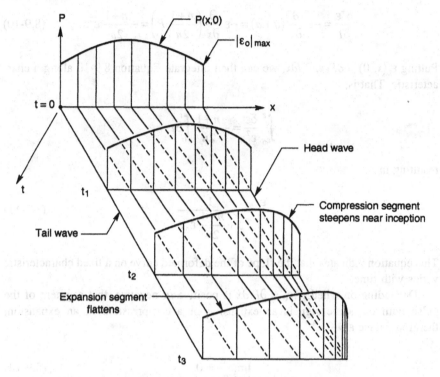

FIGURE 8.9-2 Steepening of a finite P-wave propagating to the right.

acteristics tend to converge. Furthermore, the slope at the head of the deformed pulse envelope tends to steepen with time. Eventually, two characteristics intersect, indicating a signularity, which is the signal for a shock beginning to form.

INCEPTION TIME FOR SHOCK FORMATION

We start with the derivative of an arbitrary Riemann parameter $P(x, t)$ along a C^+-characteristic:

$$\frac{\delta^+ P}{\delta t} \equiv \frac{\partial P}{\partial t} + (u + a)\frac{\partial P}{\partial x} = 0. \tag{8.9-9}$$

Next take $\partial/\partial x$ of Equation 8.9-9 and put $\varepsilon\,(x,\,t) \equiv \partial P/\partial x$. If the function $P(x)$ is sufficiently regular that we can interchange the order of differentiation, then

$$\frac{\delta^+ \varepsilon}{\delta t} = \frac{\partial \varepsilon}{\partial t} + (u + a)\frac{\partial \varepsilon}{\partial x} + \varepsilon\frac{\partial(u + a)}{\partial x} = 0.$$

We identify the sum of the first two terms after the first equals sign as $\delta^+\varepsilon/\delta t$. By solving,

$$\frac{\delta^+\varepsilon}{\delta t} = -\varepsilon\frac{\partial}{\partial x}(u+a) = -\varepsilon\frac{\partial}{\partial x}\left(\frac{n+1}{2n}P\right) = -\frac{n+1}{2n}\varepsilon^2. \qquad (8.9\text{-}10)$$

Putting $\varepsilon_0(x, 0) \equiv \partial P(x, 0)/\partial x$, we can then integrate Equation 8.9-10 along a characteristic. That is,

$$\int_{\varepsilon_0}^{\varepsilon}\frac{\delta\varepsilon}{\varepsilon^2} = -\frac{n+1}{2n}\int_0^t \delta t,$$

resulting in

$$\frac{\varepsilon}{\varepsilon_0} = \frac{1}{1+\dfrac{n+1}{2n}\varepsilon_0 t}. \qquad (8.9\text{-}11)$$

This equation indicates that the slope of the deformed curve on a fixed characteristic varies with time.

Depending on whether $\varepsilon_0 = \partial P/\partial x$ is positive or negative, that segment of the pulse with the same sign is an expansion or a compression. In an expansion, therefore, since $\varepsilon > 0$,

$$\lim_{t\to\infty}\frac{\varepsilon}{\varepsilon_0} = 0, \qquad (8.9\text{-}12)$$

which confirms the fact that the curve flattens out and that the C^+-characteristics diverge from each other.

For a compression, however, ε_0 is negative, in which case the denominator of Equation 8.9-12 eventually becomes zero, creating a singularity. The inception time t_i to reach the singular state on a particular characteristic is

$$t_i = -\frac{2n}{(n+1)\varepsilon_0}. \qquad (8.9\text{-}13)$$

For a given compression pulse the shortest time will occur on the characteristic on which $|\varepsilon_0|$ is a maximum. This signals the beginning of a shock wave formation at which the distribution of $P(x, t)$ becomes discontinuous. This point is actually the intersection of two characteristics, originally arbitrarily close to each other, and is an indication that at the intersection P has become doubly valued.

DISCUSSION

The analysis for the development of a shock in the general case is complicated and requires a numerical treatment. One way to obtain some insight to the process is to consider the wave diagram, this time for a flow generated on the compression side of a moving piston. We cannot choose an impulsively accelerated piston because,

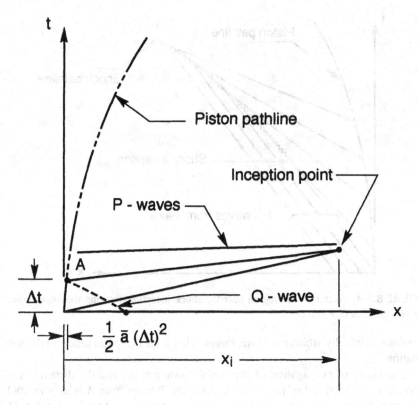

FIGURE 8.9-3 Sketch of a wave diagram for a compression generated by a piston impulsively accelerated to the right showing the formation of a shock.

as it turns out, such motion generates a shock created "instantly," which precedes the piston, resulting in a continuously increasing body of gas contiguous to the piston, which moves to the right at the piston velocity.

Instead, we consider a piston that accelerates from rest at a uniform acceleration $\bar{a}$. The wave diagram of Figure 8.9-3 is a rough schematic of the starting process. The piston moves toward a gas at rest in which the sound speed is a_1. The piston pathline is a parabola whose equation is $x_P = \frac{1}{2}\bar{a}t^2$. After a short time increment Δt, the velocity of the piston is $u_P = \bar{a}(\Delta t)$, and its x-coordinate is $\frac{1}{2}\bar{a}(\Delta t)^2$. Once again, this is a flow that involves changes only in the family of P-characteristics; i.e., the C^- characteristics are straight and parallel throughout the field. Likewise $Q = na - u = $ constant throughout. The initial data line is the positive x-axis. Thus, for the C^- characteristic through point A we can write $na_A - u_A = na_1$, since $u_1 \equiv 0$. Now, since the flow speed at A must be the same as that of the piston, we can solve for $a_A = a_1 + \bar{a}(\Delta t)/n$.

The slope of any C^+-characteristic is the inverse of the propagation speed of a wave along the same line. The slope of the P-wave from A, $\delta^+t/\delta x = 1/[a_1 + \bar{a}(\Delta t)(1 + 1/n)]$, is less than that of the P-wave from the origin $\delta^+t/\delta x = 1/a_1$; these two lines must intersect. We seek the time of intersection, in the limit as $\Delta t \to 0$, where the

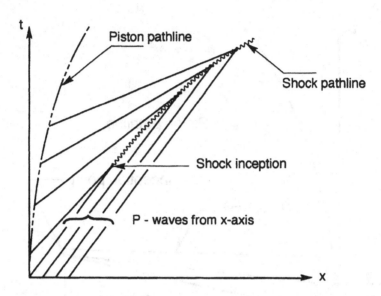

FIGURE 8.9-4 Sketch of the region near the shock inception showing the intersection of the two families of P-waves.

two lines, originally arbitrarily close, merge. This is the inception time t_i of the shock creation.

The speed of propagation of the initial P-wave is a_1, and the distance it will travel is $a_1 t_i$. The speed of propagation along the P-wave from A is $a_1 + \overline{a}(\Delta t)(1 + 1/n)$, and the distance the wave travels in the increment $t_i - \Delta t$ is $[a_1 + \overline{a}(1 + 1/n)]$ $(t_i - \Delta t)$. The difference of these two distances is the length $\frac{1}{2}\overline{a}t_i^2$ that the piston has moved in the interval Δt. Therefore,

$$a_1 t - \left[a_1 + \overline{a}(\Delta t)\left(1 + 1/n\right)\right]\left(t - \Delta t\right) = \tfrac{1}{2}\overline{a}t^2.$$

The term $a_1 t_i$ cancels. Then, letting $\Delta t \to 0$, $t_i = na_1^2/(n + 1)\overline{a}$. The space coordinate of the shock inception point is $x_i = na_1^2/(n + 1)\overline{a}$.

It takes a formidable rate of acceleration of the piston to initiate a shock formation in a short distance. In air, with $a_1 = 1116$ ft/sec, $n = 5$, and if $\overline{a} = 10$ ft/sec², then $x_i = 19.7$ mi; if $\overline{a} = 1000$ ft/sec² (≈ 31 g!), then $x_i = 1038$ ft.

The process of the shock formation itself must be handled by numerical techniques. Figure 8.9-4 — plotted to a different hypothetical scale than Figure 8.9-3 — gives a hint of the situation. Starting at the inception point, the shock path is determined, point by point, by matching conditions that $P = na + u$ must satisfy on either side of the shock, on their respective C^+-characteristics, while ensuring that the Rankine-Hugoniot equations across the shock are satisfied. Since the shock location is not known in advance, the computation must be iterative. The process is greatly eased by the availability of a computer especially programmed for the purpose.

8.10 AN APPROACH TO ACOUSTICS

INTRODUCTION

Having developed the method of characteristics with several applications involving only one of the Riemann variables as variables, the groundwork to examine elementary applications in acoustical theory has been laid. The theory of acoustics had already been well developed in the 19th century, particularly by Lord Rayleigh (John W. Strutt) in England and Helmholtz in Germany. As Rayleigh (1877) made clear in the introduction to his classic work, the subjects of acoustics* and vibrations are inextricably intertwined. We have the space to examine a mere single application of the former.

The general equations of compressible flow had been derived in the late 18th and early 19th centuries. However, except for one-dimensional applications, few solutions had been obtained by mid-19th century, due to the inherent nonlinearity of the equations. The acoustical equations are derived by imposing rather drastic restrictions on the equations of compressible flow, thereby linearizing them.

We take for our point of departure the exact equations of one-dimensional nonsteady flow, and repeat Equations 8.9-1b, 8.9-2, and 8.9-3:

- Differential equations for $C^{\pm}$-characteristics:

$$\delta^{\pm}x/\delta t = u \pm a; \tag{8.10-1}$$

- Equations for the Riemann parameters:

$$P, Q \equiv na \pm u = \text{constants}, \quad \text{on } C^{\pm}; \tag{8.10-2}$$

- Sound speed and velocity:

$$a = \tfrac{1}{2n}(P+Q), \quad u = \tfrac{1}{2}(P-Q). \tag{8.10-3}$$

It is important to understand these restrictions and their physical significance. If the ambient speed of sound is a_0, it is not merely that the flow velocity $u(x, t)$, and the sound speed deviation from ambient $a(x, t) - a_0$, are small but, in fact, that both are *negligible* compared with the ambient value. More properly, $u/a_0 \ll 1$ and $|a - a_0|/a_0 \ll 1$. In consequence, the acoustical version of Equation 8.10-1 becomes

$$\frac{\delta^{\pm}x}{\delta t} = \pm a_0. \tag{8.10-4a}$$

These can be immediately integrated to obtain the equations for the characteristic lines:

---

* For a short history of acoustics see Stephens and Bate (1966), *Historical Introduction*, pp. 1–10. For general treatments of acoustical theory, see Stephens and Bate (1966) or Kinsler et al. (1982).

$$\xi, \eta = x \mp a_0 t, \quad \text{on } C^{\pm}, \tag{8.10-4b}$$

where ξ, η are the initial space-coordinates (i.e., constants of integration) on the $C^{\pm}$-characteristics and will be used as parameters to designate the specific characteristics, which for each family are obviously all straight, parallel lines.

It follows that the derivative operators along the characteristic lines simplify to

$$\frac{\delta^{\pm}}{\delta t} = \frac{\partial}{\partial t} + (u \pm a) \frac{\partial}{\partial x} \rightarrow \frac{\partial}{\partial t} \pm a_0 \frac{\partial}{\partial x}. \tag{8.10-5}$$

The expressions for the Riemann parameters cannot be linearized lest they be reduced to a triviality. In other words we do *not* put $P, Q \rightarrow \pm n a_0$ in Equation 8.10-3. Similarly, Equation 8.10-3 remains unchanged.

The restriction of a homentropic fluid remains in effect. Therefore, the differential equations for the Riemann parameters, Equation 8.7-17, become

$$\frac{\delta^{\pm}}{\delta t} \begin{pmatrix} P \\ Q \end{pmatrix} = \left[\frac{\partial}{\partial t} \pm a_0 \frac{\partial}{\partial x} \right] \begin{pmatrix} P \\ Q \end{pmatrix} = 0, \tag{8.10-6a}$$

where the plus/minus signs are used with P/Q, respectively, or

$$\frac{\partial P}{\partial t} + a_0 \frac{\partial P}{\partial x} = 0, \quad \frac{\partial Q}{\partial t} - a_0 \frac{\partial Q}{\partial x} = 0. \tag{8.10-6b}$$

We next operate on the first equation in Equation 8.10-6b with the following expression:

$$\frac{\partial}{\partial t} - a_0 \frac{\partial}{\partial x}.$$

The cross-derivatives cancel, resulting in

$$\frac{\partial^2 P}{\partial t^2} - a_0^2 \frac{\partial^2 P}{\partial x^2} = 0. \tag{8.10-7a}$$

Also, a similar procedure applied to the second equation produces

$$\frac{\partial^2 Q}{\partial t^2} - a_0^2 \frac{\partial^2 Q}{\partial x^2} = 0. \tag{8.10-7b}$$

The functions P and Q, therefore, both satisfy the *wave equation*, a second-order, linear, partial differential equation, whose solutions are known to be

$$P = P(\xi) = P(x - a_0 t), \quad \text{on } x - a_0 t = \xi,$$
$$Q = Q(\eta) = Q(x + a_0 t), \quad \text{on } x + a_0 t = \eta. \qquad (8.10\text{-}8)$$

It can be verified by substitution that the functions of Equation 8.10-8 do indeed satisfy Equations 8.10-7. The most general form of solution is $P(\xi) + Q(\eta)$, where P, Q are arbitrary functions, which are determined from the boundary and/or initial conditions. If we can solve for $P = P(\xi)$, $Q = Q(\eta)$ in some domain, then we can find $u = u[\xi(x, t), \eta(x, t)]$, $a = a[\xi(x, t), \eta(x, t)]$ in the same domain by use of Equations 8.10-3.

APPLICATION

At $t = 0+$ we assume that a pulse, consisting of both P- and Q-waves, has been created within a constant-area tube over a segment of length π. We need not concern ourselves about how such a pulse can be created. The pulse is defined as follows:

$$u(x, 0+) = u_0 \sin x, \quad 0 \le x \le \pi,$$
$$= 0, \qquad\qquad x < 0, \, x > \pi. \qquad (8.10\text{-}9a)$$

$$a(x, 0+) \equiv a_0 = \text{constant} \qquad (8.10\text{-}9b)$$

Therefore, from Equation 8.10-2,

$$P(x, 0+) = n a_0 + u_0 \sin x. \qquad (8.10\text{-}10)$$

To the uninitiated the next step may seem like black magic, but here goes. Since $P(x, t) = P(x - a_0 t) = P(\xi)$, on C^+ and, since $\xi = x - a_0 t \to x$, for $t \to 0$, and since $\xi = \text{constant}$ on C^+, we can replace x in Equation 8.10-10 by ξ to obtain

$$P(\xi) = n a_0 + u_0 \sin \xi, \quad \text{on } C^+, \quad 0 \le \xi \le \pi. \qquad (8.10\text{-}11a)$$

Similarly,

$$Q(\eta) = n a_0 - u_0 \sin \eta, \quad \text{on } C^-, \quad 0 \le \eta \le \pi. \qquad (8.10\text{-}11b)$$

Therefore, in the triangle ABC, the domain within which the interaction between the P- and the Q-waves takes place, we can obtain expressions for $u(\xi, \eta)$, $a(\xi, \eta)$ from Equation 8.10-3

$$u(\xi, \eta) = \tfrac{1}{2} u_0 (\sin \xi + \sin \eta) \qquad (8.10\text{-}12a)$$

and

$$a(\xi, \eta) - a_0 = \left(u_0/2n\right)\left(\sin \xi - \sin \eta\right). \tag{8.10-12b}$$

The latter equation indicates that the deviations of the sound speed from the ambient value are of the same order as the velocity function.

If desired, it is straightforward to convert from the parametric form of the solution involving the parameters ξ, η to the original independent variables x, y. For example, Equation 8.10-12a becomes

$$u/u_0 = \tfrac{1}{2}\left(\sin \xi + \sin \eta\right) = \sin \tfrac{1}{2}\left(\xi - \eta\right) \cos \tfrac{1}{2}\left(\xi + \eta\right) = \sin \times \cos a_0 t.$$

It turns out that the former form is superior for purposes of calculation.

The solutions in the other domains are obtained just as easily. We have

- Within the semi-infinite rhomboid ABDE,

$$P = na_0 + u_0 \sin \xi,$$

$$Q = na_0,$$

$$u/u_0 = \tfrac{1}{2} \sin \xi, \qquad \text{on } C^+, \ \ 0 \le \xi \le \pi; \qquad (8.10\text{-}13a)$$

$$a - a_0 = \frac{u_0}{2n} \sin \xi,$$

- Within the semi-infinite rhomboid ABCD,

$$P = na_0,$$

$$Q = na_0 - u_0 \sin \eta,$$

$$u/u_0 = \tfrac{1}{2} \sin \eta, \qquad \text{on } C^-, \ \ 0 \le \eta \le \pi. \qquad (8.10\text{-}13b)$$

$$a - a_0 = \frac{u_0}{2n} \sin \eta,$$

Everywhere else $P = Q = na_0$, $u = 0$, $a = a_0$.

The solution for the dimensionless velocity distribution is shown in Figure 8.10-1, plotted isometrically to an arbitrary scale. As previously noted, the most convenient method is to use ξ, η as independent variables. The isometric grid is formed by the characteristics themselves. As t increases, we see that the pulse shape changes continuously in the domain ABC until $a_0 t = \pi$. At this station the velocity distribution corresponding to the original pulse has decomposed into separate sine waves of half-amplitude corresponding to the decomposed P- and Q-pulses, the first propagating to the right, the other to the left.

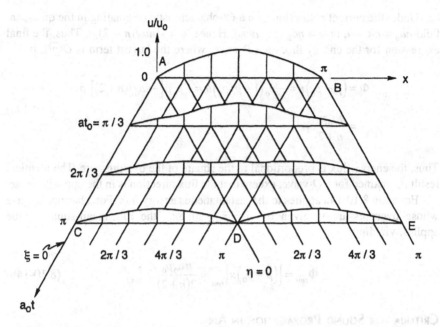

FIGURE 8.10-1 Sketch of the velocity distribution of two interacting acoustic pulses.

According to elementary acoustical theory, there is neither any dissipative effect nor any tendency for the characteristics to converge creating singularities, and the waves would propagate unchanged indefinitely. Anyone familiar with the voice tubes (not necessarily straight) sometimes installed in older homes, old-fashioned department stores, and old ships, knows that the ordinary voice can travel therein, essentially unimpeded, for surprisingly long distances and with amazing clarity.

WAVE ENERGY TRANSMISSION

We again follow Kantrowitz. In Section 3.19 flow work in a duct was defined as puA; the flow work per unit area is called the *power* $P = pu$. In a constant-area duct of quiescent fluid at ambient pressure p_0 and sound speed a_0, subject to superimposed isentropic wave motion, it is only the *overpressure* $p - p_0$ (in most references referred to simply as the *sound pressure*) that contributes to the term $\Phi \equiv (p - p_0)u$, which Kantrowitz calls the *wave energy transmission* or *energy flux*.

We have the identity $p = p_0 + (p - p_0)$. We can define a dimensionless overpressure as $\bar{p} \equiv (p - p_0)/p_0$, so that $p/p_0 = 1 + \bar{p}$, where $\bar{p} \ll 1$. For an isentropic wave propagation, we have

$$\frac{a}{a_0} = \left(\frac{p}{p_0}\right)^{1/(n+2)} = (1 + \bar{p})^{1/(n+2)} = 1 + \tfrac{1}{n+2}\,\bar{p} + O(\bar{p}^2),$$

after expanding in a binomial series. Therefore, to first order in $\bar{p}$, $1 - a/a_0 = -[1/(n + 2)]\bar{p}$.

Under the current restrictions, on a C^+-characteristic originating in the quiescent fluid $na_0 = na + u$, or $u = na_0 (1 - a/a_0)$. Hence, $u = -[na_0/(n + 2)]\bar{p}$. Thus, the final expression for the energy flux on a P-wave, where the largest term is $O(\bar{p}^2)$, is

$$\Phi = \left(p - p_0\right)u = -p_0\left(1 - p/p_0\right)u = -p_0\bar{p}\left[-na_0/(n+2)\right]\bar{p},$$

$$= \frac{na_0p_0}{n+2}\,\bar{p}^2. \tag{8.10-14a}$$

Thus, the energy flux is proportional to the square of the overpressure. The identical result is obtained for a Q-wave, except that the flux direction is in the opposite sense.

Equation 8.10-14a applies to the instantaneous energy flux. For a harmonic wave whose overpressure is given by $\bar{p} = \bar{p}_{max} \sin \omega t,$, the root-mean-square value applies, yielding

$$\Phi_{rms} = \left[\left(p - p_0\right)u\right]_{rms} = \frac{na_0p_0}{2(n+2)}\,\bar{p}_{max}^2. \tag{8.10-14b}$$

CRITERIA FOR SOUND PROPAGATION IN AIR

The average human can detect pure harmonic tones in the frequency range of 16 to 20,000 Hz. According to Harris (1991), at 1000 Hz the rms amplitude of the weakest audible sound has an overpressure* of $(p - p_0)_{ref} \equiv 20 \times 10^{-6}$ Pa. Since 1 atm = 1.013 $\times 10^5$ Pa, the reference value is a tiny pressure indeed. Although not directly related, there has also been defined a reference energy flux of $\Phi_{ref} \equiv 10^{-12}$ W/m^2.

It is the practice to indicate the level of overpressure as a ratio to the reference value. Further, since experimental data indicate that the ability of the ear to detect various sound intensities (which depends on the energy flux rather than the over-pressure) seems to follow a logarithmic scale, the unit bel** was defined such that

$$1\,bel = \log_{10}\frac{\left(p - p_0\right)}{\left(p - p_0\right)_{ref}}, \quad \text{when} \quad \frac{\left(p - p_0\right)}{\left(p - p_0\right)_{ref}} = 10. \tag{8.10-15a}$$

A finer subdivision was also created, the decibel (dB); there are 10 dB to one bel. Therefore,

$$dB = 10\,\log_{10}\frac{\left(p - p_0\right)}{\left(p - p_0\right)_{ref}}. \tag{8.10-15b}$$

---

* In the standard acoustics literature, e.g., Harris (1991), what we have called overpressure is defined as the *sound pressure*. Similarly, what Kantrowitz termed the wave energy transmission is the *sound intensity*. When the word *level* is used (as in *sound intensity level*), a ratio is implied, usually in the form of $\log_{10}$ of the ratio. The denominator of the ratio is a reference value adopted by international agreement.
** Named for Alexander Graham Bell, inventor of the telephone.

TABLE 8.10-1
Examples of Sound on the Decibel Scale

Subjective Effects	Example	Sound Pressure Level (dB)	Φ/Φ_{ref}
Lethal	Rocket launching pad	180	10^{18}
Hearing loss, short exposure	Turbojet	150	10^{15}
Pain threshold	Riveting, chipping	130	10^{13}
Discomfort[a]	Discotheque	120	10^{12}
Safe	Street traffic	70	10^{7}
Intrusive	Average conversation	60	10^{6}
Quiet	Private office	50	10^{5}
Faint	Whisper	30	10^{3}
Silent (youth)		0	10^{0}

[a] Presumably not for the occupants.

Source: Abstracted from Gehring (1980).

Acousticians get around this awkward notation, effectively, by discarding the ambient pressure p_0 in the formula and using the symbol p as the sound pressure. Furthermore, as noted, the decibel rating is referred to as the sound pressure level. By doubling the pressure ratio in Equation 8.10-15b the pressure level is increased by one decibel unit. Associated with this is the decibel rating of energy flux or, equivalently, the sound intensity level, for which

$$dB = 10 \log_{10} \left(\Phi/\Phi_{ref} \right) = 10 \log_{10} \left[\frac{(p - p_0)}{(p - p_0)_{ref}} \right]^2 = 20 \log_{10} \frac{(p - p_0)}{(p - p_0)_{ref}}. \quad (8.10\text{-}15c)$$

To put the decibel scale into physically meaningful terms Table 8.10-1 is abstracted from Gehring (1980).

SOME FLOW-PROPERTY MAGNITUDES IN ACOUSTICS

We can make some estimates of the flow behavior in an acoustical pulse. Assume for the current purpose that a harmonic wave at 1000 Hz is propagated at the reference superpressure that generates the reference energy flux, all values being rms. Then, using the preceding formulas, the rms velocity of a particle is $u_{rms} = \Phi_{ref}/(p - p_0)_{ref} = 10^{-12}/(2 \times 10^{-5}) = 5 \times 10^{-8}$ m/s. Thus, we see that the particle velocity is proportional to the amplitude of the overpressure or, equivalently, to the square root of the energy flux. In our numerical example the velocity is obviously negligible compared with the local speed of sound. For the particle displacement, we must have $x = x_{rms} \sin \omega t$; the corresponding expression for the velocity is obtained by differentiating $\dot{x} = \omega x_{rms} \cos \omega t$. We can then equate ωx_{rms} to the value just computed to find that the amplitude of the particle displacement is $x_{rms} = 0.8 \times 10^{-11}$ m, a value much smaller than the sea-level mean-free-path length.

For a pulse of the same frequency at the discomfort level, the displacement and velocity would both be multiplied by 10^6, but would still remain small quantities. Alternatively, we see that the ear is an extraordinarily sensitive organ, responding as it does to extremely small stimuli, yet can function over a wide range of magnitudes of sound pressure levels before it is endangered.

The problem of analyzing voice, or sound, transmissions in air is significantly more complicated since each of these involves a spectrum of frequencies of varying intensities. This field of research is a specialty of its own.

References

Abramovitz, M. and I.A. Stegun (1964), *Handbook of Mathematical Functions*, NBS AMS 55, U.S. Government Printing Office, reprinted by Dover, New York.

Ackeret, J. (1927), Gasdynamik, Hand buch der Physia, Vol. 7, Chapter 5, Springer, Berlin.

Ackeret, J. (1967), Aspects of internal flow in *Fluid Mechanics of Internal Flow*, Elsevier, New York, 1–26.

Adams, J.Q. (1821), *Report on Weights and Measures*, from the Secretary of State to the Congress of the United States.

Adamson, A.W. (1982), *Physical Chemistry of Surfaces*, 4th ed., John Wiley & Sons, New York.

Alder, B.J. and T. Wainwright (1959), Studies in molecular dynamics I, general method, *J. Chem. Phys.*, 31(2), 459–466.

Alder, G.M. (1979), The numerical solution of choked and supercritical ideal gas flow through orifices and convergent conical nozzles, *J. Mech. Eng. Sci.*, 21(3), 197–203.

Ames Research Staff (1953), Equations, Tables and Charts for Compressible Flow, NACA Rep. 1135, Ames Aeronautical Laboratory, Moffett Field, CA, printed by the U.S. Goverement Printing Office.

Anderton, P. and P.H. Bigg (1965), *Changing to the Metric System*, Her Majesty's Stationery Office, London.

Anonymous (1988), Flow of Fluids through Valves, Fittings, and Pipe, Technical Paper No. 410, 24th Printing, Crane Co., 800 Third Ave., King of Prussia, PA.

ASME (1981) *ASME Orientation and Guide for Use of SI (Metric) Units*, 8th ed., ASME Guide SI-1, New York.

ASTM (1980), *Standard for Metric Practice*, E 380-79, American Society for Testing Materials, Philadelphia.

Baumeister, T. (1961), Turbomachinery, in *Handbook of Fluid Dynamics*, V.L. Streeter, ed., McGraw-Hill, New York, Section 19.

Bean, H.S. (1971), *Fluid Meters — Their Theory and Application*, Report of the ASME Research Committee on Fluid Meters, New York.

Becker, E. (1968), *Gas Dynamics*, Academic Press, New York, translated from the German edition of 1968.

Bernoulli, D. (1738), *Hydrodynamica*, translated from the Latin by Thomas Carmody and Helmut Kobus; preface by H. Rouse. Published together with *Hydraulica* by J. Bernoulli, Dover, New York, 1968.

Bikerman, J.J. (1958), *Surface Chemistry, Theory and Applications*, 2nd ed., Academic Press, New York.

Bird, G.A. (1976), *Molecular Gas Dynamics*, Oxford University Press, New York.

Bird, R.B., W.E. Stewart, and E.N. Lightfoot (1963), *Transport Phenomena*, John Wiley & Sons, New York.

Birkhoff, G. (1960), *Hydrodynamics*, rev. ed., Princeton University Press, Princeton, NJ. The first edition was published in 1950.

Biswas, A.K. (1970), *History of Hydrology*, North-Holland, Amsterdam.

Blasius, H. (1913), *Das Ähnlichkeitgesetz bei Reibungsvorgängen in Flussigkeiten*, Mitteilungen uber Forschungarbeiten, Verein deutscher Ingenieure, Julius Springer, Berlin.

Bloch, E. (1968), Numerical solution of free boundary problems by the method of steepest descent, in *High-Speed Computing in Fluid Dynamics, IUTAM Symposium*, F.N. Frenkiel and K. Stewartson, Eds., The Physics of Fluids Supplement II, 1969, pp. 129–132, American Institute of Physics.

Boys, C.V. (1959), *Soap Bubbles, Their Colors and Forces Which Mold Them*, reprint of the 1911 edition, with a new introduction by S.Z. Levin, Dover, New York.

Bridgman, P.W. (1931), *Dimensional Analysis*, rev. ed., Yale University Press, New Haven, CT.

Brougham, H.L. and E.J. Routh (1855), *Analytical View of Sir Isaac Newton's Principia*, Longman, Brown, Green, and Longman, London.

Brower, W.B., Jr. and J.L. Way (1964), On Non-Steady Liquid Discharge from Reservoirs, RPI TR AE 6407, Rensselaer Polytechnic Institute, Troy, NY.

Brower, W.B., Jr. (1990), *Theory, Tables, and Data for Compressible Flow*, Hemisphere, New York.

Brower, W.B., Jr., E. Eisler, E.J. Filkorn, J. Gonenc, C. Plati, and J. Stagnitti (1993), On the compressible flow through an orifice, *J. Fluids Eng.*, 115, 660–664.

Browne, M.W. (1996), Kinder, gentler push for metric inches along, *New York Times*, Science section, June 4, 1996.

Buckingham, E. (1914), On physically similar systems, *Phys. Rev.*, 4(2), 345–376.

Burgreen, D. (1960), Development of flow in tank draining, *J. Hydraulics Div., ASCE*, 13–28.

Busemann, A. (1937), Hodographen Methods der Gasdynamik, Z.A.M.M., Vol. 17, p. 73.

Cascetta, F. and P. Vigo (1990), *Flow Meters — A Comprehensive Survey and Guide to Selection*, rev. 2nd printing, ISA, Research Triangle Park, NC.

Chapman, S. and T.G. Cowling (1953), *The Mathematical Theory of Non-Uniform Gases*, 2nd ed., Cambridge University Press, New York, reprint of the 1939 edition.

Cole, A.E., A. Court, and A.J. Kantor (1965), Model atmosphere, in *Handbook of Geophysics and Space Environments*, McGraw-Hill, New York, Chap. 2.

Cormier, R.C., J.C. Ulwick, J.A. Klobacher, W. Pfister, and T.J. Keneshea (1965), Ionospheric physics, in *Handbook of Geophysics and Space Environments*, McGraw-Hill, New York, Chap. 12.

Courant, R. and K.O. Friedrichs (1948), *Supersonic Flow and Shock Waves*, Interscience, New York.

Courant, R. and D. Hilbert (1962), *Methods of Mathematical Physics*, Vol. II, *Partial Differential Equations*, Interscience, New York.

Cousins, R.R. (1969), Shear Flow Past a Sphere, NPL Report Ma 80.

Dean, R.C., Jr. (1959), On the necessity of nonsteady flow in fluid machines, *Trans. ASME, J. Basic Eng.*, Series D, 81(1), 24–28.

De Carlo, J.P. (1984), *Fundamentals of Flow Measurement*, ISA, Research Triangle Park, NC.

de Saint-Venant, B. and L. Wantzel (1839), Mémoir et expériences sur l'éncoulement de l'air, *J. Éc. R. Polytech.*, 16, 85–122.

Diehl, W.S. (1925), *Standard Atmosphere Tables and Data*, NACA Rep. 218, U.S. Government Printing Office, Washington, D.C.

Dowdell, R.B., Ed. (1974), *Flow, Its Measurement and Control in Science and Industry*, 3 vol., Instrument Society of America, Research Triangle Park, NC.

Dublin, M., A.R. Hull, and K.S.W. Champion, Eds. (1976), *U.S. Standard Atmosphere, 1976*. U.S. Government Printing Office, Washington, D.C.

Dwight, H.B. (1961), *Tables of Integrals and Other Mathematical Data*, 4th ed., Macmillan, New York.

Eck, B. (1957), *Technische Stromungslehre*, 5th ed., Springer-Verlag, Heidelberg.

Fermi, E. (1956), *Thermodynamics*, Dover, New York, reprinted from the original 1937 edition.

Feynman, R.P., R.B. Leighton, and M. Sands (1963), *Lectures on Physics*, Vol. I, Addison-Wesley, Reading, MA, 1–2.

Feynman, R.P., R. Leighton, and E. Hutchings (1985), *Surely You're Joking Mr. Feynman*, W.W. Morton, New York.

Foa, J.V. (1960), *Elements of Flight Propulsion*, John Wiley & Sons, New York.

Folsom, R.G. (1956), Review of the pitot tube, *Trans. ASME*, 78.

Fourier, J.B.J. (1822), *La Théorie Analytique de la Chaleur*. English translation by A. Freeman, published in 1872 by Cambridge University Press, reprinted by Dover, New York, 1955.

Freeman, J.R. (1888), The discharge of water through fire hose and nozzles, *Trans. ASCE*, 21, 308–482.

Garabedian, P.R. (1956), Calculation of axially symmetric cavities and jets, *Pacific J. Math.*, 6, 611–684.

Gehring, W.H. (1980), *Reference Data for Acoustic Noise Control*, 2nd printing, Ann Arbor Services, Ann Arbor, MI.

General Electric Co. (1981), *Fluid Flow Data Book*, GE Corporate Research and Development, Schenectady, NY.

Glass, I.I. (1952), On the speed of sound in gases, Readers forum, *J. Aero. Soc.*, 29(4), 286.

Goldman, D.T. and R.J. Bell, Eds. (1981), *The International System of Units*, NBS Special Publication 330.

Gronksy, R. (1984), *Electron Microscopy at Atomic Resolution, Materials Research Society Symposium*, Vol. 31, Elsevier Science, New York.

Hagen, G. (1839), Über die Bewegung des Wassers in engen cylindrischen Röhren, *Poggendorf's Ann. Phys. Chem.*, 46, 423–442, reprinted in 1933 by L. Schiller in *Die Klassiker der Stromonglehre: Hagen, Poiseuille Hagenbach*, Academische Verlagsgesellschaft M.B.H., Leipzig. The Poiseuille reprint is a translation into German of the 1846 paper.

Hagen, G. (1854), *Uber die Einfluss der Temperatur auf die Bewegung des Wassers in Rohren*, Mathematische Abhandlungen der Akademie der Wissenschaften, Berlin.

Haltiner, J. and F.L. Martin (1957), *Dynamical and Physical Meteorology*, McGraw-Hill, New York.

Harris, C.M., Ed. (1991), *Handbook of Acoustical Measurements and Noise Control*, 3rd ed., McGraw-Hill, New York.

Hayes, W. (1958), The basic theory of gas dynamic discontinuities, in *Fundamentals of Gas Dynamics*, H.W. Emmons, Ed., Vol. III of *High Speed Aerodynamics and Jet Propulsion*, Princeton University Press, Princeton, NJ, article C.

Heath, T.L., Ed. (1897), *The Works of Archimedes*, Dover, New York.

Henry, J.R. (1944), Design of Power Plant Installations — Pressure Loss Characteristics of Duct Components, NACA ARR No. L4F26.

Herschel, C. (1887), The Venturi water meter, *Trans. ASCE*, 17.

Herschel, C. (1913), *The Two Books on the Water Supply of Rome of Sextus Julius Frontinus*, translated from the Latin.

Hildebrand, J.H. (1963), *An Introduction to Molecular Kinetic Theory*, Reinhold, New York.

Hilsenrath, J. (1955), *Tables of Thermodynamic and Transport Properties*, NBS Circular 564, reprinted by Pergamon Press, New York, 1960.

Hinch, E.J. (1991), *Perturbation Methods*, Cambridge University Press, New York.

Hirschfelder, J.O., C.F. Curtiss, and R.B. Bird (1954), *Molecular Theory of Gases and Liquids*, John Wiley & Sons, New York.

Holt, M. (1961), Dimensional analysis, in *Handbook of Fluid Dynamics*, V.L. Streeter, Ed., McGraw-Hill, New York, section 15.

Holton, J.R. (1972), *An Introduction to Dynamic Meteorology*, Academic Press, New York.

Howarth, L., Ed. (1953), *Modern Developments in Fluid Dynamics*, Vol. I, Clarendon Press, Oxford, U.K.

Hugoniot, H. (1889), Sur la propagation du mouvement dans les corps et spécialement dans les gaz parfaits, *J. Ec. Polytech.*, 58, 1–125.

Idelchik, I.E. (1986), *Handbook of Hydraulic Resistance*, Hemisphere, Harper and Row, New York.

ISO (1983) Measurement of Water Flow in Enclosed Conduits — Meters for Cold Potable Water — Part 3: Test Method and Equipment, 1st ed. 4064-3-1983. International Organization for Standardization, Geneva, Switzerland. Copies of ISO publications can be obtained from the American National Standards Institute (ANSI), New York.

Jammer, M. (1962), *Concepts of Force*, Harper and Brothers Torchbook, reprinted from the 1957 edition, Harvard University Press, Cambridge, MA.

Jeans, J. (1952), *An Introduction to the Kinetic Theory of Gases*, Cambridge University Press, New York, reprint of the 1940 edition.

Johansen, F.C. (1930), Flow through pipe orifices at low Reynolds number, *Proc. Roy. Soc. A*, 126(801), 231.

Jones, F.E. (1995), *Techniques and Topics in Flow Measurement*, CRC Press, Boca Raton, FL.

Joule, J.P. and W. Thomson (1854), On the thermal effect of fluids in motion, *Proc. Roy. Soc. London*, VIII, 178–185. This is reprinted both in the collected works of Joule and of Thomson.

Judson, L.V. (1976), Weights and Measures Standards of the United States, A Brief History, NBS Special Publication 447.

Jursa, A., Scientific Ed. (1985), *Handbook of Geophysics and the Space Environment*, U.S. Air Force Geophysics Laboratory, 4th ed., available from N.T.I.S., Springfield, VA, ADA 167000.

Kantrowitz, A. (1958), One-dimensional treatment of nonsteady gas dynamics, in *Fundamentals of Gas Dynamics*, H.W. Emmons, Ed., Vol. III of *High Speed Aerodynamics and Jet Propulsion*, Princeton University Press, Princeton, NJ, section C.

Kaufman, W. (1963), *Fluid Mechanics*, McGraw-Hill, New York. Translated from the German.

King, H.W. and E.F. Brater (1954), *Handbook of Hydraulics*, 4th ed., McGraw-Hill, New York.

Kinsler, L.E., A.R. Frey, A.B. Coppens, and J.V. Sanders (1982), *Fundamentals of Acoustics*, John Wiley & Sons, New York.

Kirchhoff, G. (1869), Zur Theorie freier Flussigkeitsstrahlen, *J. Reine Angewandt Math.*, 70, 289.

Kline, S.J. (1986), *Similitude and Approximation Theory*, 2nd ed., Springer-Verlag, Heidelberg.

Kozeny, J. (1954), *Hydraulik*, Springer-Verlag, Vienna (in German).

Knudsen, M. (1933), *The Kinetic Theory of Gases*, Methuen, London.

Lamb, H. (1932), *Hydrodynamics*, Dover reprint in 1945 of the 6th ed. The first edition dates from 1874.

Langhaar, H.L. (1951), *Dimensional Analysis and Theory of Models*, John Wiley & Sons, New York.

Laplace, M. (1816), Sur la vitesse du son dans l'air et dans l'eau, *Ann. Chim. Phys.*, 3, 238.

Liepmann, H.W. and A. Roshko (1957), *Elements of Gas Dynamics*, John Wiley & Sons, New York.

Lin, C.C. and L. Segel (1974), *Mathematics Applied to Deterministic Problems in the Natural Sciences*, Macmillan, New York.

Loeb, L.B. (1934), *The Kinetic Theory of Gases*, reprint of the 3rd ed., Dover, New York, 1961.

McCarty, R.D. (1980), Interactive Fortran IV Computer Programs for the Thermodynamic and Transport Properties of Selected Cryogens (Fluid Pack), NBS Technical Note 1025, U.S. Department of Commerce, Washington, D.C.

Maxwell, J.C. (1911), Capillary action, in *Encyclopedia Brittanica*, 11th ed., Vol. 5, 256–275, originally published in the 9th ed., 1875–1878, and updated by Lord Rayleigh.

Meriam, J.L. and L.G. Kraige (1986) *Dynamics*, John Wiley & Sons, New York.

Middleton, W.E.K. (1968), *The History of the Barometer*, John Hopkins Press, Baltimore.

Milne-Thomson, L.M. (1968), *Theoretical Hydrodynamics*, 5th ed., The Macmillian Press, London.

Mirels, H. (1958), Comments on characteristics and sound speed in nonisentropic gas flows with nonequilibrium thermodynamic states, Readers Forum, *J. Aero. Sci.*, 25, 460.

Moody, L.F. (1944), Friction factors for pipe flow, *Trans. ASME*, 66, 671–684.

NACA TM 952, *Standard for Discharge Measurement*, 1940. Translated from the German.

Newton, I. (1686), *Principia*, see the Dover reprint of the Motte translation of 1729, edited by F. Cajori, published by University of California Press, Berkeley, 1934, p. 338.

Nikuradse, J. (1932), Gesetzmäsigkeit der turbulenten strömung in glatten Röhren, *Forsch. Arbeit Ing. Wes.*, 356.

Nikuradse, J. (1933), Strömungesetze in rauhen Röhren, *Forsch. Arbeit Ing. Wes.*, 361.

Olien, N.A. (1979), Present and future sources of fluid property data, ASME paper 79-WA/HT-19, New York.

Olmstead, J.M.H. (1961), *Advanced Calculus*, Appleton-Century-Crafts, New York.

Partington, J. (1955), *An Advanced Treatise on Physical Chemistry*, Vol. 2, *The Properties of Liquids*, corrected from the 1951 edition, Longmans Green, London.

Patterson, G.N. (1956), *Molecular Flow of Gases*, John Wiley & Sons, New York.

Patterson, G.N. (1971), *Introduction to the Kinetic Theory of Gas Flows*, University of Toronto Press, Toronto, Canada.

Pigott, R.J.S. (1933), The flow of fluids in closed conduits, *Mech. Eng.*, 55, 497–501.

Planck, M. (1926), *Treatise on Thermodynamics*, Dover, New York, 3rd rev., English ed. based on the 7th German ed. The first German ed. was published in 1897.

Poiseuille, J. (1846), Récherches expérimentales sur le mouvement des liquides dans les tube de très petites diamètres. *Mem. Savants Etrangers*, 9, pp. 433–543. See also *C.R.* 11, 961–1041, 1840, and 12, 112, 1841.

Poisson, S.D. (1833), Traité de Mécanique, 2nd vol., 3rd ed., translated into English by H.H. Harte, Longman & Co., London, 1842.

Polyanin, A.D. and V.F. Zaitsev (1995), *Handbook of Exact Solutions for Ordinary Differential Equations*, CRC Press, Boca Raton, FL.

Prandtl, L. (1927), Uber den Reibungswiderstand strömender Luft, *Ergeb. Aerodyn. Versuchanstalt Göttingen*, 3.

Prandtl, L. and O.G. Tietjens (1934), *Applied Hydro- and Aeromechanics*, Vol. II, based on the lectures of L. Prandtl, Engineering Societies Monograph, reprinted by Dover, New York, 1957, translated from the German.

Prandtl, L. (1952), *Essentials of Fluid Mechanics*, Hafner, translation of the 2nd German ed. of 1949, New York.

Prudnikov, A.P., Yu.A. Brichkov, and O.I. Marichev (1986), *Integrals and Series*, Vol. 1 *Elementary Functions*, Vol. 2 *Special Functions*, Vol. 3 *More Special Functions* (Vol. 3 publ. in 1990), translated from the Russian by S. McQueen, Gordon and Breach Science Publishing, New York.

Rankine, W.J.M. (1865), On the mechanical principles of the action of propellers, *Trans. Inst. Naval Archit.*, 6, 13–30, London.

Rankine, W.J.M. (1870), On the thermodynamic theory of waves of finite longitudinal disturbance, *Phil. Trans.*, 160, 277–288, London.

Rasmussen, H.J., Test Track Division, Hollomon Air Force Base (1976), Survey of braking techniques for high speed track vehicles, paper prepared for 27th meeting of the Aeroballistic Range Association.

Rayleigh, J.W.S., Lord (1877), *Theory of Sound*, 2 Vol., Macmillan, London.

Reynolds, O. (1883), An experimental investigation of the circumstances which determine whether the motion of water shall be direct or sinuous and of the law of resistance in parallel channels, *Philos. Trans. Roy. Soc.*, 174, 935–982, London.

Reynolds, O. (1885), On the flow of gases, *Philos. Mag.*, 21(130); *Proc. Manch. Lit. Philos. Soc.*, 185–199.

Reynolds, O. (1894), On the dynamical theory of incompressible fluids and the determination of the criterion, *Philos. Trans. Roy. Soc.*, 186. See also *Collected Papers*.

Riemann, B. (1858), Über die Fortpflanzung ebener Luftwellen von endlicher Schwingungsweite, *Göttingen Abh.*, 8, 43–65.

Rouse, H. and S. Ince (1957), *History of Hydraulics*, Iowa Institute of Hydraulic Research, State University of Iowa, Iowa City, Iowa.

Rouse, H. (1961), *Fluid Mechanics for Hydraulic Engineers*, Dover reprint of the original 1938 ed., New York.

Ruark, A.E. (1935), Inspectional analysis: a method which supplements dimensional analysis, *J. Elisha Mitchell Sci. Soc.*, 51, 127–133.

Rudinger, G. (1955), *Wave Diagrams for Nonsteady Flow in Ducts*, Van Nostrand.

Saph, V. and E.H. Schoder (1903), An experimental study of the resistance to the flow of water in pipes, *Trans. Am. Soc. Civ. Eng.*, 51, 944.

Schlichting, H. (1979), *Boundary Layer Theory*, 7th ed., McGraw-Hill, New York.

Shapiro, A. (1953), *The Dynamics and Thermodynamics of Compressible Fluid Flow*, Vol. I, The Ronald Press Co., Vol. II published in 1954, New York.

Smith, R.W. (1958), *The Federal Basis for Weights and Measures*, U.S. Department of Commerce, NBS Circular 593.

Smits, A.J. and M.V. Zagarola (1997), Design of a high Reynolds number testing facility using compressed air, AIAA 97-1917, presented at the 28th Fluid Dynamics Conference, June 29–July 2, 1997, Snowmass Village, CO.

Sommerfeld, A. (1952), *Mechanics*, Academic Press, New York; translated from the 4th German edition.

Stanton, T.E. (1911), The mechanical viscosity of dense fluids, *Proc. R. Soc. London., A.*, 85, 366.

Stary, F. (1962), On the discharge time of a tank with a horizontal bottom opening, *ZAMM*, 42, in German.

Stephan, K. and K. Lucas (1979), *Viscosity of Dense Fluids*, Plenum Press, New York.

Stephens, R.W.B. and A.E. Bate (1966), *Acoustics and Vibrational Analysis*, St. Martins Press, New York.

Stevenson, B. (1948), The MacMillian Book of Proverbs, Maxims and Famous Phrases, New York, p. 1247.

Stodola, A. and L.C. Lowenstein (1945), *Steam and Gas Turbines*, Peter Smith, reprinted from the McGraw-Hill 1927 ed., which is a translation of the 6th German ed.

Stokes, G. (1845), On the theories of the internal friction of fluids in motion, and of the equilibrium and motion of elastic solids, *Trans. Cambridge Philos. Soc.*, VIII, 387. See also *Collected Works*.

Tabor, D. (1991), *Gases, Liquids and Solids and Other States of Matter*, 3rd ed., Cambridge University Press, New York. The first edition was published in 1969.

Taylor, G.I. (1950), The formation of a blast wave by a very intense explosion, *Proc. R. Soc.,* A201, 159–174.

Thompson, P.A. (1972), *Compressible Fluid Dynamics*, McGraw-Hill, New York.

Thompson, P.A., T.W. Strock, and D.S. Lim (1983), Estimate of shock thickness based on entropy production, *Phys. Fluids*, 21(1), 48–49.

Torricelli, E. (1643), *De Motu Gravium Naturaliter Accelerato*, Firenze, Italia.

Touloukian, Y.S., S.C. Saxena, and P. Hestermans (1975), *Viscosity*, Vol. 11 in Thermophysical Properties of Matter – The TRPC Data Series (13 Volumes), Plenum, New York.

Truesdell, C. (1951), On the velocity of sound in fluids, Readers Forum, *J. Aero. Sci.*, 18, 501.

Truesdell, C. (1962), Reactions of the history of mechanics upon modern research, in *Proc. of the 4th U.S. Natl. Congress of Applied Mechanics*, also published in *Journal of Applied Mechanics*, Vol. 29.

Truesdell, C. (1967), Reactions of late baroque mechanics to success, conjecture, error, and failure in Newton's *Principia, Texas Q.*, Autumn issue.

Van Dyke, M.D. (1975), *Perturbation Methods in Fluid Mechanics*, annotated ed., Parabolic Press, Stanford, CA. The original version was published in 1964.

Van Wylen, G.J. and R.E. Sonntag (1976), *Fundamentals of Classical Thermodynamics*, 2nd ed., John Wiley & Sons, New York.

Vargaftik, N.B. (1980), *Tables on the Thermophysical Properties of Liquids and Gases*, Hemisphere, Washington, D.C., translated from the Russian edition.

Vincenti, W.G. and C.H. Kruger, Jr. (1965), *Introduction to Physical Gas Dynamics*, John Wiley & Sons, New York.

Viswanath, D.S. and G. Natarajan (1989), *Data Book on the Viscosity of Liquids*, Hemisphere, New York.

von Kármán, Th. (1946), The similarity law of transonic flow, *J. Math. Phys.*, 24, 182–190. Also reprinted in volume IV of the *Collected Works of Theodore von Kármán*, Butterworths Scientific Publications, London, 1956.

von Kármán, Th. (1954), *Aerodynamics, Selected Topics in the Light of Their Historical Development*, Cornell University Press, Ithaca, NY, 80–81.

Ward-Smith, A.J. (1980), *Internal Fluid Flow*, Clarendon Press, Oxford.

Weber, M. (1919), *Die Grundlagen der Ähnlichkeits mechanik und ihre Verwertung Bei Modellversuchen*, Jahrbuch der Schiffbautechnischen Gasellschaft.

Weinbaum, S. and R.W. Garvine (1969), The viscous counterpart of the sonic throat, *J. Fluid Mech.*, 39(1), 57–85.

Weisbach, J. (1855), *Die Experimental-Hydraulik*, J.G. Englehardt, Freiberg.

White, F.M. (1974), *Viscous Fluid Flow*, McGraw-Hill, New York.

Wilde, H. (1885), On the velocity with which air rushes into a vacuum, *Proc. Manch. Lit. Philos. Soc.*, 25(2), 17.

Willis, A.P. (1931), *Vector Analysis with an Introduction to Tensor Analysis*, Reprinted by Dover Publications, New York, 1958.

Winspear, A.D. (1955), De Rerum Natura, p. 28 of the didactic poem of Lucretius (Titus Lucretius Carus), set in English verse by Allen Dewes Winspear, The Harbor Press, Ltd., Cambridge, MA.

Young, T. (1809), On the functions of the heart and arteries, *Philos. Trans. R. Soc. London*, 99, 1–31.

Zagarola, M.V. (1996), Mean-Flow Scaling of Turbulent Pipe Flow, Ph.D. thesis, Princeton University, available from UMI, Ann Arbor, MI.

Zagarola, M.V. and A.J. Smits (1997), Reynolds number dependence of the mean flow in a circular pipe, AIAA 97-0649, presented at the 35th Aerospace Sciences Meeting, January 6–10, 1997, Reno, Nev.

Index

Printed in the United States
by Baker & Taylor Publisher Services

Printed in the United States
by Baker & Taylor Publisher Services